One Complex Variable from the Several Variable Point of View

Traditionally speaking, those who study the function theory of one complex variable spend little or no time thinking about several complex variables. Conversely, experts in the function theory of several complex variables do not consider one complex variable. One complex variable is the inspiration and testing ground for several complex variables, and several complex variables are the natural generalization of one complex variable.

The authors' thesis here is that these two subject areas have much in common. These subject areas can gain a lot by learning to communicate with each other. These two fields are logically connected, and each can be used to explain and put the other into context. This is the purpose of this book.

The point of view and the methodology of the two subject areas are quite different. One complex variable is an aspect of traditional hard analysis. Several complex variables are more like algebraic geometry and differential equations, with some differential geometry thrown in. The authors intend to create a marriage of the function theory of one complex variable and the function theory of several complex variables, leading to a new and productive dialogue between the two disciplines.

The hope is for this book to foster and develop this miscegenation in a manner that leads to new collaborations and developments. There is much fertile ground here, and this book aims to breathe new life into it.

It is a pleasure to thank Pavel Dovbush for producing technical illustrations specifically for this book. We are also most grateful to Elias Wegert of the Bergakademie Freiberg for creating the graphic for our book cover.

Peter V. Dovbush Dr. habil., Associate Professor, in Moldova State University, Institute of Mathematics and Computer Science. He received his Ph.D. from Lomonosov Moscow State University in 1983 and Doctor of Sciences in 2003. He has published over 50 scholarly articles.

Steven G. Krantz is a Professor of Mathematics at Washington University in St. Louis. He has previously taught at UCLA, Princeton University, and Penn State University. He received his Ph.D. from Princeton University in 1974. Krantz has directed 20 Ph.D. students and eight Masters students. He has published over 130 books and over 300 scholarly articles. He is the holder of the Chauvenet Prize, the Beckenbach Book Award, and the Kemper Prize. He is a Fellow of the American Mathematical Society.

Textbooks in Mathematics
Series editors:
Al Boggess, Kenneth H. Rosen

Business Process Analytics
Modeling, Simulation, and Design
Manuel Laguna and Johan Marklund

Quantitative Literacy Through Games and Gambling
Mark Hunacek

Measure and Integral
Theory and Practice
John Srdjan Petrovic

Contemporary Abstract Algebra, Eleventh Edition
Joseph A. Gallian

Student Solutions Manual for Gallian's Contemporary Abstract Algebra, Eleventh Edition
Joseph A. Gallian

Algebra, Second Edition
Groups, Rings, and Fields
Louis Halle Rowen and Uzi Vishne

Functional Analysis for the Applied Mathematician
Todd Arbogast and Jerry L. Bona

Exploring Linear Algebra, Second Edition
Labs and Projects with Mathematica®
Crista Arangala

Measure Theory and Fine Properties of Functions, Second Edition
Lawrence Craig Evans and Ronald F. Gariepy

Set Theory
An Introduction to Axiomatic Reasoning
Robert André

Introduction to Differential and Difference Equations Through Modeling
William P. Fox and Robert E. Burks

Abstract Algebra, Third Edition
An Interactive Approach
William Paulsen

Elements of Algebraic Topology, Second Edition
James R. Munkres, Steven G. Krantz, and Harold R. Parks

One Complex Variable from the Several Variable Point of View
Peter V. Dovbush and Steven G. Krantz

https://www.routledge.com/Textbooks-in-Mathematics/book-series/CANDHTEX-BOOMTH

One Complex Variable from the Several Variable Point of View

Peter V. Dovbush and Steven G. Krantz

CRC Press
Taylor & Francis Group
Boca Raton London New York

CRC Press is an imprint of the
Taylor & Francis Group, an **informa** business
A CHAPMAN & HALL BOOK

First edition published 2025
by CRC Press
2385 Executive Center Drive, Suite 320, Boca Raton, FL 33431, U.S.A.

and by CRC Press
4 Park Square, Milton Park, Abingdon, Oxon, OX14 4RN

CRC Press is an imprint of Taylor & Francis Group, LLC

Library of Congress Cataloging-in-Publication Data
Names: Dovbush, Peter V. author | Krantz, Steven G. (Steven George), 1951- author
Title: One complex variable from the several variable point of view / Peter V. Dovbush and Steven G. Krantz.
Description: First edition. | Boca Raton : CRC Press, 2025. | Series: Textbooks in mathematics | Includes bibliographical references and index.
Identifiers: LCCN 2024060040 (print) | LCCN 2024060041 (ebook) | ISBN 9781041020141 hbk | ISBN 9781041025061 pbk | ISBN 9781003619581 ebk
Subjects: LCSH: Functions of complex variables
Classification: LCC QA331.7 .D68 2025 (print) | LCC QA331.7 (ebook) | DDC 515/.9--dc23/eng/20250310
LC record available at https://lccn.loc.gov/2024060040
LC ebook record available at https://lccn.loc.gov/2024060041

ISBN: 978-1-041-02014-1 (hbk)
ISBN: 978-1-041-02506-1 (pbk)
ISBN: 978-1-003-61958-1 (ebk)

DOI: 10.1201/9781003619581

Typeset in Latin Modern font
by KnowledgeWorks Global Ltd.

Publisher's note: This book has been prepared from camera-ready copy provided by the authors.

Contents

Preface

Traditionally speaking, those who study the function theory of one complex variable spend little or no time thinking about several complex variables. And, conversely, experts in the function theory of several complex variables do not consider one complex variable. But this is in some sense wrong-headed, because one complex variable is the inspiration and testing ground for several complex variables, and several complex variables are the natural generalization of one complex variable.

Of course, the point of view and the methodology of the two subject areas are quite different. One complex variable is an aspect of traditional hard analysis. Several complex variables are more like algebraic geometry and differential equations, with some differential geometry thrown in. But our thesis here is that, nonetheless, these two subject areas have much in common. They can gain a lot by learning to communicate with each other.

What is true is that these two fields are logically connected, and each can be used to explain and put the other into context. That is the purpose of this book. We intend to create a marriage of the function theory of one complex variable and the function theory of several complex variables. We hope that this will lead to a new and productive dialogue between the two disciplines. The subject of one complex variable and the subject of several complex variables have a lot to learn from each other.

Just to illustrate these ideas we give now a few examples.

In several complex variables, there is always a C_c^∞ solution of the equation $\overline{\partial} u = f$ when f has C_c^∞ coefficients. In one complex variable, this is never the case. This technical distinction in fact explains a lot about the difference between the two disciplines. Indeed, the Hartogs extension phenomenon is a manifestation of this distinction. Thanks to Hörmander, there is an elegant proof of the one-variable corona theorem using the $\overline{\partial}$ problem. No such solution has been found in several complex variables.

In fact there is no domain in one complex variable for which the corona problem is known to be false. And there is no domain in several complex variables for which the corona problem is known to be true.

In one complex variable, every domain is a domain of holomorphy. In several complex variables, some domains are domains of holomorphy and others are not. One of the big vectors of activity of twentieth-century research in several complex variables was to come up with an intrinsic characterization of domains of holomorphy in complex n-space. This was the so-called *Levi problem.* Its solution, by Oka and many others, helped to shape the subject.

In one complex variable, it is a great classical result that every H^p function has a factorization into an inner function and an outer function. There is no such factorization in several complex variables. In this sense, the function theory of several complex variables is not as rich as that of one complex variable.

The zero set of a holomorphic function of one complex variable is always discrete. The zero set of a holomorphic function of several complex variables is never discrete—in fact, it is always a codimension 1 variety. But, in fact, that is also what it is in one complex variable—if we look at things in the right way.

As the reader can see, there is a common *lingua franca* for these two disciplines. They really can speak to each other—in a productive and useful manner. We hope in this book to foster and develop this miscegenation in a manner that leads to new collaborations and developments. There is much fertile ground here, and we plan to breathe new life into it.

We look forward to hearing from our readership and to continue contributing to this new dialogue.

Peter V. Dovbush

Steven G. Krantz

1

Introduction

The theory of functions of several complex variables found its roots early in the 20th century, as a natural follow-up to the mathematical work conducted by Augustin Louis Cauchy (1789–1857), Georg Friedrich Bernhard Riemann (1826–1866), and Karl Theodor Wilhelm Weierstrass (1815–1897) on holomorphic functions of one complex variable; and in the course of time it has achieved a broader autonomy and relevance. Despite the large number of texts written in recent years, some of the older classics remain the best. A few that are worth looking at are A. Hurwitz and R. Courant, *Vorlesungen ü Allgemeine Punktionentheorie und Elliptischc Funktionen* (Berlin: Julius Springer, 1925); E.T. Whittaker and G.N. Watson, *A Course in Modern Analysis*, Fourth Edition (London: Cambridge University Press, 1927); E.T. Titchmarch, *The Theory of Functions*, 2d ed. (NewYork: Oxford University Press. 1939, reprinted 1985); and K. Knopp, *Theory of Functions* (New York: Dover. 1947).

Certainly, the theory of several complex variables is now much too broad a field so that we have no presumption of giving a complete outline of the main developments as well as of the interface with other fields of mathematical research. For them we refer to [212–218] and also to [24, 46, 52, 60]. The reader who wishes further information on various of the more advanced topics can profitably consult E. Hille, *Analytic Function Theory*, 2 volumes, (Boston: Ginn, 1959); L.V. Ahlfors, *Complex Analysis* (New York: McGraw-Hill, 1966); W. Rudin, *Real and Complex Analysis* (New York: McGraw-Hill, 1969); and P.M. Morse and H. Feshbach, *Methods of Theoretical Physics* (New York: McGraw-Hill, 1953). Some additional references are given throughout the text.

From time immemorial teaching has been adapted to the evolution of knowledge. But this adaptation has sometimes lagged behind, to the great detriment of both science and teaching. During the last fifty years or so scientific progress has been so rapid that delay in adaptation was inevitable.

After the fundamental papers by Levi on holomorphic convexity for domains in $\mathbb{C}^n$, in the early 1950's, the theory received a new impetus from the algebraic and cohomological methods. It develops in the more general setting of the complex spaces with singularities and gives rise to a series of results culminating with the fundamental theorems on the cohomology of complex spaces with values in coherent sheaves. At the same time, the methods of the theory of plurisubharmonic functions and PDEs applied to the $\bar{\partial}$-equation, make now possible a new approach which will, within a few years, give many important contributions. An authentic turning point occurred in the theory around the

DOI: 10.1201/9781003619581-1

1970s with the establishment of representation formulas by Henkin, Ramirez, Grauert, and others. These developments have produced a great change in the field. In a synthetic and evocative way, we can say that we have passed from "existence problems" to "existence problems with control of growth."

The denomination itself has been modified as the methods have changed and the research has covered broader mathematical areas.[1]

It is imperative that we clear the way for the younger generation. In order for they may assimilate mathematics more easily, we must show them clearly the great simplifying ideas, and teach them how to handle complex situations by teaching them the unifying theories, which create bridges between the various fields. This will require sacrifice and agreeing to abandon certain elegant theories which, though polished by centuries of work, are now seen as isolated branches. On the other hand, the new mathematics has brought simplifications and economy of thought to all fields from which the physicists and the engineers can benefit as well as future mathematicians. It is obvious that the text books for such a renewal are still wanting.

It is no longer possible to divide the teaching of mathematics into its classical divisions of algebra, geometry, PDE and analysis.

It becomes therefore essential to think of a teaching whose main fabric will be the fundamental structures. Algebra and geometry will support each other, algebra bringing its symbolism and its operations, geometry its language loaded with intuitions. Geometry will provide analysis with its topological framework, the tool of convexity, and a convenient interpretation of integration and differentiation; analysis in turn will produce for algebra a rich collection of groups and vector spaces.

This book is endowed with an ample bibliography consisting of more than 200 items whose aim is to give both the reference for topics in the book but also the introduction to some areas of complex analysis which are not treated here.

The purpose of the function theory of one complex variable is to construct holomorphic functions with (*) certain properties. The purpose of the function theory of several complex variables is to understand the geometry of domains. Our goal in this book is to create a marriage and understanding between these two ostensibly different subjects. Traditionally one complex variable and several complex variables have different methodologies and study different questions. Our goal here is to show, by way of many examples, that the two different subjects really speak the same language. And each can shed light on the other.

---

[1] The current name of Complex Analysis first appeared at the International Congress of Mathematicians (IMC), held in Vancouver in 1974, where a section of the Congress was specifically reserved for research in the field.

2

Complex numbers

A satisfactory discussion of the main concepts of analysis (such as convergence, continuity, differentiation, and integration) must be based on an accurately defined number concept. In the course of real analysis the so-called "real number system" is constructed. We shall not, however, enter into any discussion of the axioms that govern the arithmetic of the numbers but assume familiarity with the theory of natural numbers. The reader will have met complex numbers in his course on elementary algebra. It is fundamental that real and complex numbers obey the same basic laws of arithmetic. We begin our study of complex function theory by stressing and implementing this analogy. There are many ways in which such a field can be constructed we give the simplest. Many readers will already know the contents of this chapter: they should read it through rapidly to check such items as notation and pass it on at once to the next.

2.1 The Field of Complex Numbers

The function

$$f(x) = \frac{1}{1+x^2}$$

is well defined on the real number axis, since the equation

$$x^2 + 1 = 0 \tag{2.1}$$

does not have any solution in the field of real numbers.

The function $f(x)$ it is equally good (infinitely differentiable) at all points of the real axis, and its Taylor series

$$f(x) = 1 - x^2 + x^4 - x^6 + \cdots$$

ceases to converge when $|x| > 1$. The reason for this cannot be understood in the context of the real numbers: after all, the points $x = \pm 1$ separate the sets of convergence and divergence of the series.

One can say that this is because $f(x)$ at the points $x = \pm\sqrt{-1}$ goes to infinity.

DOI: 10.1201/9781003619581-2

The inexperienced reader may find these thoughts perplexing, even challenging. Complex numbers were introduced to express the roots of an equation as numbers, a step into a completely new branch of mathematics that includes elliptic functions and Riemann surfaces.

Of course, one could simply agree to write the solutions of (2.1) in the form $\pm\sqrt{-1}$ as a formal notational convention. But this is not a satisfactory answer to the problem; we would like to attach some meaning to the notation. We shall discuss this matter on several levels.

The following discussion will assume some familiarity with the basic ideas and notions of algebra.

It is impossible to define division in the ring of integers, and this makes it necessary to extend the integers to the field of rational numbers. Similarly, since it is impossible to extract square roots from negative numbers in the field of real numbers, we are forced to extend this field to a larger one called the field of complex numbers.

To understand better what the field of complex numbers is, we should first try to understand the nature of the field of real numbers. Its rigorous construction is usually covered in a real analysis course. We will not consider all the details here; however, we remark that there are several such constructions (e.g., infinite decimal fractions, Dedekind cuts of rationals, etc). Formally speaking, each method leads to a different field.

Which one is the "real" field of reals? The answer is that they are all isomorphic, and we should simply view them as different models of the same object called the field of real numbers. Put in other words, there is only one field of real numbers because they are determined by the *axioms* of the reals.

In such a situation, the most satisfactory approach is always the axiomatic one: first, we formulate as axioms the properties that our object must have, and then prove that these properties define it uniquely up to isomorphism. Finally, we prove the existence by constructing a model. In the case of the field of real numbers such as a choice of axioms (in addition to the axioms of a field) may be the order axioms, the Archimedean postulate, and the axiom of continuity.

It is not difficult to prove that not only every two models of the field of real numbers are just isomorphic, but that the isomorphism between them is unique. (The proof comes down to demonstrating that an isomorphism from $\mathbb{R}$ to itself is the identity map. This demonstration, in turn, relies on the observation that nonnegative numbers in $\mathbb{R}$ are squares, and hence must be mapped to nonnegative numbers under any isomorphism.) This means that every element of $\mathbb{R}$ is specific, i.e., in any model, we can identify the numbers $10, \sqrt{2}, \pi$, etc.

2.1.1 An Auxiliary Construction

We would like to extend the field of real numbers $\mathbb{R}$ in such a way that the equation (2.1) has a solution in the new field.

As a model (example) of such a field extension, let us take the set P of all square matrices

$$\begin{pmatrix} a & b \\ -b & a \end{pmatrix} \in M_2(\mathbb{R}). \tag{2.2}$$

We claim that P is a field. First of all, P contains the zero

$$0 = \begin{pmatrix} 0 & 0 \\ 0 & 0 \end{pmatrix}$$

and the identity

$$E = \begin{pmatrix} 1 & 0 \\ 0 & 1 \end{pmatrix}$$

of the ring $M_2(\mathbb{R})$. Next, the relations

$$\begin{pmatrix} a & b \\ -b & a \end{pmatrix} + \begin{pmatrix} c & d \\ -d & c \end{pmatrix} = \begin{pmatrix} a+c & b+d \\ -(b+d) & a+c \end{pmatrix} \tag{2.3}$$

$$-\begin{pmatrix} a & b \\ -b & a \end{pmatrix} = \begin{pmatrix} -a & -b \\ -(-b) & -a \end{pmatrix}, \tag{2.4}$$

$$\begin{pmatrix} a & b \\ -b & a \end{pmatrix} \begin{pmatrix} c & d \\ -d & c \end{pmatrix} = \begin{pmatrix} ac-bd & ad+bc \\ -(ad+bc) & ac-bd \end{pmatrix} \tag{2.5}$$

imply that P is closed under addition and multiplication. The associativity of these operations follows from the associativity of addition and multiplication in the ring $M_2(\mathbb{R})$. The same goes for the distributive laws. Thus P is a subring of $M_2(\mathbb{R})$ and P is commutative by (2.5). It remains to show that any matrix (2.2) with a and b, not both zero has an inverse in P. (Note that the determinant of the matrix (2.2) is $a^2 + b^2 \neq 0$). By solving the linear system which comes from the condition

$$\begin{pmatrix} a & b \\ -b & a \end{pmatrix} \begin{pmatrix} x & y \\ -y & x \end{pmatrix} = \begin{pmatrix} 1 & 0 \\ 0 & 1 \end{pmatrix},$$

we find that

$$\begin{pmatrix} a & b \\ -b & a \end{pmatrix}^{-1} = \begin{pmatrix} c & d \\ -d & c \end{pmatrix} \qquad \text{where } c = \frac{a}{a^2+b^2}, \quad d = \frac{-b}{a^2+b^2}.$$

Thus P is a field.

Using the rule for multiplying matrices by scalars, we can write any element of P in the form

$$\begin{pmatrix} a & b \\ -b & a \end{pmatrix} = aE + bJ, \qquad \text{where } a, b \in \mathbb{R}, \quad J = \begin{pmatrix} 0 & 1 \\ -1 & 0 \end{pmatrix}. \tag{2.6}$$

The field P contains the subfield $\{aE \mid a \in \mathbb{R}\} \cong \mathbb{R}$, and the relation

$$J^2 + E = 0$$

shows that "up to isomorphism" the element $J \in P$ is a solution of equation (2.1). So there is no need for any mysticism about J being an "imaginary quantity".

But it is not the field P which is called the *field of complex numbers*, but rather a certain field isomorphic to it whose elements are identified with the points of a plane.

It is natural to want to think of P in terms of a geometrical realization; after all, the real number field $\mathbb{R}$ is inseparable in our minds from the number line, i.e., with a line having a point denoted 0 and a fixed scale of distance to the point 1.

2.2 The Complex Plane $\mathbb{C}$

Thus we would like to construct a field $\mathbb{C}$ whose elements are the points of the plane $\mathbb{R}^2$, having addition and multiplication operations which satisfy the field axioms, and solving our problem of being able to solve equation (2.1). On the cartesian plane, take the usual rectangular coordinate system with x-and y-axes. We write (a, b) for the point with x-coordinate a and y-coordinate b. We define the sum and product of two points (a, b) and (c,d) according to the rules

$$(a, b) + (c, d) = (a + c, b + d) \tag{2.7}$$

$$(a, b)(c, d) = (ac - bd, ad + bc) \tag{2.8}$$

(we use the same symbols + and $\cdot$ as in the field $\mathbb{R}$, but this should not cause confusion). We could verify by a straightforward but tedious series of calculations that these operations satisfy all of the field axioms, i.e., the operations (2.7) and (2.8) make the set of pairs (points in the plane) into a field. But happily, there is no need for this verification, because if we let each point of the plane $\mathbb{C}$ correspond to an element of our earlier field P as follows

$$(a, b) \longmapsto \begin{pmatrix} a & b \\ -b & a \end{pmatrix}$$

by comparing (2.3–2.5) and (2.7–2.8) we see that the set $\mathbb{C}$ is a field isomorphic to P. The set $\mathbb{C}$ is called the *field of complex numbers*. Because of the geometrical realization of $\mathbb{C}$, it is also often called the *complex plane*.

The points on the x-axis, i.e., the set of points of the form $(a, 0)$ have the same properties as the real number line, so we set $(a, 0) = a$. Then the zero $(0, 0)$ and the identity $(1, 0)$ remain the same as on the real number line. The point $(0, 1)$ on the y-axis is traditionally denoted i (the "imaginary" unit); it is a root of equation (2.1), since $i^2 = (0, 1) \cdot (0, 1) = (-1, 0) = -1$. An arbitrary complex number $z = (x, y)$ can now be written in the customary form

$$z = x + iy, \quad x, y \in \mathbb{R},$$

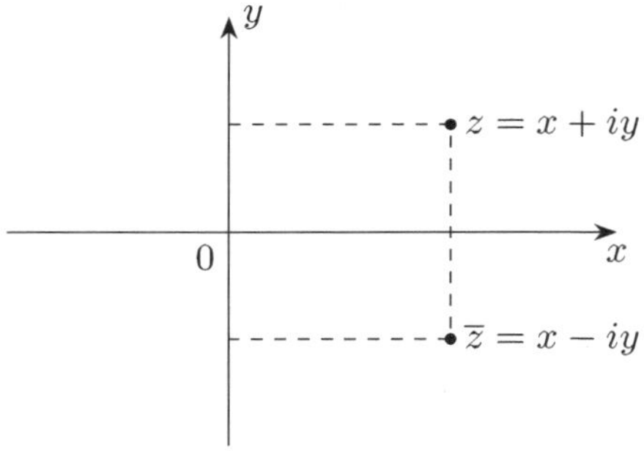

FIGURE 2.1

which is very close to the form (2.6) for the elements of the field P. Note that $\mathbb{Q} \subset \mathbb{R} \subset \mathbb{C}$. Hence, $\mathbb{C}$ is a field of characteristic zero.

Theorem 2.2.1. *The field of complex numbers $\mathbb{C}$ is unique up to an isomorphism that maps all real numbers to themselves. Every complex number can be uniquely written as $a + ib$, where $a, b \in \mathbb{R}$ and i is a (fixed) element such that $i^2 = -1$.*

Proof. Let $\mathbb{C}$ is the field constructed above. Consider its subset

$$K = \{a + bi \mid a, b \in \mathbb{R}\}.$$

Properties of the field operations and the identity $i^2 = -1$ imply that

$$(a_1 + b_1 i) + (a_2 + b_2 i) = (a_1 + a_2) + (b_1 + b_2)\, i \tag{2.9}$$

$$(a_1 + b_1 i)\,(a_2 + b_2 i) = (a_1 a_2 - b_1 b_2) + (a_1 b_2 + b_1 a_2)\, i \tag{2.10}$$

By solving appropriate equations, we also obtain

$$-(a + bi) = (-a) + (-b)i \tag{2.11}$$

and

$$(a + br)^{-1} = \frac{a}{a^2 + b^2} + \left(-\frac{b}{a^2 + b^2}\right) i, \qquad \text{whenever } a^2 + b^2 \neq 0\,. \tag{2.12}$$

Equations (2.9–2.12) show that K is a subfield of $\mathbb{C}$. Since K obviously contains i and $\mathbb{R}$, $K = \mathbb{C}$.

Every element of $\mathbb{C}$ can be presented in the form $a + bi$, where $a, b \in \mathbb{R}$. We have to show that such a presentation is unique. Let $a_1 + b_1 i = a_2 + b_2 i$, with $a_1, b_1, a_2, b_2 \in \mathbb{R}$. Then

$$a_1 - a_2 = (b_2 - b_1)\, i.$$

Taking squares of both sides, we get

$$(a_1 - a_2)^2 = -(b_2 - b_1)^2$$

which implies

$$a_1 - a_2 = b_2 - b_1 = 0,$$

as required.

Now, let $\mathbb{C}'$ be another field of complex numbers and $i' \in \mathbb{C}'$ an element such that $(i')^2 = -1$. As equations (2.9) and (2.10) remain valid when i is replaced with i', the map

$$f : \mathbb{C} \to \mathbb{C}', \quad a + bi \mapsto a + b' \quad (a, b \in \mathbb{R})$$

is an isomorphism from $\mathbb{C}$ to $\mathbb{C}'$. □

It is remarkable that the field so created contains not only square roots of each of its elements, but even n-th roots for every positive integer n. All the more remarkable is that adjoining a solution of the single equation $x^2 + 1 = 0$ to $\mathbb{R}$ results in an algebraically closed field: every nonconstant polynomial with coefficients in $\mathbb{C}$ can be factored over $\mathbb{C}$ into linear factors. This is the "Fundamental Theorem of Algebra," first established by C.F. Gauss (1777–1855) in 1799. Despite the theorem's name, it is in the final analysis a theorem in analysis.

However, it took nearly three hundred years until the work of Casper Wessel (ca. 1797), Jean Robert Argand (1806), Karl Friedrich Gauss (1831), Sir William R. Hamilton (1837), and others, when "imaginary" numbers were recognized as legitimate mathematical objects, and it was realized that there is nothing "imaginary" about them at all (although this term is still used).

Geometrical interpretation of operations with complex numbers. Since a complex number is nothing but an ordered pair of real numbers, it can be put into one-to-one correspondence with the points of a plane oriented by a rectangular coordinate system. The plane is then called "the (Gaussian or complex) number plane" or, more briefly, "the z-plane."

To say it another way, each point in the plane can be labeled by a complex number. The real numbers correspond to points on the horizontal axis, which is thus referred to in this context as the real axis. Complex numbers of the form iy with y real, so-called purely imaginary numbers, correspond to points on the vertical axis, which is thus referred to as the imaginary axis.

When wishing to emphasize this geometric interpretation, we shall refer to $\mathbb{C}$ as the "complex plane."

In geometric terms, the addition of two complex numbers is just vector addition according to the parallelogram law (see Fig. 2.2).

It is useful to call special attention to the following simple facts which follow from this equivalence of point and number.

Triangle inequality. If z_1 and z_2 are complex numbers, then

$$|z_1 + z_2| \leq |z_1| + |z_2|.$$

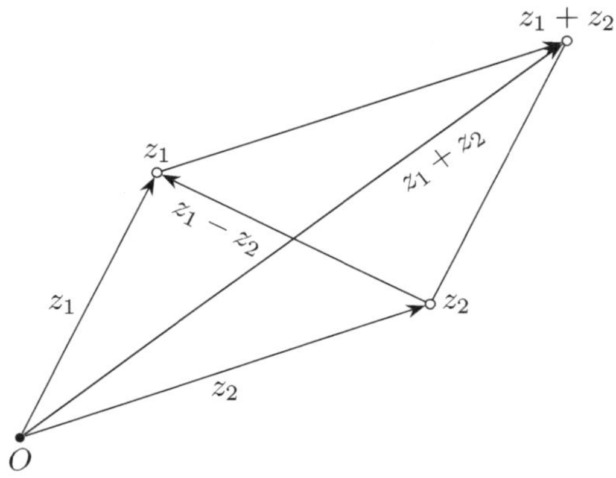

FIGURE 2.2

The inequality is strict unless one of z_1 and z_2 is zero, or z_1 and z_2 are positive multiples of each other.

In view of the interpretation of addition in $\mathbb{C}$ as vector addition, the inequality expresses the geometric fact that the length of any side of a triangle does not exceed the sum of the lengths of the other two sides.

The decomposition of a complex number $z \in \mathbb{C}$ as $a + bi$, $a, b \in \mathbb{R}$, is called its *algebraic form;* a is called the *real part* of z and b is the *imaginary part* of z. Notation: $a = \text{Re } z$, $b = \text{Im } z$. Complex numbers that are not real are called *imaginary*; those of the form bi, $b \in \mathbb{R}$, are called *purely imaginary.*

If we substitute $\mathbb{C}$ for $\mathbb{C}'$ and $-i$ for i' in the first part of the proof of Theorem 2.2.1, we see that the map

$$z = a + bi \mapsto \overline{z} = a - bi \quad (a, b \in \mathbb{R})$$

is an isomorphism of $\mathbb{C}$ into itself. This map is called *complex conjugation.* In general, an isomorphism of an algebraic structure into itself is called an automorphism. Thus, complex conjugation $z \mapsto \overline{z}$ is an automorphism of the field of complex numbers. It is clear that $\overline{\overline{z}} = z$.

The real numbers are precisely those complex numbers which are equal to their conjugates. It follows that for any $z \in \mathbb{C}$, $z + \overline{z}$ and $z\overline{z}$ are real. Indeed,

$$\overline{z + \overline{z}} = \overline{z} + z = z + \overline{z}, \qquad \overline{z\overline{z}} = \overline{z}z = z\overline{z}$$

It is easy to see that if $z = a + bi$, $a, b \in \mathbb{R}$, then

$$z + \overline{z} = 2a, \quad z - \overline{z} = 2ib, \quad |z|^2 := z\overline{z} = a^2 + b^2 .$$

Parallelogram equality. If z_1 and z_2 are complex numbers, then

$$|z_1 + z_2|^2 + |z_1 - z_2|^2 = 2\left(|z_1|^2 + |z_2|^2\right)$$

In geometric terms equality says that the sum of the squares of the lengths of the diagonals of a parallelogram equals the sum of the squares of the lengths of the sides. To establish it we note that the left side equals

$$(z_1 + z_2)(\bar{z}_1 + \bar{z}_2) + (z_1 - z_2)(\bar{z}_1 - \bar{z}_2)$$

which, when multiplied out, becomes

$$z_1\bar{z}_1 + z_1\bar{z}_2 + z_2\bar{z}_1 + z_2\bar{z}_2 + z_1\bar{z}_1 - z_1\bar{z}_2 - z_2\bar{z}_1 + z_2\bar{z}_2$$

After making the obvious cancellations, we obtain the right side.

Polar form. The argument of a complex number $z = a + ib$ is the angle formed by the corresponding vector with the positive real axis. The argument is determined up to 2π and is not defined for 0. The notation is $\arg z$.

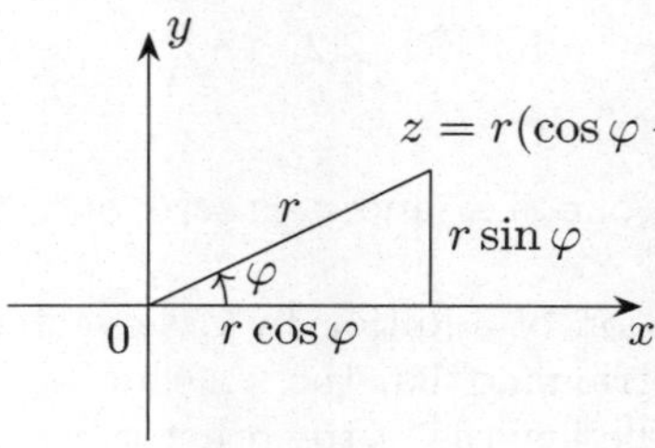

FIGURE 2.3

Let r and φ be the absolute value and argument of a complex number z. Obviously,

$$a = r\cos\varphi, \quad b = r\sin\varphi$$

hence

$$z = r(\cos\varphi + i\sin\varphi).$$

The above representation of a complex number z is called its *polar form.*

We will use a shorthand notation

$$e^{i\varphi} := \cos\varphi + i\sin\varphi.$$

Since the polar form of a complex number is determined up to addition to φ a multiple of 2π,

$$\begin{aligned} r_1(\cos\varphi_1 + i\sin\varphi_1) &= r_2(\cos\varphi_2 + i\sin\varphi_2) \\ &\iff \{r_1 = r_2, \varphi_1 = \varphi_2 + 2\pi k, k \in \mathbb{Z}\}, \qquad \text{for } r_1, r_2 > 0. \end{aligned}$$

Expressing complex numbers in polar form is useful for performing such operations as multiplication, division, raising to a power, and root extraction. Namely, formulas for the sine and cosine of a sum of two angles imply

$$\begin{aligned} &r_1(\cos\varphi_1 + i\sin\varphi_1) \cdot r_2(\cos\varphi_2 + i\sin\varphi_2) \\ &= r_1 r_2(\cos(\varphi_1 + \varphi_2) + i\sin(\varphi_1 + \varphi_2)) \end{aligned}$$

i.e., when multiplying complex numbers, we multiply their absolute values and add their arguments. Thus we have the following formulas for the ratio of complex numbers:

$$\frac{r_1(\cos\varphi_1 + i\sin\varphi_1)}{r_2(\cos\varphi_2 + i\sin\varphi_2)} = \frac{r_1}{r_2}(\cos(\varphi_1 - \varphi_2) + i\sin(\varphi_1 - \varphi_2))$$

and for a positive integer power of a complex number:

$$[r(\cos\varphi + i\sin\varphi)]^n = r^n(\cos n\varphi + i\sin n\varphi) \quad \text{(De Moivre's formula).}$$

To extract a root of the nth degree from a complex number $z = r(\cos\varphi + i\sin\varphi)$ is to solve the equation $z^n = c$. Put $|c| = s, \arg z = \psi$. Then $s^n = r$, $n\psi = \varphi + 2\pi k, k \in \mathbb{Z}$. Hence,

$$s = \sqrt[n]{r} \quad \text{(positive root)}, \qquad \psi = \frac{\varphi + 2\pi k}{n}.$$

Combining these formulas, we get

$$c = \sqrt[n]{r}\left(\cos\frac{\varphi + 2\pi k}{n} + i\sin\frac{\varphi + 2\pi k}{n}\right)$$

With the above formula, we obtain the same answer for different values of k if and only if they are congruent modulo n. It follows that for $z \neq 0$, the equation $z^n = c$ has exactly n solutions corresponding to, say, $k = 0, 1, \ldots, n-1$. Represented geometrically, these numbers lie at the vertices of a regular n-gon with its center at the origin.

$\mathbb{C}$ *as a metric space.* Since $\mathbb{C}$ as a set coincides with the Euclidean plane, $\mathbb{R}^2$, becomes a metric space if we endow it with the Euclidean metric. In this metric, the distance between the two points z_1 and z_2 equals $|z_1 - z_2|$. We can thus use in $\mathbb{C}$ all of the standard notions from the theory of metric spaces, such as open and closed sets, compactness, connectedness, convergence, and continuity. For example, a sequence $\{z_j\}$ in $\mathbb{C}$ will be said to converge to the complex number z provided

$$\lim_{j\to\infty} |z - z_j| = 0.$$

Stereographic projection. To obtain another geometric picture of $\mathbb{C}$, we identify $\mathbb{C}$ with the horizontal coordinate plane in $\mathbb{R}^3$. Let $\mathbb{S}$ denote the unit sphere (the sphere of unit radius with center $(0,0,0)$) in $\mathbb{R}^3$. A given point $z = x + iy$ in $\mathbb{C}$ and the point $(0,0,1)$ (the north pole of $\mathbb{S}$) determine a line in $\mathbb{R}^3$. That line intersects $\mathbb{S}$ at one point other than the north pole. We denote that other point by (ξ, η, ζ) and map $\mathbb{C}$ to $\mathbb{S}$ by sending z to (ξ, η, ζ). The map is clearly one-to-one, and each point of the sphere other than the north pole is an image point.

To represent the map analytically we note that the line determined by z and $(0,0,1)$ is given parametrically by the function

$$(tx, ty, 1-t), \quad -\infty < t < \infty$$

($t = 0$ corresponds to the north pole and $t = 1$ to z). The point $(tx, ty, 1 - t)$ lies on $\mathbb{S}$ provided

$$t^2x^2 + t^2y^2 + (1 - t)^2 = 1$$

which has the two solutions $t = 0$ (giving the north pole) and $t = \frac{2}{(x^2+y^2+1)}$ (giving the image point (ξ, η, ζ)). From this one sees that our map is given by

$$z \mapsto \frac{(2 \operatorname{Re} z, 2 \operatorname{Im} z, |z|^2 - 1)}{|z|^2 + 1}$$

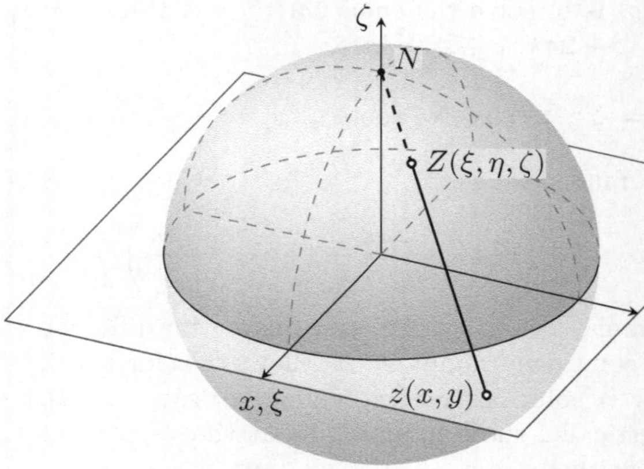

FIGURE 2.4: The stereographic projection

The reader will easily verify that the inverse map is given by

$$(\xi, \eta, \zeta) \mapsto \frac{\xi + i\eta}{1 - \zeta}.$$

The latter map is the stereographic projection of cartographers.

Both the map from $\mathbb{C}$ to $\mathbb{S}$ and its inverse are continuous and so preserve topological properties such as convergence and compactness. For example, if a sequence in $\mathbb{C}$ converges, then so does the corresponding sequence on $\mathbb{S}$, and if a sequence on $\mathbb{S}$ converges to a point other than the north pole, then the corresponding sequence in $\mathbb{C}$ converges.

Spherical metric. For z_1 and z_2 in $\mathbb{C}$, let $\rho(z_1, z_2)$ denote the Euclidean distance between the corresponding points on $\mathbb{S}$. This gives a metric in $\mathbb{C}$, the spherical metric, equivalent to the Euclidean metric (i.e., yielding the same family of open sets). Establish the following formula for the spherical metric:

$$\rho(z_1, z_2) = \frac{2\,|z_1 - z_2|}{\sqrt{|z_1|^2 + 1}\sqrt{|z_2|^2 + 1}}.$$

The *spherical arc length element* ds on the Riemann sphere $\mathbb{S}$ works out to be

$$ds = \frac{|dz|}{1 + |z|^2}$$

and the *spherical area element* dA is given by

$$dA = \frac{dxdy}{(1 + |z|^2)^2} \qquad (z = x + iy)$$

The *spherical length*

$$L(\gamma) = \int_\gamma \frac{|dw|}{1 + |w|^2}$$

of a curve γ on the Riemann sphere S induces a metric in the following manner. Given distinct points a, b on the Riemann sphere, define

$$\sigma(a, b) = \inf\{L(\gamma)\},$$

where the infimum is taken over all differentiable curves on S which join a with b. Then $\sigma(a, b)$ is the Euclidean length of the shortest arc of the great circle on S joining a and b and defines a metric on the sphere known as the *spherical metric.*

Indeed,

$$\rho(z_1, z_2) \le \sigma(z_1, z_2) \le \frac{\pi}{2}\rho(z_1, z_2),$$

so that the two metrics are uniformly equivalent and generate the same open sets on $\mathbb{S}$. Thus, from a topological point of view, the metrics ρ and σ can be treated as one and the same.

Extended complex plane. In many situations it is convenient to work with the enlargement of $\mathbb{C}$ that one obtains by appending a "point at infinity," corresponding to the north pole of $\mathbb{S}$. We denote this added point by ∞. (Set-theoretic purists can take ∞ to be any object not already in $\mathbb{C}$.) By the extended complex plane, we shall mean the set $\overline{\mathbb{C}} = \mathbb{C} \cup \{\infty\}$, to which we extend the spherical metric by defining $\rho(z, \infty)$, for z in $\mathbb{C}$, to be the Euclidean distance between the point on $\mathbb{S}$ corresponding to z and the north pole. Thus, for example, a sequence $\{z_j\}$ in $\mathbb{C}$ converges to ∞ in this metric if and only if $|z_j| \to \infty$ (in the standard sense).

Establish the formula

$$\rho(z, \infty) = \frac{2}{\sqrt{|z|^2 + 1}}.$$

It is also true that $\rho(\cdot, \cdot)$ is a metric on $\overline{\mathbb{C}}$, with $\rho(z_1, z_2) \le |z_1 - z_2|$ in $\mathbb{C}$. The virtue of the chordal metric is that it allows $z = \infty$ to be treated like any other point.

The extended complex plane is often referred to as the Riemann sphere, after G.F.B. Riemann (1826–1866), a pioneer in our subject whose ideas had and continue to have a profound influence on its development.

2.3 The Complex Space $\mathbb{C}^n$

We consider even-dimensional Euclidean space $\mathbb{R}^{2n}$, whose points are ordered sets of $2n$ real numbers $(x_1, \dots, x_{2n})$. We introduce a complex structure on it by setting $z_\nu = x_\nu + ix_{n+\nu}(\nu = 1, \dots, n)$. We shall often write $x_{n+\nu} = y_\nu$, so that $z_\nu = x_\nu + iy_\nu(\nu = 1, \dots, n)$. We shall call the space whose points are ordered sets of n complex numbers

$$z = (z_1, \dots, z_n),$$

the n-dimensional complex Euclidean space and we denote it by the symbol $\mathbb{C}^n$. In particular, for $n = 1$ we obtain $\mathbb{C}^1 = \mathbb{C}$, the complex plane. The space $\mathbb{C}^n$ is the Cartesian product of n planes

$$\mathbb{C}^n = \underbrace{\mathbb{C} \times \cdots \times \mathbb{C}}_{n \text{ times}}.$$

Thus the points of n-dimensional complex Euclidean space $\mathbb{C}^n$ are the points of $2n$-dimensional real Euclidean space $\mathbb{R}^{2n}$. However, the introduction of a complex structure in $\mathbb{R}^{2n}$ immediately introduces an asymmetry into this space—the coordinates in it are not all equivalent (for example, x_1 and x_{n+1} are combined into the complex number z_1, but x_1 and x_2 are not combined).

One can naturally introduce the structure of a vector space over the field $\mathbb{C}$ in $\mathbb{C}^n$; addition and multiplication by a complex scalar λ are defined coordinate-by-coordinate: $z + w = (z_1 + w_1, \dots, z_n + w_n)$ and $\lambda z = (\lambda z_1, \dots, \lambda z_n)$. A vector $z \in \mathbb{C}^n$ is sometimes written in the form $z = x + iy$, where $x = (x_1, \dots, x_n)$ and $y = (y_1, \dots, y_n)$ are vectors belonging to $\mathbb{R}^n$. The space $\mathbb{R}^n(x)$, consisting of vectors of the form $z = x + i0$, is called the real subspace of $\mathbb{C}^n$ (for $n = 1$ this is the real axis).

A set $P \subseteqq \mathbb{C}^n$ is said to be an *open (closed) polydisc* if there are open (closed) disc $P_1, \dots, P_n$ in $\mathbb{C}$ such that $P = P_1 \times \cdots \times P_n$. If ξ_j is the center of P_j, then the point $(\xi_1, \dots, \xi_n)$ is called the center of P. If $r_j > 0$ is the radius of P_j, then $(r_1, \dots, r_n)$ is called the multi-radius of P. Denote by $P(a; r)$ the polydisc with the center a and the multi-radius r.

Considered only as a topological space (or as a real vector space), $\mathbb{C}^n$ is of course just the same as $\mathbb{R}^{2n}$, the ordinary Euclidean space of In dimensions. Thus we can impose on $\mathbb{C}^n$ in a natural manner any of the structures of $\mathbb{R}^{2n}$; for instance, the Lebesgue measure on R2n becomes a measure on $\mathbb{C}^n$, which will be denoted by dV.

In $\mathbb{C}^n$ one has usual Euclidian metric, according to which the distance between the points a and b is equal to $|a - b|$, the Hermitian form of this metric is

$$ds^2 = \sum_{k=1}^{2n} dx_k^2 = \sum_{k=1}^{n} |dz_k|^2,$$

where $|dz_k|^2 = dx_k^2 + dy_k^2$.

3

The Cauchy-Riemann Equations

The first place where the Cauchy-Riemann equations appear is in d'Alembert's work from 1752, Essai d'une nouvelle théorie de la résistance des fluides. The functions for which d'Alembert establishes these equations are the horizontal and vertical components of the force acting on a fluid. He also shows that there is a relationship with the functions of a complex variable.

The Cauchy-Riemann equations are one way of looking at the condition for a function to be differentiable in the sense of complex analysis: in other words, they encapsulate the notion of the function of a complex variable by means of conventional differential calculus. In the theory there are several other major ways of looking at this notion, and the translation of the condition into other language is often needed.

3.1 The Cauchy-Green Formula

A complex-valued function f on a subset $U \subset \mathbb{C}$ is merely a mapping from U into the complex plane; the value of the function f at a point $z \in U$ will be denoted by $f(z)$, as usual.

Let $\mathbb{C}$ be the complex plane. By x_1, x_2 we denote the real coordinates in $\mathbb{C}$ such that $\mathbb{C} \ni z = x_1 + ix_2$. For every complex-valued continuous function f in an open domain in $\mathbb{C}$ we define (in the sense of distributions)

$$\frac{\partial f}{\partial z} := \frac{1}{2}\left(\frac{\partial f}{\partial x_1} + \frac{1}{i}\frac{\partial f}{\partial x_2}\right), \quad \frac{\partial f}{\partial \overline{z}} := \frac{1}{2}\left(\frac{\partial f}{\partial x_1} - \frac{1}{i}\frac{\partial f}{\partial x_2}\right),$$

$$\partial f := \frac{\partial f}{\partial z} dz \quad \text{and} \quad \overline{\partial} f := \frac{\partial f}{\partial \overline{z}} d\overline{z},$$

where $\overline{z} := x_1 - ix_2$. Then the differential df of f can be expressed as $df = \partial f + \overline{\partial} f$. We also write $d_z f$, $\partial_z f$ and $\overline{\partial}_z f$ instead of df, ∂f, and $\overline{\partial} f$ to make clear that the differentiation is with respect to z if f depends on other variables, too.

Recall that a C^1-function f is holomorphic if and only if $\overline{\partial} f = 0$ (the Cauchy-Riemann equations). The set of all holomorphic functions in domain $U \subset \mathbb{C}$ will be denoted by $\mathcal{O}(U)$.

DOI: 10.1201/9781003619581-3

Viewed properly, the Cauchy integral formula in one complex variable is *not* a result of holomorphic functions. Rather, it is a result of *arbitrary smooth functions.* And it is a corollary of Stokes's theorem. We now explicate this point of view. Cauchy integral formula *not* about holomorphic functions.

If $\omega = udz + vd\bar{z}$ is a 1-form, then

$$d\omega = \frac{\partial u}{\partial \bar{z}} d\bar{z} \wedge dz + \frac{\partial v}{\partial z} dz \wedge d\bar{z} = \left(\frac{\partial v}{\partial z} - \frac{\partial u}{\partial \bar{z}} \right) dz \wedge d\bar{z} . \tag{3.1}$$

Of course, 1-forms can be integrated over 1-manifolds, such as the circle, and 2-forms can be integrated over 2-manifolds, such as the sphere or torus.

Finally, we recall the exterior differential operator d. In the complex setting we can decompose $d = \partial + \bar{\partial}$, where

$$\partial \omega = \frac{\partial v}{\partial z} dz \wedge d\bar{z}$$

and

$$\bar{\partial} \omega = \frac{\partial u}{\partial \bar{z}} d\bar{z} \wedge dz .$$

Refer back to equation (3.1).

Theorem 3.1.2 (Stokes). *Let $U \subset \mathbb{C}$ be a bounded open domain with C^1 boundary. Let ω be a differential form of degree 1 with coefficients in $C^1(\overline{U})$. Then*

$$\int_{\partial U} U = \int_U d\omega = \int_U \partial \omega + \bar{\partial} \omega .$$

Standard references for Stokes's theorem are [187] and [59]. The Stokes's theorem that we have recorded here is the standard one, simply expressed in complex notation.

Cauchy—Green's formula, plays a fundamental role in the theory of functions of a complex variable, together with the Cauchy-Riemann equations is now the usual way to prove Cauchy's theorem.

Now we have

Theorem 3.1.3 (Cauchy-Green formula). *Let $U \subset\subset \mathbb{C}$ be an open domain with C^1 boundary ∂D, and let f be a complex-valued continuous function on $\overline{U}$ such that $\bar{\partial} f$ is also continuous on $\overline{U}$.*[1] *Then*

$$f(z) = \frac{1}{2\pi i} \int_{\partial U} \frac{f(\zeta) d\zeta}{\zeta - z} - \frac{1}{2\pi i} \int_{\partial U} \frac{\bar{\partial} f(\zeta) \wedge d\zeta}{\zeta - z} , \quad z \in U. \tag{3.2}$$

Proof. Fix $z \in U$. Then, for $\zeta \in \overline{U} \backslash z$, we see that

$$d_\xi [(f(\zeta) d\zeta)/(\zeta - z)] = \bar{\partial} f(\zeta) \wedge d\zeta / (\zeta - z)$$

[1] This means $\partial f / \partial \bar{z}$, which is defined in the sense of distributions in U, is continuous in U and admits a continuous extension to $\overline{D}$.

and, therefore, by Stokes's formula, for sufficiently small $\varepsilon > 0$

$$\frac{1}{2\pi i}\int_{|\zeta-z|=\varepsilon}\frac{f(\zeta)d\zeta}{\zeta-z} = \frac{1}{2\pi i}\int_{\zeta\in\partial U}\frac{f(\zeta)d\zeta}{\zeta-z} - \frac{1}{2\pi i}\int_{\zeta\in U,|\zeta-z|>\varepsilon}\frac{\overline{\partial}f(\zeta)\wedge d\zeta}{\zeta-z}.$$

It is clear that the right-hand side of this relation tends to the right-hand side of (3.2) when $\varepsilon \to 0$. Consider the left-hand side. Since $d\overline{\zeta}\wedge d\zeta/2i$ is the Lebesgue measure on $\mathbb{C}$, Stokes's formula gives

$$\int_{|\zeta-z|=\varepsilon}\frac{d\zeta}{\zeta-z} = \frac{1}{\varepsilon^2}\int_{|\zeta-z|=\varepsilon}(\overline{\zeta}-\overline{z})d\zeta = \frac{1}{\varepsilon^2}\int_{|\zeta-z|<\varepsilon}d\overline{\zeta}\wedge d\zeta = 2\pi i.$$

Furthermore,

$$\left|\int_{|\zeta-z|=\varepsilon}\frac{f(\zeta)-f(z)}{\zeta-z}d\zeta\right| \leqq 2\pi \max_{|\zeta-z|=\varepsilon}|f(\zeta)-f(z)| \to 0 \quad \text{for} \quad \varepsilon\to 0\,.$$

Hence

$$\lim_{\varepsilon\to 0}\int_{|\zeta-z|=\varepsilon}\frac{f(\zeta)d\zeta}{\zeta-z} = 2\pi i f(z)\,.$$

□

It is important to note now that if U is as in the statement of the Theorem 3.1.3 and $f\in C^1(\overline{U})$ and holomorphic on the interior, then equation (3.2) simplifies to

$$f(z) = \frac{1}{2\pi i}\int_{\partial U}\frac{f(\zeta)}{\zeta-z}\,d\zeta\,.$$

This, of course, is the Cauchy integral formula that one can find in any complex analysis text.

Since the first integral on the right-hand side of the Cauchy-Green formula (3.2) depends holomorphically on $z\in U$, and since $\overline{\partial}f = 0$ for holomorphic functions, we obtain:

Corollary 3.1.4. *Let f be a continuous complex-valued function in an open domain $U\subseteqq\mathbb{C}$. Then f is holomorphic in U if and only if $\overline{\partial}f = 0$ in U.*

Corollary 3.1.5. *If $g\in C_0^1(\mathbb{C})$, i.e., g is C^1 with compact support, then for all $z\in\mathbb{C}$ the following formula holds*

$$g(z) = \frac{1}{2\pi i}\int_{\mathbb{C}}\frac{g_{\overline{\zeta}}(\zeta)}{\zeta-z}d\zeta\wedge d\overline{\zeta}$$

Proof. Apply (3.2) to a disc $\mathbb{D}(0,r)\supset \operatorname{supp} g\cup\{z\}$. □

Theorem 3.1.6. *Let $U\subset\subset\mathbb{C}$ be a bounded open domain, and let f be a bounded continuous complex-valued function in U. Then the continuous function*

$$u(z) := -\frac{1}{2\pi i}\int_U\frac{f(\zeta)d\overline{\zeta}\wedge d\zeta}{\zeta-z}, \quad z\in U$$

is a solution of $\frac{\partial u}{\partial\overline{z}} = f$ in U.

Proof. First we consider the case that f is continuously differentiable in U. Fix $\xi \in U$. It is sufficient to prove that $\partial u/\partial \bar{z} = f$ in some neighborhood of ξ. Choose a $\mathcal{C}^\infty$-function χ on $\mathbb{C}$ such that $\chi = 1$ in a neighborhood $V_\xi \subset\subset U$ of ξ and $\chi = 0$ in some neighborhood of $\mathbb{C}\backslash U$. Then $u = u_1 + u_2$, where

$$u_1(z) := -\frac{1}{2\pi i}\int_U \frac{\chi(\zeta)f(\zeta)d\bar{\zeta}\wedge d\zeta}{\zeta - z},$$

$$u_2(z) := -\frac{1}{2\pi i}\int_U \frac{(1-\chi(\zeta))f(\zeta)d\bar{\zeta}\wedge d\zeta}{\zeta - z}.$$

Since $1-\chi = 0$ in V_ξ, u_2 is holomorphic in V_ξ, that is, $\bar{\partial} u_2 = 0$ in V_ξ. Therefore it remains to prove that $\partial u_1/\partial \bar{z} = f$ in V_ξ. Since $\chi = 0$ in a neighborhood of $\mathbb{C}\backslash U$, the function χf can be continued by zero to $\mathbb{C}$, and we obtain

$$u_1(z) = -\frac{1}{2\pi i}\int_{\mathbb{C}} \frac{\chi(\zeta)f(\zeta)d\bar{\zeta}\wedge d\zeta}{\zeta - z} = -\frac{1}{2\pi i}\int_{\mathbb{C}} \frac{\chi(\zeta+z)f(\zeta+z)d\bar{\zeta}\wedge d\zeta}{\zeta}.$$

Taking into account that $\partial/\partial \bar{z}[\chi(\zeta+z)f(\zeta+z)]d\bar{\zeta} = \bar{\partial}_\zeta[\chi(\zeta+z)f(\zeta+z)]$, we conclude that

$$\frac{\partial u_1(z)}{\partial \bar{z}} = \frac{1}{2\pi i}\int_{\mathbb{C}} \frac{\bar{\partial}_\zeta[\chi(\zeta+z)f(\zeta+z)]\wedge d\zeta}{\zeta}$$
$$= \frac{1}{2\pi i}\int_{\mathbb{C}} \frac{\bar{\partial}_\zeta[\chi(\zeta)f(\zeta)]\wedge d\zeta}{\zeta - z}, \quad z \in U.$$

Application of Theorem 3.1.3 with f replaced by χf and U equal to a disc containing the support of χ now gives that $\partial u_1/\partial \bar{z} = f$ in V_ξ.

Now we consider the case that f is an arbitrary continuous and bounded function on U. Choose a sequence f_k of continuously differentiable functions which converge uniformly on U to f. Then the sequence

$$u_k(z) := -\frac{1}{2\pi i}\int_U \frac{f_k(\zeta)d\bar{\zeta}\wedge d\zeta}{\zeta - z}$$

converges uniformly on U to u, and therefore we have in the sense of distributions $\partial u/\partial \bar{z} = \lim \partial u_k/\partial \bar{z} = \lim f_k = f$. □

3.1.1 The Compact Support Case

Suppose that g is compactly supported. Starting from (3.4), we prove the following existence theorem:

Theorem 3.1.7. *Let $g \in C_0^k(\mathbb{C}), 1 \le k \le \infty$. The function*

$$u_0(z) = \frac{1}{2\pi i}\int_{\mathbb{C}} \frac{g(\zeta)}{\zeta - z} d\zeta \wedge d\bar{\zeta} \tag{3.3}$$

is $C^k(\mathbb{C})$ and it is a solution of the Cauchy-Riemann equation

$$\frac{\partial u}{\partial \overline{z}} = g$$

Proof. Setting $w = \zeta - z$, (3.3) becomes

$$u_0(z) = \frac{1}{2\pi i}\int_{\mathbb{C}} \frac{g(w+z)}{w} dw \wedge d\overline{w}$$

Hence, we may differentiate under the integral sign, obtaining that $u_0 \in C^k(\mathbb{C})$. In particular,

$$\frac{\partial u_0}{\partial \overline{z}} = \frac{1}{2\pi i}\int_{\mathbb{C}} \frac{g_{\overline{\zeta}}}{\zeta - z} d\zeta \wedge d\overline{\zeta} = g$$

by Corollary 3.1.5. □

Remark 3.1.8. In general, the equation

$$\frac{\partial u}{\partial \overline{z}} = g$$

with $g \in C_0^\infty(\mathbb{C})$ does not admit compactly supported solutions. Indeed, if $v \in C_0^\infty(\mathbb{C})$ is a solution and u_0 is given by (3.3), then $\varphi = u_0 - v$ is entire and, since

$$|u_0(z)| \leq \frac{C}{d(z, \operatorname{supp} g)}$$

(C is a constant), we have $|\varphi(z)| \to 0$ when $|z| \to +\infty$. It follows that $\varphi \equiv 0$ and $u_0 = v$. Thus u_0 has compact support, consequently, in view of (3.3), g must fulfill the condition

$$\int_{\mathbb{C}} \frac{g(\zeta)}{\zeta - z} d\zeta \wedge d\overline{\zeta} = 0 \tag{3.4}$$

for $|z| \gg 0$.

Example 3.1.9. The function

$$g = \begin{cases} e^{-\frac{1}{1-|z|^2}}, & \text{for } |z| < 1 \\ 0, & \text{for } |z| \geq 1 \end{cases}$$

does not satisfy the condition (3.4).

3.1.2 The General Case

The proof of the existence theorem for

$$\frac{\partial u}{\partial \overline{z}} = g$$

when $g \in C^\infty$ is more involved and requires an approximation theorem.

Theorem 3.1.10 (Runge). *Let $K \subset U \subset \mathbb{C}$ be a compact set. Then $\mathcal{O}(U)$ is dense in $\mathcal{O}(K)$, with respect to the sup norm on K if and only if one of the following equivalent conditions holds*

(I) *$U \backslash K$ has no relatively compact component in D;*

(II) *K coincides with its $\mathcal{O}(U)$-envelope $\widehat{K}_U$.*

In particular, if $U = \mathbb{C}$ and $\mathbb{C} \backslash U$ are connected, $\mathbb{C}[z]$ is dense in $\mathcal{O}(K)$.

For the proof we refer to [111].

Remark 3.1.11. The Runge Theorem was generalized by Mergelyan [155] (see also [208]): if condition **(I)** holds, then $\mathcal{O}(U)$ is dense, with respect to the sup norm on K, in $A(K)$, the algebra of holomorphic functions on the interior of K, continuous on K.

Theorem 3.1.12. *Let U be a domain in $\mathbb{C}$ and $g \in C^\infty(U)$. The equation*

$$\frac{\partial u}{\partial \bar{z}} = g$$

admits a solution $u \in C^\infty(U)$.

Proof. We consider preliminary two particular cases, namely $U = \mathbb{C}$ and $U \subsetneq \mathbb{C}$ simply connected. Let

$$U_j = \{z \in \mathbb{C} \mid |z| < j\}$$

$n \in \mathbb{N}$. For each $n \in \mathbb{N}$ let $\rho_j \in C_0^\infty(U_{j+1})$ with $\rho_j \equiv 1$ on a neighborhood of $\overline{U}_j$. Let $v_j \in C^\infty(\mathbb{C})$ be such that

$$\frac{\partial v_j}{\partial \bar{z}} = \rho_j g, \quad n \in \mathbb{N}$$

Our goal is to modify the solutions v_j in order to have a sequence $\{v_j\}_{j \in \mathbb{N}}$ which converges to a smooth solution of the problem. Let $u_1 = v_1$. The function $v_2 - u_1$ is holomorphic on a neighborhood of $\overline{U}_1$ and so there is a polynomial $p = p(z)$ such that

$$\|(v_2 - u_1) - p\|_{\overline{U}} < \frac{1}{2}$$

Define $u_2 = v_2 - p$. Obviously, in a neighborhood of $\overline{U}_2$ we have

$$\frac{\partial u_2}{\partial \bar{z}} = g$$

We may repeat the argument for $v_3 - u_2$ and so on. In this way we construct a sequence of functions $u_j \in C^\infty(\mathbb{C})$ with the properties

$$\frac{\partial u_j}{\partial \bar{z}} = g$$

in a neighborhood of $\overline{U}_j$ and

$$\|u_{n+1} - u_j\|_{\overline{U}_j} < \frac{1}{2^j} \tag{3.5}$$

for all $n \in \mathbb{N}$. The sequence $\{u_j\}$ converges in $C^\infty(\mathbb{C})$ to a solution u. Indeed, on each disc U_k

$$u = \lim_{j \to +\infty} u_j = u_k + \sum_{j=k}^{\infty} (u_{j+1} - u_j)$$

and the series of holomorphic functions

$$\sum_{j=k}^{\infty} (u_{j+1} - u_j)$$

converges uniformly on $\overline{U}_k$ in view of (3.5). Thus, on U_k, for every $k \in \mathbb{N}$ we have

$$\frac{\partial u}{\partial \overline{z}} = \frac{\partial u_k}{\partial \overline{z}} = g$$

The case when $U \subsetneq \mathbb{C}$ is simply connected reduces to the case of the unit disc, in view of the Riemann mapping theorem. Indeed, let $f : \mathbb{D} \to U$ be a biholomorphism from the unit disc to the domain U and define

$$\tilde{g}(z') = g(f(z')) \overline{\frac{\partial f}{\partial z'}}$$

If $\tilde{u}$ is a solution of

$$\frac{\partial \tilde{u}}{\partial \overline{z}} = \tilde{g}$$

on the unit disc, then the function $u = \tilde{u} \circ f^{-1}$ is a solution of

$$\frac{\partial u}{\partial \overline{z}} = g$$

on U. In order to solve the problem for the unit disc, we use the same method as in the case of $\mathbb{C}$.

As for the general case, we consider an exhaustive sequence $\{K_j\}$ of compact sets such that $K_j \subset \overset{\circ}{K}_{j+1}$ and $K_j = \widehat{K}_{j\mathcal{O}(U)}$. Let $\rho_j \in C_0^\infty\left(\overset{\circ}{K}_{j+1}\right)$ such that $\rho_j \equiv 1$ on a neighborhood of K_j. The proof runs as in the first two cases using Theorem 3.1.10. □

3.1.3 Why the Cauchy-Riemann Equations?

Very often problems in the theory of several complex variables reduce to solving Cauchy-Riemann equations. This happens even in the most simple cases, as shown by the following examples.

3.1.3.1 Examples in $\mathbb{C}$

Example 3.1.13. *Construction of a meromorphic function with given poles.*
Given a discrete set $F = \{z_j\}_{j\in\mathbb{N}}$ in a domain $\Omega \subset \mathbb{C}$ and a sequence $\{n_j\}_{j\in\mathbb{N}}$ of positive integers, we ask for a meromorphic function f on Ω, whose poles are the points z_j, with respective orders n_j. Consider the meromorphic functions

$$\varphi_j = \frac{1}{(z - z_j)^{n_j}}$$

$j \in \mathbb{N}$. Each φ_j is holomorphic on $\tilde{V}_j = \mathbb{C}\setminus\{z_j\}$; $\{V_j = \tilde{V}_j \cap \Omega\}_{j\in\mathbb{N}}$ is an open covering of Ω, and $\varphi_{jk} = \varphi_j - \varphi_k$ is holomorphic on $V_{jk} = V_j \cap V_k$. We may choose a locally finite refinement $\{U_j\}_{j\in\mathbb{N}}$ of $\{V_j\}_{j\in\mathbb{N}}$. Let $\{\rho_j\}_{j\in\mathbb{N}}$ be a partition of unity associated to $\{U_j\}_{j\in\mathbb{N}}$ and let

$$f_j = \sum_{k\in\mathbb{N}} \rho_k \varphi_{jk}$$

$f_j \in C^\infty(U_j)$ and $f_j - f_k = \varphi_{jk}$ is holomorphic on U_{jk}. Thus the local functions

$$g_j = \frac{\partial f_j}{\partial \bar{z}}$$

define a function $g \in C^\infty(\Omega)$ and, since the Cauchy-Riemann equation for g is solvable in Ω, there exists $u \in C^\infty(\Omega)$ such that

$$\frac{\partial u}{\partial \bar{z}} = g.$$

Then

(I) $f_j - u \in \mathcal{O}(U_j)$;

(II) on U_{jk}, $(f_j - u) - (f_k - u) = \varphi_{jk} \in \mathcal{O}(U_{jk})$, so the local functions $\varphi_j + u - f_j$ define a meromorphic global function f. On U_j we have

$$f - \varphi_j = f_j - u \in \mathcal{O}(U_j)$$

i.e., z_j is the only pole in U_j of f (of order n_j).

Example 3.1.14. *Construction of a holomorphic function with given zeros.*
Consider a discrete set $F = \{z_j\}_{j\in\mathbb{N}}$ in a domain $\Omega \subset \mathbb{C}$ and a sequence $\{n_j\}_{j\in\mathbb{N}}$ of positive integers. The problem is to find a holomorphic function $f : \Omega \to \mathbb{C}$, whose zeros are the points z_j, with respective orders n_j.

Consider a locally finite covering $\{U_j\}_{j\in\mathbb{N}}$ of Ω with the following properties:

(I) $U_j \cap F = \{z_j\}$;

(II) $U_{jk} = U_j \cap U_k$ is simply connected for all j and k.

In each U_j we define the holomorphic function

$$\varphi_j(z) = (z - z_j)^{n_j}$$

Following the lines of proof in the previous example, we must find invertible holomorphic functions $h_j : U_j \to \mathbb{C}$, such that

$$\frac{\varphi_j}{\varphi_k} = \frac{h_j}{h_k} \tag{3.6}$$

on $U_{jk} \neq \emptyset$.

Since U_{jk} is simply connected and

$$\varphi_{jk} = \frac{\varphi_j}{\varphi_k} \neq 0$$

there exists a logarithm $\theta'_{jk} = \log \varphi_{jk}$ of φ_{jk}. If $U_{jkl} = U_j \cap U_k \cap U_l \neq \emptyset$

$$\theta'_{jk} + \theta'_{kl} + \theta'_{lj} = 2\beta_{jkl}\pi i$$

with $\beta_{jkl} \in \mathbb{Z}$. From the homology theory (see [33]), we know that for every pair of indexes j and k such that $U_{jk} \neq \emptyset$, there is a $\beta_{jk} \in \mathbb{Z}$ such that

$$\beta_{jl} = -\beta_{lj} \quad \beta_{jk} = \beta_{jt} + \beta_{tj} + \beta_{jl}.$$

Define $\theta_{jk} = \theta'_{jk} - 2\beta_{jk}\pi i$. It follows that $\log \varphi_{jk} = \beta_{jk}$ and

$$\theta_{jk} + \theta_{kl} + \theta_{lj} = 0$$

Arguing as in Example 3.1.13, there exist holomorphic functions g_j, $j \in \mathbb{N}$, such that $\theta_{jk} = g_j - g_k$ for every j, k, hence

$$\varphi_{jk} = e^{\theta_{jk}} = \frac{e^{g_j}}{e^{g_k}}.$$

Thus (3.6) holds with $h_j = e^{g_j}$.

From Examples 3.1.13 and 3.1.14, we derive the following

Proposition 3.1.15. *Let Ω be a domain in $\mathbb{C}$. Let $\{z_j\}$ and $\{w_j\}$ be disjoint discrete sets in Ω and $\{n_j\}$ and $\{m_j\}$ sequences of positive integers. Then there exists a meromorphic function f on Ω such that*

(I) $\{z_j\}_{j\in\mathbb{N}}$ is the set of zeros with respective orders $\{n_j\}_{j\in\mathbb{N}}$;

(II) $\{w_j\}_{j\in\mathbb{N}}$ is the set of poles with respective orders $\{m_j\}_{j\in\mathbb{N}}$;

(III) $f(z) \in \mathbb{C}^*$, $\mathbb{C}^* = \mathbb{C} \setminus \{0\}$, for $z \neq z_j$, $z \neq w_j$, $j \in \mathbb{N}$.

3.2 Holomorphic Functions in $\mathbb{C}^n$

Now we fix once and for all a definition of holomorphic function in $\mathbb{C}^n$ and develop some elementary properties of these functions.

A complex-valued function f on a subset $\Omega \subset \mathbb{C}^n$ is merely a mapping from Ω into the complex plane; the value of the function f at a point $z \in \Omega$ will be denoted by $f(z)$, as usual.

For $E \subseteq \mathbb{C}^n$, let $C^0(E)$ be the space of continuous complex-valued functions on E. If $\Omega \subseteq \mathbb{C}^n$ is open, then $C^k(\Omega)(k = 1, 2, \ldots, \infty)$ is the space of k-times continuously differentiable (with respect to the real coordinates x_j) complex-valued functions in Ω, and $C_0^k(\Omega)$ is the subspace of all $f \in C^k(\Omega)$ vanishing outside a compact subset of Ω.

Let $\mathbb{C}^n (n = 1, 2, \ldots)$ be the space of all n-tuples $z = (z_1, \ldots, z_n)$ of complex numbers z_j. The components $z_1, \ldots, z_n$ of $z \in \mathbb{C}^n$ will be called the canonical (complex) coordinates of z. By $x_1, \ldots, x_{2n}$ we denote the real coordinates in $\mathbb{C}^n$ such that $z_y = x_y + ix_{j+n}$.

We introduce the differential operators

$$\frac{\partial}{\partial z_j} := \frac{1}{2}\left(\frac{\partial}{\partial x_j} + \frac{1}{i}\frac{\partial}{\partial x_{j+n}}\right) \quad \text{and} \quad \frac{\partial}{\partial \bar{z}_j} := \frac{1}{2}\left(\frac{\partial}{\partial x_j} - \frac{1}{i}\frac{\partial}{\partial x_{j+n}}\right)$$

These operators $\partial/\partial z_j$ and $\partial/\partial \bar{z}_j$ will also be denoted by ∂_j and $\bar{\partial}_j$, respectively.

A multi-index is an n-tuple $k = (k_1, \ldots, k_n)$ of non-negative integers. We define

$$|k| = k_1 + \cdots + k_n$$
$$k! = k_1! \cdot \cdots \cdot k_n!$$

For every multi-order k we write $z^k := z_1^{k_1} \ldots z_n^{k_n}$ if $z \in \mathbb{C}^n$,

$$\partial^k := \partial_1^{k_1} \circ \ldots \circ \partial_n^{k_n} \text{ and } \bar{\partial}^k := \bar{\partial}_1^{k_1} \circ \ldots \circ \bar{\partial}_n^{k_n}.$$

If $\Omega \cong \mathbb{C}^n$ is open and $f \in C^0(\Omega)$, then we define

$$\partial f := \sum_{j=1}^{n} \frac{\partial f}{\partial z_j} dz_j \quad \text{and} \quad \bar{\partial} f := \sum_{j=1}^{n} \frac{\partial f}{\partial \bar{z}_j} d\bar{z}_j \text{ in } \Omega.$$

Then $df = \partial f + \overline{\partial f}$ in Ω.

3.2.1 Power Series

Many of the more important functions studied in real analysis, such as the exponential and trigonometric functions, are most conveniently defined using power series.

Power series are even more important when we move from real to complex functions. In the 1820s Cauchy made considerable use of the power series $\sum_{j=1}^{\infty} a_j z^j$ of a complex variable z. In particular, any real function having a power series development automatically gives rise to a complex function with the corresponding power series. This provides a natural method for extending real functions to the complex case. In the 1840s Weierstrass showed how to base the theory of complex functions on power series.

We are going to show now that the holomorphic functions in an open set $\Omega \subset \mathbb{C}^n$ can be characterized as those functions that are locally representable by convergent power series.

Let

$$\sum_{k_1,\dots,k_n \in \mathbb{N}} a_{k_1 \dots k_n} \left(\xi_1 - \xi_1^0\right)^{k_1} \cdots \left(\xi_n - \xi_n^0\right)^{k_n} \tag{3.7}$$

be a power series in $\xi_1 - \xi_1^0, \dots, \xi_n - \xi_n^0$, also denoted

$$\sum_{k \in \mathbb{N}^n} a_k \left(z - \xi^0\right)^k$$

If $n > 1$, there does not exist a natural way of summing (3.7). To each increasing sequence $\mathcal{I} = \{I_m\}_{m \geq 1}$ of finite subsets of $\mathbb{N}^n$, such that $\bigcup_{m \geq 1} I_m = \mathbb{N}^n$, corresponds a way of summing the series; setting

$$S_m(z) = \sum_{\alpha \in I_m} a_\alpha \left(z - \xi^0\right)^\alpha \tag{3.8}$$

if there exists $S(z) = \lim_{m \to +\infty} S_m(z)$, it is called the sum of the series with respect to the sequence $\mathcal{I}$.

In general, $S(z)$ depends on $\mathcal{I}$. This does not occur if the series (3.8) converges absolutely, i.e., if the series

$$\sum_{k_1,\dots,k_n \in \mathbb{N}} |a_{k_1 \dots k_n}| \cdot \left|(\xi_1 - \xi_1^0)\right|^{k_1} \cdots \left|(\xi_n - \xi_n^0)\right|^{k_n}$$

is convergent.

The following result, known as Abel's Lemma, gives the basic general result about convergence of power series.

Lemma 3.2.1. *Let a_k be complex numbers defined for all multi-orders k. Suppose that for some $\xi \in \mathbb{C}^n$ $(\xi_j \neq 0$ for all $j)$ the power series $\sum_k a_k \xi^k$ converges. Then the series $\sum_k a_k z^k$ converges normally in the polydisc $|z_j| < |\xi_j|, j = 1, \dots, n$, that is, for all $r < 1$*

$$\sum_k \sup_{|z_j| \leq r |\xi_j|} \left|a_k z^k\right| < \infty. \tag{3.9}$$

Proof. By hypothesis there is $C < \infty$ such that $\left|a_k \xi^k\right| \leqq C$ for all k. Consequently,

$$\sum_k \sup_{|z_j| \leq r |\xi_j|} \left|a_k z^k\right| \leqq C \sum_k r^{k_1 + \dots + k_n} < \infty$$

□

Theorem 3.2.2. *Let $\Omega \subseteqq \mathbb{C}^n$ be an open domain and f a complex-valued function in Ω. Then the following conditions are equivalent:*

(I) $f \in C^0(\Omega)$ *and* $\overline{\partial} f = 0$ *in* Ω.

(II) $f \in C^0(\Omega)$ *and f is holomorphic in each variable z_j when the other variables are kept fixed.*

(III) $f \in C^0(\Omega)$ *and, for every polydisc* $P = P_1 \times \cdots \times P_n \subset\subset \Omega$,

$$f(z) = \frac{1}{(2\pi i)^n} \int_{\partial P_1 \times \cdots \times \partial P_n} \frac{f(\zeta) d\zeta_1 \wedge \cdots \wedge d\zeta_n}{(\zeta_1 - z_1) \cdot \cdots \cdot (\zeta_n - z_n)}, \quad z \in P$$

(IV) *f admits local power series expansions, that is, for every point $\xi \in \Omega$ there are complex numbers a_k defined for every multi-order k such that, for all z in some neighborhood of ξ,*

$$f(z) = \sum_k a_k (z - \xi)^k \tag{3.10}$$

If these equivalent conditions are fulfilled, then, moreover:

(V) $f \in C^\infty(\Omega)$.

(VI) *For every polydisc $P = P_1 \times \cdots \times P_n \subset\subset \Omega$ and every multi-order k,*

$$\partial^k f(z) = \frac{k!}{(2\pi i)^n} \int_{\partial P_1 \times \cdots \times \partial P_n} \frac{f(\zeta) d\zeta_1 \wedge \cdots \wedge d\zeta_n}{(\zeta_1 - z_1)^{k_1+1} \ldots (\zeta_n - z_n)^{k_n+1}}, \quad z \in P \tag{3.11}$$

(VII) *The coefficients in the power series expansion (3.11) are uniquely determined, where*

$$a_k = \frac{\partial^k f(\xi)}{k!} \tag{3.12}$$

(VIII) *The power series expansion (3.10) converges uniformly in each polydisc $P \subset\subset \Omega$ centered at ξ.*

Proof. **(I)** $\Leftrightarrow$ **(II)** according to Corollary 3.1.4. By repeated use of (3.2), we obtain the implication **(I)** $\Rightarrow$ **(III)**.

We prove that **(III)** $\Rightarrow$ **(IV)**. Let $\xi \in \Omega$ and $P = P_1 \times \cdots \times P_n \subset\subset \Omega$ be a polydisc centered at ξ. Since for $z \in P$ and $\zeta \in \partial P_1 \times \cdots \times \partial P_n$

$$\frac{1}{(\zeta_1 - z_1) \ldots (\zeta_n - z_n)} = \sum_{k_1, \ldots, k_n = 0}^{\infty} \frac{(z_1 - \xi_1)^{k_1} \ldots (z_n - \xi_n)^{k_n}}{(\zeta_1 - \xi_1)^{k_2+1} \ldots (\zeta_n - \xi_n)^{k_n+1}}$$

with uniform convergence in ζ, it follows from (3.11) that

$$f(z) = \sum_k \left[\frac{1}{(2\pi i)^n} \int_{\partial P} \frac{f(\zeta) d\zeta_1 \wedge \cdots \wedge d\zeta_n}{(\zeta_1 - \xi_1)^{k_1+1} \ldots (\zeta_n - \xi_n)^{k_n+1}} \right] (z-\xi)^k, \quad z \in P \tag{3.13}$$

Now we prove that **(IV)** $\Rightarrow$ **(II)**. It follows from (3.10) and Lemma 3.2.1 that f is locally the uniform limit of polynomials in $z_1, \ldots, z_n$. Since such polynomials fulfill **(II)**, and since the uniform limit of continuous functions is continuous and the uniform limit of holomorphic functions of one complex variable is holomorphic, we conclude that f satisfies condition **(II)**.

Now we suppose that the equivalent conditions **(I)**–**(IV)** are fulfilled. Then **(V)** follows from **(III)** and **(VI)** can be obtained by differentiation under the sign of integration in (3.11). Property **(VII)** follows, having chosen a sufficiently small polydisc $P = P_1 \times \cdots \times P_n$ centered at ξ, by the following computation, which is permissible in view of (3.11) and Lemma 3.2.1:

$$\begin{aligned} \partial^k f(\xi) &= \frac{k!}{(2\pi i)^n} \sum_l a_l \int_{\partial P_1 \times \cdots \times \partial P_n} \frac{(\zeta - \xi)^l d\zeta_1 \wedge \cdots \wedge d\zeta_n}{(\zeta_1 - \xi_1)^{k_1+1} \ldots (\zeta_n - \xi_n)^{k_n+1}} \\ &= \frac{k!}{(2\pi i)^n} \sum_l a_l \frac{(2\pi i)^n}{l!} \delta_\zeta^k (\zeta - \xi)^l \Big|_{\zeta = \xi} = k! a_k. \end{aligned}$$

(VIII) follows from **(VII)**, (3.13) and Lemma 3.2.1. □

Definition 3.2.3. Let $\Omega \subseteqq \mathbb{C}^n$ be an open domain. A complex-valued function f in Ω is said to be holomorphic (or analytic) if the equivalent conditions in Theorem 3.2.2 are fulfilled. The set of all holomorphic functions in Ω will be denoted by $\mathcal{O}(\Omega)$.

Corollary 3.2.4. *For every open domain* $D \subseteqq \mathbb{C}^n$*, the space* $\mathcal{O}(\Omega)$ *is a ring; that is, if* $f, g \in \mathcal{O}(\Omega)$*, then* $f+g \in \mathcal{O}(\Omega)$ *and* $fg \in \mathcal{O}(\Omega)$*. Further, if* $f \in \mathcal{O}(\Omega)$ *and* $f(z) \neq 0$ *for all* $z \in \Omega$*, then* $1/f \in \mathcal{O}(\Omega)$*.*

Proof. This follows from condition **(II)** in Theorem 3.2.2 and the corresponding properties of holomorphic functions of one complex variable. □

Corollary 3.2.5 (Maximum principle). *Let* $D \subseteqq \mathbb{C}^n$ *be an open domain and* $f \in \mathcal{O}(\Omega)$*. Suppose that there exists a point* $\xi \in \Omega$ *such that* $|f(z)| \leqq |f(\xi)|$ *for all* $z \in \Omega$*. Then* f *is constant in* Ω *if* Ω *is connected.*

Proof. This follows from condition **(II)** in Theorem 3.2.2 and the maximum principle for holomorphic functions of one complex variable. □

Corollary 3.2.6. *Let* $\Omega \subseteq \mathbb{C}^n$ *be an open set and* $f_j \in \mathcal{O}(\Omega)$*. If* $f_f \to f$ *when* $j \to \infty$*, uniformly on every compact subset of* Ω*, then* $f \in \mathcal{O}(\Omega)$*.*

Proof. This follows from condition **(II)** in Theorem 3.2.2 and the corresponding property of holomorphic functions of one complex variable. Clearly, it can also be obtained from condition **(III)** in Theorem 3.2.2. □

Corollary 3.2.7 (uniqueness of holomorphic continuation). *Let $D \subseteqq \mathbb{C}^n$ be an open domain and $f \in \mathcal{O}(\Omega)$. If there is a point $\xi \in \Omega$ where $\partial^k f(\xi) = 0$ for all multi-indices k, then $f = 0$ in Ω if Ω is connected.*

Proof. This follows from condition **(IV)** and relation (3.12) in Theorem 3.2.2. □

Corollary 3.2.8. *For every open domain $\Omega \subseteqq \mathbb{C}^n$ and all multi-indices k, we have $\partial^k \mathcal{O}(\Omega) \subseteq \mathcal{O}(\Omega)$.*

Proof. This follows from property **(VI)** in Theorem 3.2.2. □

Corollary 3.2.9 (Cauchy's inequality). *If f is holomorphic in the polydisc $P := \{z \in \mathbb{C}^n \mid |z_i| < r_i, i = 1, \dots, n\}$, where $0 < r_i < \infty$, $i = 1, \dots, n$, then for all multi-indices k*

$$\left|\hat{\partial}^k f(0)\right| \leqq k! r^{-k} \sup_{z \in P} |f(z)| .$$

Proof. This follows from property **(VI)** in Theorem 3.2.2. □

Theorem 3.2.10. *Let $\Omega \subseteq \mathbb{C}^n$ be an open set. Then, for every compact set $K \subset\subset \Omega$ and every multi-order k, there exists a constant $C = C(K, k)$ such that*

$$\max_{z \in K} \left|\partial^k f(z)\right| \leqq C \int_\Omega |f| d\sigma_{2n} \quad \text{for all} \quad f \in \mathcal{O}(\Omega)$$

where $d\sigma_{2n}$ is the Lebesgue measure in $\mathbb{C}^n$.

Proof. First let $n = 1$. Choose $\chi \in C_0^\infty(\Omega)$ such that $\chi = 1$ in a neighborhood U_E of K. Then, by (3.2), for every $f \in \mathcal{O}(\Omega)$

$$\chi(z) f(z) = \frac{-1}{2\pi i} \int_{\Omega \setminus U_K} \frac{f(\zeta) \bar{\partial} \chi(\zeta) \wedge d\zeta}{\zeta - z}$$

and differentiation leads to the required estimate. Repeated use of this gives the theorem for polydiscs Ω of arbitrary dimension. The general case follows by use of a covering of K by a finite number of polydiscs $\subset\subset \Omega$. □

Corollary 3.2.11 (Stieltjes-Vitali). *Let $\Omega \subseteq \mathbb{C}^n$ be an open set, and let f_k be a sequence of holomorphic functions in Ω which is uniformly bounded on every compact subset of Ω. Then there is a subsequence f_k converging uniformly on every compact subset of Ω to a limit in $\mathcal{O}(\Omega)$.*

Proof. Since $\bar{\partial} f_k = 0$, it follows from Theorem 3.2.10 that all first-order derivatives (with respect to the real coordinates) of f_k are uniformly bounded on any compact subset of Ω. Hence the corollary follows from Ascoli's theorem and Corollary 3.2.6. □

Remark 3.2.12. Suppose the function f is continuous in $\overline{P}$ and is holomorphic with respect to each variable. Then it can be represented by a Cauchy integral (see (**III**)), and then by (**IV**) also by a power series. Since by (**VIII**) these series converge uniformly on compact subsets of P, and their terms are continuous with respect to all the variables jointly, then any of the derivatives is $\mathbb{R}$-differentiable in P and therefore since this derivative is holomorphic in each variable it is holomorphic in U. Hence we see that the function f is $\mathbb{C}$ differentiable, and hence, is also holomorphic in P. Thus, in order to prove the fundamental theorem of Hartogs in Section 3.2.4 it suffices to prove that if f is holomorphic in each variable separately in $\overline{P}$ then it is continuous in $\overline{P}$.

Remark 3.2.13. The hypothesis that $f \in C^1(\Omega)$ in the definition of holomorphic function was not used in the proof of the Cauchy integral formula for polydiscs (**III**). However, to differentiate under the integral sign or to compute the power series in (**VII**), Fubini's theorem must be applied. Therefore, one needs to assume, for instance, that f is bounded on compact sets. Our hypothesis is that $f \in C^0$ is much stronger than is needed.

Lemma 3.2.14 (Cauchy's Integral Formula). *If f is a function which is defined in a neighborhood U of the closed polydisc $\bar{P}(a,r)$ and holomorphic in each variable z_i at each point of U, then*

$$f(z) = \left(\frac{1}{2\pi i}\right)^n \int_{|\zeta_n - a_n| = r_n} \cdots \int_{|\zeta_1 - a_1| = r_1} \frac{f(\zeta_1, \ldots, \zeta_n)\, d\zeta_1 \ldots d\zeta_n}{(\zeta_1 - z_1) \ldots (\zeta_n - z_n)}$$

for each $z = (z_1, z_2, \ldots, z_n) \in P(a,r)$.

This follows immediately from the repeated application of the one-variable Cauchy formula.

Note that, at this point, we do not assume that the function f of the above lemma is continuous or bounded on $\bar{P}(a,r)$ or even integrable as a function on the product of the circles $|\zeta_i - a_i| = r_i$. It is measurable since a separately continuous function on $\mathbb{R}^k$ is a Borel function. However, the integral of the lemma makes perfectly good sense, as an iterated integral, though, as far as we know at this point, not necessarily as an integral relative to the corresponding product measure.

Notice that polynomials in the functions $z_1, \ldots, z_n$ are holomorphic in all of $\mathbb{C}^n$. From Abel's lemma (Lemma 3.2.1) follows that a power series expansion of the form (3.13) is absolutely uniformly convergent in all suitably small open polydiscs $P(w;r)$ centered at the point w. A first consequence of this observation is that the function f is continuous in such polydiscs $P(w;r)$; hence, any function holomorphic in Ω is also continuous in Ω (see (**II**)).

A second consequence is that the power series (3.13) can be rearranged arbitrarily and will still represent the function f. In particular, if the coordinates $z_1, \dots, z_{j-1}, z_{j+1}, \dots, z_n$ are given any fixed values $a_1, \dots, a_{j-1}, a_{j+1}, \dots, a_n$, then this power series can be arranged as a convergent power series in the variable z_j alone, for z_j sufficiently close to w_j; and this holds for any values a_i sufficiently near w_i. That is to say, the function f is holomorphic in **each variable separately** throughout the domain in which it is analytic; thus the ordinary complex derivative with respect to one of the variables z_j is well-defined, and will be denoted by $\partial/\partial z_j$. A converse to the boldface statement is also true, as follows.

Theorem 3.2.15 (Osgood's Lemma). *If f is a function on an open set $U \subset \mathbb{C}^n$, which is bounded on each compact subset of U, and holomorphic in each variable separately on U, then, for each compact polydisc $\overline{P}(a, s) \subset U$, there is a power series of the form*

$$\sum_{i_1,\dots,i_n=0}^{\infty} c_{i_1\dots i_n} (z_1 - a_1)^{i_1} \dots (z_n - a_n)^{i_n}$$

which converges uniformly to f on $\overline{P}(a, s)$.

Proof. Since $\overline{P}(a, s)$ is compact in U, we may choose numbers $r_i > s_i$ such that the compact polydisc $\overline{P}(a, r)$ with polyradius $r = (r_1, \dots, r_n)$ is also contained in U. Then f is bounded on $\overline{P}(a, r)$. We substitute into the integrand of (3.11) the series expansion:

$$\frac{1}{(\zeta_1 - z_1)\dots(\zeta_n - z_n)} = \sum_{i_1,\dots,i_n=0}^{\infty} \frac{(z_1 - a_1)^{i_1} \dots (z_n - a_n)^{i_n}}{(\zeta_1 - a_1)^{i_1+1} \dots (\zeta_n - a_n)^{i_n+1}}.$$

This series converges uniformly in ζ and z for $|z_i - a_i| \le s_i < r_i = |\zeta_i - a_i|$ and, thus, can be integrated termwise. The resulting series in z is uniformly convergent for $z \in \overline{P}(a, s)$ and yields a series expansion for f of the form

$$f(z) = \sum_{i_1,\dots,i_n=0}^{\infty} c_{i_1\dots i_n} (z_1 - a_1)^{i_1} \dots (z_n - a_n)^{i_n}$$

uniformly convergent on $\overline{P}(a, s)$, with

$$c_{i_1\dots i_n} = \left(\frac{1}{2\pi i}\right)^n \int_{|\zeta_n - z_n| = r_n} \dots \int_{|\zeta_1 - z_1| = r_1} \frac{f(\zeta_1, \dots, \zeta_n)\, d\zeta_1 \dots d\zeta_n}{(\zeta_1 - a_1)^{i_1+1} \dots (\zeta_n - a_n)^{i_n+1}}.$$

This completes the proof. □

Thus, Osgood's lemma says that a function on Ω, which is holomorphic in each variable separately at each point of Ω, is holomorphic on Ω, provided it is bounded on each compact subset of Ω. In particular, it is holomorphic if it is holomorphic in each variable separately and is continuous.

One of the most useful theorems of basic complex analysis is the following result, first noted by Giacinto Morera.

Morera's theorem which is stated below may be considered as a characterization of holomorphic functions among the class of continuous functions; in fact, it is the converse to Cauchy's theorem.

A closed path T is said to be *triangular* if it is polygonal and has three sides.

Theorem 3.2.16 (Morera's Theorem, [55, p. 86]). *Let U be a region and let $f : U \to \mathbb{C}$ be a continuous function such that $\int_T f = 0$ for every triangular path T in U; then f is analytic in U.*

Proof. First observe that f will be shown to be analytic if it can be proved that f is analytic on each open disk contained in U. Hence, without loss of generality, we may assume U to be an open disk; suppose $U = B(a; R)$.

Use the hypothesis to prove that f has a primitive. For z in U define $F(z) = \int_{[a,z]} f$. Fix z_0 in U; then for any point z in U the hypothesis gives that $F(z) = \int_{[a,z_0]} f + \int_{[z_0,z]} f$. Hence

$$\frac{F(z) - F(z_0)}{z - z_0} = \frac{1}{z - z_0} \int_{[z_0,z]} f$$

This gives

$$\begin{aligned} \frac{F(z) - F(z_0)}{z - z_0} - f(z_0) &= \frac{1}{(z - z_0)} \int_{[z_0,z]} f - f(z_0) \\ &= \frac{1}{(z - z_0)} \int_{[z_0,z]} [f(w) - f(z_0)]\, dw \end{aligned}$$

But by taking absolute values

$$\left| \frac{F(z) - F(z_0)}{z - z_0} - f(z_0) \right| \leq |f(z) - f(z_0)|$$

which shows that

$$\lim_{z \to z_0} \frac{F(z) - F(z_0)}{z - z_0} = f(z_0)$$

□

Morera's Theorem enables one to establish the analyticity of functions in situations where resort to the definition and the attendant calculation of difference quotients would lead to hopeless complications. Applications of this sort occur, for instance, in the proofs of the Schwarz Reflection Principle and other theorems on the extension of analytic functions. Nor is its usefulness limited to this circle of ideas; the important fact that the uniform limit of analytic functions is again analytic is an immediate consequence (observed already by Morera himself, as well as by Osgood [166], who had rediscovered Morera's theorem).

Theorem 3.2.17. *If $\{f_j\}$ is a sequence of holomorphic functions on $\Omega \subset \mathbb{C}^n$, which is uniformly convergent on each compact subset of Ω, then the limit is also holomorphic on Ω.*

Proof. For holomorphic functions of one variable, this is a standard result. Its proof is a simple application of Morera's theorem. In the several variable case we simply apply this result in each variable separately (with the other variables fixed) to conclude that the limit of a sequence of holomorphic functions, converging uniformly on compacta, is holomorphic in each variable separately. Such a limit is also obviously continuous, and so is holomorphic, by Osgood's lemma. □

3.2.2 Chains and Cycles

The winding number of a closed curve γ in the complex plane can be expressed in terms of the complex coordinate $z = x+iy$. Specifically, if we write $z = re^{i\theta}$, then

$$dz = e^{i\theta} dr + ire^{i\theta} d\theta$$

and therefore

$$\frac{dz}{z} = \frac{dr}{r} + id\theta = d[\ln r] + id\theta$$

As γ is a closed curve, the total change in $\ln(r)$ is zero, and thus the integral of $\frac{dz}{z}$ is equal to i multiplied by the total change in θ. Therefore the winding number of a closed path γ about the origin is given by the expression

$$\frac{1}{2\pi i} \oint_\gamma \frac{dz}{z} .$$

More generally, if γ is a closed curve parameterized by $t \in [\alpha, \beta]$, the winding number of γ about a, also known as the index of a with respect to γ, is defined for complex $a \notin \gamma([\alpha, \beta])$ as

$$\operatorname{Ind}_\gamma (a) = \frac{1}{2\pi i} \oint_\gamma \frac{d\zeta}{\zeta - a} = \frac{1}{2\pi i} \int_\alpha^\beta \frac{\gamma'(t)}{\gamma(t) - a} dt .$$

Then the cycle γ is *homologous to zero* in U if the winding number $\operatorname{Ind}_\gamma (a) = 0$ for all $a \notin U$. If γ is a closed path homotopic to zero, then it is homologous to zero.

Suppose that $\gamma_1, \ldots, \gamma_n$ are paths in the plane, and put $K = \gamma_1^* \cup \ldots \cup \gamma_n^*$. We let $\gamma^\star$ denote the set of all points $\gamma(t)$, for t in the parameter interval of γ.

Each γ_i induces a linear functional $\widetilde{\gamma}_i$ on the vector space $C(K)$ by the formula

$$\widetilde{\gamma}_i(f) = \int_{\gamma_i} f(z) dz .$$

Define

$$\widetilde{\Gamma} = \widetilde{\gamma}_1 + \cdots + \widetilde{\gamma}_m \tag{3.14}$$

Explicitly, $\widetilde{\Gamma}(f) = \widetilde{\gamma}_1(f) + \cdots + \widetilde{\gamma}_n(f)$ for all $f \in C(K)$. The relation (3.14) suggests that we introduce a "formal sum"

$$\Gamma = \gamma_1 \dot{+} \ldots \dot{+} \gamma_m \tag{3.15}$$

and define

$$\int_\Gamma f(z)dz = \widetilde{\Gamma}(f)\,.$$

Then (3.15) is merely an abbreviation for the statement

$$\int_\Gamma f(z)dz = \sum_{i=1}^{n} \int_{\gamma_i} f(z)dz \quad (f \in C(K))\,. \tag{3.16}$$

Note that (3.16) serves as the definition of its left side.

The objects Γ so defined are called *chains*. If each γ_j in (3.15) is a closed path, then Γ is called a *cycle*.

If each γ_j in (3.15) is a path in some open set Ω, we say that Γ is a chain in Ω. If (3.15) holds, we define

$$\Gamma^* = \gamma_1^* \cup \ldots \cup \gamma_n^*\,.$$

If Γ is a cycle and $\alpha \notin \Gamma^*$, we define the index of α with respect to Γ by

$$\mathrm{Ind}_\Gamma(\alpha) = \frac{1}{2\pi i} \int_\Gamma \frac{dz}{z-\alpha}\,.$$

Obviously, (3.15) implies

$$\mathrm{Ind}_\Gamma(\alpha) = \sum_{i=1}^{n} \mathrm{Ind}_{\gamma_i}(\alpha)\,.$$

Suppose that each γ_i in (3.15) is replaced by its opposite path. (However, suppose that [0, 1] is the parameter interval of a path γ_i, and $\gamma_i^- = \gamma_i(1-t)$, $0 < t < 1$.) We call γ_i^- the *path opposite* to γ_i — the resulting chain will be denoted by $-\Gamma$. Then

$$\int_{-\Gamma} f(z)dz = -\int_\Gamma f(z)dz \quad (f \in C(\Gamma^*))\,.$$

In particular, $\mathrm{Ind}_{-\Gamma}(\alpha) = -\mathrm{Ind}_\Gamma(\alpha)$ if Γ is a cycle and $\alpha \notin \Gamma^*$.

Chains can be added and subtracted in the obvious way, by adding or subtracting the corresponding functionals: The statement $\Gamma = \Gamma_1 \dot{+} \Gamma_2$ means

$$\int_\Gamma f(z)dz = \int_{\Gamma_1} f(z)dz + \int_{\Gamma_2} f(z)dz$$

for every $f \in C(\Gamma_1^* \cup \Gamma_2^*)$.

Finally, note that a chain may be represented as a sum of paths in many ways. To say that

$$\gamma_1 \dot{+} \ldots \dot{+} \gamma_n = \delta_1 \dot{+} \ldots \dot{+} \delta_k$$

means simply that

$$\sum_i \int_{\gamma_i} f(z)dz = \sum_j \int_{\delta_j} f(z)dz$$

for every f that is continuous on $\gamma_1^* \cup \ldots \cup \gamma_n^* \cup \delta_1^* \cup \ldots \cup \delta_k^*$. In particular, a cycle may very well be represented as a sum of paths that are not closed.

Definition 3.2.18. A chain Γ is said to bound the region U if and only if $Ind_\Gamma(a)$ is defined and equal to 1 for all points a ∂U and either undefined or equal to zero for all points not in U.

3.2.3 Cauchy's Theorem

Let $f(x)$ be a complex valued function defined and continuous in a region U of the complex plane. We define the function

$$h(x,y) = \frac{f(x) - f(y)}{x - y}, \quad \text{for} \quad x, y \in U, x \neq y$$
$$h(x,x) = f'(x), \quad \text{for} \quad x \in U, \quad \text{when } f'(x) \text{ exists at } x.$$

Thus $h(x,y)$ is defined at all points of the cartesian product space $U \times U$ except at points (x,x) of the diagonal Δ of this space where f fails to be differentiable at x. Further, $h(x,y)$ is continuous in (x,y) at all points of $U \times U - \Delta$ and is continuous in x (and in y) separately at points (x,x) of Δ where it is defined, i.e., such that $f'(x)$ exists. We note also that $h(x,y)$ is symmetric in x and y.

Moreover, let $w = f(x)$ be continuous in a region U and differentiable at all points of $U \setminus F$ where F is a finite set of points. Then $h(x,y)$ is defined, and continuous at all points (x,y) in $U \times U$. Thus $f'(x)$ exists and is continuous at all points of U [209, (4.2) Theorem, p. 75].

Therefore we see that the following result holds:

Lemma 3.2.19. *If $f \in \mathcal{O}(U)$ and g is defined in $U \times U$ by*

$$g(z,w) = \begin{cases} \frac{f(z)-f(w)}{z-w} & \text{if } w \neq z \\ f'(z) & \text{if } w = z \end{cases}$$

then g is continuous in $U \times U$.

The proof of this result is more simple than that in [209, (4.2) Theorem, p. 75].

Proof. The only points $(z,w) \in U \times U$ at which the continuity of g is possibly in doubt have $z = w$.

Fix $a \in U$. Fix $\epsilon > 0$. There exists $r > 0$ such that $D(a;r) \subset U$ and $|f'(\zeta) - f'(a)| < \epsilon$ for all $\zeta \in D(a;r)$. If z and w are in $D(a;r)$ and if

$$\zeta(t) = (1-t)z + tw$$

then $\zeta(t) \in D(a;r)$ for $0 \le t \le 1$, and

$$g(z,w) - g(a,a) = \int_0^1 [f'(\zeta(t)) - f'(a)]\, dt$$

The absolute value of the integrand is less than ϵ, for every t. Thus $|g(z,w) - g(a,a)| < \epsilon$. This proves that g is continuous at (a,a). □

Theorem 3.2.20 (Cauchy's Theorem)**.** *Suppose that $f \in \mathcal{O}(U)$, where U is an arbitrary open set in the complex plane. If Γ is a cycle in U that satisfies*

$$\operatorname{Ind}_\Gamma(\alpha) = 0 \quad \textit{for every } \alpha \textit{ not in } U \tag{3.17}$$

then

$$f(z) \cdot \operatorname{Ind}_\Gamma(z) = \frac{1}{2\pi i} \int_\Gamma \frac{f(w)}{w-z} dw \quad \textit{for } z \in U \setminus \Gamma^* \tag{3.18}$$

and

$$\int_\Gamma f(z) dz = 0\,. \tag{3.19}$$

If Γ_0 and Γ_1 are cycles in U such that

$$\operatorname{Ind}_{\Gamma_0}(\alpha) = \operatorname{Ind}_{\Gamma_1}(\alpha) \quad \textit{for every } \alpha \textit{ not in } U \tag{3.20}$$

then

$$\int_{\Gamma_0} f(z) dz = \int_{\Gamma_1} f(z) dz \tag{3.21}$$

Proof. The function g defined in $U \times U$ by

$$g(z,w) = \begin{cases} \frac{f(w)-f(z)}{w-z} & \text{if } w \neq z \\ f'(z) & \text{if } w = z \end{cases}$$

is continuous in $U \times U$ (see Lemma 3.2.19). Hence we can define

$$h(z) = \frac{1}{2\pi i} \int_\Gamma g(z,w) dw \quad (z \in U)$$

For $z \in U \setminus \Gamma^*$, the Cauchy formula (3.18) is clearly equivalent to the assertion that

$$h(z) = 0 \tag{3.22}$$

To prove (3.22), let us first prove that $h \in \mathcal{O}(U)$. Note that g is uniformly continuous on every compact subset of $U \times U$. If $z \in U$, $z_n \in U$, and $z_n \to z$, it follows that $g(z_n, w) \to g(z,w)$ uniformly for $w \in \Gamma^*$ (a compact subset of

U). Hence $h(z_n) \to h(z)$. This proves that h is continuous in U. Let Δ be a closed triangle in U. Then

$$\int_{\partial\Delta} h(z)dz = \frac{1}{2\pi i}\int_{\Gamma}\left(\int_{\partial\Delta} g(z,w)dz\right)dw \qquad (3.23)$$

For each $w \in U, z \to g(z,w)$ is holomorphic in U. (The singularity at $z = w$ is removable.) The inner integral on the right side of (3.23) is therefore 0 for every $w \in \Gamma^*$. Morera's theorem (Theorem 3.2.16) shows now that $h \in \mathcal{O}(U)$.

Next, we let U_1 be the set of all complex numbers z for which $\mathrm{Ind}_\Gamma(z) = 0$, and we define

$$h_1(z) = \frac{1}{2\pi i}\int_{\Gamma}\frac{f(w)}{w-z}dw \qquad (z \in U_1)$$

If $z \in U \cap U_1$, the definition of U_1 makes it clear that $h_1(z) = h(z)$. Hence there is a function $\varphi \in H(U \cup U_1)$ whose restriction to U is h and whose restriction to U_1 is h_1.

Our hypothesis (3.17) shows that U_1 contains the complement of U. Thus φ is an entire function. U_1 also contains the unbounded component of the complement of Γ^*, since $\mathrm{Ind}_\Gamma(z)$ is 0 there. Hence

$$\lim_{|z|\to\infty}\varphi(z) = \lim_{|z|\to\infty} h_1(z) = 0$$

Liouville's theorem implies now that $\varphi(z) = 0$ for every z. This proves (3.22), and hence (3.18).

To deduce (3.19) from (3.18), pick $a \in U\backslash\Gamma^*$ and define $F(z) = (z-a)f(z)$. Then

$$\frac{1}{2\pi i}\int_{\Gamma} f(z)dz = \frac{1}{2\pi i}\int_{\Gamma}\frac{F(z)}{z-a}dz = F(a)\cdot \mathrm{Ind}_\Gamma(a) = 0$$

because $F(a) = 0$.

Finally, (3.21) follows from (3.20) if (3.19) is applied to the cycle $\Gamma = \Gamma_1 - \Gamma_0$. This completes the proof. □

To paraphrase Cauchy's theorem: if the integral $\int_\gamma f(z)dz$ vanishes for every function of the form $\frac{1}{z-a}$, where $a \notin U$, then it vanishes for every holomorphic function on U.

Corollary 3.2.21. *Let U be a simply connected domain. Then*

$$\int_\gamma f(z)dz = 0$$

for every holomorphic function f on U and every cycle γ.

Proof. Every cycle is a linear combination of closed paths. Any closed path in a simply connected domain is homotopic to zero, hence homologous to zero (winding number argument). Thus, Cauchy's theorem applies. □

Theorem 3.2.22 (argument principle). *Let* $\gamma \subset \mathbb{C}$ *be a simple closed Jordan curve. Let* U *be the region enclosed by* γ. *If* $f(z)$ *is a holomorphic function on* U, *and* f *has no zeros on* γ, *then*

$$\frac{1}{2\pi\sqrt{-1}}\int_\gamma \frac{f'(z)}{f(z)}dz = N(0)$$

where $N(0)$ *denotes the number of zeros inside the contour* γ, *with multiplicities counted.*

Proof. By change of variables $z \mapsto w := f(z)$, we have that

$$\frac{1}{2\pi\sqrt{-1}}\int_\gamma \frac{f'(z)}{f(z)}dz = \frac{1}{2\pi\sqrt{-1}}\int_{f(\gamma)} \frac{1}{w}dw = N(0),$$

where the last equation is due to a direct computation. This completes the proof of the argument principle. □

If γ is a closed curve in the complex plane which does not pass through a zero of $f(z)$ and R is the interior of γ, it is well known from the theory of complex variables that

$$s_N = \frac{1}{2\pi i}\int_C z^N \frac{f'(z)}{f(z)}dz = \sum_{i=1}^{\nu} {z_i}^N$$

where $z_i (i = 1, 2, \ldots, \nu)$ are all the zeros of $f(z)$ which lie in R. (A multiple zero is counted according to its multiplicity in this formula.)

We are not satisfied by knowing how many zeros are inside a contour γ. Roughly speaking, for a fixed q in the image of f, we want to compute the number of z so that $f(z) = q$. For this, we analyze $f(z) - q$ and obtain the following generalized argument principle.

Corollary 3.2.23 (generalized argument principle). *Let* $\gamma \subset \mathbb{C}$ *be a closed contour, i.e., a simple closed Jordan curve. Let* Ω *be the region enclosed by* γ. *Let* $f(z)$ *be a holomorphic function on* Ω. *Fix* $q \in f(\Omega)$ *and assume there is no* $z \in \gamma$ *so that* $f(z) = q$. *Then*

$$\frac{1}{2\pi\sqrt{-1}}\int_\gamma \frac{f'(z)}{f(z)-q}dz = N(q)$$

where $N(q)$ *denotes the number of* z *for which* $f(z) = q$ *inside the contour* γ, *with multiplicities counted.*

Proof. We observe that

$$\frac{1}{2\pi\sqrt{-1}}\int_\gamma \frac{f'(z)}{f(z)-q}dz \in \mathbb{Z}^+$$

for all holomorphic functions f. Observe that $\frac{f'(z)}{f(z)-q}$ is continuous on γ if and only if $f(z) \neq q$ on γ. Thus

$$\frac{1}{2\pi\sqrt{-1}} \int_\gamma \frac{f'(z)}{f(z)-q} dz$$

is continuous. But a continuous function with image in the integers is a constant function. This completes the proof.

Moreover, the generalized argument principle says that $N(q)$ remains constant on each component of $f(\Omega)\backslash f(\gamma)$ as long as no points in $f^{-1}(q)$ cross the boundary. □

Remark 3.2.24.

(a) The last part of Theorem 3.2.20 shows under what circumstances integration over one cycle can be replaced by integration over another, without changing the value of the integral.

For example, let Ω be the plane with three disjoint closed discs D_i removed. If $\Gamma, \gamma_1, \gamma_2, \gamma_3$ are positively oriented circles in Ω such that Γ surrounds $D_1 \cup D_2 \cup D_3$ and γ_i surrounds D_i but not D_j for $j \neq i$, then

$$\int_\Gamma f(z)dz = \sum_{i=1}^{3} \int_{\gamma_i} f(z)dz$$

for every $f \in \mathcal{O}(\Omega)$.

(b) In order to apply Theorem 3.2.20, it is desirable to have a reasonably efficient method of finding the index of a point with respect to a closed path. The following theorem does this for all paths that occur in practice.

It says, essentially, that the index increases by 1 when the path is crossed "from right to left." If we recall that $\mathrm{Ind}_\gamma(\alpha) = 0$ if α is in the unbounded component of the complement W of γ^*, we can then successively determine $\mathrm{Ind}_\gamma(\alpha)$ in the other components of W, provided that W has only finitely many components and that γ traverses no arc more than once.

Theorem 3.2.25. *If $\{f_j\} \subset \mathcal{O}(U)$ and $\{f_j\}$ converges uniformly on all compact subsets of U, then the limit function f is holomorphic on U.*

Proof. We already know that $f \in \mathcal{C}(U)$. Let γ be any closed curve homotopic to a point in U. Then, by Cauchy's theorem,

$$\int_\gamma f_j(z)dz = 0$$

By uniform convergence it follows that

$$\int_\gamma f(z)dz = \lim_{j\to\infty} \int_\gamma f_j(z)dz = 0$$

and then, by Morera's theorem (Theorem 3.2.16), f is holomorphic on U. □

Theorem 3.2.26. *Let* $\{f_j\}$ *be a sequence of holomorphic functions on* U *such that* $f_j \to f$ *uniformly on all compact subsets of* U.

If $f_j(z) \neq 0$ *for all* $z \in U$ *and all* $j \in \mathbb{N}$, *then either*

(a) f *is identically zero, or*

(b) $f(z) \neq 0$ *for all* $z \in U$.

Proof. Assume that there is a $a \in U$ with $f(a) = 0$ and that f is not identically zero. Then there exists a disc $\mathbb{D}(a)$ with center a such that $\mathbb{D}(a) \subset U$ and $f(z) \neq 0$ for all $z \in \mathbb{D}(a) \setminus \{a\}$. Set $\gamma = \partial\mathbb{D}(a)$.

Therefore the number N of zeros of f in $\mathbb{D}(a)$ satisfies

$$N = \frac{1}{2\pi\imath} \int_\gamma \frac{f'(z)}{f(z)} dz \geq 1$$

But

$$\int_\gamma \frac{f'(z)}{f(z)} dz = \lim_{n\to\infty} \int_\gamma \frac{f_j'(z)}{f_j(z)} dz = 0$$

□

Green's Theorem [187, Theorem 10.45, p. 282]: Let $\Omega \subseteq \mathbb{C}$ be a domain with C^1 boundary. Suppose that $F(x, y) = u(x, y) + iv(x, y)$ is a function that is in $C^1(\overline{\Omega})$. Then

$$\oint_C udx + vdy = \iint_\Omega (v_x - u_y)\, dxdy$$

In notation that is more closely allied with complex function theory, Green's theorem may be expressed (with $F = u + iv$) as

$$\oint_C Fdz = 2i \iint_\Omega \frac{\partial F}{\partial \overline{z}} dxdy$$

If the complex line integral of F around any closed curve in Ω is 0 then by Green's theorem the integral of $\partial F/\partial\overline{z}$ on any small region is zero. Therefore $\partial F/\partial\overline{z}$ is identically 0.

The vanishing of $\oint_\gamma Fdz$ for all closed γ is sometimes referred to as "path independence."

A proof of Morera's theorem based on other ideas is contained in the book of Greene/Krantz [87].

The next consequence of uniform convergence on compact sets is that uniform convergence of a sequence of holomorphic functions on compact subsets implies uniform convergence of the derivatives on compact subsets. This is another feature of holomorphic functions not shared by real differentiable functions: It is easy to construct a sequence of differentiable functions converging uniformly on a closed interval and such that the sequence of derivatives does not converge uniformly.

Theorem 3.2.27 (Weierstrauss's theorem). *If $\{f_j\} \subset \mathcal{O}(\Omega)$ and $f_j \to f$ uniformly on all compact subsets of $\Omega \subset \mathbb{C}^n$, then for any $k = (k_1, \ldots, k_n)$,*

$$\frac{\partial^{|k|} f_j}{\partial z^k} \to \frac{\partial^{|k|} f}{\partial z^k}$$

on any compact subset of Ω.

Proof. First we give proof that $f \in \mathcal{O}(\Omega)$. Let $\Omega \subset \mathbb{C}$. We already know that $f \in \mathcal{C}(\Omega)$. Let γ be any closed curve homotopic to a point in Ω. Then, by Cauchy's theorem,

$$\int_\gamma f_n(z)dz = 0$$

By uniform convergence it follows that

$$\int_\gamma f(z)dz = \lim_{j\to\infty} \int_\gamma f_j(z)dz = 0$$

and then, by Morera's (Theorem 3.2.16), f is holomorphic on Ω.

In the several variable case we simply apply this result in each variable separately (with the other variables fixed) to conclude that the limit of a sequence of holomorphic functions, converging uniformly on compacta, is holomorphic in each variable separately. Such a limit is holomorphic, by Hartogs's Theorem 3.2.31.

It suffices to prove the second part of the theorem for a neighborhood of an arbitrary point $z^0 \in \Omega$ and the derivative with respect to a single variable z_μ. We take the polydisc $P = P\left(z^0, r\right) \subset\subset \Omega$ and use Cauchy's formula, according to which at any point $z \in P$

$$\frac{\partial f_j}{\partial z_\mu} - \frac{\partial f}{\partial z_\mu} = \frac{1}{(2\pi i)^n} \int_\Gamma \frac{f_j(\zeta) - f(\zeta)}{(\zeta - z)\left(\zeta_\mu - z_\mu\right)} d\zeta \tag{3.24}$$

where Γ is the skeleton of P. Since $f_j \to f$ uniformly on Γ, it follows that for any $\varepsilon > 0$ there is a j_0 such that $\|f_j - f\|_\Gamma < \varepsilon$ for all $j \geq j_0$. If $K \subset\subset U$, then from (3.24) we will have, for all $z \in K$ and $j \geq j_0$,

$$\left| \frac{\partial f_j}{\partial z_\mu} - \frac{\partial f}{\partial z_\mu} \right| < \frac{\varepsilon}{(2\pi)^n} \frac{(2\pi)^n r_1 \cdots r_n}{\min_{z\in K, \zeta\in\Gamma} |\zeta - z| \left|\zeta_\mu - z_\mu\right|}$$

from this it follows that $\partial f_j / \partial z_\mu \to \partial f / \partial z_\mu$ uniformly on K. □

Theorem 3.2.27 has no analog in real variables: It is easy to see that the absolute value function on $\mathbb{R}$, which has no derivative at 0, can be approximated uniformly by differentiable functions. A more extreme example was constructed by Weierstrass:

$$F(\theta) = \sum_{j=0}^{\infty} 2^{-j} \exp^{-i2^j \theta}.$$

Function $F(\theta)$ defined on $[0,1]$ is nowhere differentiable and uniformly approximated by polynomials.

We will need the following theorem which is an n-dimensional version of Hurwitz's theorem. Observe that its proof uses nothing other than the amximum modulus theorem.

Theorem 3.2.28 (Hurwitz). *Let Ω be a domain in $\mathbb{C}^n$ and $\{f_j\}$ a sequence of non-vanishing holomorphic unctions f_j which converges uniformly on compact subsets to a holomorphic function f on Ω. Then f vanishes either everywhere or nowhere.*

Proof. Let $n = 1$. If the limit function is constant, our assertion is evident. If $f(z) \not\equiv$ const, then one can find for any point $z_0 \in \Omega$ a circle $\gamma_r = \{z \in \mathbb{C} \mid |z - z^0| = r\}$ on which $f(z) \neq 0$. Then there exists a number $\varepsilon > 0$ such that $|f(\zeta)| > \varepsilon$ and $|f_j(\zeta)| > \varepsilon/2$ when $\zeta \in \gamma_r$ and $j > j_0$. Here j_0 is some natural number chosen appropriately. Since $f_j(z) \neq 0$ for $z \in \Omega$, the maximum modulus theorem implies that for all z such that $|z - z^0| \leq r$ we have $|f_j(z)| > \varepsilon/2$. Hence it follows that $f(z^0) \neq 0$.

Let $n > 1$, $z^0 \in \Omega$ be an arbitrary point and the function $f(z) \not\equiv$ const in the domain $\Omega \subset \mathbb{C}^n$. Then one can choose a complex line $l_{z^0}(t) = \{z \in \mathbb{C}^n \mid z = z^0 + ta, t \in \mathbb{R}\}$ through z^0, where $|a| = 1$ and $t \in \mathbb{C}$, in such a way that the function $\phi(t) = f(l_{z^0}(t))$ is not constant in a neighborhood of the point $t = 0$. Set $\phi_j(t) = f_j(z^0 + at)$; then in some neighborhood of 0 all of the functions $\phi_j(t) \neq 0$ and $\lim_{j\to\infty} \phi_j(t) = \phi(t)$ is reached uniformly. Hence, as we have just seen, it follows that $\phi(0) = f(z^0) \neq 0$. The lemma is proved. □

Thus far our development of the theory of analytic functions of more than one complex variable has paralleled the one-variable theory very closely.

3.2.4 Hartogs's Theorem

Historically, holomorphic functions on $\Omega \subseteq \mathbb{C}^n$ were defined to be functions that are holomorphic in each variable separately and bounded on compact sets; then the Cauchy integral for polydiscs implies that they are infinitely differentiable. It was Hartogs who proved, in 1906, that the hypothesis of boundedness on compacta is superfluous. His proof, using a lemma of Osgood and some subharmonic function theory, is very ingenious and a bit difficult to fathom. To this day there is not an essentially simpler proof.

In what follows, the notation $|U|$ will be used to denote the volume of a set $U \subset \mathbb{C}^n$ and dV will denote volume measure on $\mathbb{C}^n$.

Theorem 3.2.29 (Jensen's Inequality). *If f is holomorphic in a neighborhood of $\overline{P} = \overline{P}(a,r)$, then*

$$\log|f(a)| \leq \frac{1}{|P|}\int_P \log|f(z)|dV(z)\,.$$

Proof. We recall the single variable version of Jensen's inequality:

$$\log|g(a)| \leq \frac{1}{2\pi}\int_0^{2\pi} \log\left|g\left(a+\rho e^{i\theta}\right)\right| d\theta$$

and apply it to f, considered as a function of z_1 only, with the other variables fixed at $a_2, \ldots, a_n$. For $0 \leq \rho_1 \leq r_1$, this yields:

$$\log|f(a)| \leq \frac{1}{2\pi}\int_0^{2\pi} \log\left|f\left(a_1+\rho_1 e^{i\theta_1}, a_2, \ldots, a_n\right)\right| d\theta_1 .$$

We now apply Jensen's inequality to the integrand of this expression, where f is considered as a function of z_2 with z_1 fixed at $a_1+\rho_1 e^{i\theta_1}$ and the remaining variables fixed at $a_3, \ldots, a_n$ to obtain, for $0 \leq \rho_2 \leq r_2$:

$$\log|f(a)| \leq \frac{1}{(2\pi)^2}\int_0^{2\pi}\int_0^{2\pi} \log\left|f\left(a_1+\rho_1 e^{i\theta_1}, a_2+\rho_2 e^{i\theta_2}, a_3, \ldots, a_n\right)\right| d\theta_1 d\theta_2 .$$

Continuing in this way, we obtain:

$$\log|f(a)| \leq \frac{1}{(2\pi)^n}\int_0^{2\pi}\cdots\int_0^{2\pi} \log\left|f\left(a_1+\rho_1 e^{i\theta_1}, \ldots, a_n+\rho_n e^{i\theta_n}\right)\right| d\theta_1 \ldots d\theta_n .$$

Finally, we multiply both sides of this expression by the radial variables $\rho_1 \ldots \rho_n$ and integrate with respect to $d\rho_1 \ldots d\rho_n$ over the set

$$\{0 \leq \rho_i \leq r_i; \quad i = 1, \ldots, n\}$$

to obtain the inequality of the theorem. □

In the next lemma we will write a point $z \in \mathbb{C}^n$ as $z = (z', z_n)$, with $z' \in \mathbb{C}^{n-1}$. Similarly we will write a polyradius r as $r = (r', r_n)$, so that a polydisc $P(a, r)$ can be written as $\Delta(a', r') \times \Delta(a_n, r_n)$.

Lemma 3.2.30 (Hartogs's lemma). *Let f be holomorphic in $P(0, r)$ and let*

$$f(z) = \sum_k f_k(z')\, z_n^k$$

be the power series expansion of f in the variable z_n, where the f_k are holomorphic in $P(0, r')$. If there is a number $c > r_n$ such that this series converges in $\overline{P}(0, c)$, for each $z' \in P(0, r')$, then it converges uniformly on each compact subset of $P(0, r') \times P(0, c)$. Thus, f extends to be holomorphic in this larger polydisc.

Proof. Choose a point $a' \in P(0, r')$, a closed polydisc $\overline{P}(a', s') \subset P(0, r')$, and some positive $b < r_n$. Since f is holomorphic in $P(0, r)$, there is an upper bound $M > 1$ for $|f|$ on the compact polydisc $\overline{P}(a', s') \times \overline{P}(0, b) \subset P(0, r)$. It follows from Cauchy's inequalities that $|f_k(z')| \leq Mb^{-k}$ for all $z' \in \overline{P}(a', s')$. Hence

$$k^{-1} \log |f_k(z')| \leq k^{-1} \log M - \log b \leq \log M - \log b = M_0$$

for all $z' \in \overline{P}(a', s')$ and all k. If the above series converges at (z', c), then $|f_k(z')| c^k$ converges to 0. Hence

$$\limsup_k k^{-1} \log |f_k(z')| \leq -\log c$$

for all $z' \in \overline{P}(a', s')$.

The functions $k^{-1} \log |f_k(z')|$ are bounded and measurable on $\overline{P}(a', s')$, and so, with $P = P(a', s')$, Fatou's lemma implies

$$\begin{aligned} \limsup_k \int_P k^{-1} \log |f_k(z')| \, dV(z') &\leq \int_P \limsup_k k^{-1} \log |f_k(z')| \, dV(z') \\ &\leq -|P| \log c \,. \end{aligned}$$

We conclude that if $0 < c_0 < c$, then there is a k_0 and a $\delta > 0$ such that

$$\int_P k^{-1} \log |f_k(z')| \, dV(z') \leq -|P| \log c_0 - \delta$$

for all $k \geq k_0$. Now, since $k^{-1} \log |f_k(z')|$ is bounded by M_0 for all k and all $z' \in P$, we may choose ϵ sufficiently small that, if $P = P(a', s')$ is replaced by $P_\epsilon = P(w', s' - \hat{\epsilon})$, where $w' \in P(a', \hat{\epsilon})$ and $\hat{\epsilon} = (\epsilon, \dots, \epsilon)$, we will have

$$\int_{P_\epsilon} k^{-1} \log |f_k(z')| \, dV(z') \leq -|P_\epsilon| \log c_0 \,.$$

By Jensen's inequality, this yields

$$k^{-1} \log |f_k(w')| \leq -\log c_0$$

or

$$|f_k(w')| \, c_0^k \leq 1$$

for all $k \geq k_0$ and $w' \in P_\epsilon$. This implies that our series is uniformly convergent on compact subsets of $P_\epsilon \times P(0, c_0)$. Since a' is an arbitrary point of $P(0, r')$ and c_0 is an arbitrary positive number less than c, our power series serves to extend f to a function which is continuous on $P(0, r') \times P(0, c)$ and holomorphic in each variable. It follows from Osgood's lemma that this extension of f is, in fact, holomorphic. □

Theorem 3.2.31 (Hartogs's Theorem). *If a complex-valued function f is holomorphic in each variable separately in a domain $U \subset \mathbb{C}^n$, then it is holomorphic in U.*

Proof. We prove this by induction on the dimension n. There is nothing to prove if $n = 1$. Thus, we suppose that $n > 1$ and that the theorem is true for dimension $n - 1$.

Let a be a point in U and let $\overline{P}(a, r)$ be a closed polydisc contained in U. As in the previous lemma, we write $z = (z', z_n)$ for points of $\mathbb{C}^n$ and write $\overline{P}(a, r) = \overline{P}(a', r') \times \overline{P}(a_n, r_n)$.

Consider the sets

$$X_k = \left\{z_n \in \overline{P}(a_n, r_n/2) \mid |f(z', z_n)| \le k \quad \forall z' \in \overline{P}(a', r')\right\}.$$

For each k, this set is closed, since $f(z', z_n)$ is continuous in z_n for each fixed z'. On the other hand, by the induction hypothesis, $f(z', z_n)$ is also continuous in z' and, hence, bounded on $\overline{P}(a', r')$, for each fixed $z_n \in \overline{P}(a_n, r_n/2)$. We conclude that $\overline{P}(a_n, r_n/2) \subset \bigcup_k X_k$. It follows from the Baire category theorem that, for some k, the set X_k contains a neighborhood $P(b_n, P)$ of some point $b_n \in P(a_n, r_n/2)$.

We now know that f is separately holomorphic and uniformly bounded in the polydisc $P(a', r') \times \overline{P}(b_n, P)$. We conclude from Osgood's lemma that f is holomorphic on $P(a', r') \times P(b_n, P)$ and, in fact, has a power series expansion about (a', b_n) which converges uniformly on compact subsets of this polydisc.

We choose $s_n > r_n/2$ so that $P(b_n, s_n) \subset P(a_n, r_n)$. Then $f(z', z_n)$ is holomorphic in z_n on $P(b_n, s_n)$ for each $z' \in P(a', r')$ and, hence, its power series expansion about (a', b_n) converges as a power series in z_n on $P(b_n, s_n)$ for each fixed point $z' \in P(a', r')$. It follows from Hartogs's lemma that f is actually holomorphic on all of $P(a', r') \times P(b_n, s_n)$. Since a is in this set and a was an arbitrary element of U, the proof is complete. □

A function is holomorphic in each variable $z_j = x_j + iy_j$ separately in a domain U if and only if, for each j, it is differentiable as a function of the pair of real variables x_j, y_j and satisfies the equation $\partial f/\partial \bar{z}_j = 0$ in U. Thus, by Hartogs theorem, we have:

Corollary 3.2.32. *A function defined in an open set $U \subset \mathbb{C}^n$ is holomorphic in U provided, for $j = 1, \ldots, n$, it is differentiable in x_j, y_j and $\partial f/\partial \bar{z}_j = 0$ in U.*

Remark 3.2.33. A much stronger result is true: A distribution ψ, defined in a domain in $\mathbb{C}^n$, which satisfies the Cauchy-Riemann equations, $\partial\psi/\partial\bar{z}_j = 0, j = 1, \ldots, n$ in the distribution sense, is actually a holomorphic function. We will not prove this here, but it follows from regularity theorems for elliptic PDE's.

Remark 3.2.34. The "real variable analogue" of Hartogs's theorem is that a function

$$f : \mathbb{R}^N \to \mathbb{R}$$

such that

$$t \mapsto f(x_1, \ldots, x_{j-1}, t, x_{j+1}, \ldots, x_N)$$

is in $C^\infty(\mathbb{R})$ for each j, and for each fixed set of $x_1, \ldots, x_{j-1}, x_{j+1}, \ldots, x_N$, is continuous as a function on $\mathbb{R}^N$. This assertion is false (it is even false if we replace $\mathcal{C}^\infty$ by real analytic—see Krantz and Parks [145]). The reader should construct an example to see that this is the case.

3.2.5 Another Proof of Hartogs's Theorem

Lemma 3.2.35 (The Schwarz lemma). *Let a function f be holomorphic in the unit disk $\mathbb{D}$, satisfying $|f(z)| \leq 1$ for all $z \in \mathbb{D}$ and $f(0) = 0$. Then we have*

$$|f(z)| \leq |z| \tag{3.25}$$

for all $z \in \mathbb{D}$. Moreover, if equality holds in (3.25) for at least one $z \neq 0$ then it holds everywhere in $\mathbb{D}$ and in this case $f(z) = e^{i\alpha}z$, where α is a real constant.

Proof. Consider the function $\phi(z) = f(z)/z$. It is holomorphic in $\mathbb{D}$ since $f(0) = 0$. Let $\mathbb{D}_r = \{z \in \mathbb{C} \mid |z| < r\}$, $r < 1$ be an arbitrary disk centered at zero. The function $\phi(z)$ attains its maximum in $\mathbb{D}_r$ on its boundary $\gamma_r = \{z \in \mathbb{C} \mid |z| = r\}$. However, we have $|\phi| \leq 1/r$ on γ_r since $|f| \leq 1$ by assumption. Therefore we have

$$|\phi(z)| \leq 1/r \tag{3.26}$$

everywhere in $\mathbb{D}_r$. We fix $z \in \mathbb{D}$ and observe that $z \in \mathbb{D}_r$ for $r > |z|$. Therefore (3.26) holds for any given z with all $r > |z|$. We let $r \to 1$ and, passing to the limit $r \to 1$, we obtain $|\phi(z)| \leq 1$ or $|f(z)| \leq |z|$. This proves the inequality (3.25).

Let us assume that equality in (3.25) holds for some $z \in U$; then $|\phi|$ attains its maximum equal to 1 at this point. Then ϕ is equal to a constant so that $\phi(z) = e^{i\alpha}$ and $f(z) = e^{i\alpha}z$. $\square$

Suppose that the function φ is holomorphic in the disc

$$\mathbb{D}_r = \{z \in \mathbb{C} \mid |z| < r\} \subset \mathbb{C},$$

where $\varphi = 0$ at some point $z_0 \in U_r$ and $|\varphi| \leq M$ everywhere in $\mathbb{D}_r$. Then everywhere in $\mathbb{D}_r$ we have the estimate

$$|\varphi(z)| \leq Mr\frac{|z - z_0|}{|r^2 - \overline{z}_0 z|} \tag{3.27}$$

(for $r = M = 1$ and $z_0 = 0$ we obtain the usual statement).

For the proof we take a linear-fractional mapping of $\mathbb{D}_r$ onto the unit disc $\mathbb{D}$:

$$\lambda : z \to r\frac{z - z_0}{r^2 - \overline{z}_0 z}\,.$$

We denote by λ^{-1} the inverse mapping $\mathbb{D} \to \mathbb{D}_r$. Consider the function $\psi = \frac{1}{M}\varphi \circ \lambda^{-1}$. It satisfies the hypotheses of the usual Schwarz lemma, and by this lemma $|\psi(z)| \leq |z|$ everywhere in $\mathbb{D}$. Replacing z here by $\lambda(z)$, we obtain the inequality (3.27).

Lemma 3.2.36. *If the function f is holomorphic with respect to each variable z_k in the polydisc $P = P(a, r)$ and bounded in P, then it is continuous at each point of P with respect to the set of all the variables.*

Proof. Let $z^0, z \in P$, be arbitrary points. Then

$$f(z) - f\left(z^0\right) =$$
$$\sum_{k=1}^{n}\{f\left(z_1^0, \ldots, z_{k-1}^0, z_k, \ldots, z_n\right) - f\left(z_1^0, \ldots, z_k^0, z_{k+1}, \ldots, z_n\right)\} \tag{3.28}$$

and consider the k th summand as a function φ_k of the variable z_k with the remaining values of the argument being fixed. If $|f| \leq M/2$ in P, then the function φ_k satisfies the hypotheses of Schwarz's lemma in the form just given, and, applying inequality (3.27) to each term of the sum (3.28), we find that

$$\left|f(z) - f\left(z^0\right)\right| \leq M \sum_{k=1}^{n} r_k \frac{\left|z_k - z_k^0\right|}{\left|r_k^2 - \overline{z_k^0} z_k\right|}.$$

The assertion follows from this inequality. □

Lemma 3.2.36 shows that a function that is separately holomorphic and locally bounded on an open domain Ω is in fact C^∞, and hence holomorphic. Thus, to prove Hartogs's theorem it is necessary to prove the boundedness in some polydisc with center at a of a function that is holomorphic with respect to each variable.

Now the proof of Hartogs's theorem is by induction on dimension. When $n = 1$ there is, of course, nothing to prove. Assume inductively (and for the rest of the section) that the result has been proved in dimensions $j = 1, \ldots, n-1$. (However, it should be mentioned that Lemma 3.2.40 stands alone and has been formulated accordingly.) The following result, sometimes called Hartogs's lemma, is a key step in the proof of the Hartogs Theorem 3.2.31.

Proposition 3.2.37. *Let f_j be a sequence of subharmonic functions on Ω. Suppose that for each $K \subset\subset \Omega$ there is a finite $M_K > 0$ such that $f_j \leq M_K$ on K for every j. Suppose further that* $\limsup_{j\to\infty} f_j(x) \leq C < \infty$ *for all $x \in \Omega$. Then for each $\epsilon > 0$ and each $K \subset\subset \Omega$, there is a $J = J(\epsilon, K)$ such that*

$$f_j(x) \leq C + \epsilon \quad \textit{for all } j \geq J, \textit{ all } x \in K.$$

Proof. Fix $K \subset\subset \Omega$. Choose K_0 such that $K \subset\subset K_0 \subset\subset \Omega$ and K_0 is the closure of an open set. We may suppose that $f_j \leq -1$ on K_0, all j. Let $0 < r < \operatorname{dist}(K, {}^cK_0)/3$. Choose, for $x \in K, \epsilon > 0$, a natural number j_0 so large that (by Fatou's lemma)

$$\int_{B(x,r)} f_j(t)dV(t) \leq V(B(0,1)) \cdot r^N \cdot (C + 2\epsilon), \quad j \geq j_0$$

(Note that the negativity of the functions involved reverses the usual inequality of Fatou.) Let

$$\delta = r \cdot \left[\left(\frac{C + 2\epsilon}{C + \epsilon} \right)^{1/N} - 1 \right].$$

Then $|x - x'| < \delta$ implies that

$$\begin{aligned} V(B)(r+\delta)^N f_j(x') &\leq \int_{B(x', r+\delta)} f_j(t)dV(t) \\ &\leq \int_{B(x,r)} f_j(t)dV(t) \\ &\leq V(B)r^N(C + 2\epsilon) \end{aligned}$$

or $f_j(x') \leq C + \epsilon$, all $x' \in B(x, \delta)$. Now we invoke the compactness of K to complete the proof. □

Lemma 3.2.38. *Let $\Omega \subseteq \mathbb{C}^n$ be a domain and let $f : \Omega \to \mathbb{C}$ be holomorphic in each variable separately. Let*

$$\overline{\mathcal{D}} \equiv \overline{D}(z_1, r_1) \times \cdots \times \overline{D}(z_n, r_n) \subseteq \Omega$$

with $r_1, \ldots, r_n > 0$. Then there exist $\overline{D}(z_j', r_j') \subseteq \overline{D}(z_j, r_j), j = 1, \ldots, n$, with each $r_j' > 0$ and with $\overline{D}(z_n', r_n') = \overline{D}(z_n, r_n)$, such that f is bounded on

$$\overline{\mathcal{D}}' \equiv \overline{D}(z_1', r_1') \times \cdots \times \overline{D}(z_n', r_n').$$

In particular, f is C^∞ and holomorphic on $\mathcal{D}' = D(z_1', r_1') \times \cdots \times D(z_n', r_n')$.

Proof. We write $\zeta = (\zeta_1, \ldots, \zeta_n) = (\zeta', \zeta_n)$. For $M = 1, 2, \ldots$, define

$$S_M \equiv \left\{ \zeta' \in \prod_{j=1}^{n-1} \overline{D}(z_j, r_j) \;\middle|\; |f(\zeta', \zeta_n)| \leq M, \text{ all } \zeta_n \in D(\zeta_n, r_n) \right\}.$$

Then

$$\prod_{j=1}^{n-1} \overline{D}(z_j, r_j) = \bigcup_{M=1}^{\infty} S_M.$$

By the inductive hypothesis, each $f(\cdot, \zeta_n)$ is continuous as a function of ζ', so each S_M is closed. By Baire's theorem, one of the S_M's, say S_{M_0}, has interior. The proof is completed by selecting $\overline{D}(z_1', r_1') \times \cdots \times \overline{D}(z_{n-1}', r_{n-1}') \subseteq S_{M_0}$. □

Remark 3.2.39. Notice that Lemma 3.2.38 does not yet yield Hartogs's theorem. All we know so far is that every polydisc in Ω contains a polydisc (of full size in the last variable) on which f is holomorphic. Thus there may still be a closed set without interior on which f is ill behaved. This possibility is obviated by the following calculation, which is the heart of the proof.

Lemma 3.2.40.

Let $0 < r < R$, and suppose that $g : D^n(P, R) \to \mathbb{C}$ satisfies

1. *For each $\zeta_n \in D(P_n, R)$, the function $\zeta' \mapsto g(\zeta', \zeta_n)$ is holomorphic on $D^{n-1}(P', R)$,*

2. *The function g is holomorphic and bounded on*

$$\hat{D} \equiv D^{n-1}(P', r) \times D(P_n, R)\,.$$

Then g is holomorphic on $D^n(P, R)$.

Proof. Assume for simplicity that $P = 0$. For fixed ζ_n, we expand $g(\cdot, \zeta_n)$ in a power series expansion on $D^{n-1}(0, R)$ (we use here hypothesis *1.*):

$$g(\zeta', \zeta_n) = \sum_{\alpha \in (\mathbb{Z}^+)^{n-1}} a_\alpha(\zeta_n)(\zeta')^\alpha\,. \tag{3.29}$$

Here

$$a_\alpha(\zeta_n) = \left(\frac{\partial}{\partial \zeta'}\right)^\alpha g(0, \zeta_n)/\alpha!\,.$$

In particular, a_α is a holomorphic function of ζ_n by hypothesis *2.* It is our goal to see that the series (3.29) converges uniformly on compact subsets of $D^n(0, R)$. Let M be a bound for g on $\overline{D}$. Let $0 < R_1 < R_2 < R$.

Since $\sum_\alpha a_\alpha z^\alpha$ converge absolutely on a neighborhood of a polydisc of the form $\overline{D}(0, R_2)$ then, for $\epsilon > 0$ sufficiently small, by Abel's test and the Cauchy estimates on this polydisc, we obtain

$$|a_\alpha| \le \frac{C_\epsilon}{\prod_{j=1}^n (R_2 + \epsilon)^{\alpha_j}}\,.$$

It follows that

$$|a_\alpha(\zeta_n)|\, R_2^{|\alpha|} \to 0 \quad \text{when} \quad |\alpha| \to \infty$$

and hypothesis *2.* with the Cauchy estimates yields that

$$|a_\alpha(\zeta_n)|\, r^{|\alpha|} \le M \quad \text{when} \quad |\zeta_n| < R\,.$$

This last line tells us that the function

$$\zeta_n \mapsto \frac{1}{|\alpha|} \log |a_\alpha(\zeta_n)| \equiv h_\alpha(\zeta_n)$$

is subharmonic and bounded above (by $\log(M + 1)$) on the disc $\{\zeta_n \in \mathbb{C} \mid |\zeta_n| < R\}$. And the line that precedes it tells us that

$$\limsup_{|\alpha| \to \infty} h_\alpha (\zeta_n) \le \log (1/R_2) .$$

Thus we may apply Hartogs's Proposition 3.2.37 to see that, for $|\alpha|$ sufficiently large,

$$\frac{1}{|\alpha|} \log |a_\alpha (\zeta_n)| \le \log \frac{1}{R_1}, \quad |\zeta_n| < R_1 .$$

In conclusion,

$$|a_\alpha (\zeta_n)| \cdot (R_1)^{|\alpha|} \le 1, \text{ all } |\zeta_n| < R_1.$$

This trivially yields that the series (3.29) converges normally on $\{|\zeta_n| < R_1\}$. Since R_1 was an arbitrary positive number less than R, we have that the series (3.29) converges normally on $\{|\zeta_n| < R\}$. We conclude that the series (3.29) converges to a harmonic, hence locally bounded, hence (by Cauchy theory) holomorpic function on $D^n(P, R)$. □

Final Argument in the Proof of Theorem 3.2.31. Let $w \in \Omega$, $R > 0$ satisfy $\overline{D}^n(w, 2R) \subseteq \Omega$. Apply Lemma 3.2.40 to find a $P \in D^n(w, R)$ and $r = \min_{j=1,\ldots,n} \{r'_j\}$ so that the hypotheses of Lemma 3.2.40 are satisfied with $g = f$. Then, by 3.2.40, f is holomorphic in a neighborhood of w. Since $w \in \Omega$ was arbitrary, we are finished. □

The proof of Hartogs's theorem that we have presented is not essentially different from the original proof of F. Hartogs [95]. In spite of the intervening years, no essentially simpler argument has been found. An old conjecture of Hervé [106] was that a separate subharmonic function is subharmonic. This implication would give an easy and natural way to see that a separate analytic function is locally bounded above, hence is holomorphic. Unfortunately, the Hervé conjecture is false, as was discovered by Wiegerinck [210].

It would be quite interesting to have a truly new proof of Hartogs's theorem.

Hartogs's original proof of this result is quite tricky. The difficulty is that one does not know that the function f is locally bounded, or even locally integrable. In fact early treatments (those that preceded Hartogs) of the function theory of several complex variables would define a holomorphic function of n variables to be a locally bounded function that is holomorphic in each variable separately. The advantage of this particular definition is that one can apply the Cauchy theory (on a polydisc in the domain, let us say) and (changing the order of integration—which is justified because of the local boundedness) then prove that the function is infinitely differentiable, indeed real analytic.

But the Hartogs result is in the spirit, if not the literal detail, of the Forelli theorem. So it is worth discussing here. We offer now a slightly weaker form of the Hartogs theorem, but one that is very natural in terms of modern developments in analysis. And it is much easier to prove.

This is an alternative form of Hartogs's theorem. Since Hartogs did not have the theory of distributions available to him, he could not have conceived of nor formulated such a result. But it is quite natural in the context of modern analysis. We shall also see that the proof of this new result is very natural.

Theorem 3.2.41. *Let f be a function on a domain $\Omega \subseteq \mathbb{C}^n$ that is holomorphic in each variable separately. Further, assume that f induces a distribution—by way of integration. Then f is holomorphic as a function of n complex variables.*

Proof. For simplicity we take Ω to be $\mathbb{C}^n$.

Let φ be a function of one complex variable ζ that is nonnegative, radial, and has a total integral equal to 1. Then, for any $z \in \mathbb{C}^n$,

$$f(z) = \iint \cdots \int f(\zeta_1, \zeta_2, \ldots, \zeta_n)\, \varphi(z_1 - \zeta_1)\, \varphi(z_2 - \zeta_2) \cdots \varphi(z_n - \zeta_n)\, dA(\zeta_1)\, dA(\zeta_2) \cdots dA(\zeta_n)$$

Here dA is the standard area measure in the complex plane. We are invoking here (a total of n times) the mean-value property for harmonic functions of one complex variable.

Now a standard structure theorem for distributions [140] allows us to write

$$f = \sum_{\alpha,\beta} \left(\frac{\partial^{\alpha+\beta}}{\partial z^\alpha \partial \bar{z}^\beta} \right) u_{\alpha,\beta}$$

where the sum is finite and each $u_{\alpha,\beta}$ is a continuous function. Thus

$$f(z) = \iint \cdots \int \sum_{\alpha,\beta} \left(\left(\frac{\partial^{\alpha+\beta}}{\partial z^\alpha \partial \bar{z}^\beta} \right) u_{\alpha,\beta}(\zeta_1, \zeta_2, \ldots, \zeta_n) \right) \times \varphi(z_1 - \zeta_1)\, \varphi(z_2 - \zeta_2) \cdots \varphi(z_n - \zeta_n)\, dA(\zeta_1)\, dA(\zeta_2) \cdots dA(\zeta_n).$$

By the definition of distribution, we may, in each summand, throw the derivative over onto the testing function φ. So we end up with an expression which is a sum of integrals of continuous functions multiplied times smooth functions. Now it is clearly possible to differentiate under the integral sign to see that f is C^∞, indeed real analytic.

The theorem is proved. □

We shall say that a function f is holomorphic at a point $a \in \mathbb{C}^n$ *in the sense of Riemann* if

(R) f is holomorphic with respect to each variable z_ν in some polydisc $P(a, r)$.

We shall say that a function f is *holomorphic at the point a in the sense of Weierstrass* if

(W) f can be expanded in some polydisc $P(a, r)$ in a power series

$$f(z) = \sum_{|k|=0}^{\infty} c_k (z-a)^k.$$

The implication **(W)** $\Rightarrow$ **(R)** is obvious and the implication **(R)** $\Rightarrow$ **(W)** is the content of the theorem of Hartogs. Consequently, Hartogs theorem can be formulated as:

Theorem 3.2.42. *The concepts of holomorphy in the sense of Riemann and holomorphy in the sense of Weierstrass are equivalent.*

3.3 Cauchy-Green Equation: The Case $n > 1$

3.3.1 Local Solvalibity of the Cauchy—Riemann Equations

Now we consider the particular case of the polydisc.

Theorem 3.3.1. *Let $P = P(0; r) \subset \mathbb{C}^n, r > 0$, and $g = \sum_{k=1}^{n} g_k d\overline{z}_k$ a smooth $\overline{\partial}$-closed $(0,1)$-form on a neighborhood U of $\overline{P}$. Then there exist a polydisc $P' = P(0; r')$, $r' > r$, and $u \in C^\infty(P')$ such that*

$$\overline{\partial} u = g$$

on P'. In particular, the Cauchy-Riemann equation is always locally solvable.

Proof. There exists $r' > r$ such that g is defined on a neighborhood of $\overline{P}' = \overline{P(0; r')}$ The function

$$v(z) = \frac{1}{2\pi i} \int_{|\zeta_1| \leq r'} \frac{g_1(\zeta_1, z_2, \ldots, z_n)}{\zeta_1 - z_1} d\zeta_1 \wedge d\overline{\zeta}_1$$

is in $C^\infty(P')$ and

$$\frac{\partial v}{\partial \overline{z}_1} = g_1$$

on P' (see Corollary 3.1.3). It follows that

$$g - \overline{\partial} v = h_2 d\overline{z}_2 + \cdots + h_n d\overline{z}_n = h$$

on P', where h is a $\overline{\partial}$-closed $(0,1)$-form of class $\mathcal{C}^\infty$ on (P'), independent of $d\overline{z}_1$. If u satisfies $\overline{\partial} u = h$, then $w = u + v$ satisfies $\overline{\partial} w = g$. Arguing by induction on the number of variables, it is sufficient to prove the theorem for forms g which depend only on one differential, say $d\overline{z}_1$, i.e., $g = g_1 d\overline{z}_1$. The condition $\overline{\partial} g = 0$ now implies that g_1 is holomorphic in $z_2, \ldots, z_n$ thus, in order to solve $\overline{\partial}u = g$, we have to find u of class $\mathcal{C}^\infty$ in a neighborhood of $\overline{P}$, which is holomorphic in the variables $z_2, \ldots, z_n$ and such that

$$\frac{\partial u}{\partial \overline{z}_1} = g_1$$

The function $u : P' \to \mathbb{C}$ given by

$$u(z_1, \ldots, z_n) = \frac{1}{2\pi i} \int_{|\zeta_1| \le r'} \frac{g_1(\zeta_1, z_2, \ldots, z_n)}{\zeta_1 - z_1} d\zeta_1 \wedge d\overline{\zeta}_1$$

$u \in C^\infty(P')$, is holomorphic in the variables $z_2, \ldots, z_n$, and satisfies

$$\frac{\partial u}{\partial \overline{z}_1} = g_1$$

□

Arguing as in Theorem 3.1.12 we obtain

Corollary 3.3.2. *On an open polydisc P the Cauchy-Riemann equation it always solvable. In particular, the Cauchy-Riemann equation is solvable in $\mathbb{C}^n$.*

The system of differential equations

$$\frac{\partial u}{\partial \overline{z}_j} = f_j, \quad j = 1, \ldots, n, \qquad (\overline{\partial}u = \sum_{j=1}^n f_j d\overline{z}_j) \tag{3.30}$$

is also called the $\overline{\partial}$-*equation* Cauchy—Rieman equations. It is clear that the following compatibility condition is necessary for the existence of a solution:

$$\frac{\partial f_j}{\partial \overline{z}_k} = \frac{\partial f_k}{\partial \overline{z}_j}, \quad j, k = 1, \ldots, n. \tag{3.31}$$

In general, the equation (3.30) is not solvable, as shown by the following example.

Example 3.3.3. Let $n = 2$ and consider the Bochner-Martinelli Kernel:

$$\omega = -\frac{\overline{z}_2}{|z|^4} d\overline{z}_1 + \frac{\overline{z}_1}{|z|^4} d\overline{z}_2 \,.$$

Here ω is a $C^\infty \overline{\partial}$-closed 1-form in $\mathbb{C}^2 \setminus \{0\}$, which is not $\overline{\partial}$-exact.

Indeed, the Bochner-Martinelli formula (Theorem 10.6.5) applied to the constant function $f \equiv 1$ on the unit ball gives

$$1 = \frac{1}{(2\pi i)^2} \int_{\mathbb{S}^3} dz_1 \wedge dz_2 \wedge \omega \tag{3.32}$$

If ω were exact on $\mathbb{C}^2\backslash\{0\}$, $\omega = \overline{\partial}u$, then, on the unit sphere $\mathbb{S}^3$ we would obtain $dz_1 \wedge dz_2 \wedge \omega = dz_1 \wedge dz_2 \wedge \overline{\partial}(u) = dz_1 \wedge dz_2 \wedge d(u) = d\left(udz_1 \wedge dz_2\right)$ and thus, in view of (3.32),

$$
\begin{aligned}
1 &= \frac{1}{(2\pi i)^2}\int_{\mathbb{S}^3} dz_1 \wedge dz_2 \wedge \omega = \frac{1}{(2\pi i)^2}\int_{\mathbb{S}^3} d\left(udz_1 \wedge dz_2\right) \\
&= \frac{1}{(2\pi i)^2}\int_{\partial\mathbb{S}^3} udz_1 \wedge dz_2 = 0
\end{aligned}
$$

a contradiction.

We now present a formula that will solve the inhomogeneous $\overline{\partial}$-equation for compactly supported data in $\mathbb{C}^n$. Although we shall later learn that this is in fact not the most interesting type of inhomogeneous Cauchy-Riemann equation to solve, it still has some short-term utility.

Theorem 3.3.4. *Let $f_j \in C_0^k\left(\mathbb{C}^n\right)$, $n \geq 2$, $j = 1,\ldots,n$, and let $k \geq 1$ be a positive integer such that (3.31) is satisfied. Then there is a function $u \in C_0^k\left(\mathbb{C}^n\right)$ satisfying (3.30). When $k = 0$, if (3.31) is satisfied in the distribution sense, then there exists a function $u \in C_0\left(\mathbb{C}^n\right)$ such that u satisfies (3.30) in the distribution sense.*

This theorem repeats ideas already presented in Section 3.1 on the Cauchy—Green formula.

The compatibility conditions on f are necessary, but the compactness is not. However, in that case the boundary of the domain where the equation lives would come into play. Let us not worry about this, and prove that this simple compactly supported version always has a solution. Without the compact support condition the solution is clearly not unique. Given any holomorphic g, $\overline{\partial}(u+g) = f$. But since the difference between any two solutions u_1 and u_2 is holomorphic, and the only holomorphic compactly supported function is 0, then the compactly supported solution u is unique.

Proof. For $k \geq 1$, set

$$
\begin{aligned}
u(z) &= \frac{1}{2\pi i}\int_{\mathbb{C}} \frac{f_1\left(\zeta, z_2, \ldots, z_n\right)}{\zeta - z_1} d\zeta \wedge d\overline{\zeta} \\
&= \frac{1}{2\pi i}\int_{\mathbb{C}} \frac{f_1\left(\zeta + z_1, z_2, \ldots, z_n\right)}{\zeta} d\zeta \wedge d\overline{\zeta}
\end{aligned}
$$

It is easily seen that $u \in C^k\left(\mathbb{C}^n\right)$ from differentiation under the integral sign. We also have $u(z) = 0$ when $|z_2|+\cdots+|z_n|$ is sufficiently large, since f vanishes on the set. By Theorem 3.1.3, we have

$$
\frac{\partial u}{\partial \overline{z}_1} = f_1(z).
$$

For $j > 1$, using the compatibility condition (3.31), we obtain

$$\frac{\partial u}{\partial \bar{z}_j} = \frac{1}{2\pi i}\int_{\mathbb{C}} \frac{\frac{\partial f_j}{\partial \bar{\zeta}}(\zeta, z_2, \ldots, z_n)}{\zeta - z_1} d\zeta \wedge d\bar{\zeta} = f_j(z)$$

Hence, $u(z)$ is a solution to the $\bar{\partial}$-equation (3.30). In particular, u is holomorphic on the unbounded component of the complement of the support of f. Since $u(z) = 0$ when $|z_2| + \cdots + |z_n|$ is sufficiently large, we see from the Identity Theorem for holomorphic functions that u must be zero on the unbounded component of the complement of the support of f. This completes the proof of the theorem for $k \geq 1$.

When $k = 0$, define $u(z)$ by the same equation. We see that $u \in C^0(\mathbb{C}^n)$, and that $u(z) = 0$ when $|z_2| + \cdots + |z_n|$ is sufficiently large. By Theorem 3.1.3,

$$\frac{\partial u}{\partial \bar{z}_1} = f_1(z)$$

For $j > 1$, let $\phi \in C_0^\infty(\mathbb{C}^n)$. Then, using $(\cdot, \cdot)_{\mathbb{C}}^n$ to denote the pairing between distributions and test functions, we have

$$\begin{aligned}
\left(\frac{\partial u}{\partial \bar{z}_j}, \phi\right)_{\mathbb{C}}^n &= \left(u, -\frac{\partial \phi}{\partial \bar{z}_j}\right)_{\mathbb{C}}^n \\
&= -\int_{\mathbb{C}}^n \left(\frac{1}{2\pi i}\int_{\mathbb{C}} \frac{f_1(\zeta + z_1, z_2, \ldots, z_n)}{\zeta} d\zeta \wedge d\bar{\zeta}\right) \frac{\partial \phi}{\partial \bar{z}_j} d\lambda(z) \\
&= \frac{-1}{2\pi i}\int_{\mathbb{C}} \frac{1}{\zeta}\left(\int_{\mathbb{C}^n} f_1(\zeta + z_1, z_2, \ldots, z_n) \frac{\partial \phi}{\partial \bar{z}_j} d\lambda(z)\right) d\zeta \wedge d\bar{\zeta} \\
&= \frac{1}{2\pi i}\int_{\mathbb{C}} \frac{1}{\zeta}\left(\frac{\partial f_1}{\partial \bar{z}_j}(\zeta + z_1, z_2, \ldots, z_n), \phi\right)_{\mathbb{C}^n} d\zeta \wedge d\bar{\zeta} \\
&= \frac{1}{2\pi i}\int_{\mathbb{C}} \frac{1}{\zeta}\left(\frac{\partial f_j}{\partial \bar{z}_1}(\zeta + z_1, z_2, \ldots, z_n), \phi\right)_{\mathbb{C}^n} d\zeta \wedge d\bar{\zeta} \\
&= \frac{-1}{2\pi i}\int_{\mathbb{C}} \frac{1}{\zeta}\left(\int_{\mathbb{C}^n} f_j(\zeta + z_1, z_2, \ldots, z_n) \frac{\partial \phi}{\partial \bar{z}_1} d\lambda(z)\right) d\zeta \wedge d\bar{\zeta} \\
&= -\int_{\mathbb{C}}^n \left(\frac{1}{2\pi i}\int_{\mathbb{C}} \frac{f_j(\zeta + z_1, z_2, \ldots, z_n)}{\zeta} d\zeta \wedge d\bar{\zeta}\right) \frac{\partial \phi}{\partial \bar{z}_1} d\lambda(z) \\
&= \left(\frac{\partial}{\partial \bar{z}_1}\left(\frac{1}{2\pi i}\int_{\mathbb{C}} \frac{f_j(\zeta, z_2, \ldots, z_n)}{\zeta - z_1} d\zeta \wedge d\bar{\zeta}\right), \phi\right)_{\mathbb{C}}^n \\
&= (f_j, \phi)_{\mathbb{C}^n}
\end{aligned}$$

where the last equality is again guaranteed by Theorem 3.1.3. Hence,

$$\frac{\partial u}{\partial \bar{z}_j} = f_j(z)$$

for $j > 1$ in the distribution sense. One shows, similarly, that u must vanish on the unbounded component of the complement of the support of f. This proves the theorem. □

Example 3.3.5. *Extension of holomorphic functions.*

Let $\Omega \subset \mathbb{C}^n$, $X = \{z \in \mathbb{C}^n \mid h(z) = 0\}$, where $h \in \mathcal{O}(\Omega)$ and $dh \neq 0$ on X. X is a complex submanifold of Ω of dimension $n-1$. Thus every holomorphic function $f \in \mathcal{O}(X)$ locally extends in the neighborhood of each point $x \in X$ to a holomorphic function. In general, the restriction homomorphism

$$\mathcal{O}(\Omega) \longrightarrow \mathcal{O}(X)$$

is not surjective. For instance, let

$$\Omega = \left\{ z \in \mathbb{C}^n \;\middle|\; 0 < |z_1|^2 + |z_2|^2 < 1 \right\} \subset \mathbb{C}^2$$

and

$$X = \{z \in \mathbb{C}^n \mid 0 < |z_1| < 1, z_2 = 0\} \subset \Omega$$

X is a closed subvariety, the function $f = z_1^{-1} \in \mathcal{O}(X)$ cannot be holomorphically extended on Ω.

As in the previous examples, the problem we are interested in can be reduced to a Cauchy–Riemann equation. Indeed, given $f \in \mathcal{O}(X)$, there is a locally finite covering $\{U_j\}_{j \in J}$ of Ω such that U_{jk} is simply connected and a family $\{f_j\}_{j \in J}$ of holomorphic functions, $f_j \in \mathcal{O}(U_j)$ with the property

$$f_j|_{U_j \cap X} = f|_{U_j \cap X}$$

for $j \in J$. If $U_{jk} \cap X \neq \emptyset$, $f_j - f_k = 0$ on U_{jk}, hence

$$f_j - f_k = g_{jk} h$$

with $g_{jk} \in \mathcal{O}(U_{jk})$, $h \in \mathcal{O}(\Omega)$, and $g_{jk} + g_{kl} + g_{lj} = 0$. Arguing as in Example 3.1.13, we get a Cauchy—Riemann equation

$$\overline{\partial} u = g \tag{3.33}$$

where $g = \{\overline{\partial} g_j\}$ and $g_j = \sum_k \rho_k g_{jk}$. If (3.33) has a solution u, then $u - g_j$ is holomorphic, $(g_j - u) - (g_k - u) = g_{jk}$ and so

$$f_j - f_k = (g_j - u)\, h - (g_k - u)\, h$$

i.e., the local holomorphic functions $f_j - (g_j - u)\, h$ glue together and define a holomorphic function on Ω, $\widehat{f}$, such that $\widehat{f_{|X}} = f$.

In particular, this is true if Ω is a polydisc or $\Omega = \mathbb{C}^n$.

3.3.2 Some Applications

In this section, we start investigating phenomena peculiar to analytic functions of two or more complex variables.

One major difference between one and several complex variables is the Hartogs phenomenon, which states that if a bounded domain Ω in $\mathbb{C}^n$, $n \geq 2$, has a connected boundary, then any holomorphic function $f(z)$ defined in some open neighborhood of the boundary $\partial\Omega$ can be holomorphically extended to the entire domain Ω. This is no longer true in one complex variable. For instance, $f(z) = 1/z$ is holomorphic on the entire complex plane except the origin, but there is no way to extend it as an entire function.

Theorem 3.3.6 (Hartogs's Phenomenon).
Let $f(z,t)$ be analytic on $P^2\backslash P(r)^2$, with $r < 1$. Then f extends to an analytic function on P^2.

Proof. Define $F(z,t) = (2\pi i)^{-1} \int_{S^1} f(\zeta,t)/(z-\zeta)d\zeta$. Since the integrand is holomorphic as a function of (z,t), so is F. But $F(z,t) = f(z,t)$ for $|t| > r$, so it provides the desired extension. □

Let us prove a stronger version of this fact, to illustrate the new phenomena that arise in several complex variables.

Theorem 3.3.7 (Hartogs's Phenomenon). *Let $\Omega \subseteq \mathbb{C}^n$ be a bounded domain, $n > 1$. Let K be a compact subset of Ω with the property that $\Omega\backslash K$ is connected. If f is holomorphic on $\Omega\backslash K$, then there is a holomorphic F on Ω such that $F|_{\Omega\backslash K} = f$.*

The idea of the proof is to extend the function in some fashion and then to use the solution of the $\bar\partial$-problem to correct the result to make it holomorphic.

Proof. Let ϕ be a function in $C_c^\infty(\Omega)$ that is identically 1 on a neighborhood of K. Define

$$\widetilde{f}(z) = \begin{cases} (1-\phi(z))\cdot f(z) & \text{if} \quad z \in \Omega\backslash K \\ 0 & \text{if} \quad z \in K \end{cases}$$

Then $\widetilde{f} \in C^\infty(\Omega)$. Finally, set

$$\psi(z) = \bar\partial \widetilde{f}(z)$$

Then ψ satisfies the following crucial properties:

1. ψ has C^∞ coefficients;
2. $\bar\partial\psi = \bar\partial^2 \widetilde{f} \equiv 0$;
3. $\operatorname{supp}\psi$ is a compact subset K_0 of Ω.

The first two of these properties are obvious, and the last follows because $\widetilde{f}$ is holomorphic in $\Omega\cap$ (neighborhood of $\partial\Omega$). By Theorem 3.3.4, there is a function $u \in C_c^\infty(\mathbb{C}^n)$, with support compact in Ω, so that $\bar\partial u = \psi$. In particular, the function u is identically 0 in a neighborhood W of $\partial\Omega$. We define $F = \widetilde{f} - u$. Then

$$\bar\partial F = \bar\partial\widetilde{f} - \bar\partial u = \psi - \psi = 0$$

so F is holomorphic on Ω. Also, shrinking W if necessary,

$$F|_U = (\widetilde{f} - u)\Big|_W = \widetilde{f}\Big|_W = f|_W$$

Therefore, F agrees with f near $\partial\Omega$. Since $\Omega \backslash K$ is connected, we may conclude by the uniqueness of analytic continuation that $F = f$ on $\Omega \backslash K$. The proof is complete. □

Remark 3.3.8. Let us examine this proof and see why it works. The idea is that $\widetilde{f}$ is a $\mathcal{C}^\infty$ extension of f to Ω. It could not, in general, be holomorphic, because it is identically 0 on K (which could have interior). What we hope is that we can subtract a function u from $\widetilde{f}$ to make it holomorphic. That is, we want $\overline{\partial}(\widetilde{f} - u) = 0$ or $\overline{\partial}u = \psi$.

The equation $\overline{\partial}u = \psi$ has many solutions because $\overline{\partial}$ is a linear operator with a large null space (all holomorphic functions). One solution is the function $\widetilde{f}$ itself; but this would be a poor choice since then $\widetilde{f} - u$ would be identically zero. This is why it is so important that we be able to find a solution u with support compact in Ω. Since $\widetilde{f}$ has support near $\partial\Omega$, then $\widetilde{f} - u$ is not the zero function and agrees with the original function f near $\partial\Omega$.

The hypothesis $n \geq 2$ made in Theorem 3.3.4 is crucial. We cannot solve the equation $\partial u / \partial \overline{z} = f$ for a solution with compact support in $\mathbb{C}$ when the given function f has compact support.

Corollary 3.3.9 (Non-existence of isolated singularities). *Let $n \geqq 2$ and let $\Omega \subseteq \mathbb{C}^n$ be an open set. If $z \in \Omega$, then every holomorphic function in $\Omega \backslash \{z\}$ can be holomorphically continued to z.*

One of many consequences of the Hartogs phenomenon is that the zero set of a holomorphic function f is never compact in dimension 2 or higher. If it were compact, $1/f$ would provide a contradiction.

Corollary 3.3.10. *Let $n \geqq 2$, and let $\Omega \cong \mathbb{C}^n$ be a connected open set. If $f \in \mathcal{O}(\Omega)$ and $N(f) := \{z \in \Omega \mid f(z) = 0\}$, then*

(I) $\Omega \backslash N(f)$ is connected,

(II) $N(f)$ is not compact.

Proof.

(I) We only have to prove that for every open ball $B \subseteq \Omega$ the set $B \backslash N(f)$ is connected. To do this we consider two arbitrary points $z, w \in B \backslash N(f)$ and show that z and w belong to the same component of $B \backslash N(f)$. Let l be the complex line which contains z and w. Then the restriction of f to $l \cap B$ can be considered as a holomorphic function of one complex variable. Since $f \neq 0$ on $l \cap B(f(z) \neq 0$ and $f(w) \neq 0)$, $l \cap B \cap N(f)$ is discrete in $l \cap B$. Since $l \cap B$ is a disc and therefore connected, this implies that $(B \cap l) \backslash N(f)$ is connected. Consequently, z and w belong to the same component of $B \backslash N(f)$.

(II) Assume that $N(f)$ is compact. Since, by part (I), $\Omega \backslash N(f)$ is connected, then it follows from Theorem 3.3.7 that $1/f$ can be continued holomorphically to $N(f)$. This is a contradiction because $f = 0$ on $N(f)$.

□

From Theorem 3.3.7 we see that holomorphic functions of $n \geq 2$ variables cannot have isolated singular points; the singularities of such functions (if they do not separate the domain) must go beyond the boundary of the domain or extend to infinity. The function $f = 1/z_1$ on the z_1-plane has the singular point $\{0\}$, and in $\mathbb{C}^2$ it has the singular complex line $\{z_1 = 0\}$, which extends to infinity.

Next, we prove another version of the holomorphic extension theorem which is an easy application of the Cauchy integral formula.

Theorem 3.3.11. *Let f be a continuous function on a domain Ω in $\mathbb{C}^n$, and let S be a smooth real hypersurface in $\mathbb{C}^n$. Suppose that f is holomorphic in $\Omega \backslash S$. Then f is holomorphic on Ω.*

Proof. It suffices to show that f is holomorphic near each $p \in \Omega \cap S$. Without restriction we may assume that p is the origin and that locally near p, the hypersurface S is realized as a graph which can be represented as

$$S = \left\{ ('z, z_n = x + iy) \in \mathbb{C}^{n-1} \times \mathbb{C} \mid y = \phi\left(z', x\right) \right\}$$

for some smooth function ϕ such that $\phi(0) = 0$ and $d\phi(0) = 0$.

Hence, for any small $\beta > 0$, there exists a $\delta_\beta > 0$ and a polydisc U_β in $\mathbb{C}^{n-1}$ centered at the origin, such that

$$|\phi\left('z, x\right)| < \beta$$

for all $|x| < \delta_\beta$ and $'z \in U_\beta$. Let $\beta_1 > 0$ be sufficiently small, and let $\beta_2 > \beta_1$ be a positive number sufficiently close to β_1, then we may assume that, where

$$V_0 = \{z_n \in \mathbb{C} \mid |x| < \delta_{\beta_1} \text{ and } \beta_1 < y < \beta_2\}$$

is contained in $\Omega \backslash S$. Thus, $f \in \mathcal{O}\left(U_{\beta_1} \times V_0\right)$.

Next, for each $'z \in U_{\beta_1}$, $f\left('z, z_n,\right)$ is continuous on

$$V = \{z_n \in \mathbb{C} \mid |x| < \delta_{\beta_1} \text{ and } |y| < \beta_2\}$$

and holomorphic on V except for the smooth curve $y = \phi\left('z, x\right)$. By Morera's theorem in one complex variable, $f\left('z, z_n\right)$ is holomorphic in z_n on V. Now choose a contour of integration Γ in U_{β_1}. Namely, let $\Gamma = \Gamma_1 \times \cdots \times \Gamma_{n-1}$, where $\Gamma_j = \{z_j \in \mathbb{C} \mid |z_j| = r_j\}$, for $1 \leq j \leq n-1$, so that $\Gamma \subset U_{\beta_1}$.

Define

$$F\left('z, z_n\right) = \frac{1}{(2\pi i)^{n-1}} \int_\Gamma \frac{f\left('\zeta, z_n\right)}{(\zeta_1 - z_1) \cdots (\zeta_{n-1} - z_{n-1})} d\zeta_1 \cdots d\zeta_{n-1}$$

It is easily seen that $F('z, z_n)$ is holomorphic on $U \times V$, where

$$U = \Omega_1 \times \cdots \times \Omega_{n-1} \text{ and } \Omega_j = \{z_j \in \mathbb{C} \mid |z_j| < r_j\} .$$

Since for $('z, z_n) \in U \times V_0$, we have

$$F('z, z_n) = f('z, z_n) .$$

Since f is continuous, it follows from the Identity Theorem that f is holomorphic on $U \times V$. The proof of the theorem is now complete. □

Another application of Theorem 3.3.4 deals with the boundary value problem for holomorphic functions

$$\begin{cases} \overline{\partial} u = 0 & \text{in } \Omega \\ u = h & \text{on } \partial\Omega \end{cases} \tag{3.34}$$

where Ω is a bounded domain of $\mathbb{C}^n$ with smooth connected boundary and h is C^1.

We observe that the problem (3.34) is not always solvable. The boundary value h cannot be arbitrary, but it must satisfy a differential condition, the so-called *tangential Cauchy-Riemann condition* (see conditions (3.35), (3.36) below).

Indeed, suppose that problem (3.34) has a (obviously unique) solution $u \in C^1(\overline{\Omega})$. Then,

$$du \wedge dz_1 \wedge \ldots \wedge dz_n = \partial u \wedge dz_1 \wedge \ldots \wedge dz_n = 0$$

on $\overline{\Omega}$ hence, by restriction to $\partial\Omega$,

$$0 = (du \wedge dz_1 \wedge \ldots \wedge dz_n)|_{\partial\Omega} = dh \wedge (dz_1 \wedge \ldots \wedge dz_n)|_{\partial\Omega} . \tag{3.35}$$

Condition (3.35) goes back to Severi (see [150, 193]). If $h \in C^2(\partial\Omega)$, it is equivalent to the following: for each C^2 extension $\widehat{h}$ of h, at all points of $\partial\Omega$

$$\overline{\partial}\widehat{h} \wedge \overline{\partial}\rho = 0 \tag{3.36}$$

where $\rho = 0$ is a local equation for $\partial\Omega$.

The equivalence between (3.35) and (3.36) is left as an exercise to the reader.

Functions h satisfying the equivalent conditions (3.35) and (3.36) are called *CR functions.*

Theorem 3.3.12 (Bochner). *If $\partial\Omega$ is connected, then each CR function $h \in C^k(\partial\Omega)$, $k \geq 4$, extends holomorphically to Ω.*

Proof. Let $\partial\Omega$ be defined as the zero set of a smooth function (C^4 suffices) ρ in a neighborhood of $\overline{\Omega}$. Let $\widehat{h}$ be a C^k extension of h. Since, in view of (3.36), $\overline{\partial}\widehat{h}$ and $\overline{\partial}\rho$ are proportional at each point of $\partial\Omega$, $\overline{\partial}\widehat{h} = \lambda_0\overline{\partial}\rho$, where $\lambda_0 \in C^{k-1}(\partial\Omega)$, hence, in a neighborhood W of $\overline{\Omega}$,

$$\overline{\partial}\widehat{h} = \lambda_0\overline{\partial}\rho + \rho\lambda_1 \tag{3.37}$$

where $\lambda_0 \in C^{k-1}(W)$ and λ_1 is a $(0,1)$-form on W of class C^{k-2}. We claim that, up to a change of the extension $\widehat{h}$, $\overline{\partial}\widehat{h}$ is zero of order at least 2 on $\partial\Omega$. Indeed, (3.37) implies

$$\overline{\partial}\widehat{h} = \overline{\partial}\left(\lambda_0\rho\right) - \rho\overline{\partial}\lambda_0 + \rho\lambda_1$$

i.e.,

$$\overline{\partial}\left(\widehat{h} - \lambda_0\rho\right) = \rho\lambda_2 \tag{3.38}$$

with $\lambda_2 \in C^{k-2}(W)$. Differentiating we obtain

$$0 = \rho\overline{\partial}\lambda_2 + \overline{\partial}\rho \wedge \lambda_2$$

hence

$$\overline{\partial}\rho \wedge \lambda_2 = 0$$

on $\partial\Omega$. So

$$\lambda_2 = \lambda_3\overline{\partial}\rho + \rho\lambda_4$$

where $\lambda_3 \in C^{k-2}(W)$ and $\lambda_4 \in C^{k-3}(W)$. Substituting in (3.38), we get

$$\begin{aligned}\overline{\partial}\left(\hat{h} - \lambda_0\rho\right) &= \lambda_3\rho\overline{\partial}\rho + \lambda_4\rho^2\\ &= \lambda_3\frac{1}{2}\overline{\partial}\left(\rho^2\right) + \rho^2\lambda_4\\ &= \overline{\partial}\left(\frac{\lambda_3}{2}\rho^2\right) - \frac{\rho^2}{2}\overline{\partial}\lambda_3 + \rho^2\lambda_4\end{aligned}$$

and thus

$$\overline{\partial}\left(\widehat{h} - \lambda_0\rho - \frac{\lambda_3}{2}\rho^2\right) = \rho^2\left(\lambda_4 - \frac{1}{2}\overline{\partial}\lambda_3\right)$$

The function $u = \widehat{h} - \lambda_0\rho - \frac{\lambda_3}{2}\rho^2$ is the claimed extension of h satisfying $\overline{\partial}u = O\left(\rho^2\right)$ on $\partial\Omega$. Now consider the $(0,1)$-form

$$g = \begin{cases}\overline{\partial}u & \text{on } \Omega\\ 0 & \text{on } \mathbb{C}^n\backslash\Omega\end{cases}$$

Since $\overline{\partial}u = O\left(\rho^2\right)$ on $\partial\Omega$, g is a $(0,1)$-form, with compact support, of class $C^{k-3}(k \geq 1)$ on $\mathbb{C}^n$. By Theorem 3.3.4 and Theorem 3.1.7, there is a function $v \in \overline{C}_0^{k-3}(\mathbb{C}^n)$ such that $\overline{\partial}v = g$. The function v is holomorphic on $\mathbb{C}^n\backslash\overline{\Omega}$ and zero for $|z| >> 0$. Hence v is zero on $\mathbb{C}^n \backslash \Omega$, since it is connected by hypothesis. On $\overline{\Omega}$ we have $\overline{\partial}(u - v) = 0$, i.e.,

$$w = u - v$$

is in $\mathcal{O}(\Omega) \cap C^{k-3}(\overline{\Omega})$ and $w|_{\partial\Omega} = h$. □

Remark 3.3.13. The proof shows also that if $h \in C^\infty(\partial\Omega)$ is CR, then its extension u is $\mathcal{C}^\infty$ on $\overline{\Omega}$.

Remark 3.3.14. The boundary value problem for holomorphic functions is a particular case of the following one:

$$\begin{cases} \overline{\partial} u = f & \text{in } \Omega \\ u = h & \text{on } \partial\Omega \end{cases}$$

where f is a $\overline{\partial}$-closed regular $(0,1)$-form on $\overline{\Omega}$. Condition (3.35) becomes

$$dh \wedge (dz_1 \wedge \ldots \wedge dz_n)|_{\partial\Omega} = f \wedge (dz_1 \wedge \ldots \wedge dz_n)|_{\partial\Omega}$$

Remark 3.3.15. The proof shows also that if $h \in C^\infty(\partial\Omega)$ is CR, then its extension u is $\mathcal{C}^\infty$ on $\overline{\Omega}$.

4

Convergent Power Series

The idea of power series dates back at least to Isaac Newton. It is an attractive way to represent a function that is not a polynomial. It is also a powerful tool in the study of differential equations and other aspects of mathematics and physics.

In this chapter we develop some elementary properties of complex power series of complex numbers, mostly by direct analogy with the real case, and then specialize to a deeper study of holomorphic functions.

4.1 Reinhardt Domains

We want to give a characterization of domains of convergence of complex power series of several variables. Consider the power series of one complex variable given by

$$\sum_{j=0}^{\infty} a_j z^j$$

for $z \in \mathbb{C}$. It is elementary to see that there is an r, $0 \leq r \leq \infty$, such that the series converges absolutely and uniformly for $|z| < r$ and diverges for $|z| > r$. Observe that the domain of convergence is always a disc.

By analogy with functions of one variable one wants to expect in several variables that this set of convergence will be a polydisc, completed by some set of boundary points. However, very simple examples show that the situation is quite different.

Reinhardt domains appear naturally when one considers power series:

Definition 4.1.1. *Reinhardt domains* with a center at a point $a \in \mathbb{C}^n$ are defined as domains with the following property: for any point $z^0 = \left\{z_\nu^0\right\}$ that belongs to the domain, the domain also contains any point

$$z = \left\{a_\nu + \left(z_\nu^0 - a_\nu\right) e^{i\theta_\nu}\right\}, \quad 0 < \theta_\nu < 2\pi.$$

A Reinhardt domain with center a is said to be *complete* if along with each point z^0 it contains all points $z = \{z_\nu\}$ for which $|z_\nu - a_\nu| \leq \left|z_\nu^0 - a_\nu\right|$, $\nu = 1, \ldots, n$.

DOI: 10.1201/9781003619581-4

Observe that a complete Reinhardt domain in $\mathbb{C}$ (centered at 0) is an open disc with center 0. The ball $B(0, r)$ and polydisc $P(0, r)$ are complete Reinhardt domains also.

Whereas in one complex variable the domain of convergence of a power series is always a disc, in several complex variables the domain of convergence does not have a simple geometric description.

The set of convergence of the power series

$$\frac{1}{1 - z_1 z_2} = \sum_{k=0}^{\infty} z_1^k z_2^k$$

in $\mathbb{C}^2$ is the complete Reinhardt domain $\{|z_1 z_2| < 1\}$.

The set of convergence in $\mathbb{C}^2$ for the series

$$\frac{z_1}{(1 - z_1)(1 - z_2)} = \sum_{|k|=0}^{\infty} z_1^{k_1+1} z_2^{k_2}$$

is the bidisc $\{|z_1| < 1, |z_2| < 1\}$, completed by the complex lines $\{z_1 = 0\}$ and $\{z_2 = 0\}$.

The situation becomes more simple if instead of the set of convergence, we consider its interior. The domains of convergence of power series in several variables exhibit a much greater variety than in one variable. In order to see this, it is enough to calculate the domain of convergence of the following power series:

$$\sum_{\substack{k_1 \geq 0 \\ k_2 > 0}} z_1^{k_1} z_2^{k_2}, \qquad \sum_{k=0}^{\infty} (z_1 z_2)^k, \qquad \sum_{\substack{k_1 \geq 0 \\ k_2 > 0}} \frac{k_1!}{k_2!} z_1^{k_1} z_2^{k_2}.$$

Definition 4.1.2. The domain of convergence of the power series

$$\sum_{|k|=0}^{\infty} c_k (z - a)^k$$

is the interior $\overset{\circ}{S}$ of the set S of points $z \in \mathbb{C}^n$ at which this series converges for some ordering of its terms.

Notice that (power-series) always converges for $z = 0$, but if $n > 1$, S may be empty even if (the power-series) converges at some point $z \neq 0$. For example, the power series

$$\sum_{\substack{k_1 \geq 0 \\ k_2 > 0}} k_1! z_1^{k_1} z_2^{k_2}$$

converges for $z = (z_1, 0)$, but not for $z = (z_1, z_2)$ if $z_1, z_2 \neq 0$; hence its domain of convergence is empty.

The domain of convergence Ω of the power series $\sum c_j z^v$ is a (possibly empty) complete Reinhardt domain and Ω is the interior of the set of points

$w \in \mathbb{C}^n$ which satisfies (3.9). The convergence is normal in Ω. Not every complete Reinhardt domain is the precise domain of convergence of some power series (except in case $n = 1$, of course).

From Abel's lemma (Lemma 3.2.1) follows

Corollary 4.1.3. *The domain of convergence S of the power series $\sum c_k z^k$ is a (possibly empty) complete Reinhardt domain, and Ω is the interior of the set of points $w \in \mathbb{C}^n$ which satisfies (3.9). The convergence is normal in S.*

Theorem 4.1.4. *Let Ω be a complete Reinhardt domain. Then the Taylor series of $f \in \mathcal{O}(\Omega)$ at the center p of Ω*

$$f(z) = \sum_{|k|=0}^{\infty} c_k (z-p)^k. \tag{4.1}$$

converges on Ω.

Proof. Let p be an arbitrary point of Ω. Then the polydisc

$$\overline{P} = \{|z_\nu - p_\nu| \leq |z_\nu^0 - p_\nu|\} \subset\subset \Omega$$

and the function f is represented in P by a Taylor expansion centered at p. The coefficients of this expansion are computed via the derivatives of f at p, and hence, coincide with c_k, i.e., this expansion is the same as (4.1). □

Thus complete Reinhardt domains play the same role in the case of functions of several variables as discs do in the case of one variable. Not every complete Reinhardt domain is the precise domain of convergence of some power series (except in case $n = 1$, of course).

4.2 Laurent Series

Functions of one complex variable that are holomorphic on an annulus have Laurent series expansions, an analog of power series expansions for holomorphic functions on disks.

We see that holomorphic functions in two or more variables, in contrast to the situation in one variable, cannot have isolated singularities.

The extension of holomorphic functions can often be obtained by first expanding into a Laurent series and then showing that, under suitable geometric hypotheses, the Laurent series is actually a power series which converges on a larger domain.

Proposition 4.2.1. *Suppose* $0 < R_j < \infty$ *and* $0 \leq r_j \leq R_j$ *for* $1 \leq j \leq n$, *and let* $A(r, R) = \{z \in \mathbb{C}^n \mid r_j \leq |z_j| \leq R_j, 1 \leq j \leq n\}$. *Then every* $f \in \mathcal{O}(A)$, *has a unique representation*

$$f(z) = \sum_{k \in \mathbb{Z}^n} c_k z^k \qquad \text{for } z \in A(r, R) \tag{4.2}$$

The series (4.2) converges uniformly and absolutely on $A(r, R)$. *Moreover, for any* $\rho = (\rho_1, \ldots, \rho_n) > 0$ *with* $r \leq \rho \leq R$,

$$c_k = (2\pi i)^{-n} \int_{\partial_0 P(0,\rho)} f(\zeta)\zeta^{-k-1} d\zeta_1 \ldots d\zeta_n \qquad \text{for } k \in \mathbb{Z}^n \tag{4.3}$$

and

$$\text{if } r_l = 0 \text{ for some } l, \text{ then } c_k = 0 \text{ whenever } k_l < 0 \tag{4.4}$$

Remark 4.2.2. The series (4.2) is called the *Laurent series* of f on $A(r, R)$. It is uniquely determined by f and $A(r, R)$. It is a particular type of multiple series, with index set $\mathbb{Z}^n$ rather than $\mathbb{N}^n$.

Proof. We first prove that uniform convergence in (4.2) implies (4.3) and (4.4); in particular, this implies the uniqueness of the Laurent expansion. Given ρ as in the Theorem, the uniform convergence of (4.2) implies that

$$\begin{aligned} \int_{\partial_0 P(0,\rho)} f(\zeta)\zeta^\alpha d\zeta &= \sum_{v \in \mathbb{Z}^n} c_k \int_{\partial_0 P(0,\rho)} \zeta^v \zeta^\alpha \, d\zeta \\ &= \begin{cases} (2\pi i)^n c_k & \text{if } v + \alpha = (-1, \ldots, -1) \\ 0 & \text{otherwise.} \end{cases} \end{aligned}$$

Thus (4.3) follows. Furthermore, if $r_l = 0$ for some $1 \leq l \leq n$ and if $v_l < 0$, the integrand in (4.4) is holomorphic in ζ_l for $0 \leq |\zeta_l| \leq R_l$; so integration in ζ_l gives zero by Cauchy's Theorem, and (4.4) holds.

In order to prove the existence part we use induction on the number of variables n. The classical case $n = 1$ is known (see [3, p. 184]), so we assume that $n \geq 2$ and that the theorem holds for $n-1 > 0$ variables. Given $f \in \mathcal{O}(A(r, R))$ we choose $\lambda \in \mathbb{R}$ with $0 < \lambda < 1$ so that f is defined and holomorphic on the larger set $A_\lambda = A(\lambda r, R/\lambda)$. Define $'A_\lambda = \{'z \in \mathbb{C}^{n-1} \mid \lambda r_j \leq |z_j| \leq R_j/\lambda\}$ for $1 \leq j \leq n-1$ and $A_{n,\lambda} = \{z_n \in \mathbb{C} \mid \lambda r_n \leq |z_n| \leq R_n/\lambda\}$, and set $M = |f|_{A_\lambda} < \infty$. For each $z_n \in A_{n,\lambda}$ one has $f(\cdot, z_n) \in \mathcal{O}('A_\lambda)$; so, by inductive hypothesis,

$$f('z, z_n) = \sum_{k' \in \mathbb{Z}^{n-1}} c_{k'}(z_n)('z)^k \qquad \text{for } ('z, z_n) \in {}'A_\lambda \times A_{n,\lambda} \tag{4.5}$$

and, according to (4.3) in case of $n-1$ variables,

$$c_{k'}(z_n) = (2\pi i)^{n-1} \int_{\partial_0 P(0,\rho')} f('\zeta, z_n)('\zeta)^{-v'-1} d\zeta_1 \ldots d\zeta_{n-1} \tag{4.6}$$

for any $\rho' \in \mathbb{R}^{n-1}$ with $\rho' > 0$ and $\lambda r' \leq \rho' \leq R'/\lambda$. Equation (4.6) implies that $c_{k'} \in \mathcal{O}(A_{n,\lambda})$; so the case $n = 1$ gives

$$c_{k'}(z_n) = \sum_{k_n \in \mathbb{Z}} c_{v',k_n} z_n^{v_n} \quad \text{for } z_n \in A_{n,\lambda}$$

and

$$c_{k',k_n} = (2\pi i)^{-1} \int_{|\zeta|=\rho_n} c_{k'}(\zeta)\zeta^{-k_n-1} d\zeta \tag{4.7}$$

for $\rho_n > 0$ with $\lambda r_n \leq \rho_n \leq R_n/\lambda$. Substituting (4.7) into (4.5) and writing $(k', k_n) = k \in \mathbb{Z}^n$, one obtains

$$f(z) = \sum_{k' \in \mathbb{Z}^{n-1}} \left(\sum_{k_n \in \mathbb{Z}} c_k z^k \right) \text{ for } z \in A_\lambda \,. \tag{4.8}$$

We now claim that

$$\left| c_k z^k \right| \leq M \lambda^{|k_1| + \cdots + |k_n|} \quad \text{for } z \in A(r, R) \,. \tag{4.9}$$

Since $\lambda < 1$, (4.9) readily implies that the series $\sum c_k z^k$ converges uniformly and absolutely on $A(r, R)$, and hence its limit agrees with the iterated series (4.8), i.e., with $f(z)$. This will complete the proof of the theorem.

In order to prove (4.9), notice that by estimating (4.6) and (4.7) it follows that

$$|c_k| \leq M/\rho^k \tag{4.10}$$

for all $\rho \in \mathbb{R}^n, \rho > 0$, with $\lambda r \leq \rho \leq R/\lambda$. Given $z \in A(r, R)$ with $z_j \neq 0$ for $1 \leq j \leq n$ and $v \in \mathbb{Z}^n$, we choose $\rho_1, \ldots, \rho_n$ as follows: for j with $k_j \geq 0$, set $\rho_j = |z_j|/\lambda$, and if $k_j < 0$, set $\rho_j = \lambda |z_j|$. Then (4.10) implies

$$\left| c_k z^k \right| \leq M \frac{|z_1|^{k_1 \cdots} \; |z_n|^{k_n}}{\rho^k} = M \lambda^{|k_1| + \cdots + |k_n|}$$

Finally, if $z \in K(r, R)$ satisfies $z_l = 0$ for some l, then one must have $r_l = 0$; so, by letting $\rho_l \to 0$ in (4.10), it follows that $c_k = 0$ if $v_l < 0$, and if $v_l \geq 0$, (4.9) is obvious. So (4.9) holds for all $z \in A(r, R)$. □

Theorem 4.2.3. *Let Ω be a connected Reinhardt domain with center 0. Then every $f \in \mathcal{O}(\Omega)$ has a Laurent series representation*

$$f(z) = \sum_{k \in \mathbb{Z}^n} c_k z^k \quad \textit{for } z \in \Omega \tag{4.11}$$

which converges normally on Ω. Moreover, if $\Omega \cap \{z \in \mathbb{C}^n \mid z_l = 0\} \neq \varnothing$ for some $1 \leq l \leq n$, then $c_k = 0$ for $k_l < 0$.

Proof. Suppose $f \in \mathcal{O}(\Omega)$. For $a \in \Omega$, we choose $A(r,R) \subset \Omega$ so that a lies in the interior of $A(r,R)$. By Proposition 4.2.1, $f(z) = \sum c_k(a) z^k$ uniformly and absolutely on $A(r,R)$, and the coefficients $c_k(a)$, $k \in \mathbb{Z}^n$, are independent of the particular choice of r and R. Moreover, if a and a^* are two points in the interior of $A(r,R)$, then $c_k(a) = c_k(a^*)$. So, for each $k \in \mathbb{Z}^n$, $a \to c_k(a)$ is locally constant on Ω, and hence globally constant since Ω is connected. This proves the existence of the representation (4.11) valid on Ω. The uniform absolute convergence on each $A(r,R)$ and a standard compactness argument show that (4.11) converges normally in Ω. Finally, if for some l, $1 \le l \le n$, there is $a \in \Omega$ with $a_l = 0$, then we can choose r and R with $r_l = 0$, so that $a \in A(r,R) \subset \Omega$; hence, by Proposition 4.2.1, $c_k = c_k(a) = 0$ for $k_l < 0$. □

4.3 Series in Homogeneous Polynomials

A polynomial $p(z)$ is said to be *homogeneous of degree k* if $p_\nu(\lambda z) = \lambda^k p_\nu(z)$ for all $\lambda \in \mathbb{C}$ and $z \in \mathbb{C}^n$. Series in such polynomials are obtained from power series by regrouping the terms:

$$\sum_{|k|=0}^{\infty} c_k z^k = \sum_{\nu=0}^{\infty} \left(\sum_{|k|=\nu} c_k z^k \right) = \sum_{\nu=0}^{\infty} p_\nu(z)$$

(ν is a scalar index). This regrouping may lead to an extension of the domain of convergence; we see now that series in homogeneous polynomials converge in a wider class of domains than power series:

Theorem 4.3.1. *Let $f \in \mathcal{O}(\Omega)$, where Ω is a complete circular domain with center 0. Then f can be expanded in this domain in a series in homogeneous polynomials*

$$f(z) = \sum_{j=0}^{\infty} p_j(z) \tag{4.12}$$

where p_j is formed from the terms $c_k z^k$ of the Taylor expansion of f at 0 for which $|k| = j$, and the series converges uniformly on compact subsets of Ω.

Proof. Suppose $z \in \Omega \setminus \{0\}$ and $\omega = z/|z|$. The complex line $\zeta = \lambda\omega$, $\lambda \in \mathbb{C}$, passes through z and intersects Ω in the disc $\{\lambda \mid |\lambda| < R(\omega)\}$. Since $0 \in \Omega$, the series

$$f(\zeta) = \sum_{|k|=0}^{\infty} c_k \zeta^k = \sum_{j=0}^{\infty} p_j(\zeta)$$

converges in some neighborhood of the origin and restriction to the complex line $\{\zeta = \lambda\omega\}$ gives

$$\varphi(\lambda) = f(\lambda\omega) = \sum_{j=0}^{\infty} p_j(\lambda\omega) = \sum_{j=0}^{\infty} p_j(\omega)\lambda^j \tag{4.13}$$

Since $f \in \mathcal{O}(\Omega)$, φ is holomorphic in the disc $\{|\lambda| < R(\omega)\}$ and (4.13) is its Taylor expansion; it converges in this disc, and since we assumed $|z| < R(\omega)$ and $z = |z|\omega$, then the series (4.12) converges at the point z. Now if $K \subset\subset \Omega$, then there exist functions $r(\omega)$, $0 < r(\omega) < R(\omega)$, and a number q, $0 < q < 1$, such that for all $z \in K$ we have $|z| \leq qr(\omega)$, where $\omega = z/|z|$. Therefore for all $z \in K$

$$|p_j(z)| = |p_j(\omega)|\,|z|^{\nu} \leq |p_j(\omega)|\, r^j(\omega)q^j$$

But by the integral formula for the Taylor coefficients

$$p_j(\omega) = \frac{1}{2\pi i}\int_{|\lambda|=r(\omega)} \frac{f(\lambda\omega)}{\lambda^{j+1}}d\lambda$$

from which, using the fact that the set $\{\lambda\omega \mid |\lambda| = r(\omega)\} \subset\subset \Omega$ and hence $|f(\lambda\omega)| \leq M$ there, we find that $|p_j(\omega)| \leq M/r^j(\omega)$ (Cauchy's inequalities for the Taylor coefficients). Substituting this in (4.12), we obtain that $|p_j(z)| \leq Mq^j$ for all $z \in K$. Since $0 < q < 1$, this proves the uniform convergence of (4.12) on K. □

4.4 The Domain of Convergence

We now discuss a characterization of domains of convergence of power series in terms of a simple geometric condition which, for complete Reinhardt domains, is equivalent to holomorphic convexity.

Definition 4.4.1. We denote by

$$z \to \tau(z) = (\ln|z_1|, \ldots, \ln|z_n|)$$

a mapping of the set $\{z \in \mathbb{C}^n \mid z_1 \cdots z_n \neq 0\}$ into the space $\mathbb{R}^n$. The logarithmic image of a set $M \subset \mathbb{C}^n$ is the set $M^* = \tau(M_0)$, where $M_0 = \{z \in M \mid z_1 \cdots z_n \neq 0\}$. The set M is said to be *logarithmically convex* if its logarithmic image M^* is a convex set in $\mathbb{R}^n$.

Obviously, if $\Omega \subset \mathbb{C}^n$ is a complete Reinhardt domain, then

$$\xi \in \log\tau(\Omega) \text{ implies } \{\eta \in \mathbb{R}^n \mid \eta_j \leq \xi_j \text{ for } 1 \leq j \leq n\} \subset \log\tau(\Omega). \tag{4.14}$$

Lemma 4.4.2. *If $\Omega \subset \mathbb{C}^n$ is the region of convergence of a power series $\sum_k c_k z^k$, then $\log\tau(\Omega)$ is a convex subset of $\mathbb{R}^n$.*

Proof. Suppose ξ and η are points in $\log\tau(\Omega)$; we must show that $t\xi+(1-t)\eta\in\log\tau(\Omega)$ for $0\le t\le 1$. Choose $p,q\in\Omega$, and $\lambda>1$, so that $\xi_j=\log|p_j|$ and $\eta_j=\log|q_j|$ for $1\le j\le n$, and $\lambda p,\lambda q\in\Omega$. Since $\sum c_k z^k$ converges at λp and λq, there is $M<\infty$ such that

$$|c_k|\,\lambda^{|k|}\left|p^k\right|\le M \quad\text{and}\quad |c_k|\,\lambda^{|k|}\left|q^k\right|\le M$$

for all k. It follows that

$$|c_k|\,\lambda^{|k|}\left|p^k\right|^t\left|q^k\right|^{1-t}\le M \quad \text{for } 0\le t\le 1$$

According to Abel's Lemma (Lemma 3.2.1), this implies that $\sum c_k z^k$ converges in a neighborhood of

$$a_t=\left(|p_1|^t\,|q_1|^{1-t},\ldots,|p_n|^t\,|q_n|^{1-t}\right),$$

i.e., $a_t\in\Omega$. So $t\xi+(1-t)\eta\in\log\tau(\Omega)$ for all $0\le t\le 1$. □

Lemma 4.4.3. *Let Ω be a complete Reinhardt domain. If $\log\tau(\Omega)$ is convex, then Ω is convex with respect to the family $\mathcal{M}=\left\{z^k,k\in\mathbb{N}^n\right\}$ of holomorphic monomials on $\mathbb{C}^n$.*

Proof. We must show that $\hat{K}_{\mathcal{M}}\cap\Omega\subset\subset\Omega$ for every compact set $K\subset\Omega$, where $\widehat{K}_{\mathcal{M}}=\{z\in\mathbb{C}^n \mid |m(z)|\le|m|_K \text{ for all } m\in\mathcal{M}\}$. Since $\widehat{K}_{\mathcal{M}}$ is obviously compact, this will follow once we show that $\widehat{K}_{\mathcal{M}}\cap\partial\Omega=\varnothing$. Given a compact set $K\subset\Omega$, we cover K by finitely many polydiscs $P\left(0,\tau\left(q^{(l)}\right)\right)\subset\Omega$, $l=1,\ldots,k$, where $q^{(l)}\in\Omega$ and $q_j^{(l)}\ne 0$ for $1\le j\le n$. Let $Q=\left\{q^{(1)},\ldots,q^{(k)}\right\}$. Then $\widehat{K}_{\mathcal{M}}\subset\hat{Q}_{\mathcal{M}}$, and we shall show that $\hat{Q}_{\mathcal{M}}\cap\partial\Omega=\varnothing$, completing the proof of the Lemma.

Let $p\in\partial\Omega$. Assume first that $p_j\ne 0$ for $1\le j\le n$. Then $p^*=(\log|p_1|,\ldots,\log|p_n|)\in b[\log\tau(\Omega)]$, and since $\log\tau(D)$ is assumed convex, there is a linear function $L(\xi)=\sum_{j=1}^n\mu_j\xi_j,\mu_j\in\mathbb{R}$, so that $L(\xi)<L(p^*)$ for all $\xi\in\log\tau(D)$. Since $\log\tau(D)$ satisfies (4.14), we must have $\mu_j\ge 0$. Let Q^* be the finite set of points in $\log\tau(D)$ which corresponds to Q. One can find rational numbers $\alpha_j>\mu_j\ge 0$ sufficiently close to $\mu_j, 1\le j\le n$, so that for $\tilde{L}(\xi)=\sum\alpha_j\xi_j$ one has

$$\tilde{L}(\xi)<\tilde{L}(p^*) \qquad \text{for all } \xi\in Q^* \tag{4.15}$$

Equation (4.15) remains true after multiplication with the (positive) common denominator of $\alpha_1,\ldots,\alpha_n$, so we may assume that $\alpha_j\in\mathbb{N}^+$. The monomial $m_\alpha(z)=z_1^{\alpha_1}\ldots z_n^{\alpha_n}$ then satisfies $|m_\alpha|_Q<|m_\alpha(p)|$, i.e., we have shown that $p\notin\hat{Q}_{\mathcal{M}}$. For any of the remaining points $p\in bD$, we may renumber the coordinates so that for some $1\le l<n$ one has $p_1\cdot\cdots\cdot p_l\ne 0$, while $p_{l+1}=\cdots=p_n=0$. If $\pi_l:\mathbb{C}^n\to\mathbb{C}^l$ is the projection $z\mapsto(z_1,\ldots,z_l)$, then $\log\tau(\pi_l(D))\subset\mathbb{R}^l$ is convex and satisfies (4.14). The preceding argument applied to $\pi_l(p)$ now gives a monomial m in $z_1,\ldots,z_l$ with $|m|_Q<|m(p)|$, so $p\notin\widehat{Q}_{\mathcal{M}}$ in this case also. □

We now have all the necessary ingredients for the main result of this section.

Theorem 4.4.4. *The following are equivalent for a complete Reinhardt domain Ω in $\mathbb{C}^n$ with center 0.*

(I) *Ω is the region of convergence of a power series.*

(II) $\log \tau(\Omega) \subset \mathbb{R}^n$ *is convex.*

(III) *Ω is $\mathcal{M}$-convex.*

(IV) *Ω is holomorphically convex.*

(V) *Ω is a domain of holomorphy.*

Proof. The implications **(I)** $\Rightarrow$ **(II)** $\Rightarrow$ **(III)** are proved in Lemma 4.4.2 and Lemma 4.4.3. **(III)** $\Rightarrow$ **(IV)** is trivial, since $\mathcal{M} \subset \mathcal{O}(\Omega)$, and **(IV)** $\Rightarrow$ (V) (A holomorphically convex domain Ω in $\mathbb{C}^n$ is a domain of holomorphy). To see that **(V)** implies **(I)**, let $f \in \mathcal{O}(\Omega)$ be completely singular at every boundary point of Ω. By Cauchy estimates, the Taylor series of f converges on Ω, and it obviously cannot converge on any region strictly larger than Ω. □

Many important function theoretic properties of domains in $\mathbb{C}^n$ are characterized by convexity with respect to an appropriate family of functions.

4.5 Forelli's Theorem

We give three proofs of a radial analogue of the classical theorem of Hartogs, proved in 1978 by F. Forelli [74]. A number of generalizations variants and applications of Forelli's theorem are presented in [143].

For Forelli's proof of his theorem, we need the following

Theorem 4.5.1. *Any function f that is holomorphic in the unit ball $B \subset \mathbb{C}^n$ can be expanded in B in a series of homogeneous polynomials*

$$f(z) = \sum_{j=0}^{\infty} p_j(z), \tag{4.16}$$

where p_j is formed from the terms $c_k z^k$ of the Taylor expansion of f at 0 for which $|k| = j$, and the series converges uniformly on compact subsets of B.

Proof. Suppose $z \in B \backslash \{0\}$ and $\omega = z/|z|$. The complex line $\zeta = \lambda\omega$, $\lambda \in \mathbb{C}$, passes through z and intersects B in the disc $\{\lambda \mid |\lambda| < 1\}$. Since $0 \in B$, the series

$$f(\zeta) = \sum_{|k|=0}^{\infty} c_k \zeta^k = \sum_{j=0}^{\infty} p_j(\zeta)$$

converges in some neighborhood of the origin and restriction to the complex line $\{\zeta = \lambda\omega\}$ gives

$$\varphi(\lambda) = f(\lambda\omega) = \sum_{j=0}^{\infty} p_j(\lambda\omega) = \sum_{j=0}^{\infty} p_j(\omega)\lambda^j. \tag{4.17}$$

Since $f \in \mathcal{O}(B)$, φ is holomorphic in the disc $\{|\lambda| < 1\}$ and (4.17) is its Taylor expansion; it converges in this disc, and since we assumed $|z| < 1$ and $z = |z|\omega$, then the series (4.16) converges at the point z. Now let K be a compact subset of B, then there exist numbers r, $0 < r < 1$, and q, $0 < q < 1$, such that for all $z \in K$ we have $|z| \le qr$. Therefore, for all $z \in K$,

$$|p_j(z)| = |p_j(\omega)|\, |z|^j \le |p_j(\omega)|\, q^j. \tag{4.18}$$

We have

$$p_j(\omega) = \frac{1}{2\pi i}\int_{|\lambda|=r} \frac{f(\lambda\omega)}{\lambda^{j+1}}\, d\lambda\,.$$

Since $|f(\lambda\omega)| \le M$ on the compact set $\{\lambda\omega \mid |\lambda| = r\} \subset B$, this implies $|p_j(\omega)| \le M/r^j$. Substituting this into (4.18), we obtain that $|p_j(z)| \le Mq^j$ for all $z \in K$. Since $0 < q < 1$, this proves the uniform convergence of (4.16) on K. □

Theorem 4.5.2 (Forelli's Theorem). *If f is a function defined in the unit ball $B \subset \mathbb{C}^n$, holomorphic on the intersection of B with every complex line l passing through the point $z = 0$, and if f is of class $\mathcal{C}^\infty$ in a neighborhood of this point, then it is holomorphic in B.*

Proof. [**Forelli's Proof**] Let $z \in B \setminus \{0\}$ be an arbitrary point; from the condition that f is holomorphic on the complex line passing through $z = 0$ and this point, it follows that

$$f(\lambda z) = f(0) + \sum_{k=1}^{\infty} F_k(z)\lambda^k \tag{4.19}$$

and the series converges absolutely and uniformly in the closed disc $\{|\lambda| \le 1\}$. From this it follows that the series of the $|F_k(z)|$ converges, and hence for $|z| < 1$ the function

$$F(z) = \sum_{k=1}^{\infty} F_k(z)$$

is well-defined.

Further, from (4.19) for any λ and μ from the unit disc $U \subset \mathbb{C}$ we will obtain two expansions

$$f(\lambda\mu z) = f(0) + \sum_{k=1}^{\infty} F_k(\lambda z)\mu^k = f(0) + \sum_{k=1}^{\infty} F_k(z)\lambda^k\mu^k$$

from which by the uniqueness theorem for power series expansion we get the identities

$$F_k(\lambda z) = \lambda^k F_k(z), \quad k = 1, 2, \ldots. \tag{4.20}$$

By Cauchy's theorem for the coefficients of (4.19),

$$F_k(z) = \frac{1}{2\pi i} \int_{\{|\lambda|=1\}} \frac{f(\lambda z)}{\lambda^{k+1}} d\lambda,$$

and since by hypothesis the function f is of class $\mathcal{C}^\infty$ in some ball $B_r = \{|z| < r\}$, then it follows from this formula that $F_k \in C^\infty(B_r)$. In particular, these functions are bounded in B_r and then from (4.19) it follows that $F_k(z) = O\left(|z|^k\right)$ as $z \to 0$.

By Taylor's formula for smooth functions, we will obtain from this that for $z \in B_r$

$$F_k(z) = \sum_{\mu+\nu=k} P_{\mu\nu}(z) + |z|^k \gamma(z), \tag{4.21}$$

where $P_{\mu\nu}$ is a polynomial in z and $\overline{z}$ of general degree μ in z and ν in $\overline{z}$, and $\gamma(z) \to 0$ as $z \to 0$ (the lower-order terms vanish because $F_k(z) = O\left(|z|^k\right)$). Substituting this result in (4.20), we obtain the identity

$$\sum_{\mu+\nu=k} P_{\mu\nu}(z)\lambda^\mu \overline{\lambda}^\nu + |\lambda|^k |z|^k \gamma(\lambda z) = \sum_{\mu+\nu=k} P_{\mu\nu}(z)\lambda^k + \lambda^k |z|^k \gamma(z), \tag{4.22}$$

valid for all $z \in B_r$ and $|\lambda| \leq 1$. In particular, for positive $\lambda < 1$ it follows from this that $\gamma(\lambda z) = \gamma(z)$, and hence as $\lambda \to 0$ we find that $\gamma(z) \equiv 0$ in B_r. But then from (4.22), again by the theorem on the uniqueness of power series expansions, we see that $P_{\mu\nu} \equiv 0$ for $\nu > 0$, and then it follows from (4.21) that $F_k(z) = P_{k0}(z)$ are polynomials in z, i.e., are holomorphic functions, and not only in B_r but everywhere in $\mathbb{C}^n$.

The series of the F_k by Theorem 4.5.1 converges uniformly on compact subsets of B, and from the theorem of Weierstrass we conclude that $F \in \mathcal{O}(B)$. But from (4.19) for $\lambda = 1$ it follows that $f(z) = f(0) + F(z)$, and hence $f \in \mathcal{O}(B)$.

Krantz's new proof of Forelli's theorem Let f satisfy the hypotheses of Theorem 4.5.2. For convenience, we assume in addition that f is continuous as a function of several variables (some of the aforementioned references also make this assumption).

Let $B(P, r)$ be a Euclidean ball contained in the unit ball B. Then the intersection of any of the complex lines $l_{\mathbf{v}}$ with $B(P, r)$ is a complex disc $\mathbf{d_v}$ and, since f is harmonic on the line $l_{\mathbf{v}}$, it satisfies the mean-value property on $\mathbf{d_v}$. In case $B(P, r)$ does not contain the origin, then the discs $\mathbf{d_v}$ are pairwise disjoint and exhaust $B(P, r)$. So we may apply the coarea formula (in [67, Theorem 3.2.12]) to conclude that f satisfies the mean-value property on $B(P, r)$. If instead $B(P, r)$ does contain the origin, then the discs $\mathbf{d_v}$ are not pairwise disjoint. So the coarea formula does not apply. We may assume that $B(P,r)$ lies in the neighborhood U of the origin on which f is $\mathcal{C}^\infty$. Thus we can exploit this smoothness—by way of the Taylor expansion, for instance—to again conclude that f satisfies the mean value property on $B(P,r)$. In conclusion, f is harmonic. So, in particular (because the Poisson kernel on balls is real analytic), f is real analytic.

So f has a convergent power series centered at the origin. And the holomorphicity of f along each of the complex lines $l_{\mathbf{v}}$ now guarantees—just using a little linear algebra—that this power series will have no conjugate-holomorphic terms in it. We conclude then that f is holomorphic.

Krantz's second new proof of Forelli's theorem We continue to assume that f is continuous. Since f is holomorphic on the complex line $l_{\mathbf{v}}$, with $\mathbf{v} = (1, 0, 0, \ldots, 0)$, it is analytic in the variable z_1. Thus we may conclude that

$$\left| \frac{\partial^k f}{\partial z_1^k}(0) \right| \leq C \cdot \frac{k!}{R^k}$$

for some constant $C > 0$ and for some number R with $0 < R < 1$ and for k a positive integer. A similar argument shows, for $j = 2, 3, \ldots, n$, that

$$\left| \frac{\partial^k f}{\partial z_j^k}(0) \right| \leq C \cdot \frac{k!}{R^k}$$

for some constant $C > 0$ and for some number R with $0 < R < 1$ and for k a positive integer. Some elementary Fourier analysis now allows us to conclude that

$$\left| \frac{\partial^{\alpha+\beta} f}{\partial z^\alpha \partial \overline{z}^\beta}(0) \right| \leq C \cdot \frac{\alpha!\beta!}{R^{|\alpha|+|\beta|}}$$

for multi-indices α and β. If we set

$$a_{\alpha,\beta} = \frac{1}{\alpha!\beta!} \cdot \frac{\partial^{\alpha+\beta} f}{\partial z^\alpha \partial \overline{z}^\beta}(0)$$

then the power series

$$\sum_{\alpha,\beta} a_{\alpha,\beta} z^\alpha \overline{z}^\beta \tag{4.23}$$

converges to a real analytic function g on a nontrivial polydisc centered at the origin. But we cannot assert that this power series converges to f. That requires an extra argument. First, we note that, by the argument presented in the first new proof of Forelli, all the summands in (4.23) with $\overline{z}$ vanish. So the series (4.23) defines a holomorphic function. We now show that that function is f. Let us compare f and g on the complex line $l_{\mathbf{v}}$ with $\mathbf{v} = (1, 0, 0, \ldots, 0)$. Both functions have a power series expansion centered at the origin along this complex line. Both these power series have the same coefficients—namely those given by

$$\frac{\partial^k f}{\partial z_1^k}(0)$$

So f and g agree on this particular complex line. But in fact the same reasoning applies to all the complex lines $l_{\mathbf{v}}$. So in fact f and g agree everywhere. We conclude that f is holomorphic as a function of several complex variables. □

We conclude this section by noting that the hypothesis that f is C^∞ in a neighborhood of the origin cannot be weakened. The example (see [114] for the source)

$$f(z_1, z_2) = \begin{cases} \frac{\bar{z}_2}{\bar{z}_1} \cdot z_1^{k+2} & \text{if } z_1 \neq 0 \\ 0 & \text{if } z_1 = 0 \end{cases}$$

is a function that is C^k (for k a finite, positive integer) and satisfies the condition (II) of the hypothesis of Forelli's theorem, but it is plainly not holomorphic in the several variable sense.

4.6 Domains of Holomorphy

At a quick glance, one might be tempted to think of the analysis of several complex variables (or several real variables, for that matter) as being essentially one variable theory with the additional complication of multi-indices. This perception turns out to be incorrect. Deep new phenomena and profound (and as yet unsolved) problems present themselves in the theory of several variables. The purpose of this section is to give the reader an appreciation for some of these new features.

The kind of regions without Hartogs phenomenon is called domains of holomorphy. The study of domains of holomorphy is one of the main topics in the research of the function theory of several complex variables in the last century. The reader can refer to books about the function theory of several complex variables such as S.G. Krantz [139] and R. Narasimhan [161].

Here we give the definition of a domain of holomorphy. Let Ω be a domain in $\mathbb{C}^n, n \geq 1$. We first make the following definition.

Definition 4.6.1. Ω is called a domain of holomorphy if there is a holomorphic function f defined on Ω that is singular at every boundary point, that is, f cannot be extended holomorphically across any boundary point.

Remark 4.6.2. We do not assume that a domain of holomorphy is connected.

According to Hartogs' theorem (Theorem 3.3.7), if we remove a compact subset K from the unit ball $B(0;1) \in \mathbb{C}^n$, $n \geq 2$, such that $B(0;1) \setminus K$ is connected, then the remaining set $B(0;1) \setminus K$ is not a domain of holomorphy.

In the theory of holomorphic functions of one complex variable, every open set is a domain of holomorphy. Here is a simple proof that the unit disc $D \subseteq \mathbb{C}$

is a domain of holomorphy. Let

$$f_0(\zeta) = \sum_{j=0}^{\infty} 2^{-j} \zeta^{2^j}$$

Then the series defining f_0 converges absolutely and uniformly on $\overline{D}$ by the Weierstrass M-test. Therefore, it is continuous on $\overline{D}$ and holomorphic on D. However, the boundary function

$$\theta \mapsto \sum_{j=0}^{\infty} 2^{-j} e^{i2^j \theta}$$

is the Weierstrass nowhere differentiable function (see Y. Katznelson [116]), so that f_0 cannot be extended even differentiably, much less analytically, to a larger open set. Since there exists one analytic function on D that cannot be continued to a larger open set, it follows that D is a domain of holomorphy. Now if Ω is a simply connected, smoothly bounded domain then, by the Riemann mapping theorem, there is a conformal mapping $\Phi : \Omega \to D$. By an old result of Painlevé (see M. Tsuji [204] or S. Bell and S.G. Krantz [22]), the mapping Φ continues to be continuously differentiable from $\overline{\Omega}$ to $\overline{D}$. Likewise the mapping Φ^{-1} extends to be continuously differentiable from $\overline{D}$ to $\overline{\Omega}$. Then $f_0 \circ \Phi$ is holomorphic on Ω and continuous on $\overline{\Omega}$, yet could not continue to be even differentiable on any strictly larger open set. It follows that Ω is a domain of holomorphy.

For the general case, let U be any connected open subset of $\mathbb{C}$. Let $\{z_j\} \subseteq U$ be a sequence of points that has no accumulation point in U but that accumulates at every boundary point of U (the reader may construct such a sequence by exploiting the distance function to the boundary). By Weierstrass's theorem there is a nonconstant analytic function h on U that vanishes at each z_j and nowhere else. If h were to analytically continue to an analytic function $\hat{h}$ on some strictly larger open set $\hat{U}$, then it would necessarily hold that $\hat{U}$ contains a boundary point P of U. But then $\hat{h}$ would have zero set with interior accumulation point P. It follows that $\hat{h} \equiv 0$, which is a contradiction. Thus U is a domain of holomorphy.

The next famous example tells us that there exists such a power scries that every point on its circle of convergence is a singularity.

Example 4.6.3. Let

$$f(z) = z^{1!} + z^{2!} + \cdots + z^{n!} + \cdots \tag{4.24}$$

Then every point on its circle of convergence is a singularity.

Proof. Since

$$a_j = \begin{cases} 1, & \text{if } j = k! \\ 0, & \text{if } j \neq k! \end{cases}$$

we have

$$\varlimsup_{j\to\infty} \sqrt[n]{|a_j|} = 1$$

Thus, the radius of convergence of (4.24) is $R = 1$. That is $f(z)$ is holomorphic on $D(0,1)$. If $z_0 \in D(0,1)$ and $|z_0| = 1/2$, then $f(z)$ has a Taylor expansion

$$g(z) = \sum_{j=0}^{\infty} \frac{f^{(j)}(z_0)}{j!}(z - z_0)^j$$

A circle $\mathbb{D}(z_0, 1/2)$ intersects the circle $|z| = 1$ at a point ζ_0. If we can show that the radius of convergence of $g(z)$ is $1/2$, then $f(z)$ cannot be analytically continued from ζ_0. Suppose the contrary. If ρ is the radius of convergence of $g(z)$ and $\rho > 1/2$, then $\mathbb{D}(\zeta_0, \rho) \cap \mathbb{D} \neq \emptyset$. Thus, $\mathbb{D}(\zeta_0, \rho)$ contains an arc σ of $|z| = 1$ where $\zeta_0 \in \sigma$. Since the set of points $\exp\{(2\pi ip)/q\}$ (where p, q are integers and p/q is a reduced fraction) is dense on $|z| = 1$, there exists a point $\zeta_1 = \exp\{(2\pi ip)/q\}$ on σ and $\lim_{r\to 1} g(r\zeta_1) = g(\zeta_1)$, $(0 < r < 1)$. Since $g(z) = f(z)$ for $z \in \mathbb{D}$, we have $\lim_{r\to 1} f(r\zeta_1) = g(\zeta_1)$. Since

$$f(r\zeta_1) = \sum_{j=1}^{q-1} r^{j!}\xi_1^{j!} + \sum_{j=q}^{\infty} r^{j!}$$

it is easy to see that

$$\sum_{j=q}^{\infty} r^{j!} > \sum_{j=q}^{N} r^{j!} > (N-q)r^{N!}$$

Thus, $\sum_{j=q}^{\infty} r^{j!}$ is greater than any positive integers as $r \to 1$. Therefore $\lim_{r\to 1} |f(r\zeta_1)| = \infty$ and this contradicts with $\lim_{r\to 1} f(r\zeta_1) = g(\zeta_1)$. This proves that ζ_0 is a singularity of $f(z)$. □

The situation in several complex variables is decidedly different. Hartogs's Phenomenon (Theorem 3.3.7) shows that, for $n \geqq 2$, there are examples of open sets $\Omega \subsetneqq \widetilde{\Omega} \subseteq \mathbb{C}^n$ such that every holomorphic function in Ω admits a holomorphic extension to $\tilde{\Omega}$.

Let

$$\Omega \equiv P(0,3)\backslash\overline{P}(0,1) \subseteq \mathbb{C}^2.$$

We claim that every holomorphic function on Ω analytically continues to the domain $\hat{\Omega} \equiv P(0,3)$. Of course, this fact follows from Theorem 3.3.7, but we give a simple proof using only elementary one-variable facts about power series, of the continuation result.

Now let h be holomorphic on Ω. For z_1 fixed, $|z_1| < 3$, we write

$$h_{z_1}(z_2) = h(z_1, z_2) = \sum_{j=-\infty}^{\infty} a_j(z_1) z_2^j \tag{4.25}$$

where the coefficients of this Laurent expansion are given by

$$a_j(z_1) = \frac{1}{2\pi i}\oint_{|\zeta|=2} \frac{h(z_1,\zeta)}{\zeta^{j+1}} d\zeta$$

In particular, $a_j(z_1)$ depends holomorphically on z_1 (by Morera's theorem, for instance). But $a_j(z_1) = 0$ for $j < 0$ and $1 < |z_1| < 3$. Therefore, by analytic continuation, $a_j(z_1) \equiv 0$ for $j < 0$. But then the series expansion (4.25) becomes

$$\sum_{j=0}^{\infty} a_j(z_1) z_2^j$$

and this series defines a holomorphic function $\hat{h}$ on all of $P(0,3)$ that agrees with the original function h on Ω. Thus Ω is not a domain of holomorphy; holomorphic functions on Ω continue to the larger domain $P(0,3)$. This completes the proof of the "Hartogs extension phenomenon."

In one complex variable, every domain is a domain of holomorphy. One way to see this fact is as follows.

Let U be a domain on the complex plane, and let $p \in \partial D$ be a boundary point of U. Then $g(z) = 1/(z-p)$ is a holomorphic function on U, but singular at p. To construct a holomorphic function f on U that is singular at every boundary point, we use the following fact.

Theorem 4.6.4. *Let K be a compact subset of U in $\mathbb{C}$ and let $z_0 \in U\backslash K$. If the component of $U\backslash K$ that contains z_0 is not relatively compact in U, then there is a holomorphic function $h(z)$ on U such that $|h(z_0)| > \sup_{z\in K}|h(z)|$.*

The proof of Theorem 4.6.4 uses the Runge approximation theorem. Actually, we can make $|h(z_0)|$ arbitrarily large and $\sup_{z\in K}|h(z)|$ arbitrarily small. For instance, if we set $m = (|h(z_0)| + \sup_{z\in K}|h(z)|)/2$, then $(h/m)^l$, for large $l \in \mathbb{N}$, will do the job.

Now, let $\mathcal{P}$ be the set containing all points in U with rational coordinates. Clearly, $\mathcal{P}$ is countable and dense in U. Let $\{\zeta_i\}_{i=1}^{\infty}$ be a sequence of points in U such that every point belonging to $\mathcal{P}$ appears infinitely many times in the sequence. Next, we exhaust U by a sequence of increasingly compact subsets $\{K_j\}_{j=1}^{\infty}$ of U defined as follows:

$$K_j = \left\{z \in U \mid \text{dist}(z, U^c) \geq j^{-1}\right\} \cap \overline{\mathbb{D}(0;j)}, \quad j \in \mathbb{N}$$

where $\text{dist}(z, u^c)$ denotes the distance from z to the complement of U and $\mathbb{D}(0;j) = \{z \in \mathbb{C} \mid |z| < j\}$. Hence, we have $K_j \subset \overset{\circ}{K}_{j+1}$, where $\overset{\circ}{K}_{j+1}$ is the interior of K_{j+1}. For each i, denote by B_{ζ_i} the largest disc centered at ζ_i and contained in U. Then, inductively choose a subsequence $\{K_{m_j}\}$ of $\{K_j\}$ such that for each j, pick a $z_j \in \left(B_{\zeta_j}\backslash K_{m_j}\right) \cap \overset{\circ}{K}_{m_j+1}$, and a $f_j(z) \in \mathcal{O}(D)$ satisfying

$$|f_j(z)| < \frac{1}{2^j}, \quad z \in K_{m_j}$$

and

$$|f_j(z_j)| \geq \sum_{i=1}^{j-1} |f_i(z_j)| + j + 1$$

It follows that

$$h(z) = \sum_{j=1}^{\infty} f_j(z)$$

defines a holomorphic function on U and that

$$|h(z_j)| \geq |f_j(z_j)| - \sum_{i=1}^{j-1} |f_i(z_j)| - \sum_{i=j+1}^{\infty} |f_i(z_j)| \geq j$$

which implies $h(z)$ is singular at every boundary point of U. For if $h(z)$ extends holomorphically across some boundary point, then $h(z)$ would be bounded on $\bar{B}_{\zeta_i}$ for some ζ_i. Obviously, it contradicts the construction of h.

We now move to higher-dimensional Euclidean spaces. We see that the phenomenon is totally different. Let us start with the following simple example in $\mathbb{C}^2$. We consider the domain Ω defined by

$$\Omega = \left\{ (z_1, z_2) \in \mathbb{C}^2 \,\middle|\, \frac{1}{2} < |z_2| < 1 \text{ and } |z_1| < 1 \right\}$$
$$\cup \left\{ (z_1, z_2) \in \mathbb{C}^2 \,\middle|\, |z_2| \leq \frac{1}{2} \text{ and } |z_1| < \frac{1}{2} \right\}$$

For any $f \in \mathcal{O}(\Omega)$, set

$$F(z) = \frac{1}{2\pi i} \int_\Gamma \frac{f(z_1, w)}{w - z_2} dw$$

where $\Gamma = \{w \in \mathbb{C} \mid |w| = 3/4\}$. Then it is not hard to see that $F|_\Omega = f$ on D and $F \in \mathcal{O}(\Omega_1)$, where $\Omega_1 = \{(z_1, z_2) \mid |z_1| < 1, |z_2| < 1\}$. This implies that Ω is not a domain of holomorphy since any holomorphic function f on Ω can be holomorphically extended to a strictly larger set Ω_1 via applying the Cauchy integral formula in z_2. It also indicates that the analog of Theorem 4.6.4 in higher-dimensional cases in general need not hold.

Thus, it becomes fundamental to determine whether a given domain Ω is a domain of holomorphy or not.

When $n = 1$, the key in the above construction of a holomorphic function singular at every boundary point is Theorem 4.6.4. We observe that one of the crucial hypotheses stated in Theorem 4.6.4 is to exclude the points sitting in the components, which are relatively compact in Ω, of the complement of K due to the maximum modulus principle. Technically, to avoid such a difficulty one may enlarge K by considering the holomorphically convex hull of K. Define

$$\widehat{K}_\Omega = \left\{ z \in \Omega \,\middle|\, |f(z)| \leq \sup_K |f| \text{ for all } f \in \mathcal{O}(\Omega) \right\} \tag{4.26}$$

From equation 4.26, it is clear that if $z \in \Omega \backslash \widehat{K}_\Omega$, then there is $f \in \mathcal{O}(\Omega)$ such that $|f(z)| > \sup_{\widehat{K}_\Omega} |f|$. This is exactly the property required in Theorem 4.6.4. If $K = \widehat{K}_\Omega$, we shall call K holomorphically convex.

For $n = 1, \widehat{K}_\Omega = K \cup \cup_j K_j$, where K_j is a component, relatively compact in Ω, of the complement of K in $\mathbb{C}$. Obviously, if K is a compact subset of Ω, so is $\widehat{K}_\Omega$.

When $n > 1$, $\widehat{K}_\Omega$ is still a closed subset of Ω if K is a compact subset of Ω. But $\widehat{K}_\Omega$ in general need not be compact in Ω. Again, we use the domain Ω defined above to illustrate this effect.

Let $K = \left\{ (3/4, (3/4)e^{i\theta}) \,; \theta \in [0, 2\pi] \right\}$ be a compact subset of Ω. Since any function holomorphic in Ω can be extended holomorphically to Ω_1, it is easily seen by the maximum modulus principle that

$$\widehat{K}_\Omega = \left\{ \left(\frac{3}{4}, z_2 \right) \middle| \, \frac{1}{2} < |z_2| \leq \frac{3}{4} \right\}$$

which is clearly not relatively compact in Ω. Hence, we make the following definition.

Definition 4.6.5. Let Ω be a domain in $\mathbb{C}^n$. Ω is called holomorphically convex if $\widehat{K}_\Omega$ is relatively compact in Ω for any compact subset K of Ω.

Since the holomorphically convex hull of a compact set is always contained in the geometrically convex hull of this set, it follows that we have

Lemma 4.6.6. *Any convex domain is holomorphically convex.*

It turns out that holomorphic convexity is the right condition for characterizing domains of holomorphy in $\mathbb{C}^n$.

Theorem 4.6.7. *Let Ω be a domain in $\mathbb{C}^n, n \geq 1$. Then the following statements are equivalent:*

(1) Ω is a domain of holomorphy.

(2) $\operatorname{dist}(K, \Omega^c) = \operatorname{dist}\left(\widehat{K}_\Omega, \Omega^c\right)$ for every compact subset K in Ω, where $\operatorname{dist}(K, \Omega^c)$ denotes the distance between K and $\Omega^c = \mathbb{C}^n \backslash \Omega$.

(3) Ω is holomorphically convex

Theorem 4.6.7 is due to H. Cartan and P. Thullen [39]. Thus, any convex domain is a domain of holomorphy according to Theorem 4.6.7.

Proof of Theorem 4.6.7. (2) $\Rightarrow$ (3) is obvious. The proof of (3) $\Rightarrow$ (1) is similar to the one that we adopted for $n = 1$ with compact subsets and discs being replaced by holomorphically convex compact subsets and polydiscs respectively. This is possible since by hypothesis the domain is holomorphically convex. Thus, we need to show (1) $\Rightarrow$ (2).

Let $P(0;r)$ be a polydisc centered at zero with multiradii $r = (r_1, \ldots, r_n)$. For each $z \in \Omega$, we set

$$\mathbb{D}_r(z) = \sup\{\lambda > 0 \mid \{z\} + \lambda P(0;r) \subset \Omega\}$$

To prove (1) $\Rightarrow$ (2), we first show:

Lemma 4.6.8. *Let K be a compact subset of Ω, and let $f \in \mathcal{O}(\Omega)$. Suppose that*

$$|f(z)| \le \mathbb{D}_r(z) \quad \textit{for } z \in K$$

Let ζ be a fixed point in $\widehat{K}_\Omega$. Then any $g \in \mathcal{O}(\Omega)$ extends holomorphically to $\Omega \cup (\{\zeta\} + |f(\zeta)|P(0;r))$.

Proof. For each $0 < t < 1$, the union of the polydiscs centered at $z \in K$,

$$K_t = \bigcup_{z \in K} \left(\{z\} + t|f(z)|\overline{P(0;r)}\right)$$

is a compact subset of D. Hence, there exists $M_t > 0$ such that $|g(z)| \le M_t$ on K_t. Using Cauchy's estimates of g, we obtain

$$\frac{\left|\frac{\partial^\alpha g}{\partial z^\alpha}(z)\right| t^{|\alpha|}|f(z)|^{|\alpha|} r^\alpha}{\alpha!} \le M_t \tag{4.27}$$

for $z \in K$ and all multi-indices $\alpha = (\alpha_1, \ldots, \alpha_n)$ with $|\alpha| = \alpha_1 + \cdots + \alpha_n$, where $\alpha_j \in \{0\} \cup \mathbb{N}$. Here, by definition, $\partial^\alpha g/\partial z^\alpha = \partial^{|\alpha|} g/\partial z_1^{\alpha_1} \cdots \partial z_n^{\alpha_n}$ and $r^\alpha = r_1^{\alpha_1} \cdots r_n^{\alpha_n}$. Since $(\partial^\alpha g/\partial z^\alpha)(z) f(z)^{|\alpha|}$ is holomorphic on D, by definition, 4.27 also holds for $z \in \widehat{K}_D$. Letting t tend to one, we see that $g(z)$ extends holomorphically to $D \cup (\{\zeta\} + |f(\zeta)|P(0;r))$. This proves the lemma. □

Now, we write

$$\begin{aligned}\operatorname{dist}(z, \Omega^c) = \sup\{r > 0 \mid z + aw \in \Omega \text{ for all } w \in \mathbb{C}^n,\\ |w| \le 1 \text{ and } a \in \mathbb{C}, |a| < r\} = \inf_{|w| \le 1} d_w(z),\end{aligned}$$

where

$$d_w(z) = \sup\{r > 0 \mid z + aw \in \Omega \text{ for all } a \in \mathbb{C}, |a| < r\}$$

Fix a w, and we may assume that $w = (1, 0, \ldots, 0)$. Denote by $P_j = P(0; r(j))$ the polydisc with multiradii $r(j) = (1, 1/j, \ldots, 1/j)$ for $j \in \mathbb{N}$. Then it is easily seen that

$$\lim_{j \to \infty} \delta_{r(j)}(z) = d_w(z)$$

Thus, given $\epsilon > 0$, if j is sufficiently large, we have

$$\operatorname{dist}(K, \Omega^c) \le (1 + \epsilon)\delta_{r(j)}(z), \quad z \in K \tag{4.28}$$

We let $f(z) = \text{dist}\,(K, \Omega^c)\,/(1+\epsilon)$ be the constant function. Since Ω is a domain of holomorphy, using estimate (4.28), Lemma 4.6.8 shows that

$$\text{dist}\,(K, \Omega^c) \leq (1+\epsilon)\delta_{r(j)}(\zeta) \leq (1+\epsilon) d_w(\zeta) \text{ for all } \zeta \in \widehat{K}_\Omega$$

Letting ϵ tend to zero, we get

$$\begin{aligned} \text{dist}\,(K, \Omega^c) &\leq \inf_{\zeta \in \widehat{K}_\Omega} \left(\inf_{|w| \leq 1} d_w(\zeta) \right) \\ &= \inf_{\zeta \in \widehat{K}_\Omega} \text{dist}\,(\zeta, \Omega^c) \\ &= \text{dist}\left(\widehat{K}_\Omega, \Omega^c \right) \end{aligned}$$

Since $\text{dist}\left(\widehat{K}_\Omega, \Omega^c\right) \leq \text{dist}\,(K, \Omega^c)$ is obvious, this proves (1) $\Rightarrow$ (2) and hence Theorem 4.6.7. □

It is now clear that, in order to be a domain of holomorphy, Ω must be holomorphically convex. However, such a condition is difficult to verify. Thus, we need to develop other equivalent conditions that can be actually computed. When the domain $\Omega \subset \mathbb{C}^n, n \geq 2$, has smooth boundary, we also make the following definition.

Definition 4.6.9. Let $\Omega \subset \mathbb{C}^n, n \geq 2$, be a smooth bounded domain, and let r be a smooth defining function for Ω. Ω is said to be (Levi) pseudoconvex (or strongly Levi pseudoconvex) if

$$\sum_{j,k=1}^{n} \frac{\partial^2 r}{\partial z_j \partial \bar{z}_k}(z) a_j \bar{a}_k \geq 0 \quad (\text{or } > 0) \tag{4.29}$$

for any $z \in \partial\Omega$ and $a = (a_1, \ldots, a_n) \in \mathbb{C}^n$ with $\sum_{j=1}^{n} a_j \left(\partial r / \partial z_j\right)(z) = 0$.

By a smooth defining function for Ω we mean $r : W \to \mathbb{R}$ a smooth function on some open neighborhood W of $\bar{\Omega}$ such that $\Omega = \{z \in W \mid r(z) < 0\}$ and $\partial\Omega = \{z \in W \mid r(z) = 0\}$, the boundary of Ω, and that $|dr| \neq 0$ on $\partial\Omega$. The Hermitian form (4.29) is called the Levi form of Ω. Thus, the domain Ω is pseudoconvex (or strongly pseudoconvex) if the Levi form is positive semi-definite (or positive definite) at every boundary point when applied to tangential type $(1, 0)$ vector fields on the boundary. It can also be easily verified that the semi-definiteness or definiteness of the Levi form is independent of the choice of the defining function for Ω.

The concept of pseudoconvexity is fundamental in several complex variables. It can be shown that if the domain Ω has smooth boundary, then Ω is a domain of holomorphy if and only if Ω is pseudoconvex. Condition (4.29) is computable. The Levi form can be computed easily to determine if a domain is a domain of holomorphy. However, we shall not get into the details of the proof here.

For materials of this section, the reader is referred to the books [48, 111, 141] and [177].

5

Montel's Theorem and Its Applications

Montel's theorem refers to one of two theorems about families of holomorphic functions. These are named after the French mathematician Paul Montel and give conditions under which a family of holomorphic functions is normal.

The first, and simpler, version of the theorem states that a family of holomorphic functions defined on an open subset of the complex numbers is normal if and only if it is locally uniformly bounded.

The stronger version of Montel's Theorem (occasionally referred to as the Fundamental Normality Test) states that a family of holomorphic functions, all of which omit the same two values $a, b \in \mathbb{C}$, is normal.

A heuristic principle known as Bloch's Principle (made precise by Zalcman's lemma) states that properties that imply that an entire function is constant correspond to properties that ensure that a family of holomorphic functions is normal.

Montel's circle theorems refer to a number of results about normal families of holomorphic functions.

5.1 Marty's Theorem

Definition 5.1.1. A family $\mathcal{F}$ of holomorphic functions on a domain $\Omega \subset \mathbb{C}^n$ is *normal* in Ω if every sequence of functions $\{f_j\} \subseteq \mathcal{F}$ contains either a subsequence which converges to a limit function $f \neq \infty$ uniformly on each compact subset of Ω, or a subsequence which converges uniformly to ∞ on each compact subset.

In the former case, the limit function is an analytic function by the Weierstrass Theorem. In the latter, we mean that for each compact subset $K \subset \Omega$ and constant $M > 0$, $|f_j(z)| > M$ for all $z \in K$, provided j (which may depend on K and M) is sufficiently large. For a given sequence $\{f_j\}$, both alternatives may occur.

A family $\mathcal{F}$ is said to be *normal* at a point $z_0 \in \Omega$ if it is normal in some neighborhood of z_0. [Throughout this book, the term "neighborhood of p"

DOI: 10.1201/9781003619581-5

refers to an open domain that contains the point p.] In these circumstances, the convergence is then uniform on each compact subset of Ω.

It is routine to confirm that a family of holomorphic functions $\mathcal{F}$ is normal in a domain Ω if and only if $\mathcal{F}$ is normal at each point of Ω.

For every function φ of class $C^2(\Omega)$ we define at each point $z \in \Omega$ a hermitian form

$$L_z(\varphi, v) := \sum_{k,l=1}^{n} \frac{\partial^2 \varphi}{\partial z_k \partial \overline{z}_l}(z) v_k \overline{v}_l,$$

and call it the *Levi form* of the function φ at z.

For a holomorphic function f in Ω, set

$$f^\sharp(z) := \sup_{|v|=1} \sqrt{L_z(\log(1+|f|^2), v)} \tag{5.1}$$

This quantity is well defined since the Levi form $L_z(\log(1+|f|^2), v)$ is non-negative for all $z \in \Omega$. In particular, for $n = 1$ the formula (5.1) takes the form

$$f^\sharp(z) := \frac{|f'(z)|}{1+|f(z)|^2}$$

and $z^\sharp$ coincides with the spherical metric on C.

Now, we have Marty's characterization of normal families in terms of the spherical metric.

Theorem 5.1.2. *A family $\mathcal{F}$ of functions holomorphic on Ω is normal on $\Omega \subset \mathbb{C}^n$ if and only if for each compact subset $K \subset \Omega$ there exists a constant $M(K)$ such that at each point $z \in K$*

$$f^\sharp(z) \leq M(K) \tag{5.2}$$

for all $f \in \mathcal{F}$.

Proof. Of course, it is enough to show that $\mathcal{F}$ is normal at every point $z_0 \in \Omega$. We appeal to the Theorem of Arzelà-Ascoli (Theorem 13.2.1). As the Riemann sphere $\widehat{\mathbb{C}}$ is compact, the second condition of Arzelà and Ascoli is automatically fulfilled, and our task reduces to checking that a family is equicontinuous on compacts in Ω when it is spherically bounded on such compacta.

Let us show that (5.2) implies that the family $\mathcal{F}$ is equicontinuous. Fix an arbitrary $z_0 \in \Omega$ and choose $r > 0$ such that

$$K = \{z \in \mathbb{C}^n : |z - z_0| \leq r\} \subset \Omega.$$

Let $z(t) = z_0 + t(z_1 - z_0) \subset K$, for all $t \in [0,1]$. It is elementary to see that

$$L_{z(t)}(\log(1+|f|^2), v) = \frac{|Df(z)v|^2}{(1+|f(z)|^2)^2} \geq 0,$$

where

$$Df(z)v := \sum_{k=1}^{n} \frac{\partial f}{\partial z_k}(z)v_k.$$

Hence

$$\frac{|df(z(t))/dt|^2}{(1+|f(z(t))|^2)^2} = L_{z(t)}(\log(1+|f|^2), z'(t)) \le (f^\sharp(z(t))|z'(t)|)^2.$$

Then the spherical distance satisfies

$$s(f(z_0), f(z_1)) \le \int_0^1 f^\sharp(z(t))|z'(t)|dt \le M(K) \int_0^1 |z'(t)|dt \le M(K)|z_1 - z_0|$$

for all $f \in \mathcal{F}$. Thus we see that the family $\mathcal{F}$ is equicontinuous on K—from the Euclidean metric to the spherical metric. The Ascoli-Arzelà theorem (Theorem 13.2.1) now implies that $\mathcal{F}$ is normal at z_0.

Conversely, assume that $\mathcal{F}$ is normal in Ω. If (5.2) fails for some compact set $K \subset \Omega$, then there exists a sequence $\{f_j\} \subset \mathcal{F}$ which converges locally uniformly in Ω to some $f \not\equiv \infty$ and such that the maximum of $f_j^\sharp$ over K tends to ∞. By Weierstrass's theorem (Theorem 3.2.27), $\partial f_j/\partial z_k$ converges uniformly to $\partial f/\partial z_k$ on K, therefore there exists j_0 such that for $j > j_0$

$$L_z(\log(1+|f_j|^2), v) \le L_z(\log(1+|f|^2), v) + O(|v|^2) \text{ for all } z \in K, v \in \mathbb{C}^n.$$

It follows that

$$f_j^\sharp(z) \le f^\sharp(z) + O(1) \text{ for all } j > j_0.$$

This is a contradiction since the right-hand side is bounded on K.

If $f \equiv \infty$ then consider a domain Ω' containing K and satisfying $\overline{\Omega'} \subset \Omega$. Then there exists j_0 such that the functions $1/f_j$ are holomorphic in Ω' for $j > j_0$. Since $1/f_j \to 0$ uniformly in Ω' we see as above that $(1/(f_j)^\sharp)$ are bounded on K. Since the Levi form vanishes for any pluriharmonic function,

$$\begin{aligned} L_z(\log(1+|1/f_j|^2), v) =& L_z(\log(1+|f|^2), v) - 2L_z(\log|f|, v) \\ =& L_z(\log(1+|f|^2), v)\,. \end{aligned}$$

Therefore

$$f_j^\sharp(z) = (1/f_j)^\sharp(z)$$

and we again obtain a contradiction. □

Remark 5.1.3. The proof of Marty's theorem shows that if holomorphic functions f_j converge uniformly to f in the spherical metric then $f_j^\sharp$ converges to $f^\sharp$. The converse fails: consider a sequence of constants or a sequence of rotations of a single function.

5.2 Zalcman's Lemma

By an *entire function* we shall mean an element of $\mathcal{O}(\mathbb{C}^n)$. Now Zalcman's lemma is this:

Theorem 5.2.1. *A family $\mathcal{F}$ of functions holomorphic on $\Omega \subset \mathbb{C}^n$ is not normal at some point $z_0 \in \Omega$ if and only if there exist sequences $f_j \in \mathcal{F}$, $z_j \to z_0$, $\rho_j = 1/f_j^\sharp(z_j) \to 0$, such that the sequence*

$$g_j(z) := f_j(z_j + \rho_j z) \tag{5.3}$$

converges locally uniformly in $\mathbb{C}^n$ to a non-constant entire function g satisfying $g^\sharp(z) \le g^\sharp(0) = 1$.

It is remarkable that non-normality can be described in terms of a convergent sequence. If $\{f_j\}$ were convergent, then the functions g_j given by (5.3) would converge to a constant on compact subsets of $\mathbb{C}^n$, since the radii ρ_j tend to 0. Zalcman's lemma (Theorem 5.2.1) says that arbitrarily small balls centered at z_j can be found where f_j is close to a non-trivial entire (holomorphic in $\mathbb{C}^n$) function, after rescaling.

Proof of Theorem 5.2.1. The proof will be based on Marty's normality criterion (Theorem 5.1.2). Assume $\mathcal{F}$ is not normal at a point $z_0 \in \Omega$. To simplify matters we assume that $z_0 = 0$ and all functions under consideration are holomorphic on the unit ball $\{z \in \mathbb{C}^n : |z| < 1\}$. By Marty's normality criterion (Theorem 5.1.2), $\mathcal{F}$ contains functions f_j, $j \in \mathbb{N}$, satisfying $\sup_{|z|<1/(2j)} f_j^\sharp(z) > 2j^3$. Hence there exists a ξ_j with $|\xi_j| < 1/j$ such that

$$\max_{|z|\le 1/j} (1 - j|z|) f_j^\sharp(z) = (1 - j|\xi_j|) f_j^\sharp(\xi_j) \ge j^3.$$

Put $r_j = 1/f_j^\sharp(\xi_j)$. Then from the above inequality we have $(1 - j|\xi_j|)/j^2 \ge j \cdot r_j$. Set $h_j(z) = f_j(\xi_j + r_j z)$. We claim that appropriately chosen subsequences $z_k = \xi_{j_k}$, $\rho_k = r_{j_k}$, and $g_k = h_{j_k}$ will do. First of all, h_j is defined on $|z| < (1 - |\xi_j|)/r_j = (1 - |\xi_j|) f_j^\sharp(\xi_j)$, hence on $|z| < j^3$. By the invariance of the Levi form under biholomorphic mappings, we have

$$L_z(\log(1 + |h_j|^2), v) = L_{\xi_j + r_j z}(\log(1 + |f_j|^2), r_j v)$$

and hence

$$h_j^\sharp(z) = r_j f_j^\sharp(\xi_j + r_j z) = \frac{f_j^\sharp(\xi_j + r_j z)}{f_j^\sharp(\xi_j)}.$$

Recall that $|z| < j$ and $j \cdot r_j \le (1 - j|\xi_j|)/j^2$. It follows that, for all $|z| < j$,

$$\begin{aligned}|\xi_j + r_j z| &\le |\xi_j| + r_j|z| < |\xi_j| + \frac{1}{j^2}(1 - j|\xi_j|) \\ &< \left(1 - \frac{1}{j}\right)|\xi_j| + \frac{1}{j^2} < \frac{1}{j} - \frac{1}{j^2} + \frac{1}{j^2} = \frac{1}{j}.\end{aligned}$$

Therefore

$$(1 - j|\xi_j + r_j z|) f_j^\sharp(\xi_j + r_j z) \leq (1 - j|\xi_j|) f_j^\sharp(\xi_j)$$

and

$$h_j^\sharp(z) = \frac{f_j^\sharp(\xi_j + r_j z)}{f_j^\sharp(\xi_j)} \leq \frac{1 - j|\xi_j|}{1 - j|\xi_j + r_j z|}$$

$$\leq \frac{1 - j|\xi_j|}{1 - j|\xi_j| - j \cdot r_j|z|} \leq \frac{1}{1 - 1/j} \quad (|z| < j)$$

where we again have used $j \cdot r_j \leq \frac{1}{j^2}(1 - j|\xi_j|)$. For every $m \in \mathbb{N}$ the sequence $\{h_j\}_{j>m}$ is normal in $|z| < m$ by Marty's normality criterion (Theorem 5.1.2). The well-known Cantor diagonalization process yields a subsequence $\{g_k = h_{j_k}\}$ which converges uniformly on every ball $|z| < R$. The limit function g satisfies $g^\sharp(z) \leq \limsup_{j\to\infty} h_j^\sharp(z) \leq 1 = g^\sharp(0)$. Clearly, g is non-constant because $g^\sharp(0) = 1$.

Conversely, suppose that there exist sequences $f_j \in \mathcal{F}$, $z_j \to z_0$, $\rho_j = 1/f_j^\sharp(z_j) \to 0$, such that the sequence

$$g_j(z) = f_j(z_j + \rho_j z)$$

converges locally uniformly in $\mathbb{C}^n$ to a non-constant entire function g satisfying $g^\sharp(z) \leq g^\sharp(0) = 1$, but $\mathcal{F}$ is normal. Without loss of generality, we may suppose that $z_0 = 0$. By Marty's normality criterion (Theorem 5.1.2) there exists a constant $M > 0$ such that

$$\max_{|z|<1/j_0} f_j^\sharp(z) < M.$$

Since $z_j \to 0$, $\rho_j = 1/f_j^\sharp(z_j) \to 0$, we see that for $|z| < 1/j_0$ there exists j_1 such that for all $j > j_1$

$$|z_j + \rho_j z| \leq |z_j| + \rho_j|z| \leq |z_j| + \rho_j/j_0 < 1/j_0.$$

Therefore

$$g_j^\sharp(z) = f_j^\sharp(z_j + \rho_j z)/f_j^\sharp(z_j) = f_j^\sharp(z_j + \rho_j z)\rho_j \leq M\rho_j \to 0.$$

But then $g^\sharp(0) = 0$, a contradiction, since $g^\sharp(0) = 1$. This contradiction concludes the proof of the theorem. □

5.3 Montel's Theorem

A fundamental result for the investigation of topological properties of subsets A of $\mathbb{R}^n$ is the Heine-Borel Theorem: A is compact if and only if A is closed and bounded. An analogous statement (under the name Montel's theorem) holds for $\mathcal{O}(X)$. It follows from the following theorem.

Theorem 5.3.1. *Let $\mathcal{F}$ be a family of holomorphic functions on an open domain $\Omega \subseteqq \mathbb{C}^n$ that omit two fixed complex values. Then each sequence of functions in $\mathcal{F}$ has a subsequence which converges uniformly on compact subsets.*

Proof. Since normality is a local notion, we may suppose that $\Omega = B$, the unit ball. So let $\mathcal{F}$ be as in the statement of Montel's normality criterion (Theorem 5.3.1) and suppose that $\mathcal{F}$ is not normal. Composing with a linear fractional transformation, we may assume that the omitted values are 0 and 1. This implies that if $f \in \mathcal{F}$ and $m \in \mathbb{N}$, then there exists a function g holomorphic in Ω such that $g^{2^m} = f$. Let $\mathcal{F}_m$ be the family of all such functions g. Note that

$$\left(\frac{1}{2^m}\right)^2 \left(\frac{|f|^{-1} + |f|}{|f|^{-1/2^m} + |f|^{1/2^m}}\right)^2 L_z\left(\log\left(1 + |f|^2\right), v\right) = L_z\left(\log\left(1 + |g|^2\right), v\right)$$

for all $(z, v) \in \Omega \times \mathbb{C}^n$. This implies that

$$g^\sharp(z) = \frac{1}{2^m} \frac{|f|^{-1} + |f|}{|f|^{-1/2^m} + |f|^{1/2^m}} f^\sharp(z) \geq \frac{1}{2^m} f^\sharp(z) \quad (z \in \Omega),$$

where we have used the basic inequality $a^{-1} + a \geq a^{-t} + a^t$ valid for $a > 0$ and $0 < t < 1$. By Marty's normality criterion (Theorem 5.1.2), the family $\{f^\sharp \mid f \in \mathcal{F}\}$ is not locally bounded. We deduce that, for fixed $m \in \mathbb{N}$, the family $\{g^\sharp \mid g \in \mathcal{F}_m\}$ is not locally bounded. Using Marty's normality criterion (Theorem 5.1.2) again we find that $\mathcal{F}_m$ is not normal, for all $m \in \mathbb{N}$. Note that if $g \in \mathcal{F}_m$, then g omits the values $e^{2\pi ik/2m}$ for $k, m \in \mathbb{Z}$. From the Zalcman rescalling lemma (Theorem 5.2.1), we thus deduce that there exists an entire function g_m omitting the values $e^{2\pi ik/2m}$ and satisfying $g_m(z) \leq g^\sharp_m(0) = 1$. The g_m thus forms a normal family and we have $g_{m_j} \to G$ for some subsequence $\{g_{m_j}\}$ and some nonconstant entire function G. By Hurwitz's theorem (Theorem 3.2.28), G omits the values $e^{2\pi ik/2m}$ for all $k, m \in N$. Since $G(\mathbb{C}^n)$ is open this implies that $|G(z)| \neq 1$ for all $z \in \mathbb{C}^n$. Thus either $|G(z)| < 1$ for all $z \in \mathbb{C}^n$ or $|G(z)| > 1$ for all $z \in \mathbb{C}^n$. In the first case G is bounded and thus constant by Liouville's Theorem. In the second case $1/G$ is bounded. Again $1/G$ is constant and thus G is constant. Thus we get a contradiction in both cases. □

Montel's theorem is the central device in one of the standard proofs of the Big and Little Picard Theorems.

Recall that to every a in the unit disc $\mathbb{D} := \{z \in \mathbb{C} : |z| < 1\}$ corresponds an automorphism φ_a of the disc that interchanges a and 0, namely $\varphi_a(w) := (a - w)/(1 - \overline{a}w)$. Note for later reference that $\varphi_a(0) = a$ and $\varphi'_a(0) = 1 - |a|^2$. Define $\Gamma(\mathbb{D}) := \{\varphi_a, a \in \mathbb{D}\}$.

Theorem 5.3.2 (Picard's Little Theorem). *At most one complex number is absent from the range of a non-constant entire function (= holomorphic in $\mathbb{C}$).*

Proof. Assume, to the contrary, that an entire function f omits two distinct values from its range. Composing with a linear fractional transformation, we may assume that the omitted values are 0 and 1.

The function $f(2^j\lambda)$, $\lambda \in \mathbb{D}$, where k is a positive integer, is holomorphic in $\mathbb{D}$, $f(2^j\lambda)$ takes the same values in Δ as does f in $|z| < 2^j$, and $f(2^j\lambda)$ omits the values 0 and 1.

The family $\{f(2^j\varphi_a), \varphi_a \in \Gamma(\mathbb{D}), j = 1, 2\ldots\}$, being contained in the family of functions avoiding 0 and 1, is normal in $\mathbb{D}$ by Montel's Theorem 5.3.1. By Marty's theorem, there exists $L > 0$ such that for any $f(2^j\varphi_a)$ we have

$$\frac{|f'(2^j\varphi_a)(0)|}{1 + |f(2^j\varphi_a(0))|^2} \leq L.$$

After some manipulations, using the chain rule for differentiation, and the equalities $\varphi_a(0) = a$ and $\varphi_a'(0) = 1 - |a|^2$, we find that

$$\frac{2^j\,|f'(2^j a)|\,(1 - |a|^2)}{1 + |f(2^k a)|^2} \leq L.$$

Replacing in this inequality $2^j a$ by z we get

$$\frac{2^j\,|f'(z)|\,(1 - |z/2^j|^2)}{1 + |f(z)|^2} \leq L.$$

Thus

$$\frac{|f'(z)|}{1 + |f(z)|^2} \leq \frac{2^j L}{2^{2k} - |z|^2}.$$

Keeping z fixed, let j tend to infinity to deduce that $f'(z) = 0$ for every point z in the complex plane. But this implies that $f(z)$ is constant, violating our assumption on f. The contradiction proves the theorem. □

An immediate (and easy) corollary of Picard's little theorem is Liouville's theorem:

Corollary 5.3.3 (Liouville's theorem). *Every bounded entire function is constant.*

Corollary 5.3.4 (The fundamental theorem of algebra).

Let $a_0, a_1, \ldots, a_m \in \mathbb{C}$ where $n \geq 1$ and $a_m \neq 0$, and let p denote the polynomial

$$p(z) := a_0 + a_1 z + \cdots + a_n z^m \text{ for } z \in \mathbb{C}$$

Then there exists a $z_0 \in \mathbb{C}$ such that $p(z_0) = 0$.

Proof. The proof goes by contradiction, so we assume that p never vanishes. Then the function $f(z) := 1/p(z)$ is entire and converges to 0 as $|z| \to \infty$. In particular f is bounded and so a constant by Liouville's theorem. But then p is constant, contradicting that the degree of p is ≥ 1. □

Theorem 5.3.5 (Picard's big theorem)**.** *Let f be holomorphic in*

$$\Omega = \{z \in \mathbb{C} : 0 < |z - z_0| < \delta\} .$$

If z_0 is an essential singularity of the function f, then $f(z)$ assumes an arbitrary neighborhood of z all finite complex values with one possible exception.

Proof. For the sake of simplicity suppose that f has an essential singularity at $z_0 = 0$. Suppose that there exist two finite numbers a and b and a neighborhood of zero in which $f(z)$ does not assume either the value a or the value b. Then the functions $f_j(z) = f(z/2^j)$, $j = 1, 2, \ldots$, also do not assume the values a and b in that neighborhood. Consequently, by Montel's Theorem, the family $\{f(z/2^j)\}$ is normal in the chordal metric on compact subsets of $\mathbb{C} \setminus \{0\}$. Relabeling a subsequence, we may suppose f_{j_k} converges uniformly on compact subsets of $\mathbb{C} \setminus \{0\}$ to a function g holomorphic on $\mathbb{C} \setminus \{0\}$ or to $g \equiv \infty$ by Hurwitz's theorem. If g is holomorphic, then $|g(z)| \leq M < \infty$ on $|z| = 1$ and hence $|f| \leq M + 1$ on $|z| = 1/2^{j_k}$ for $k \geq k_0$. But then, by the maximum principle, $|f| \leq M + 1$ on $1/2^{j_{k+1}} \leq |z| \leq 1/2^{j_k}$ for $k \geq k_0$. Thus $|f| \leq M + 1$ on $0 < |z| < 1/2^{j_{k_0}}$, so that, by Riemanns theorem on removable singularities, f extends to be holomorphic in a neighborhood of 0.

If $g \equiv \infty$, we can apply a similar argument to $1/f(z/2^j)$ to conclude that $1/f$ extends to be holomorphic at 0 and hence f is meromorphic in a neighborhood of 0.

But this implies that 0 is the unessential singularity of f, violating our assumption on f. The contradiction proves the theorem. □

For analytic functions, even more is true, as was proved by Gaston Julia. The above proof leads to a result somewhat more general than Picard's great theorem. It is based on the following lemma.

Lemma 5.3.6. *Let $f(z)$ be a function holomorphic in an annular neighborhood of the origin and having an essential singularity there. Then the sequence of functions $f_j(z) = f\left(z/2^j\right)$, which for all n sufficiently large are defined and holomorphic in the annulus $A = \{z \in \mathbb{C} : 1/2 < |z| < 2\}$, cannot be a normal family in A.*

Proof. Suppose that the family $\{f(z/2^j)\}$ is normal in the annulus A. Relabeling a subsequence, we may suppose f_{j_k} converges uniformly on compact subsets of $\mathbb{C} \setminus \{0\}$ to a function g holomorphic on $\mathbb{C} \setminus \{0\}$ or to $g \equiv \infty$ by Hurwitz's theorem. If g is holomorphic, then $|g(z)| \leq M < \infty$ on $|z| = 1$ and hence $|f| \leq M + 1$ on $|z| = 1/2^{j_k}$ for $k \geq k_0$. But then, by the maximum principle, $|f| \leq M + 1$ on $1/2^{j_{k+1}} \leq |z| \leq 1/2^{j_k}$ for $k \geq k_0$. Thus $|f| \leq M + 1$

on $0<|z|<1/2^{j_{k_0}}$, so that, by Riemanns theorem on removable singularities, f extends to be holomorphic in a neighborhood of 0.

If $g \equiv \infty$, we can apply a similar argument to $1/f(z/2^j)$ to conclude that $1/f$ extends to be holomorphic at 0 and hence f is meromorphic in a neighborhood of 0.

But this implies that 0 is the unessential singularity of f, violating our assumption on f. The contradiction proves the theorem. □

Theorem 5.3.7 (Julia's Theorem).

Let $f(z)$ be analytic in $D=\{z\in\mathbb{C} \mid 0<|z|<1\}$ with an essential singularity at 0. Then there exists a point $z_0 \in D$ such that, for each $\varepsilon>0, f(z)$ assumes every complex value, with at most one exception, infinitely often on the union of the homothetic discs

$$D_j=\left\{z \mid \left|z-z_0/2^j\right|<\varepsilon/2^j\right\}.$$

Proof. Again, the main tool in the proof is Montel's theorem. By Lemma 5.3.6 the functions $f_n(z)=f\left(z/2^n\right)$ do not form a normal family in A, there exists a point $\zeta \in G$ such that no matter what circle $\Delta=\Delta(\zeta;\varepsilon)$ contained in G we take the sequence $\{f_n(z)\}$ does not form a normal family in K.

Recalling the definition of $f_j(z)$ we can say that the function $f(z)$ assumes every finite value α, with the exception of one at most, in infinitely many circles $\Delta_j=\Delta\left(\zeta/2^j;\varepsilon/2^j\right)$.

In further considerations, by an angle, as the region formed by two distinct half-lines with origin at a point $z_0 \neq \infty$ (the vertex), we shall mean each of the two regions into which these two half-lines divide the plane. An angle with vertex ∞ we identify with an angle with vertex 0. Therefore, applying a translation or an inversion, we can always bring the vertex of an angle to the point 0.

Let us now consider the half-line with origin at the point 0 and passing through the point ζ[1]. Let δ be an arbitrary positive number. The angle $-\delta+\operatorname{Arg}\zeta<\operatorname{Arg} z<\delta+\operatorname{Arg}\zeta$ contains all the circles K_j, for j sufficiently large, provided that ε is taken sufficiently small. □

Theorem 5.3.8 (Casorati-Weierstrass' theorem). *Let f have an essential singularity at $z_0 \in \mathbb{C}$. Then for any $w \in \mathbb{C}$ there exists a sequence $z_1, z_2, \cdots$ such that $z_n \to z_0$ and $f\left(z_n\right) \to w$ as $n \to \infty$.*

Proof. The proof goes by contradiction: If the assertion were false then there would exist a $w_0 \in \mathbb{C}$ and an $\epsilon>0$ and a neighborhood Ω of z_0 such that

$$\left|f(z)-w_0\right| \geq \epsilon \text{ for all } z \in \Omega \backslash\left\{z_0\right\}$$

By Picard's big theorem the point z_0 can not be essential. It contradicts our assumption about z_0 being an essential singularity. □

[1]The half-line is called the *direction of Julia* or the *direction J*.

5.4 Poincaré's Theorem

A holomorphic mapping f from Ω_1 into Ω_2 in $\mathbb{C}^n$ is called a biholomorphism if f is one-to-one and onto. In this case, f^{-1} is also a one-to-one holomorphic mapping from Ω_2 onto Ω_1.

The Riemann Mapping Theorem states that a connected, simply connected domain in the complex plane is either $\mathbb{C}$ itself or else it is biholomorphic to the open unit disc. The following result shows that it is impossible to find a higher dimensional analog of Riemann's Theorem which involves only topological conditions.

Theorem 5.4.1. *There is no biholomorphic map of the bidisc $D^2(0,1)$ to the ball $B(0,1) \subseteq \mathbb{C}^2$.*

Proof. Suppose that $f = (f_1, f_2) : D^2(0,1) \to B_2$ is a biholomorphism. Let (z, w) be the coordinates in $\mathbb{C}^2$. For any point sequence $\{z_j\}$ in $D(0,1)$ with $|z_j| \to 1$ as $j \to \infty$, the sequence $g_j(w) = f(z_j, w) : D(0,1) \to B_2$ is uniformly bounded. Hence, by Montel's theorem, there is a subsequence, still denoted by $g_j(w)$, that converges uniformly on compact subsets of $D(0,1)$ to a holomorphic map $G(w) = (G_1(w), G_2(w)) : D(0,1) \to \bar{B}_2$. Since f is a biholomorphism, we must have $|G(w)| = 1$ for all $w \in D(0,1)$. Hence, $|G'(w)| = 0$ for all $w \in D(0,1)$, which implies $G'(w) \equiv 0$ on $D(0,1)$. It follows that

$$\lim_{j\to\infty} f_w(z_j, w) = G'(w) \equiv 0 \tag{5.4}$$

Equation (5.4) implies that for each fixed $w \in D(0,1)$, $f_w(z,w)$, when viewed as a function of z alone, is continuous up to the boundary with the boundary value identically equal to zero. Therefore, by the maximum modulus principle we get

$$f_w(z,w) \equiv 0 \quad \text{for all } (z,w) \in D^2(0,1)$$

This implies that f is independent of w, contradicting the fact that f is a biholomorphic map. This completes the proof of the theorem. □

The proof we presented here is based on the idea of R. Remmert and K. Stein [180], not the original one found by H. Poincaré. See also the books by R. Narasimhan [162] and R.M. Range [177].

This fact was discovered by H. Poincaré in 1907 (see [172]). Poincaré's original proof was based on computation and comparison of the groups of holomorphic automorphisms of ball and bidisc which fix the origin.

This was an early and profound use of the theory of automorphism groups. But Poincaré's theorem has had a profound influence on complex analysis. It tells us that there is no Riemann Mapping theorem in several complex variables. As a result, we seek means to compare and contrast domains in $\mathbb{C}^n$. One of the key tools in doing so is the theory of automorphism groups.

The situation in higher dimensions is thus considerably more complicated than in the plane. Theorem 5.4.1 shows that simply connected domains in dimension two or higher are much more "analytically rigid" than in the plane. Related to this theme, it has long been known that in $\mathbb{C}^2$ there exist simply connected domains whose only holomorphic automorphism is the identity (cf. [15], p. 169).

5.5 The Baire-Osgood Theorem

The Baire-Osgood theorem says that if a sequence of analytic functions from region $\Omega \subset \mathbb{C}^n$ converges pointwise to a function f, then f must be analytic at "most" points in Ω.

Proposition 5.5.1. *Let $\mathcal{F}$ be a pointwise bounded family of analytic functions on Ω. There is a (maximal) dense open set $\Omega_0 \subseteq \Omega$ such that $\mathcal{F}$ restricted to Ω_0 is locally bounded.*

Proof. For each z in Ω, let $\phi(z)$ denote the least upper bound for $\{|f(z)| \mid f \in \mathcal{F}\}$. Because each f in $\mathcal{F}$ is continuous, it is easy to see that $K_j = \{z \in \mathbb{C} \mid \phi(z) \leq j\}$ is (relatively) closed in Ω. Since $\phi(z)$ is finite for each z, Ω is the union of the K_j. The Baire Category Theorem shows that for every ball U in $\Omega, K_j \cap U$ has interior for large j. Thus, the union Ω_0 of the interiors int K_j is a dense open subset of Ω. Obviously, $\mathcal{F}$ is bounded on each int K_j, and hence $\mathcal{F}$ is locally bounded on Ω_0. Conversely, if $\mathcal{F}$ is bounded by j on an open set U, then U is contained in int K_j and thus is in Ω_0. □

Corollary 5.5.2. *Let $\{f_j\}$ be a sequence of analytic functions converging pointwise on Ω. Then the limit function f is analytic on a dense open subset of Ω.*

Proof. A sequence of functions converging pointwise is pointwise bounded. By Proposition 5.5.1, $\{f_j\}$ is locally bounded on Ω_0. By Montel's Theorem, a subsequence converges uniformly on compact subsets to f on Ω_0 and hence f is analytic on Ω_0. □

Proposition 5.5.3. *If $\{f_j\}$ is a sequence of bounded analytic functions on Ω converging pointwise to f, then f is analytic and the convergence is uniform converge on compact subsets.*

Proof. The family $\mathcal{F} = \{f_j\}$ is bounded and thus locally bounded. By Montel's Theorem, every subsequence of f_j has a sub-subsequence which converges uniform converge on compact subsets This limit is f per force, so f is analytic. If the whole sequence does not converge to f uniform converge on compact subsets, there would be a compact set K in Ω, an $\varepsilon > 0$, a subsequence f_{j_k}, and points z_k in K such that $|f_{j_k}(z_k) - f(z_k)| \geq \varepsilon$ for $k \geq 1$. Thus no subsequence of f_{j_k} could converge uniformly to f on K, which is a contradiction. □

Theorem 5.5.4 (Baire-Osgood). *Suppose that $f_1, f_2, \ldots$ are analytic in a region $\Omega \subset \mathbb{C}^n$, and that $f_j \to f$ pointwise in Ω. Then there is a dense open subset Ω_0 of Ω on which the convergence is locally uniform, and in which f is analytic.*

Here we are using the terminology that $f_j \to f$ locally uniformly in Ω if the convergence is uniform on each compact subset of Ω.

5.6 Nishino's Theorem

Of course, normal families in one and several variables are pretty much the same. But there are some nice results relating the two, such as Nishino's Theorem: if a family of functions of several variables are normal in each variable separately then they are normal in all variables.

We give a short proof of a theorem of Nishino who proved that a family $\mathcal{F}$ of analytic functions on a domain Ω in $\mathbb{C}^2$ is normal if the restriction to each coordinate line (of the form $z_1 = \lambda$ or $z_2 = \lambda$) is normal.

We begin by considering the power series

$$\sum_{|k|=0}^{\infty} c_k (z-a)^k \tag{5.5}$$

and at an arbitrary point z of its domain of convergence $\overset{\circ}{S}$ we regroup the terms of this series, arranging them in powers of one of the differences $z_\nu - a_\nu$, say the nth one (this is legal because of the absolute convergence). We obtain a series

$$\sum_{j=0}^{\infty} g_j\left('z\right)\left(z_n - a_n\right)^j \tag{5.6}$$

(j is a scalar index), whose coefficients g_j are holomorphic in $'\,\overset{\circ}{S}$, the projection of $\overset{\circ}{S}$ into the space $\mathbb{C}^{n-1}$.

Remark 5.6.1. We notice that such a regrouping of the terms of a series can lead to an extension of the domain of convergence. For example, the power series

$$\frac{1}{(1-z_1)(1-z_2)} = \sum_{|k|=0}^{\infty} z_1^{k_1} z_2^{k_2}$$

converges in the bidisc $\{|z_1| < 1, |z_2| < 1\}$. After regrouping the terms we obtain the series

$$\sum_{j=0}^{\infty}\left(\sum_{k_1=0}^{\infty} z_1^{k_1}\right) z_2^j = \sum_{j=0}^{\infty} \frac{z_2^j}{1-z_1}$$

which converges in the domain $\{z_1 \neq 1, |z_2| < 1\}$.

Definition 5.6.2. A series of the form (5.5), whose coefficients g_j are holomorphic functions of $'z = (z_1, \dots, z_{n-1})$, is called a *Hartogs series* with centers $\{z_n = a_n\}$. The domain of convergence of this series is the interior $\overset{\circ}{H}$ of the set of points $('z, z_n)$ such that the series (5.5) with coefficients $g_j('z)$ converges at the point z_n and the coefficients are holomorphic in $'z$.

Since along with each z^0 the domain of convergence $\overset{\circ}{H}$ of a Hartogs series also contains all the points $z = ('z^0, z_n)$, where $|z_n - a_n| \leq |z_n^0 - a_n|$, then $\overset{\circ}{H}$ is always a complete Hartogs domain with the plane of symmetry $\{z_n = a_n\}$, i.e., a domain of the form $\{('z, z_n) \mid 'z \in {}'\Omega, |z_n - a_n| < R('z)\}$; the function $R('z)$ is called the *Hartogs radius.*

Remark 5.6.3. The Hartogs radius $R('z)$ may not coincide with the radius of convergence $r('z)$ of the series (5.6), considered as a power series in $z_n - a_n$, since we have passed from the set of convergence H to its interior $\overset{\circ}{H}$. The function $r('z)$ is determined from the Cauchy—Hadamard formula

$$1/r('z) = \overline{\lim_{j\to\infty}} \sqrt[j]{|g_j('z)|};$$

it may not be lower semicontinuous, and then the set

$$\{'z \in {}'\Omega \mid |z_n - a_n| < r('z)\}$$

will not be open. It is not difficult to see that R is a lower semicontinuous regularization of r:

$$R('z) = \underline{\lim}_{\zeta\to 'z} r('\zeta).$$

Domains of the form $\{('z, z_n) \in \mathbb{C}^n \mid 'z \in {}'\Omega, |z_n - a_n| < R('z)\}$ play the same role for Hartogs series as Reinhardt domains do for power series. In particular, we have

Theorem 5.6.4. *Any function f that is holomorphic in a complete Hartogs domain $\Omega \subset \mathbb{C}^n$ with plane of symmetry $\{z_n = a_n\}$ is represented in Ω by an expansion*

$$f(z) = \sum_{j=0}^{\infty} g_j('z)(z_n - a_n)^j \tag{5.7}$$

with coefficients that are holomorphic in the projection $'\Omega$ of Ω into $\mathbb{C}^{n-1}$.

Proof. Along with each point $z^0 \in \Omega$ the domain Ω also contains the polydisc $U = {}'U \times U_n$, where ${}'U \subset \mathbb{C}^{n-1}$ is a sufficiently small polydisc with center at the projection ${}'z^0$ of the point z^0, and $U_n \subset \mathbb{C}$ is a disc with center a_n containing the point z_n^0. In U the function f is holomorphic with respect to z_n for any ${}'z \in {}'U$, and hence, can be expanded in a series (5.7) with coefficients

$$g_j(z) = \frac{1}{j!}\frac{\partial^j f\,({}'z, a_n)}{\partial z_n^j}$$

These coefficients, however, are defined and holomorphic not only in ${}'U$, but also in the whole projection ${}'\Omega$ of the domain Ω, so that the expansion (5.7) is also realized in the entire domain Ω. □

Remark 5.6.5. Not every complete Hartogs domain is the domain of convergence of some Hartogs series: domains of convergence are characterized by an additional property that the radius $R\,({}'z)$ must possess.

A family of holomorphic functions on the unit polidisc Δ is said to be *separately normal* if its restriction to each coordinate disk is normal. In $\mathbb{C}^2$ Nishino has proved a normal families version of the Hartogs theorem: a separately normal family of holomorphic functions is normal [164, Theoreme II, pp. 263–264].

Nishino proves a separate normality theorem not only for families of holomorphic functions but also for families of meromorphic functions.

We acknowledge that there have been prior follow-ups to Nishino's paper, notably a direct generalization by Toshiaki Terada [202]; and an extension to mappings by Theodore J. Barth [14]. Moreover, Nishino's results were anticipated by Renato Caccioppoli [36].

A family $\mathcal{F}$ is called *locally bounded* if for any p in Ω there exists an open set U containing p and a positive number M such that $|f(z)| < M$ for all $z \in U$ and all $f \in \mathcal{F}$.

Since normality is a local property (as Marty's criterion (Theorem 5.1.2) concretely illustrates), the restriction of a family $\mathcal{F}$ to any open connected subset of its normality domain is a normal family, so it suffices to consider the case when the open set Ω is a polydisc.

If F is continuous on $\Omega \subset \mathbb{C}$ and separately holomorphic in each variable when the others are kept fixed, then we know, as a consequence of Cauchy's integral formula, that it is holomorphic. We can dispense with the hypothesis of continuity.

Denote by $\Delta_k = \{z_k \in \mathbb{C} \mid |z_k| < 1\}$, $\Delta_k(0,\varepsilon) = \{z_k \in \mathbb{C} \mid |z_k| < \varepsilon\}$, $k = 1, 2$.

Hartogs's theorem is a consequence of the following result:

Theorem 5.6.6. *[211, Theorem 1.5.3, p. 23] Let f be holomorphic in $\Delta_1 \times \Delta_2(0,\epsilon)$ and separately holomorphic in $z_2 \in \Delta_2$ when z_1 is fixed in Δ_1. Then $f \in \mathrm{hol}(\Delta_1 \times \Delta_2)$.*

The next Theorem is similar in spirit to Nishino's lemma I, but in our proof we use the Baire Category theorem instead of the Poisson's integrals used in Nishino's proof. The Baire Category theorem is a very innocent-looking statement, which is, furthermore, not too hard to prove. It can, however, be used to obtain a simple proof of Nishino's theorem.

Theorem 5.6.7. *Let $\mathcal{F}$ be a family of holomorphic functions in an open polydisc $\Delta_1(0,1) \times \Delta_2(0,1)$. If $\mathcal{F}|_{\Delta_1(0,1)\times\Delta_2(0,\epsilon)}$ is a normal family in $\Delta_1(0,1) \times \Delta_2(0,\epsilon)$ and $\mathcal{F}|_{\{z_1\}\times\Delta_2(0,1)}$ is a normal family in $\Delta_2(0,1)$, for all z_1 fixed in $\Delta_1(0,1)$, then $\mathcal{F}$ is normal on $\Delta_1(0,1) \times \Delta_2(0,1)$.*

Proof. Our task is to show that from any fixed sequence $\{f_j\} \subset \mathcal{F}$ we can extract a subsequence which converges uniformly on each compact subset of $\Delta_1(0,1) \times \Delta_2(0,1)$.

(I) Suppose $\{f_j\}$ converges uniformly to a limit function f defined on $\Delta_1 \times \Delta_2(0,\epsilon)$. By Weierstrass's theorem, f is holomorphic in $\Delta_1(0,1) \times \Delta_2(0,\epsilon)$.

By hypothesis, $\mathcal{F}$ is a separately normal holomorphic family in $z_2 \in \Delta_2(0,1)$ when α is fixed in $\Delta_1(0,1)$. A choice of subsequence could change the limit function g, which by Weierstrass's theorem is holomorphic in $\Delta_2(0,1)$. Since $\Delta_2(0,\varepsilon) \subset \Delta_2(0,1)$, by the uniqueness of the limits, $g(\alpha, z_2)$ coincides with $f(\alpha, z_2)$ on $\Delta_2(0,\varepsilon)$; this means that $f(\alpha, z_2)$, previously defined on $\Delta_2(0,\varepsilon)$, can be extended to a holomorphic function on $\Delta_2(0,1)$. Denote this extension by $f(\alpha, z_2)$. We now have the hypotheses of Theorem 5.6.6 satisfied for f. We conclude that f is a holomorphic function on $\Delta_1(0,1)\times\Delta_2(0,1)$.

Take arbitrary a and b such that $0 < a < b < 1$. A holomorphic function f on $\Delta_1(0,1) \times \Delta_2(0,1)$ is bounded in modulus on a compact set $\overline{\Delta}_1(0,b) \times \overline{\Delta}_2(0,b) \subset \Delta_1(0,1) \times \Delta_2(0,1)$ by M, say.

We then consider the Hartogs series of $f - f_j$ with respect to z_2 at $z_2 = 0$: let

$$f(z) - f_j(z) = \sum_{k=0}^{\infty}[c_k(z_1) - c_k^j(z_1)]z_2^k. \tag{5.8}$$

The coefficients of this series

$$c_k(z_1) - c_k^j(z_1) = \frac{1}{k!}\frac{\partial^k[f - f_j]}{(\partial z_2)^k}(z_1, 0)$$

are holomorphic in the disc Δ_1.

By hypothesis $\{f_j(\alpha, z_2)\}$ converges uniformly to a limit function $f(\alpha, z_2)$ on $\{\alpha\} \times \Delta_2(0,\epsilon)$.

For a natural number M, let

$$S_{m,j} = \{\alpha \in \overline{\Delta}_1(0,b) \mid |f(\alpha, z_2) - f_j(\alpha, z_2)| \le M\,, \quad \text{for all } z_2 \in \Delta_2(0,b), j \ge m\}$$

and set $S_m = \bigcap_j E_{m,j}$. Since each $f - f_j$ is continuous, $S_{m,j}$ is closed, and consequently S_m is closed as intersections of the infinite collection of closed sets. It is clear that $S_m \subset S_{m+1}$ and

$$\overline{\Delta}_1(0,b) = \cup_m S_m.$$

It is at this point that Baire's theorem enters. Since $S_1, S_2, S_3, \ldots$ are closed sets, for every $x \in \overline{\Delta}_1(0,b)$ some of them must contain an open ball $B_\delta(x) \subset \overline{\Delta}_1(0,b)$. Otherwise, $\bigcup_{n=1}^{\infty} S_m$ is not the whole of $\overline{\Delta}_1(0,b)$ (see Theorem 13.3.3).

The collection of all sets $B_\delta(x) \subset \Delta_1(0,b)$ is an open cover of $\overline{\Delta}_1(0,a)$; and since $\overline{\Delta}_1(0,a)$ is compact, there is a finite set of points $x_1, \ldots, x_l$ in $\overline{\Delta}_1(0,a)$, such that

$$\overline{\Delta}_1(0,a) \subseteq \cup_{k=1}^{l} B_{\delta_k(x_k)}(x_k).$$

Using the fact that $S_m \subset S_{m+1}$, for all $m \in \mathbb{N}$, we obtain that for $\Delta_1(0,a)$ there is a natural number m_1 such that the set $S_{m_1} \supset \Delta_1(0,a)$. Since S_{m_1} is closed it follows $\overline{\Delta}_1(0,a) \subset S_{m_1}$. Hence

$$|f(\alpha, z_2) - f_j(\alpha, z_2)| \le M, \text{ for all } \alpha \in \overline{\Delta}_1(0,a), z_2 \in \Delta_2(0,b), j \ge m_1$$

It follows from Cauchy's inequalities that

$$|c_k(\alpha) - c_k^j(\alpha)| \le M b^{-k} \tag{5.9}$$

for all $\alpha \in \Delta(0,a)$, $a < b < 1$, and all k and $j > m_1$.

We see from (5.8) and (5.9) that, for all j and all $z \in \Delta_1(0,a) \times \Delta_2(0, \frac{a+b}{2})$,

$$|f(z) - f_j(z)| \le M \left[1 + \left(\frac{a+b}{2b} \right) + \left(\frac{a+b}{2b} \right)^2 + \ldots \right] \le \frac{2Mb}{b-a}.$$

This means that $\{f_j(z)\}$ is locally uniformly bounded in

$$\Delta_1(0,a) \times \Delta_2(0,(a+b)/2),$$

for arbitrary a and b, $0 < a < b < 1$. By Montel's theorem (Theorem 5.3.1) the family $\mathcal{G} = \{f_j(z_1, z_2)\}_{j > m_1}$ is normal in $\Delta_1(0,a) \times \Delta_2(0,b)$.

(II) If $\{f_j\}$ converges uniformly on $\Delta_1(0,b) \times \Delta_2(0,\epsilon)$ to ∞, then $|f_j(z)| > 1$ for all $z \in \overline{\Delta}_1(0,b) \times \overline{\Delta}_2(0,\epsilon)$, provided $j > j_0$, j_0 is sufficiently large.

A repetition of the same argument as in **(I)** gives the assertion.

Let $\alpha \in \overline{\Delta}_1(0,b)$. The hypotheses guarantee that $\{f_j(\alpha, z_2)\}$ uniformly convergent to ∞ on $\{\alpha\} \times \overline{\Delta}_2(0,b)$.

Set

$$Z_{m,j} = \left\{ \alpha \in \overline{\Delta}_1(0,b) \;\middle|\; |f_j(\alpha, z_2)| \ge 1, \text{ for all } z_2 \in \overline{\Delta}_2(0,b), j \ge m \right\}$$

and set $Z_m = \bigcap_j Z_{m,j}$.

The proofs given in **(I)** carry over verbatim to the case Z_m in place of S_m considered here, and so we shall not repeat them. A repetition of the first part of the previous argument in **(I)** shows that there is an integer m_2 such that $|f_j(z_1, z_2)| > 1$ for all $j > m_2$ and $(z_1, z_2) \in \Delta_1(0, a) \times \Delta_2(0, (a + b)/2)$, for arbitrary a and b, $0 < a < b < 1$.

By Montel's theorem (Theorem 5.3.1) the family $\mathcal{G} = \{f_j(z_1, z_2)\}_{j>m_2}$ is normal in $\Delta_1(0, a) \times \Delta_2(0, b)$.

Since a and b are taken arbitrarily we conclude that $\{f_j\}$ is locally normal in $\Delta_1(0, 1) \times \Delta_2(0, 1)$. It is routine to confirm that a family of analytic functions $\{f_j\}$ is locally normal in $\Delta_1(0, 1) \times \Delta_2(0, 1)$ if and only if $\{f_j\}$ is normal at each point of $\Delta_1(0, 1) \times \Delta_2(0, 1)$.

This completes the proof. □

Theorem 5.6.8 (cf. Nishino [164, Théorème II, p. 264]). *If $\mathcal{F}$ is a separate normal family of holomorphic functions, then $\mathcal{F}$ is a normal family.*

Remark 5.6.9. It is in fact clear that, by induction on dimension, one can prove the Nishino result in any dimension. For simplicity we restrict the proof here to the two-dimensional case.

Proof. Fix arbitrary sequence $\{f_j(z_1, z_2)\} \subset \mathcal{F}$. Take $0 < \delta < 1$ and take $(\alpha, \beta) \in \Delta_1(\alpha, \delta) \times \Delta_2(\beta, \delta)$. All that can be concluded directly from normality is that *some* subsequence converges. The choice of subsequence could change at each instance. Therefore, we need some preliminary work to find *one* sequence in what follows. We can pass to a subsequence and assume $f_j(\alpha, \beta) \to f(\alpha, \beta)$.

(III) Suppose $\{f_j(\alpha, \beta)\}$ is a bounded sequence. Convergence of any sequences of holomorphic functions on some domain Ω is always measured with respect to the euclidean metric. Directly from normality follows that $\{f_j(\alpha', \beta)\}$ and $\{f_j(\alpha, \beta)\}$ are a bounded sequence for all α, α'. Since $\{f_j(\alpha', z_2)\}$ uniformly converges on $\{\alpha'\} \times \Delta_2(0, \delta)$ to some holomorphic function $g_1(\alpha', z_2)$ and $\{f_j(z_1, \beta)\}$ uniformly converge on the set $\Delta_1(0, \delta) \times \{\beta'\}$ to some holomorphic function $g_2(z_1, \beta')$ *for all* $\alpha' \in \Delta_1(\alpha, \delta)$, $\beta' \in \Delta_2(\beta, \delta)$ (we invoke Weierstrass theorem). We see that the limit functions $g_1(\alpha', z_2)$ and $g_2(z_1, \beta')$ are both defined in (α', β') and $g_1(\alpha', \beta') = g_2(\alpha', \beta') = f(\alpha', \beta')$ for all $(\alpha', \beta') \in \Delta_1(\alpha, \delta) \times \Delta_2(\beta, \delta)$.

We obtained well-defined function f which is separately holomorphic on $\Delta_1(\alpha, \delta) \times \Delta_2(\beta, \delta)$. By Hartogs theorem f is holomorphic on $\Delta_1(\alpha, \delta) \times \Delta_2(\beta, \delta)$.

Further taking $\delta \to 1$, we can construct a well defined function $f(z_1, z_2)$ on the whole of $\Delta_1(\alpha, \delta) \times \Delta_2(\beta, \delta)$.

By the Baire-Osgood theorem, there exists an open polydisc

$$\Delta_1(a, \gamma) \times \Delta_2(b, \gamma) \subset \Delta_1(0, 1) \times \Delta_2(0, 1)$$

on which $\{f_j(z_1, z_2)\}$ is locally uniformly convergent to f. The conditions of Theorem 5.6.7 is fulfilled, and the desired result follows.

(IV) If $\{f_j(\alpha, \beta)\}$ is unbounded and assume $0 \neq f_j(\alpha, \beta) \to \infty$.

Fix arbitrary δ and b such that $0 < \delta < b < 1$.

Let $\sigma \in \overline{\Delta}_1(\alpha, \delta)$. Then $\{f_j(\sigma, z_2)\}$ uniformly converges to ∞ on $\{\sigma\} \times \overline{\Delta}_2(0, b)$.

Set

$$W_{m,j} = \{\sigma \in \overline{\Delta}_1(\alpha, \delta) \mid |f_j(\sigma, z_2)| \geq 1, \text{ for all } z_2 \in \overline{\Delta}_2(0, b), j \geq m\}$$

and set $W_m = \bigcap_j W_{m,j}$.

A repetition of the argument in **(I)** in Theorem 5.6.7 shows that there is an integer m_2 such that $|f_j(z_1, z_2)| > 1$ for all $j > m_2$ and $(z_1, z_2) \in \Delta_1(0, \delta) \times \Delta_2(0, (\delta + b)/2)$.

Put $g_j = 1/f_j$ in $\Delta_1(\alpha, \delta) \times \Delta_2(0, (\delta + b)/2)$. It is clear that $\{g_j = 1/f_j\}$ tends uniformly to 0 in $\Delta_1(\alpha, \delta) \times \Delta_2(\beta, \delta)$. By Theorem 5.6.7 $\{1/f_j\}$ tends uniformly to 0 on $\Delta_1(\alpha, \delta) \times \Delta_2(\beta, \delta)$. Since (α, β) and δ are taken arbitrarily we conclude that $\{f_j\}$ converges locally uniformly to ∞ on $\Delta_1(0, 1) \times \Delta_2(0, 1)$.

The family of analytic functions $\mathcal{F}$ is normal in $\Delta_1(0, 1) \times \Delta_2(0, 1)$ if and only if $\{f_j\}$ is normal at each point of $\Delta_1(0, 1) \times \Delta_2(0, 1)$.

This completes the proof. □

5.7 Schottky's Theorem

A collection of important theorems is associated with the Montel criterion for normality, thereby constituting the so-called Montel cycle. The famous Schottky's theorem is one of them.

We consider the class of function, analytic in the unit disc $\mathbb{D}$ for which the equations $f(z) = 0$ and $f(z) = 1$ have no solutions in $\mathbb{D}$. Schottky's theorem asserts the existence of a function $K(a, r)$ such that

$$|f(z)| \leq K(a, |z|)$$

for all functions $f(z)$ in the above class for which $|f(0)| = a$.

Schottky's Theorem can be given a completely elementary proof, based on nothing more than Montel's theorem and general properties of normal families.

Theorem 5.7.1. *Suppose $f(z)$ is holomorphic in a domain $\Omega \subset \mathbb{C}^n$, $n \geq 1$, with range omitting* 0 *and* 1. *If $p \in \Omega$ and $|f(p)| < M$, then for every $\varepsilon > 0$ sufficiently small, then we have*

$$|f(z)| < C \text{ for all } z \in \Omega, \quad \delta(z) \geq \varepsilon, \tag{5.10}$$

where $\delta(z)$ denotes the Euclidean distance from $z \in \Omega$ to $\partial\Omega$. Here C is a positive constant depending only on p, M and ε.

The proof of Theorem 5.7.1 is based on the following lemma:

Lemma 5.7.2. *Let $\mathcal{F}$ be a normal family of holomorphic functions in a domain $\Omega \subset \mathbb{C}^n$, $n \geq 1$. Let M be a positive number and $p \in \Omega$. Assume that for each function $f \in \mathcal{F}$, we have*

$$|f(p)| < M.$$

Then the family $\mathcal{F}$ is uniformly bounded on each compact subset of Ω.

Proof. Let K be a compact set of points belonging to Ω. Assume that $\mathcal{F}$ is not uniformly bounded on K. Then to each positive integer k corresponds a function $f_k \in \mathcal{F}$ such that

$$\max_{z\in K} |f_k(z)| > k.$$

The sequence $\{f_k\}$ cannot converge on compact set $\{p\} \cup K$ to a holomorphic function or to ∞ in Ω. This contradicts the hypothesis that $\mathcal{F}$ is a normal family. □

Proof of Theorem 5.7.1. Let $\mathcal{G}$ denote the family of all functions holomorphic on Ω which omit the values 0 and 1. According to Theorem 5.3.1 the family $\mathcal{G}$ is normal. By Lemma 5.7.2, the family $\mathcal{G}$ is uniformly bounded on the compact set $\overline{\Omega}_\varepsilon = \{z \in \Omega, \delta(z) \geq \varepsilon\}$, hence there is a positive constant C such that (5.10) holds for each function $f(z)$ of $\mathcal{F}$. This constant C has then the required property. □

The method of proof in Theorem 5.7.1 does not give any information about what the constant C is. In the case of the unit disc Murali Rao *et al* [179, Theorem 7.17, p. 118] gave the following bounds:

$$\sup_{|z|\leq r} |f(z)| \leq \tan(\arctan M + (c(1-r^2))^{-1}),$$

where c is some constant. The proof is based on the work of Minda and Schober [156].

We shall give an alternative argument based on the classical Montel theorem and Marty's criterion for normality.

Recall that to every z in the unit disc $\mathbb{D}$ corresponds an automorphism φ_z of the disc that interchanges z and 0, namely $\varphi_z(w) := (z-w)/(1-\overline{z}w)$. Note for later reference that $\varphi_z(0) = z$ and $\varphi_z'(0) = 1 - |z|^2$. Define $\Gamma(\mathbb{D}) := \{\varphi_z, z \in \Delta\}$.

Let $\mathcal{G}$ denote the family of all functions holomorphic on the unit disc $\mathbb{D}$ which omits the values 0 and 1. By Montel's Theorem 5.3.1 the family $\mathcal{G} \circ \Gamma(\mathbb{D})$ is normal. Letting the compact set $K \subseteq \mathbb{D}$ be the singleton $\{0\}$. Marty's Theorem 5.1.2 yields a non-negative constant L such that for any $f \circ \varphi_z \in \mathcal{G} \circ \Gamma(\mathbb{D})$

$$\frac{|(f \circ \varphi_z)'(0)|}{1 + |f(\varphi_z(0))|^2} \leq L.$$

The constant L in the inequality above does not depend on $f \circ \varphi_z$. Using the chain rule for differentiation, we obtain

$$(1 - |z|^2) f^\sharp(z) \leq L \text{ for all } z \in \mathbb{D}.$$

In what follows L will always denote

$$\sup_{z \in \Delta} (1 - |z|^2) g^\sharp(z) \text{ for all } g \in \mathcal{G}.$$

The exact value L is not known.

The following theorem is one of the main results of this section.

Theorem 5.7.3 (Schottky's theorem). *To each M and $r \in [0, 1)$ there exists a constant $C > 0$ such that the following implication holds: If $f(z)$ is holomorphic in the unit disc $\mathbb{D}$, omits 0 and 1, and $|f(0)| \leq M$, then*

$$\sup_{|z| \leq r} |f(z)| \leq \tan\left(\arctan M + 2rL/\left(1 - r^2\right)\right).$$

Proof. Since f never takes the value 0, the function $t \to |f(tz)|$ is continuously differentiable on $(0, 1)$ for any fixed $z \in \mathbb{D}_r := \{|z| < r\}$ and

$$\frac{d}{dt}\left(\arctan\left(|f(tz)|\right)\right) \leq \frac{2\,|f'(tz)|\,|z|}{1 + |f(tz)|^2} \leq \frac{2rL}{1 - r^2}.$$

Integrating this from $t = 0$ to 1, we deduce that

$$|\arctan(|f(z)|) - \arctan(|f(0)|)| \leq \int_0^1 \left|\frac{d}{dt}(\arctan(|f(tz)|))\right| dt \leq \frac{2rL}{1 - r^2}$$

and hence

$$\arctan(|f(z)|) \leq \arctan(|f(0)|) + \frac{2rL}{1 - r^2} \leq \arctan M + \frac{2rL}{1 - r^2}$$

which is the theorem with $C = \tan(\arctan M + 2rL/(1 - r^2))$. This completes the proof of Theorem 5.7.3. □

5.8 Landau's Theorem

Schottky's Theorem 5.7.3 says that the hypothesis that a function $f(z)$ holomorphic in the disc $\mathbb{D}(0; R)$ does not assume two values, e.g., 0 and 1, is already a strong limitation on the behavior of the function. The following result, which we owe to Landau, confirms this fact from another direction:

Theorem 5.8.1. *Let α and β be arbitrary complex numbers, where $\beta \neq 0$. If the function*

$$f(z) \doteq \alpha + \beta z + a_2 z^2 + a_3 z^3 + \dots \tag{5.11}$$

is holomorphic in the circle $\mathbb{D}(0; R)$ and does not assume the values 0 and 1 in it, then $R \leqslant L(\alpha, \beta)$, where $L(\alpha, \beta)$ depends only on α and β.

Proof. By Theorem 5.7.1 and Cauchy-Riemann equations, we have the inequality $|f(z)| \leqslant \Omega(\alpha, 1/2)$ for $z \in \mathbb{D}(0; R/2)$. Expressing the coefficient β in terms of the function, we get

$$|\beta| = \left| \frac{1}{2\pi i} \int_{C(0;R/2)} \frac{f(z)}{z^2} dz \right| \leqslant \frac{1}{2\pi} \cdot \frac{\Omega(\alpha, 1/2)}{(R/2)^2} \cdot 2\pi \frac{R}{2} = \frac{2\Omega(\alpha, 1/2)}{R}.$$

Consequently, $R \leqslant \frac{2\Omega(\alpha,1/2)}{|\beta|} = L(\alpha, \beta)$ and Theorem 5.8.1 is proved. □

We can state Theorem 5.8.1 in another way: For every a and $\beta \neq 0$ there exists a number $L(a, \beta)$ such that if the function (5.11) is hoiomorphic for $|z| < L(\alpha, \beta)$, then it must assume there at least one of the values $0, 1$. Since a function has at least one singular point on the circumference of the circle of convergence, we can state the matter in a still somewhat different way: For every α and $\beta \neq 0$, the function given by the series (5.11) must assume the value 0, or the value 1, in the closed circle $|z| \leqslant L(\alpha, \beta)$, or this series has a singular point in this circle.

The assumption $\beta \neq 0$ may be replaced by another more general one: Let us suppose that $\lambda \neq 0$ and that the function

$$F(z) = a + \lambda z^l + a_{l+1} z^{l+1} + a_{l+2} z^{l+2} + \dots \tag{5.12}$$

is holomorphic in the circle $\mathbb{D}(0; R)$ and does not assume the values 0 or 1 there. Then $R \leqslant L(a, \lambda, l)$, where the quantity L depends only on the arguments α, λ, l.

The proof is the same as before. From the formula

$$|\lambda| = \left| \frac{1}{2\pi i} \int_{\mathbb{D}(0;R/2)} \frac{F(z)}{z^{l+1}} dz \right| \leqslant \frac{1}{2\pi} \cdot \frac{\Omega(\alpha, 1/2)}{(R/2)^{l+1}} \cdot 2\pi \frac{R}{2} = \frac{\Omega(\alpha, 1/2)}{(R/2)^l}$$

we have $R \leqslant 2\Omega^{1/l}(a, 1/2)/|\lambda|^{1/l} = L(a, \lambda, l)$.

Every entire function $f(z)$, different from a constant, can be written in the form (5.12). Since the radius of convergence R of the series (5.12) is then infinite, the inequality $R \leqslant L(a, \lambda, l)$ is not satisfied. This proves that $f(z)$ must assume at least one of the values $0, 1$. Therefore Landau's theorem includes Picard's little theorem, even in a stronger form. For it gives an estimate of the radius R of a circle $\mathbb{D}(0, R)$ in which $f(z)$ certainly assumes either the value 0 or the value 1.

6

Boundary Behavior of Holomorphic Functions

The theorems to be developed in this chapter belong, for the most part, to an area of the theory of boundary behavior which has become classical, namely, the theory of bounded analytic functions and functions of bounded characteristic. In the first half of the chapter, we discuss Fatou's theorem and the cycle of theorems related to it. In the second half, we discuss the theory of Blashke products and invariant metrics; and we have oriented this theory to show the extent to which theorems of the type of Fatou's theorem may be generalized as far as approach to the boundary is concerned.

6.1 Fatou's Theorem on Radial Limits

The theorem, first proved by Fatou in 1906, is the following:

Theorem 6.1.1. *If $f(z)$ is analytic and bounded in $\mathbb{D}$, then the radial limits $f\left(e^{i\theta}\right) = \lim_{r\to 1} f\left(re^{i\theta}\right)$ exist for all points $e^{i\theta}$ on $\partial\mathbb{D}$, except possibly for a set of linear measure zero.*

We restrict ourselves to giving the outline of a proof based on the theory of power series which uses ideas that we shall require later on.

We show first that $\sum_{j=0}^{\infty} |a_j|^2$ is bounded. For, writing $z = re^{i\theta}$, we have

$$|f(z)|^2 = f(z)\overline{f(z)} = \sum_{j=0}^{\infty} a_j r^j e^{in\theta} \sum_{j=0}^{\infty} \bar{a}_j r^j e^{-in\theta}$$

where the bars denote conjugates, so that, multiplying the series by the usual Cauchy rule, as we may do since both are absolutely convergent, we obtain

$$|f(z)|^2 = \sum_{j=0}^{\infty} c_j(r,\theta)$$

where

$$c_j(r,\theta) = \left(a_0\bar{a}_j e^{-in\theta} + a_1\bar{a}_{j-1} e^{-i(n-2)\theta} + \cdots + a_j\bar{a}_0 e^{in\theta}\right) r^j$$

DOI: 10.1201/9781003619581-6

Now $\sum_{j=0}^{\infty} c_j(r,\theta)$ is uniformly convergent in $0 \leqslant \theta \leqslant 2\pi$ since both the series on the right are absolutely convergent. On integration term by term only the terms

$$\int_0^{2\pi} a_\nu \bar{a}_\nu r^{2\nu} d\theta = 2\pi \left|a_\nu\right|^2 r^{2\nu}$$

do not vanish so that

$$\sum_{j=0}^{\infty} \left|a_j\right|^2 r^{2n} = \frac{1}{2\pi} \int_0^{2\pi} \left|f\left(re^{i\theta}\right)\right|^2 d\theta \leqslant M^2$$

and since this holds for all $r < 1$ our assertion is proved.

Now by the Riesz-Fischer theorem

$$\sum_{j=0}^{\infty} \left|a_j\right|^2 = O(1)$$

implies that the series $\sum_{j=0}^{\infty} a_j e^{in\theta}$ is the Fourier series of a function $f(\theta)$ (Titchmarsh [203], p.423) and is therefore summable $(C,1)$ to $f(\theta)$ for almost all θ in $0 \leqslant \theta \leqslant 2\pi$. But a series which is summable $(C,1)$ is Abel summable (Hardy [93, Theorem 55, p. 108]). That is to say

$$\lim_{r\to 1} f\left(re^{i\theta}\right) = \lim_{r\to 1} \sum_{j=0}^{\infty} a_j r^j e^{in\theta}$$

exists for almost all θ in $0 \leqslant \theta \leqslant 2\pi$, and this proves the theorem.

Corollary 6.1.2. *If the function $u(z) = u(r,\theta)$ is harmonic and bounded in $\mathbb{D}$, then the radial limits $u\left(e^{i\theta}\right) = \lim_{r\to 1} u\left(r, e^{i\theta}\right)$ exist for all points $e^{i\theta}$ on $\partial\mathbb{D}$, except possibly for a set of linear measure zero.*

For $u(z)$ is the real part of a function $f(z)$ which is regular in $\mathbb{D}$. For a finite real number $A > u(z)$ for all $z \in \mathbb{D}$ the function $(f(z) - A)^{-1}$ is regular and bounded and therefore has radial limits at almost all points $e^{i\theta}$ on $\partial\mathbb{D}$. The same is therefore true of $f(z)$ and of $u(z) = \operatorname{Re} f(z)$.

6.1.1 Generalization of the Classical Fatou Theorem

Fatou's theorem can be generalized by relaxing the condition that $|f(z)|$ be bounded in $\mathbb{D}$:

Theorem 6.1.3. *Let the segment $(0,1) \subset \mathbb{R}$ be one edge of an open rectangle Q in the upper half of $\mathbb{C}$. Suppose*

(a) $f: Q \to \mathbb{C}$ is a bounded C^1-function, and

(b) $\partial f/\partial \bar{z} \in L^p(Q)$ for some $p > 1$.

Then $\lim f(x+iy)$ exists for almost all $x \in (0,1)$, as $y \to 0$.

Here, and later, L^p refers to Lebesgue measure on $\mathbb{C}$. Note that **(b)** represents a considerable weakening of the classical hypothesis that $f \in H^\infty(Q)$, i.e., that $\partial f/\partial \bar{z} = 0$.

Proof of Theorem 6.1.3. For $1 \leqq k \leqq \infty$, we shall write ${C_c}^k$ for the class of all $f : \mathbb{C} \to \mathbb{C}$ that are k times continuously differentiable and have compact support.

Lemma 6.1.4. *To every* $p, 1 < p < \infty$, *corresponds a constant* $A_p < \infty$ *such that the inequality*

$$\left\| \frac{\partial f}{\partial y} \right\|_p \leqq A_p \left\| \frac{\partial f}{\partial \bar{z}} \right\|_p$$

holds for all $f \in {C_c}^1$.

This follows from the L^p-boundedness (for $1 < p < \infty$) of the Riesz transforms on $\mathbb{R}^2$. A proof is given on p. 60 of [200].

Lemma 6.1.5. *Suppose* Ω *is a bounded open set in* $\mathbb{C}, 1 < p < \infty$, *and* $g \in L^p(\Omega)$. *If* $G \in C^1(\Omega)$ *and if*

$$G(z) = \frac{1}{2\pi i} \int_\Omega \frac{g(\zeta)}{\zeta - z} d\zeta \wedge d\bar{\zeta} \tag{6.1}$$

for almost all $z \in \Omega$, *then* $\partial G/\partial y \in L^p(\Omega)$.

Regard g as a member of $L^p(\mathbb{C})$ which is 0 off Ω. Put $k(z) = 1/\pi z$. Then k is locally L^1, and the convolution $H = g * k$, defined by

$$H(z) = \int_\mathbb{C} g(\zeta)k(z - \zeta)d\xi d\eta \quad (\zeta = \xi + i\eta) \tag{6.2}$$

exists for almost all $z \in \mathbb{C}$, as a Lebesgue integral. Moreover, a comparison of (6.1) and (6.2) show that the C^1-function G coincides with H a.e. in Ω.

Choose $\chi \in {C_c}^\infty$ so that $\chi = 1$ on Ω. Choose $\psi \in {C_c}^\infty, \psi \geqq 0$, so that $\int_\mathbb{C} \psi = 1$. For $1 \leqq t < \infty$, define $\psi_t(z) = t^2\psi(tz)$.

There is a disc $\Omega \subset \mathbb{C}$, of radius r, that contains the supports of χ and of $|g| * \psi_t$ for all $t \in [1, \infty)$. It is easily seen that

$$\int_D |k(z - \zeta)| d\xi d\eta \leqq 2r \tag{6.3}$$

for all $z \in \mathbb{C}$.

Define $H_t = H * \psi_t$. Since $H = g * k$, we have

$$H_t = (g * \psi_t) * k. \tag{6.4}$$

Since $\|\psi_t\|_1 = 1$ for all t, Hölder's inequality and (6.3) lead from (6.4) to

$$\left\{ \int_D |H_t(z)|^p \, dxdy \right\}^{1/p} \leqq 2r\|g\|_p \quad (1 \leqq t < \infty) \tag{6.5}$$

Since $g * \psi_t \in C_c^\infty$, Theorem 1.2.2 of [111] can be applied to (6.4) and shows that

$$\partial H_t / \partial \bar{z} = g * \psi_t.$$

Hence

$$\frac{\partial (\chi H_t)}{\partial \bar{z}} = H_t \cdot \frac{\partial \chi}{\partial \bar{z}} + \chi \cdot (g * \psi_t)$$

so that (6.5) gives the estimate

$$\left\| \frac{\partial (\chi H_t)}{\partial \bar{z}} \right\|_p \leqq M \|g\|_p$$

in which M is a real number that depends only on χ and r. It now follows from Lemma 6.1.4 that

$$\left\| \frac{\partial (\chi H_t)}{\partial y} \right\|_p \leqq A_p M \|g\|_p \quad (1 \leqq t < \infty). \tag{6.6}$$

To every compact $K \subset \Omega$ corresponds a $t(K)$ such that $K - \operatorname{supp} \psi_t \subset \Omega$ for all $t > t(K)$. Since $G = H$ a.e. in Ω, we have

$$H_t(z) = \int G(z - \zeta) \psi_t(\zeta) d\xi d\eta \quad (z \in K)$$

if $t > t(K)$. Since $G \in C^1(\Omega)$ and $\chi = 1$ in Ω, it follows that

$$\left(\frac{\partial G}{\partial y} \right)(z) = \lim_{t \to \infty} \frac{\partial (\chi H_t)}{\partial y}(z) \quad (z \in \Omega). \tag{6.7}$$

By (6.6) and (6.7), Fatou's lemma shows that $\partial G / \partial y \in L^p(\Omega)$. □

Remark 6.1.6. Lemma 6.1.5 would become false if, instead of (6.1), we merely assumed that $\partial G / \partial \bar{z} = g \in L^p(\Omega)$. To see this, take G holomorphic in Ω, so that $\partial G / \partial \bar{z} = 0$, but of sufficiently rapid growth near some boundary point to have $\partial G / \partial y \notin L^p(\Omega)$.

Proof of Theorem 6.1.3. To fix the notation, assume that $Q = (0,1) \times (0,h)$. Since f is bounded, there is a sequence $y_n \searrow 0$ such that the functions $x \to f(x + iy_n)$ converges weak* in $L^\infty(0,1)$, to some $\varphi \in L^\infty(0,1)$. Extend f to $(0,1) \times [0,h)$ by setting $f(x,0) = \varphi(x)$.

Choose a small $\epsilon > 0$ and define

$$Q_\epsilon = [\epsilon, 1-\epsilon] \times (0, h-\epsilon], \quad Q_{\epsilon,n} = [\epsilon, 1-\epsilon] \times [y_n, h-\epsilon]$$

For z interior to Q_ϵ and for n sufficiently large, the fact that f is C^1 on the compact set $Q_{\epsilon,n}$ shows (Theorem 1.2.1 in [111]) that

$$f(z) = \frac{1}{2\pi i} \int_{\partial Q_{\epsilon,n}} f(\zeta) \frac{d\zeta}{\zeta - z} + \frac{1}{2\pi i} \int_{Q_{\epsilon,n}} \frac{\partial f}{\partial \bar{\zeta}}(\zeta) \frac{d\zeta \wedge d\bar{\zeta}}{\zeta - z}$$

The above-mentioned weak*-convergence, combined with the fact that $\partial f/\partial\bar{\zeta} \in L^1(Q)$, shows that we can let $n \to \infty$, to obtain

$$\begin{aligned} f(z) &= \frac{1}{2\pi i}\int_{\partial Q_\epsilon} f(\zeta)\frac{d\zeta}{\zeta - z} + \frac{1}{2\pi i}\int_{Q_\epsilon} \frac{\partial f}{\partial\bar{\zeta}}(\zeta)\frac{d\zeta \wedge d\bar{\zeta}}{\zeta - z} \\ &= H(z) + G(z) \end{aligned}$$

Since H is the Cauchy integral of a bounded function, it is classical (see, for instance, Lemma 2.6 in Chap. V of [201]) that $\lim H(x+iy)$ exists, as $y \to 0$, for almost all x in $(\epsilon, 1-\epsilon)$.

Since $G = f - H, G \in C^1(Q_\epsilon)$. By Lemma 6.1.5, $\partial G/\partial y \in L^p(Q_\epsilon)$. Setting

$$M(x) = \int_0^{h-\epsilon} \left|\frac{\partial G}{\partial y}(x,y)\right|^p dy$$

it follows that $M(x) < \infty$ a.e. in $(\epsilon, 1-\epsilon)$. If $0 < y_0 < y_1 < h - \epsilon$, Hölder's inequality gives

$$|G(x+iy_1) - G(x+iy_0)| \leqq M(x)^{1/p}\,|y_1 - y_0|^{1-1/p}$$

Hence $\lim G(x+iy)$ exists, as $y \to 0$, for almost all x in $(\epsilon, 1-\epsilon)$. The arbitrariness of ϵ shows that the proof is complete. □

6.1.2 The Theorem of F. and M. Riesz

A function, continuous on the closed unit disc, holomorphic inside, and vanishing on an open subset of the boundary, vanishes identically. A simple proof of this well-known proposition is obtained by considering its Cauchy integral representation.

We have seen in Section 6.2.1 that the radial limits of a bounded analytic function exist for almost all points of $\partial\mathbb{D}$; we shall state in this section a uniqueness theorem of F. and M. Riesz [181] from which it follows that the radial limit values cannot form too sparse a set in the w-plane.

We have

Theorem 6.1.7. *If $f(z)$ is analytic and bounded, $|f(z)| < M$, in $\mathbb{D}$, and if the set E of points $e^{i\theta}$ for which $f\left(e^{i\theta}\right) = \lim_{r\to 1} f\left(re^{i\theta}\right) = 0$ has positive measure on $\partial\mathbb{D}$, then $f(z)$ is identically zero in $\mathbb{D}$.*

We remark that an immediate consequence of Theorem 6.1.7 is that two bounded analytic functions having the same radial limit values on a set of positive measure must be identical.

6.2 Privalov's Uniqueness Theorem

Let us agree on some terminology. Suppose that U is a domain bounded by a Jordan curve Γ and that a is a point on Γ at which there is a unique tangent to Γ and in a neighborhood of which the curve Γ lies on both sides of the normal. We shall refer to a continuous curve l contained in U and ending at such a point a as a *nontangential path* if that part of l in a neighborhood of a lies inside some angle of magnitude less than π with vertex at the point a and with bisector coinciding with the inner normal to Γ. When we speak of a set of points on the boundary of the domain U, we shall usually mean only points at which there are tangents to Γ possessing the property just described. Furthermore, if a function $f(z)$ that is holomorphic in U approaches a value a as z approaches a point a on Γ along every nontangential path contained in Γ, we shall say simply that $f(z)$ assumes (or has) the value a along nontangential paths.

We turn now to the investigation of functions that are holomorphic in a domain bounded by a rectifiable Jordan curve. We begin with the exceedingly important uniqueness theorem of Privalov.[1]

Theorem 6.2.1. *Let $f(z)$ denote a function that is holomorphic in a domain U bounded by a rectifiable closed Jordan curve Γ. Suppose that $f(z)$ assumes the value 0 along tangential paths on some set E of positive measure on Γ. Then $f(z) \equiv 0$ in U.*

Proof. Let us consider first the case in which U is the disk $|z| < 1$. About every point $a \in E$, let us construct a right angle with vertex at a, with rays directed into $|z| < 1$, and with bisector directed along the radius to $z = 0$. Let us cut from this angle sectors $S_a^{(j)}$, for $j = 1, 2, \cdots$, of radii $1/j$. Consider a sequence of functions $f_j(z)$, for $j = 1, 2, \cdots$, that are measurable[2] on E.

Define $f_j(z)$ at the point $a \in E$ as the maximum value of $|f_j(z)|$ in the sector $S_a^{(j)}$. Obviously, this sequence converges to zero on E. In accordance with Egorov's theorem, the set E contains a closed subset P of positive measure on which the sequence $\{f_j(z)\}$ converges uniformly to zero. The complement of P with respect to the disk $|z| = 1$ is an open set and hence consists of finitely or countably many disjoint arcs. We replace those arcs that are less than a semicircle with the sides of the right angles intercepting them from inside the disk $|z| < 1$ in such a way that the sides of the angles intersect the circle $|z| = 1$ at an angle $\pi/4$. On the other hand, we replace an arc greater than a semicircle with the two radii that pass through its endpoints. As a result,

---

[1] Privalov [1919].

[2] The measurability of $f_j(z)$ follows from the fact that we can regard $f_j(z)$ as the limit of a sequence of functions $f_{j,r}(z)$ constructed in the same way but issuing not from sectors but from the parts of them lying in $|z| < r$ as $r \to 1$. The functions $f_{j,r}(z)$ are, for r sufficiently close to unity, defined and continuous on E.

we obtain in place of the circle $|z| = 1$ a rectifiable closed Jordan curve Γ_1 bounding the domain U_1 and contained in the disk $|z| < 1$. Here, the set P lies entirely on the boundary Γ_1. The function $f(z)$ is continuous in $\bar{U}$ if we assign to it the value 0 at points of the set P. Let us now map the domain U_1 onto the disk $|\zeta| < 1$. Let $z = \omega(\zeta)$ denote the inverse of the function executing this mapping. Corresponding to the set P is a measurable set P_1 of positive measure on the circle $|\zeta| \leq 1$, The function $f_1(\zeta) = f(\omega(\zeta))$ is holomorphic in the open disk $|\zeta| < 1$, continuous on the closed disk $|\zeta| \leq 1$, and equal to zero at points of the set P_1. Then, by the F. and M. Riesz Theorem, $f_1(\zeta) \equiv 0$; that is, $f(z) \equiv 0$ in $|z| < 1$.

Thus, the theorem is proved for the case of a circle.

Let U denote an arbitrary domain defined by a rectifiable closed Jordan curve Γ. And let us map it onto the disk $|\zeta| < 1$. We denote by E' the measurable set of positive measure of $|\zeta| = 1$ into which E is mapped. The function $f(\zeta)$ is now transformed into a function $f_1(\zeta)$ that is holomorphic in the disk $|\zeta| < 1$ and that assumes the value zero along nontangential paths on the set E' after deletion of a set of measure zero at points in which this mapping does not conserve angles. In accordance with what has been proved, $f_1(\zeta) \equiv 0$ in $|\zeta| < \dot{1}$. Consequently, $f(z) \equiv 0$ in U. This completes the proof of the theorem. □

6.2.1 The Classical Lindelöf and Fatou Theorems for Angular Limits

We remark that Fatou's theorem, as we have stated it, guarantees the existence of radial limit values of $f(z)$ almost everywhere on $\partial\mathbb{D}$. This theorem can be generalized in many ways, by relaxing the condition that $|f(z)|$ be bounded in $\mathbb{D}$, for example. We shall indicate some of these possibilities, but, at this stage, we wish to show that, for a function analytic and bounded in $\mathbb{D}$, the existence of a radial limit at a point $e^{i\theta}$ implies the existence of the limit for any non-tangential approach to $e^{i\theta}$.

In the theory of functions of one complex variable, there is a theorem which states that a function that is bounded and holomorphic in a domain $\mathbb{D}$, and having a limit as $z \to \zeta \in \partial\mathbb{D}$ in the direction of the normal ν_ζ, tends to the same limit as $z \to \zeta$ in any direction not tangent to $\partial\mathbb{D}$ (Lindelöf's theorem).

As a simple example shows, the limits in the tangent directions must be excluded: the function $f(z) = e^{-1/z}$ is holomorphic and bounded in the disc $\{|z-1| < 1\}$, and at the boundary point $z = 0$ its limit along nontangential paths is equal to zero, while its limit along tangential paths does not exist.

To this end we state next a theorem of Lindelöf [154], together with two immediate corollaries, which will enable us to state Theorem 6.1.1 in a form (Theorem 6.2.6) having considerable application in the sequel. This result of Lindelöf has many applications in the theory of functions.

Theorem 6.2.2. *Let $\gamma : [0,1] \to \mathbb{D} \cup \{1\}$ be an arc with $\gamma(1) = 1$. If f is a bounded holomorphic function on $\mathbb{D}$ such that the limit $\lim_{t\to 1^-} f(\gamma(t))$ exists, call it L, then the radial limit $f^*(1) = \lim_{t\to 1^-} f(t)$ exists and is L.*

There is no assumption in this theorem about how $f(\gamma(t))$ approaches the point L as $t \to 1^-$. The approach could be highly tangential or oscillatory.

Proof. It entails no loss of generality to suppose that the limit L is 0 and that $\|f\|_{\mathbb{D}} \le 1$. Introduce the function $h \in H^\infty(\mathbb{D})$ by $h(z) = f(z)\bar{f}(\bar{z})$. We shall show that h has radial limit 0 at 1.

Suppose first that $\Im\gamma(t) > 0$ on an interval $(t_o, 1)$ for some $t_o \in (0,1)$. Let Δ be the domain in the plane bounded above by the curve $\gamma_{|[t_o,1)}$, bounded below by the curve $\bar{\gamma}(t)|_{|[t_o,1]}$, and bounded on the left by the vertical segment-which may be a single point-connecting $\gamma(t_o)$ to $\bar{\gamma}(t_o)$. The domain Δ is simply connected and is bounded by a simple closed curve. Let $\varphi : \mathbb{D} \to \Delta$ be a Riemann map. The map φ extends to a homeomorphism, still denoted by φ, of $\overline{\mathbb{D}}$ onto $\bar{\Delta}$. Without loss of generality, $\varphi(1) = 1$. Then the function $h \circ \varphi$ is bounded and holomorphic on $\mathbb{D}$, and it has continuous boundary values. Therefore $h \circ \varphi$ is continuous on $\bar{\mathbb{D}}$. (This is so, for by Fejér's theorem the Fourier series of $h \circ \varphi$ is uniformly Cesàro summable to $h \circ \varphi$. This provides a sequence of holomorphic polynomials that converges to $h \circ \varphi$ uniformly on $b\mathbb{D}$, whence $h \circ \varphi$ is in $A(\mathbb{D})$. Alternatively $h \circ \varphi$ is the Poisson integral of its continuous boundary values and so is continuous on the closed disk.) The function $h \circ \varphi$ is continuous on $\mathbb{D}$, and φ is a homeomorphism on $\overline{\mathbb{D}}$, so the function h is continuous on $\bar{\Delta}$. Consequently, the radial limit $f^*(1)$ exists and is 0.

Consider now the case of a general arc γ. Suppose without loss of generality that $\gamma(0) = 0$. If $\gamma([0,1])$ contains an interval $(t_o, 1)$ for some $t_o < 1$, we are done. In the contrary case, let $a_j, b_j \in (0,1)$ be the points such that $[0,1]\backslash\gamma([0,1])$ is the union of the open intervals (a_j, b_j). For each j, let $a_j = \gamma(\alpha_j)$ and $b_j = \gamma(\beta_j)$. Thus, for each $j, \Im\gamma > 0$ on (α_j, β_j) or else $\Im\gamma < 0$ on this interval. In either case, we denote by Δ_j the relatively compact subdomain in $\mathbb{D}$ that contains the interval (a_j, b_j) and that is bounded by the simple closed curve that is the union of the two arcs $\gamma([\alpha_j, \beta_j])$ and $\bar{\gamma}([\alpha_j, \beta_j])$. The function h is bounded on Δ_j by $\max_{t\in[a_j,b_j]} |f(\gamma(t))|^2$. The function h therefore has the radial limit 0 at the point 1, for $f(\gamma(t)) \to 0$ when $t \to 1^-$. The proof is complete. □

Let $\alpha > 1$. For $\theta \in [, 2\pi)$, define

$$\Gamma_\alpha\left(e^{i\theta}\right) = \left\{z \in \Omega \mid \left|z - e^{i\theta}\right| < \alpha(1 - |z|)\right\}$$

The set Γ_α is known as a *Stolz region* or *non-tangential approach region.* See Figure 6.1

One of the most useful strengthening of Theorem 6.1.1 is the following result, which is also often cited as Fatou's theorem.:

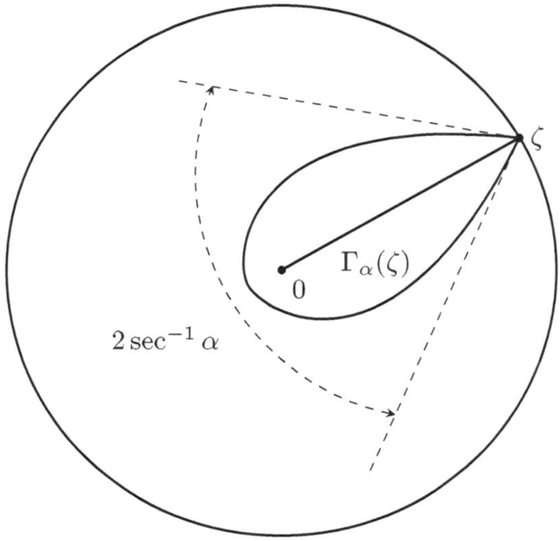

FIGURE 6.1: A Stolz region Γ_α at $\zeta \in \partial\mathbb{D}$

Theorem 6.2.3. *Let f be a bounded holomorphic function on $\mathbb{D}$. Fix a real number $\alpha > 1$. Then, for almost every $\theta \in [0, 2\pi)$, the limit*

$$f^*\left(e^{i\theta}\right) \equiv \lim_{\Gamma_\alpha(e^{i\theta}) \ni z \to e^{i\theta}} f(z)$$

exists.

Of course, we have incorporated into the statement of this result the obvious fact that the limit over the larger non-tangential approach regions will be the same as the limit taken over the radii (which they contain).

By an elementary conformal mapping, we may state Theorem 6.2.3 in the following form:

Theorem 6.2.4. *Let $f(z)$ be analytic and bounded in $\mathbb{D}$. If $f(z) \to \alpha$ as $z \to \xi \in \partial\mathbb{D}$ along some arc γ lying in $\mathbb{D}$ and terminating at ξ, then $f(z)$ tends to α uniformly as $z \to \xi$ inside any angular domain lying in $\mathbb{D}$ and having ξ as vertex.*

The theorem of Lindelöf (Theorem 6.2.2) may be easily generalized in various directions; for example, we may replace the hypothesis of boundedness by the hypothesis that $\operatorname{Re} f(z) < c$, for some finite c.

We prove here an extension of Lindelöf's theorem involving two paths:

Theorem 6.2.5. *Let $f(z) = u(z) + iv(z)$ be analytic and bounded in $\mathbb{D}$, and let γ_1 and γ_2 be two arcs lying in $\mathbb{D}$ and terminating at a point ξ of $\partial\mathbb{D}$. If $u(z) \to a$ as $z \to \xi$ on γ_1 and $v(z) \to b$ as $z \to \xi$ on γ_2, then $f(z)$ tends to $a + ib$ uniformly as $z \to \xi$ inside any angular domain of opening $\pi - \epsilon$ having vertex at ξ and bisected by the radius drawn to ξ.*

The present proof is due to Doob. By considering the function

$$g(z) = f(z) - a - ib\,,$$

we may assume that the real [resp. imaginary] part of $g(z)$ tends to 0 as $z \to \xi$ along γ_1 [resp. γ_2]; we may also assume that ξ is the point $z = 1$.

Suppose first that $\gamma_1 \cap \gamma_2$ is empty in a neighborhood of $z = 1$. Then, by a conformal mapping, the theorem can be reduced to the following. Suppose that $g(z)$ is analytic and bounded on $\mathbb{D}$, and that the Fatou boundary function $\operatorname{Re} g\left(e^{i\theta}\right)$ [resp. $\operatorname{Im} g\left(e^{i\theta}\right)$] of the real [resp. imaginary] part of g has limit 0 at $z = 1$ on approach from below [resp. above]. It is to be proved that g has radial limit 0 at $z = 1$.

Now the function $g_1(z) = \bar{g}(\bar{z})$ is analytic and bounded in $\mathbb{D}$, so that gg_1 is analytic and bounded in $\mathbb{D}$; furthermore, the real part of the Fatou boundary function has limit 0 at $z = 1$, which is a point of continuity. Therefore, by a well-known theorem of Fejér (see, for example, Titchmarsh [203, p. 414]), the Fourier series for gg_1 is summable $(C, 1)$ at $z = 1$ and its real part has radial limit 0 there; that is, $|g|^2$ has radial limit 0 at $z = 1$, as was to be proved.

If γ_1 and γ_2 intersect arbitrarily close to $z = 1$, we proceed as follows, assuming, as we may, that each of γ_1 and γ_2 is polygonal and does not intersect itself. These paths determine simple closed polygons. Let γ_i' be that part of $\gamma_i, i = 1, 2$, on one such polygon, and let z_1, z_2 be the end-points of γ_i. Let $|u - a|$ [resp. $|v - b|$] be bounded by c on γ_1' [resp. γ_2'], and let M be the supremum of $|g|$ in U.

The theorem will be proved by an application of Theorem 6.2.4 if we can exhibit a continuous path γ' in the polygon, from z_1 to z_2, on which $|g|$ is small with c, and, in fact, we shall obtain

$$|g|^2 \leqslant 2cM \tag{6.8}$$

on γ'.

We map conformally the interior of the polygon onto U in such a way that z_1 and z_2 go into ± 1, and we define γ' to be the inverse image of the diameter through ± 1. An argument based on the maximum principle for harmonic functions show that the transform of g satisfies (6.8) on the diameter through ± 1. Hence g satisfies (6.8) on γ', as was to be proved.

To prove this we observe writing $f(\xi), |\xi| \leqslant 1$, for the transform of $g(z)$ and $f_1 = \bar{f}(\bar{\xi})$, that $\operatorname{Re} f < c$ on $\partial\mathbb{D}, \operatorname{Im} \xi \leqslant 0$; $\operatorname{Im} f < c$ on $|\xi| = 1, \operatorname{Im} \xi \geqslant 0$; $|f| < M$ on $|\xi| \leqslant 1$. Hence

$$\operatorname{Re} ff_1 < 2cM \quad \text{on} \quad |\xi| = 1 \tag{6.9}$$

Now ff_1 is analytic so that $\operatorname{Re} ff_1$ is harmonic on the closed dise and hence satisfies (6.9) on the diameter through ± 1. But on this diameter

$$ff_1 = |f|^2 = \operatorname{Re} ff_1 < 2cM,$$

which proves our assertion. □

By combining Theorems 6.1.1 and 6.2.4, we obtain the following stronger form of Fatou's theorem, where the radial limits are replaced by angular limits.

Theorem 6.2.6. *Let $f(z)$ be analytic and bounded in $\mathbb{D}$. Then, for all $\xi \in \partial\mathbb{D}$, except for a possible set of linear measure zero, and for any $\epsilon > 0$, $f(z)$ tends uniformly to a limit as $z \to \xi$ inside an angular domain of opening $\pi - \epsilon$ having a vertex at ξ and bisected by the radius drawn to $e^{i\theta}$.*

6.3 Local Boundary Behavior of Bounded Holomorphic Functions of Several Complex Variables

Let $\Omega \subset\subset \mathbb{C}^n$ be a bounded domain with smooth boundary $\partial\Omega$, and let F be a bounded holomorphic function on Ω. A generalization of the classical theorem of Fatou says that the set E of points on $\partial\Omega$ at which F fails to have nontangential limits satisfies the condition $\sigma(E) = 0$, where σ denotes surface area measure. We show in the present Section that this result remains true when σ is replaced by 1-dimensional Lebesgue measure on certain smooth curves γ in $\partial\Omega$. The condition that γ must satisfy is that its tangents avoid certain directions.

6.3.1 The Domains under Consideration

To say that a bounded open set $\Omega \subset \mathbb{C}^n$ has C^k-boundary means that there is an open set $W \supset \partial\Omega$ and a k times continuously differentiable function $\rho : W \to R$ (i.e., $\rho \in C^k$) such that

$$\Omega \cap W = \{w \in W \mid \rho(w) < 0\}$$

and such that the vector

$$N(\zeta) = \left(\frac{\partial\rho}{\partial\bar{w}_1}(\zeta), \dots, \frac{\partial\rho}{\partial\bar{w}_n}(\zeta)\right)$$

is different from 0 at every $\zeta \in \partial\Omega$.

If $\rho \in C^2$ and if there is a constant $\beta > 0$ such that the inequality

$$\sum_{j,k=1}^{n} \frac{\partial^2\rho}{\partial w_j \partial\bar{w}_k}(w) z_j \bar{z}_k \geqq \beta|z|^2$$

holds for all $z \in \mathbb{C}^n$ and $w \in W$, then Ω is said to be strictly pseudoconvex. (As usual, $|z|^2 = \langle z, z\rangle^{1/2}$, where $\langle z, w\rangle = \sum z_j \bar{w}_j$ for $z \in \mathbb{C}^n, w \in \mathbb{C}^n$.)

6.3.1.1 Tangent Spaces

If Ω has C^1-boundary and $\zeta \in \partial\Omega$, the tangent space to $\partial\Omega$ at ζ is

$$T_\zeta = \{w \in \mathbb{C}^n \mid \mathrm{Re}\langle w, N(\zeta)\rangle = 0\} \tag{6.10}$$

Its maximal complex subspace is

$$T_\zeta^c = \{w \in \mathbb{C}^n \mid \langle w, N(\zeta)\rangle = 0\} \tag{6.11}$$

The directional condition mentioned in the opening paragraph is that for no $\zeta \in \gamma$ should the tangent to γ lie in T_ζ^c. To put this into a different form, let

$$\varphi : [0,1] \to \partial\Omega$$

be a C^1-parametrization of a curve γ in $\partial\Omega$, with $\varphi'(t) \neq 0$ for $0 \leqq t \leqq 1$. Then $\varphi'(t)$ is tangent to γ at $\varphi(t)$, and hence (6.10) shows that

$$\mathrm{Re}\,\langle \varphi'(t), N(\varphi(t))\rangle = 0 \quad (0 \leqq t \leqq 1) \tag{6.12}$$

According to (6.11), the tangent to γ at $\varphi(t)$ belongs to $T^c_{\varphi(t)}$ if and only if (6.12) is replaced by the stronger condition

$$\langle \varphi'(t), N(\varphi(t))\rangle = 0$$

6.3.1.2 Nontangential and Admissible Limits

If Ω has C^1-boundary and $\zeta \in \partial\Omega$, the unit outward normal at ζ is the vector

$$\nu(\zeta) = \frac{N(\zeta)}{|N(\zeta)|}$$

Following Stein [198] we let $\delta_\zeta(w)$ be the minimum of the distances from w to $\partial\Omega$ and from w to the affine tangent plane $\zeta + T_\zeta$.

For $\alpha > 0$, we define

$$\Gamma_\alpha(\zeta) = \{w \in \Omega \mid |w - \zeta| < (1+\alpha)\delta_\zeta(w)\}$$

and we let $A_\alpha(\zeta)$ be the set of all $w \in \Omega$ such that

$$A_\alpha(\zeta) = \left\{|\langle w - \zeta, \nu(\zeta)\rangle| < (1+\alpha)\delta_\zeta(w) \text{ and } |w - \zeta|^2 < \alpha\delta_\zeta(w)\right\}.$$

Since $|\,\mathrm{Re}\langle \zeta - w, \nu(\zeta)\rangle|$ is the distance from w to $\zeta + T_\zeta$, we see that, for a sufficiently small neighborhood V of $\zeta, V \cap \Gamma_\alpha(\zeta)$ lies in the cone

$$K_\alpha(\zeta) = \{w \in \mathbb{C}^n \mid |w - \zeta| < (1+\alpha)\,\mathrm{Re}\langle \zeta - w, \nu(\zeta)\rangle\}$$

and that $V \cap \Gamma_\alpha(\zeta) \supset V \cap K_\beta(\zeta)$ for some $\beta < \alpha$. Thus $\Gamma_\alpha(\zeta)$ is a nontangential approach region to ζ, and $A_\alpha(\zeta)$ is a so-called *admissible approach region* to ζ which contains

$$\Gamma_\alpha(\zeta) \cap \left\{ w \;\middle|\; |w - \zeta| < \frac{\alpha}{1+\alpha} \right\}$$

but which also contains sequences that approach ζ tangentially. (See [199, Chapter II]).

Definition 6.3.1. A function $f : \Omega \to \mathbb{C}$ is said to have a *nontangential limit* (resp. *admissible limit*) at $\zeta \in \partial\Omega$ if, for all $\alpha > 0, \lim f(w)$ exists as $w \to \zeta$ within $\Gamma_\alpha(\zeta)$ (resp. within $A_\alpha(\zeta)$).

We let $E_\Gamma(f)$ be the set of all $\zeta \in \partial\Omega$ at which f fails to have a nontangential limit, and we write $E_A(f)$ for the set where f fails to have an admissible limit. Obviously, $E_\Gamma(f) \subset E_A(f)$.

6.3.1.3 The Fatou Theorem of Korányi and Stein

This is the theorem (proved by Korányi for the ball [135] and by Stein in general [199]) that is formulated as follows:

Theorem 6.3.2. *If Ω has C^2-boundary and if $f \in H^\infty(\Omega)$ then $\sigma(E_A(f)) = 0$. Hence also $\sigma(E_\Gamma(f)) = 0$.*

(As usual, $H^\infty(\Omega)$ is the space of all bounded holomorphic functions $f : \Omega \to \mathbb{C}$, with sup-norm $\|f\|_\infty$.)

6.3.1.4 The Extension of the Fatou Theorem of Nagel and Rudin

If γ is a curve in $\partial\Omega$, parametrized by φ, we can define a measure μ on $\partial\Omega$ by setting

$$\int f d\mu = \int_0^1 f(\varphi(t))dt$$

for every continuous $f : \partial\Omega \to \mathbb{C}$. Then μ is supported by γ, and μ depends of course on the particular parametrization φ that is chosen. But the collection of sets of μ-measure 0 depends only on γ itself, and in this sense we may speak of a property holding almost everywhere on γ.

We recall that γ is said to belong to the class $\Lambda_{1+\alpha}$ if γ has a C^1-parametrization φ whose derivative φ' satisfies a uniform Lipschitz condition of order α; here $0 < \alpha < 1$.

Theorem 6.3.3. *Suppose Ω has C^1-boundary, γ is a curve in $\partial\Omega$, $\gamma \in \Lambda_{1+\alpha}$ for some $\alpha > 0$, and*

$$\langle \varphi'(t), N(\varphi(t)) \rangle \neq 0 \tag{6.13}$$

for every $t \in [0,1]$. Then $\mu(E_\Gamma(F)) = 0$ for every $F \in H^\infty(\Omega)$.

In other words, if the tangent to γ belongs nowhere to T_ζ^c (see Subsection 6.3.1.1) then every $F \in H^\infty(\Omega)$ has nontangential limits almost everywhere on γ.

Proof of Theorem 6.3.3. Referring to Subsection 6.3.1, we may of course assume that the gradient of ρ is bounded in W.

We are given $\varphi : [0,1] \to \partial\Omega$, $\varphi \in C^1$, $\varphi' \in \Lambda_\alpha$ for some $\alpha \in (0,1)$. Since (6.13) is assumed to hold; we may assume, without loss of generality, that there is a constant $\eta > 0$ such that

$$\operatorname{Im} \langle \varphi'(x), N(\varphi(x)) \rangle \geqq \eta > 0 \quad (0 \leqq x \leqq 1) \tag{6.14}$$

The proof proceeds in several steps. We extend φ to a map Φ of a rectangle Q into Ω, in such a way that each point $\varphi(x) \in \gamma$ is an end point of a nontangential curve ψ_x lying in $\Phi(Q)$. We then show that $F \circ \Phi$ and Q satisfy the hypotheses of Theorem 6.1.3, and that F therefore has limits along almost all of the curves ψ_i. The desired conclusion follows then from Chirka's extension of Lindelöf's theorem to n variables.

Step 1. The map Φ. Extend φ' to be a ($\mathbb{C}^n$-valued) function on $\mathbb{R}$, with compact support, of class Λ_α. Let $u(x,y)$ be the Poisson integral of φ', for $y \geqq 0$, and define

$$\Phi(x+iy) = \varphi(x) + iyu(x,y) \quad (0 \leqq x \leqq 1, y \geqq 0) \tag{6.15}$$

Since $(\partial\Phi/\partial y)(x) = iu(x,0) = i\varphi'(x)$, we have, by (6.14),

$$\begin{aligned}\left[\frac{\partial}{\partial y}(\rho \circ \Phi)\right](x) &= 2\,\mathrm{Re}\sum_{j=1}^{n} \frac{\partial \rho}{\partial w_j}(\Phi(x))\frac{\partial \Phi_j}{\partial y}(x) \\ &= -2\,\mathrm{Im}\,\langle \varphi'(x), N(\varphi(x))\rangle \\ &\leqq -2\eta\end{aligned}$$

If $h > 0$ is small enough; it follows that Φ maps the rectangle $Q = (0,1)\times(0,h)$ into $W \cap \Omega$, and that $(\partial/\partial y)(\rho \circ \Phi)(x+iy) \leqq -\eta$, hence

$$\rho(\Phi(x+iy)) \leqq -\eta y \quad (x+iy \in Q). \tag{6.16}$$

Standard estimates of the Poisson integral show that $|y(\partial u/\partial x)|$, $|y(\partial u/\partial y)|$, and $|u(x,y) - \varphi'(x)|$ are dominated by Cy^α, where C depends on α and on the Lipschitz constant of φ'. Hence differentiation of (6.15) yields

$$\left|\frac{\partial \Phi}{\partial \bar{z}}(x+iy)\right| \leqq C_1 y^\alpha \quad (0 \leqq x \leqq 1, y > 0) \tag{6.17}$$

Step 2. The curves ψ_x. For $0 \leqq x \leqq 1, y \geqq 0$, define

$$\psi_x(y) = \Phi(x+iy)$$

We claim that $\psi_x(y)$ tends nontangentially to $\psi_x(0) = \varphi(x) \in \gamma$ when $y \searrow 0$.
Setting $\zeta = \varphi(x)$, $w = \psi_x(y)$, (6.15) shows that

$$\zeta - w = -iyu(x,y)$$

Thus $|\zeta - w| \leqq cy$ and, by (6.14)

$$\mathrm{Re}\langle \zeta - w, N(\zeta)\rangle = y\,\mathrm{Im}\langle u(x,y), N(\varphi(x))\rangle \geqq \frac{1}{2}y\eta \geqq (2c)^{-1}\eta|\zeta - w|$$

as soon as y is small enough. Thus $\psi_x(y)$ lies in some cone $\mathfrak{D}_\alpha^\epsilon(\zeta)$ (see Subsection 6.3.1.2) for all sufficiently small y.

Step 3. Now let $F \in H^\infty(\Omega)$. Define $f : Q \to \mathbb{C}$ by $f(z) = F(\Phi(z))$. Fix $z \in Q$, for the moment. Then $w = \Phi(z)$ is the center of a ball in Ω whose radius is at least $|\rho(w)|/C_2$, where C_2 is an upper bound for the gradient of ρ in W. The one-variable Schwarz lemma, applied to restrictions of F to complex lines through w, shows therefore that

$$\left|\frac{\partial F}{\partial w_j}(w)\right| \leqq C_2|\rho(w)|^{-1}\|F\|_\infty \leqq C_2\|F\|_\infty \eta^{-1}y^{-1} \tag{6.18}$$

by (6.16), since $w = \Phi(x+iy)$. We now conclude from (6.17), (6.18), and the formula

$$\frac{\partial f}{\partial \bar{z}}(z) = \sum_{j=1}^{n} \frac{\partial F}{\partial w_j}(\Phi(z))\frac{\partial \Phi_j}{\partial \bar{z}}(z) \tag{6.19}$$

that $|(\partial f/\partial \bar{z})(x+iy)| \leqq C_3 y^{\alpha-1}$, so that $f \in L^p(Q)$ for some $p > 1$. [Observe that (6.19) depends on the fact that F is holomorphic.] The other hypotheses of Theorem 6.1.3 are obviously satisfied.

It follows that $\lim f(x+iy)$ exists, as $y \searrow 0$, for every x in a set $E \subset (0,1)$ whose complement has measure 0. In other words, F has a limit along the nontangential curve ψ_x that ends at $\varphi(x)$, for every $x \in E$. Since $F \in H^\infty(\Omega)$, Chirka's Lindelöf theorem ([206, Theorem 1]) asserts that F has a nontangential limit at $\varphi(x) \in \gamma$, for every $x \in E$. This completes the proof of Theorem 6.3.3. □

6.3.2 Review of Invariant Metrics

We begin with a quick review of the two most important invariant metrics for this discussion. In point of fact, we shall only explicitly use the Kobayashi/Royden metric in our applications in Section 6.3.10 to Fatou theorems. But we also present the Carathéodory metric just to round out the picture, and because all constructions using the Kobayashi/Royden metric can be made as well using the Carathéodory metric.

Historically the first of the metrics considered here is the one introduced by Constantin Caratheodory [38] in 1927.

If Ω_1, Ω_2 are domains in $\mathbb{C}^n$ then, we define $\Omega_1\,(\Omega_2)$ to be the collection of all mappings from Ω_2 to Ω_1.

Definition 6.3.4. If $\Omega \subseteq \mathbb{C}^n$ is open, then the infinitesimal Carathéodory metric is given by $F_C : \Omega \times \mathbb{C}^n \to \mathbb{R}$ where

$$F_C(z,\xi) = \sup_{\substack{f\in B(\Omega)\\ f(z)=0}} |f_*(z)\xi| \equiv \sup_{\substack{f\in \in(\Omega)\\ f(z)=0}} \left|\sum_{j=1}^{n} \frac{\partial f}{\partial z_j}(z)\cdot \xi_j\right|$$

Remark 6.3.5. In this definition, the $\mathbb{C}^n$ in $\Omega\times\mathbb{C}^n$ should be thought of as the tangent space to Ω at z-in other words, $\Omega\times\mathbb{C}^n$ should be canonically identified with the tangent bundle of Ω at z. We think of $F_C(z,\xi)$ as the length of the tangent vector ξ at the point $z \in \Omega$. In general, $F_C(z,\xi)$ is not given by a quadratic form $(g_{ij}(z))$, hence F_C is not a Riemannian metric.

Definition 6.3.6. Let $\Omega \subseteq \mathbb{C}^n$ be open and $\gamma : [0,1] \to \Omega$ a C^1 curve. The Carathéodory length of γ is defined to be

$$L_C(\gamma) = \int_0^1 F_C(\gamma(t), \gamma'(t))\, dt$$

This definition parallels the definition of the length of a curve in a Riemannian metric. It would be natural at this point to define the (integrated) Carathéodory distance between two points to be the infimum of lengths of all curves connecting them. One advantage of defining distance in this fashion is that it is then straightforward to verify the triangle inequality.

However, we shall not take this approach. One of the most important features of the Carathéodory metric is that it is, in a precise sense, the least metric under which holomorphic mappings are distance decreasing. The notion of distance suggested in the last paragraph is not the smallest. It is fortuitous that the following definition does result in the least distance-decreasing metric-in particular, it does satisfy the triangle inequality.

Definition 6.3.7. Let $\Omega \subseteq \mathbb{C}^n$ be an open set and $z, w \in \Omega$. The Carathéodory distance between z and w is defined to be

$$C(z,w) = \sup_{f \in B(\Omega)} \rho(f(z), f(w))$$

where ρ is the Poincare-Bergman distance on $\mathbb{B}$.

Notice that, depending on the domain Ω, the Carathéodory distance between two distinct points maybe 0.

The metric defined below, introduced by Kobayashi [121] in 1967, is in a some sense dual to the Caratheodory metric — it is defined not by maps from the domain into the disk but by maps from the disk into the domain.

Definition 6.3.8. Let $\Omega \subseteq \mathbb{C}^n$ be open. Let $e_1 = (1, 0, \ldots, 0) \in \mathbb{C}^n$. The infinitesimal form of the Kobayashi/Royden metric is given by $F_K \colon \Omega \times \mathbb{C}^n \to \mathbb{R}$, where

$$\begin{aligned} &F_K(z,\xi) \\ &\equiv \inf\left\{\alpha \,\middle|\, \alpha > 0 \text{ and } \exists f \in \Omega(B) \text{ with } f(0) = z, (f'(0))\,(e_1) = \frac{\xi}{\alpha}\right\} \\ &= \inf\left\{\frac{|\xi|}{|(f'(0))\,(e_1)|} \,\middle|\, f \in \Omega(B), (f'(0))\,(e_1) \text{ is aconstant multiple of } \xi\right\} \\ &= \frac{|\xi|}{\sup\{|(f'(0))\,(e_1)| \mid f \in \Omega(B), (f'(0))\,(e_1) \text{ is a constant multiple of } \xi\}} \end{aligned}$$

We now wish to define an integrated distance based on elements of $\Omega(B)$. The natural analogue for our definition of Carathéodory distance does not satisfy a triangle inequality. Moreover, we want the Kobayashi/Royden distance to be the greatest metric under which holomorphic mappings are distance decreasing. Therefore we proceed as follows:

Definition 6.3.9. Let $\Omega \subseteq \mathbb{C}^n$ be open and $\gamma : [0,1] \to \Omega$ a piecewise C^1 curve. The Kobayashi/Royden length of γ is defined to be

$$L_K(\gamma) = \int_0^1 F_K(\gamma(t), \gamma'(t))\, dt$$

Definition 6.3.10. Let $\Omega \subseteq \mathbb{C}^n$ be an open set and $z, w \in \Omega$. The (integrated) Kobayashi/Royden distance between z and w is defined to be

$$K(z, w) = \inf \left\{ L_K(\gamma) \;\middle|\; \gamma \text{ is a piecewise } C^1 \text{ curve connecting } z \text{ and } w \right\}$$

Recall that we did not implement a definition like this one for the (integrated) Carathéodory distance because we were able to find a smaller distance that satisfied the triangle inequality. Now we are at the other end of the spectrum: we want the Kobayashi/Royden metric to be as large as possible and to satisfy a triangle inequality as well. This definition serves that dual purpose.

We remark in passing that the use of the ball as a model domain when defining the Carathéodory and Kobayashi/Royden metrics is important, but that choice is not unique. The theory is equally successful if either the disc or the polydisc is used. However, in the current state of the theory, it is essential that the model domain have a transitive group of automorphisms.

Proposition 6.3.11 (The Distance Decreasing Properties). *If Ω_1, Ω_2 are domains in $\mathbb{C}^n$, $z, w \in \Omega_1, \xi \in \mathbb{C}^n$, and if $f : \Omega_1 \to \Omega_2$ is holomorphic, then*

$$F_C^{\Omega_1}(z, \xi) \geq F_C^{\Omega_2}(f(z), f_*(z)\xi) \qquad F_K^{\Omega_1}(z, \xi) \geq F_K^{\Omega_2}(f(z), f_*(z)\xi)$$
$$C_{\Omega_1}(z, w) \geq C_{\Omega_2}(f(z), f(w)) \qquad K_{\Omega_1}(z, w) \geq K_{\Omega_2}(f(z), f(w))$$

Proof. This is followed by an inspection from the definitions. □

Corollary 6.3.12. *If $f : \Omega_1 \to \Omega_2$ is biholomorphic then f is an isometry in both the Carathéodory and the Kobayashi/Royden metrics.*

Proof. Apply the proposition to both f and f^{-1}. □

Corollary 6.3.13. *If $\Omega_1 \subseteq \Omega_2 \subseteq \mathbb{C}^n$ then for any $z, w \in \Omega_1$, any $\xi \in \mathbb{C}^n$, we have*

$$F_C^{\Omega_1}(z, \xi) \geq F_C^{\Omega_2}(z, \xi) \qquad F_K^{\Omega_1}(z, \xi) \geq F_K^{\Omega_2}(z, \xi)$$
$$C_{\Omega_1}(z, w) \geq C_{\Omega_2}(z, w) \qquad K_{\Omega_1}(z, w) \geq K_{\Omega_2}(z, w)$$

The geometrical applications of the Kobayashy metric rely on the fact that it is invariant under biholomorphic maps and does not increase under arbitrary holomorphic maps.

In many questions of complex geometry, it is of importance to know the boundary behavior of the Carathéodory metric. Let $\Omega \subset \mathbb{C}^n$ be a bounded domain and $\delta(z, \partial\Omega)$ the Euclidean distance from the point z to the boundary of the domain. Applying Schwarz's lemma to the restriction of the map $f : \Omega \to \mathbb{D}$ to the complex line passing through the point $z \in \Omega$ in the direction of the vector $v \in \mathbb{C}^n$, it is not difficult to get the estimate

$$C_\Omega(z, v) \leqq \frac{|v|}{\delta(z, \partial\Omega)}.$$

6.3.3 Normal Functions

Recall (see [63]) that a normal function has at least two equivalent definitions.

Definition 6.3.14. Let $f : \mathbb{B} \to \overline{\mathbb{C}}$ be holomorphic (here $\overline{\mathbb{C}}$ is the Riemann sphere). We say that f is normal if, whenever $\{\varphi_j\}$ are biholomorphic self-maps of $\mathbb{B}$ then $\{f \circ \varphi_j\}$ is a normal family.

In the book [63] an important equivalent formulation was derived using ideas from invariant geometry.

Definition 6.3.15. A holomorphic function $f : \Omega \to \overline{\mathbb{C}}$ is normal if the derivative ∇f is bounded from the Kobayashi metric on Ω (equivalently, the Poincaré–Bergman metric on Ω) to the spherical metric on $\overline{\mathbb{C}}$.

Geometrically, this definition means that the spherical derivatives of $f(z)$ in various directions can be estimated in terms of holomorphic discs in Ω passing through the point z in these directions. In particular, in strictly pseudoconvex domains this means that the spherical derivatives along complex normals grow like $O(1/\delta(z))$ and along complex tangential directions, like $O(1\sqrt{\delta(z)})$.

The equivalence of these two definitions in the case Ω is the unit ball $\mathbb{B} := \{z \in \mathbb{C}^n \mid |z| < 1\}$ is a sophisticated exercise with Marty's theorem.

As a result, $f : \mathbb{B} \to \overline{\mathbb{C}}$ is normal if and only if

$$\frac{\left|\sum_{j=1}^{n} \left(\partial f/\partial z_j\right)(z) v_j\right|^2}{\left(1 + |f(z)|^2\right)^2} \leqslant M \sum_{i,j=1}^{n} \frac{\left(1 - \|z\|^2\right)\delta_{ij} + \bar{z}_i z_j}{\left(1 - \|z\|^2\right)^2} v_i \bar{v}_j$$

for all $z \in \mathbb{B}$, $v \in \mathbb{C}^n$.

It is easy to see that this is a multi-dimensional variant of the normality condition for the disc.

6.3.4 The Lindelöf Principle in Several Complex Variables

The Möbius transformations

$$\tau(z) = \frac{az+b}{cz+d}, \quad a,b,c,d \in \mathbb{C}, \quad ad-bc \neq 0$$

are the only conformal maps of $\bar{\mathbb{C}}$. They form a group with respect to composition that we denote by Möb. Möbius transformations map circles in a $\bar{\mathbb{C}}$ onto circles in $\bar{\mathbb{C}}$ where a line is considered as a circle through ∞.

The conformal maps of $\mathbb{D}$ onto $\mathbb{D}$ are of the form

$$\tau(z) = \frac{cz+z_0}{1+\bar{z}_0 cz}, \quad |z_0|<1, \quad |c|=1$$

They form a subgroup Möb($\mathbb{D}$) so that Möb($\mathbb{D}$) $= \{\tau \in \text{Möb} \mid \tau(\mathbb{D}) = \mathbb{D}\}$. It is easy to see that

$$\left(1-|z|^2\right)|\tau'(z)| = 1-|\tau(z)|^2 \quad \text{for} \quad z \in \mathbb{D}, \quad \tau \in \text{Möb}(\mathbb{D}) \tag{6.20}$$

The Poincaré metric (hyperbolic metric) is defined by

$$\lambda(z_1,z_2) = \lambda_{\mathbb{D}}(z_1,z_2) = \min_C \int_C \frac{|dz|}{1-|z|^2} \quad \text{for} \quad z_1,z_2 \in \mathbb{D} \tag{6.21}$$

where the minimum is taken over all curves C in $\mathbb{D}$ from z_1 to z_2; it is attained for the Poincaré segment S from z_1 to z_2, that is the arc of the circle through z_1 and z_2 orthogonal to $\partial\mathbb{D}$. A consequence of (6.21) is the triangle inequality

$$\lambda(z_1,z_3) \leq \lambda(z_1,z_2) + \lambda(z_2,z_3) \quad \text{for} \quad z_1,z_2,z_3 \in \mathbb{D}$$

and, together with (6.20), the invariance property

$$\lambda(\tau(z_1),\tau(z_2)) = \lambda(z_1,z_2) \quad \text{for} \quad \tau \in \text{Möb}(\mathbb{D})$$

Transforming one point to 0 we easily see that

$$\lambda(z_1,z_2) = \frac{1}{2}\log\frac{1+|z_1-z_2|/|1-\bar{z}_1z_2|}{1-|z_1-z_2|/|1-\bar{z}_1z_2|} \quad \text{for} \quad z_1,z_2 \in \mathbb{D} \tag{6.22}$$

It follows that

$$\left|\frac{z_1-z_2}{1-\bar{z}_1z_2}\right| = \frac{e^{2\lambda}-1}{e^{2\lambda}+1} = \tanh\lambda$$

Often the Poincaré metric is defined with a factor 2 in (6.22).

A *Stolz angle* at $\zeta \in \partial\mathbb{D}$ is of the form

$$\Delta(\zeta) = \left\{z \in \mathbb{D} \mid |\arg(1-\bar{\zeta}z)| < \alpha, |z-\zeta| < \rho\right\} \quad \left(0<\alpha<\frac{\pi}{2}, \rho < 2\cos\alpha\right)$$

The exact shape is not relevant. The important fact is that all points of Δ have bounded Poincaré distance from the radius $[0,\zeta]$.

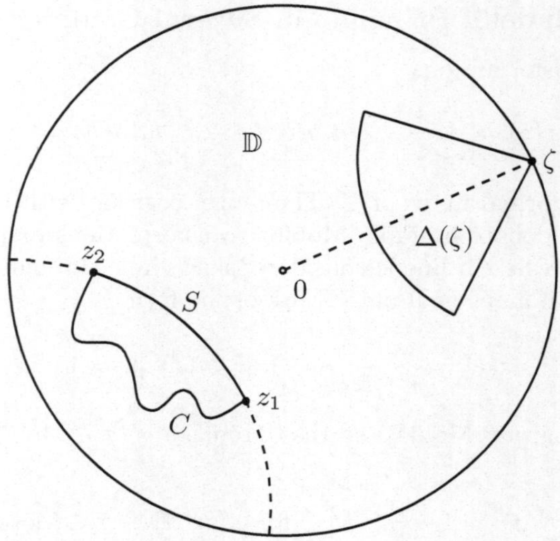

FIGURE 6.2: The non-euclidean segment S between $z_1, z_2 \in \mathbb{D}$ and a Stolz angle Δ at $\zeta \in \partial\mathbb{D}$

Definition 6.3.16. Now let f be a function from $\mathbb{D}$ to $\bar{\mathbb{C}}$. We say that f has the *angular limit* $A \in \bar{\mathbb{C}}$ at $\zeta \in \partial\mathbb{D}$ if

$$f(z) \to A \quad \text{as} \quad z \to \zeta, \quad z \in \Delta$$

for each Stolz angle Δ at ζ; the opening angle 2α of Δ can be any number $< \pi$.

The most classical formulation of the Lindelöf principle on the disc (see [154]) is as follows:

Theorem 6.3.17. *Let f be a bounded, holomorphic function on the unit disc $\mathbb{D} \subseteq \mathbb{C}$. Suppose that the radial limit*

$$\lim_{r \to 1^-} f\left(re^{i\theta}\right) \equiv \lambda \in \mathbb{C}$$

of f exists at the boundary point $e^{i\theta}$. Then in fact f has an angular limit λ at $e^{i\theta}$.

Proof. Fix $\alpha > 1$. We may assume $\lambda = 0$ and $\theta = 0$. All points of $\mathbb{D}$ have bounded Poincaré distance from the radius $[0, \zeta]$, say β. For $a \in (0,1)$ and $\beta \in (0, \infty)$ and let

$$d(a, \beta) = \{z \in \mathbb{D} \mid \varrho(a, z) < \beta\}$$

denote the disc in Poincaré metric ϱ with center a and radius β. Also, the biholomorphic map $\phi_a(z) = \frac{z-a}{1-\bar{a}z}$ send $d(a, \beta)$ into the disc $\mathbb{D}(0, r)$, where

$r = (e^\beta - 1)/(e^\beta + 1) < 1$. Set $R = (e^{2\beta} - 1)/(e^{2\beta} + 1)$ then

$$\bigcup_{a\in(0,1)} \phi_a^{-1}(\mathbb{D}(0,R)) \supseteq \triangle(1)$$

By Montel's theorem, the functions $\{f \circ \phi_a^{-1}\}$ form a normal family on $\mathbb{D}$. Choose a subsequence $\{f \circ \phi_{a_j}^{-1}\}$, $a_j \to 1$ as $j \to \infty$, that converges uniformly on $\mathbb{D}(0,R)$. Call the limit function f_0. It is easy to see that open interval $\phi_a((0,R)) \subset (0,1)$. By hypothesis, $f_0(x + i0) = 0$, all $x + i0 \in \mathbb{D}(0,R)$. Thus $f_0 \equiv 0$ uniformly on $\mathbb{D}(0,R)$. Hence

$$\lim_{\triangle(1)\ni z\to 1} f(z) = 0$$

This completes the proof of Theorem 6.3.17. □

Remark 6.3.18. The straightforward proof of this theorem was given in [139, Theorem 8.7.1, p. 362]. There

$$\Omega_j = \left\{ z \in \mathbb{C} \;\middle|\; \frac{1 - 2^{-j+1}}{2\alpha} \le \operatorname{Re} z \le \frac{1 - 2^{-j-1}}{2\alpha}, |\operatorname{Im} z| < 2^{-j+1} \right\}$$

and the function map $\phi_j : \Omega_j \to \Omega_{j_0}$ is given by $\phi_j(z) = 2^{j-j_0}(z-1) + 1$.

For $\zeta \in \partial\mathbb{D}$ and $\alpha > 1$, we define the cone

$$\Gamma_\alpha(\zeta) = \left\{ z \in \mathbb{C} \;\middle|\; |1 - z\bar{\zeta}| < \alpha(1 - |z|) \right\}.$$

The cone $\Gamma_\alpha(\zeta)$ is asymptotic to a Stolz angle with vertex ζ and angle $2\sec^{-1}(\alpha)$ that is symmetric about the radius $[0,\zeta]$. The cones $\Gamma_\alpha(\zeta)$ expand as α increases.

Now consider the domain which is the unit ball $\mathbb{B}$ in $\mathbb{C}^n$. Proceeding by analogy, we define, for $\alpha > 1$, and $P \in \partial\mathbb{B}$, new approach regions

$$\mathcal{A}_\alpha(P) = \left\{ z \in \mathbb{B} \;\middle|\; |1 - z \cdot \bar{P}| < \alpha(1 - |z|) \right\}$$

Here $z \cdot \bar{P} \equiv \sum_j z_j \bar{P}_j$.

Koranyi [135] was the first to write about these regions: for every $0 < \alpha < \infty$, he defined the admissible domain at $\zeta \in \partial\mathbb{B}$ by

$$A_\alpha(\zeta) = \left\{ z \in \mathbb{B} \;\middle|\; \frac{|\mathcal{S}(\zeta,z)|}{\mathcal{P}(\zeta,z)} < \left(\frac{1+\alpha}{2}\right)^n \right\},$$

where $\mathcal{S}$ and $\mathcal{P}$ are the Szegö and Poisson—Szegö kernels of $\mathbb{B}$.

The Szegő kernel for the unit ball $\mathbb{B}$ has the form

$$S(z,\zeta) = \frac{(n-1)!}{2\pi^n} \frac{1}{(1 - z \cdot \bar{\zeta})^n}$$

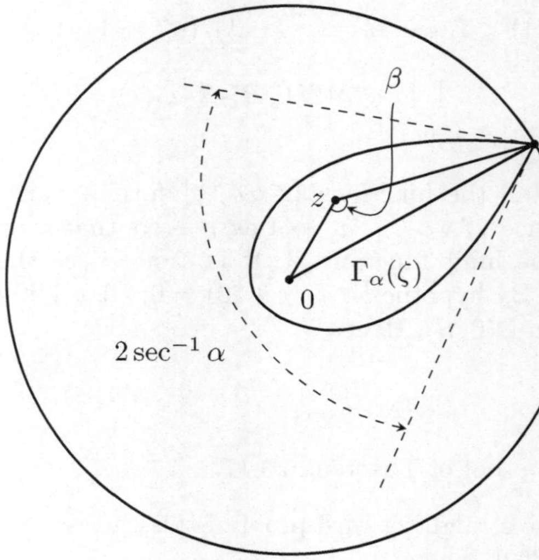

FIGURE 6.3: The cone $\Gamma_\alpha(\zeta)$

whereas the Poisson-Szegő kernel has the form

$$\mathcal{P}(z,\zeta) = \frac{(n-1)!}{2\pi^n} \frac{\left(1-|z|^2\right)^n}{|1-z\cdot\bar{\zeta}|^{2n}}$$

(see S.G. Krantz [139, Theorem 1.6.9, p. 69] for details on these matters). It is easy to see that $A_\alpha(\zeta) \equiv \mathcal{A}_{(1+\alpha)/2}(\zeta)$.

Next Korányi introduces the notion of admissible convergence of a function on $\mathbb{B}$, and proves that it is invariant under the group of holomorphic automorphisms of $\mathbb{B}$. The "admissible limit" was introduced in Korányi's article [135]: We say that F converges to f admissibly (a.e.) if, for every $\alpha > 0$, $\lim_{z\to\zeta, z\in A_\alpha(\zeta)} F(z) = f(z)$ for (almost) all $\zeta \in \partial\mathbb{B}$. Korányi proves that for $f \in L^p(\partial\mathbb{B})(p \geq 1)$ its Poisson integral converges to f admissibly a.e. This is a generalization of the classical Fatou theorem.

But Stein, in his 1972 book [199], defined an admissible region in a domain with C^2 smooth boundary. Now we will "calculate the shape" of $\mathcal{A}_\alpha(P)$. Fix the point $\mathbb{1} = (1,'0) \in \partial\mathbb{B}$. Then $\mathcal{A}_\alpha(\mathbb{1})$ is defined by the condition

$$|1 - z\cdot\bar{\mathbb{1}}| < \alpha(1-|z|)$$

More explicitly, this is

$$|1 - z_1| < \alpha(1-|z|) \tag{6.23}$$

Since $z \in \mathbb{B}$ we have

$$|z_1|^2 + |'z|^2 < 1$$

or

$$\begin{aligned}|'z|^2 &= 1 - |z_1|^2 \\ &= (1 + |z_1|)(1 - |z_1|) \\ &\leq 2|1 - z_1|\end{aligned}$$

Substituting (6.23) into this last inequality gives

$$|'z| < \sqrt{2\alpha}\sqrt{1 - |z_1|} \tag{6.24}$$

Thus the condition on $'z$ is now the square root of the condition on z_1.

A complex normal $N^c = \{(t,'0,) \mid t \in \mathbb{C}\}$ to $\partial\mathbb{B}$ through $\mathbb{1} = (1,0)$; the complex tangent plane to $\partial\mathbb{B}$ at the point $\mathbb{1}$ is defined by the equation $z_1 = 1$. A moment's thought reveals that the non-isotropic condition (6.24) forces $\mathcal{A}_\alpha$ to be conical in the z_1 direction but parabolic in the $'z$ direction.

It is natural to conjecture that, on the ball in $\mathbb{C}^n$, there is an analogous theorem with nontangential convergence replaced by admissible convergence. This is, unfortunately, false.

Example 6.3.19. Let $\mathbb{B}$ be the unit ball in $\mathbb{C}^2$. Define

$$f(z_1, z_2) = \frac{z_2^2}{1 - z_1}$$

The simple inequalities

$$|z_2|^2 < 1 - |z_1|^2 = (1 - |z_1|)(1 + |z_1|) < 2(1 - |z_1|) \leqslant 2|1 - z_1|$$

show that f is bounded.

Let $P = (1,0) \equiv \mathbb{1} \in \partial\mathbb{B}$. Notice that

$$\lim_{r \to 1^-} f(rP) = \lim_{r \to 1^-} 0 = 0$$

But if $p^j = (1 - 1/j, 1/\sqrt{j})$ for $j = 4, 5, \ldots$, then

$$f(p^j) = \frac{1/j}{1/j} = 1$$

Of course, the points p^j all lie in the admissible approach region $\mathcal{A}_2(\mathbb{1})$.

So we see that the admissible limit of f at P is 1, while the radial limit is 0. We conclude that there is no Lindelöf principle for admissible convergence.

6.3.5 Hypoadmissible Domains

Consider a bounded domain $\Omega = \{z \in \mathbb{C}^n \mid \varphi(z) < 0\}$, where $\varphi \in C^2(\bar{\Omega})$ with gradient $\nabla_z\varphi = (\partial\varphi/\partial z_1, \ldots, \partial\varphi/\partial z_n) \neq 0$ on $\partial\Omega = S$; then at a point $\xi \in S$ the unit vector $\nu_\xi = \bar{\nabla}\bar{\varphi}(\xi)/|\nabla\varphi|(\xi)$ is in the direction of the normal, while

$\tau = iv$ belongs to the usual (real) tangent plane $T_\xi(S)$ and is orthogonal to the complex tangent plane $T^c_\xi(S) = T_\xi(S) \cap iT_\xi(S)$. If v is a vector in $\mathbb{C}^n$ then $\nabla_z\varphi(\xi) \cdot v = \sum_{i=1}^n (\partial\varphi/\partial z_i)(\xi)\, v_i = v_N(\xi)$ is the (complex) normal component of v at ξ.

Denote by $\delta_\zeta(z) = |\mathrm{Re}\,(z-\zeta, \nu_\zeta)|$ the distance of the point z to the real tangent plane $T_\zeta(\partial\Omega)$, and for fixed α, $k > 0$ and $\varepsilon, 0 < \varepsilon < 1$, we consider the domain

$$\mathfrak{D}^\epsilon_\alpha = \left\{ z \in \Omega^-_a \;\middle|\; |(z-\zeta,\nu_\zeta)| < (1+\alpha)\delta_\zeta(z), |z-\zeta|^2 < k\delta_\zeta^{1+\varepsilon}(z) \right\}$$

In the direction T^c_ζ this domain is tangent to $\partial\Omega$, and in the direction T the domain forms an of acute angle with it (Figure 6.4). In fact, if z belongs to the hyperplane I on Figure 6.4, passing through ν_ζ and T^c_ζ, then $|(z-\zeta,\nu_\zeta)| = \delta_\zeta(z)$, and the first condition in the definition of $\mathfrak{D}^\epsilon_\alpha$ holds automatically, while the second gives $\xi^2 + \eta^2 < k\eta^{1+\varepsilon}$ (the meaning of ξ and η is clear from the figure); the intersection $\partial\mathfrak{D}^\epsilon_\alpha \cap \mathrm{I}$ is the surface $\xi^2 + \eta^2 = k\eta^{1+\varepsilon}$,

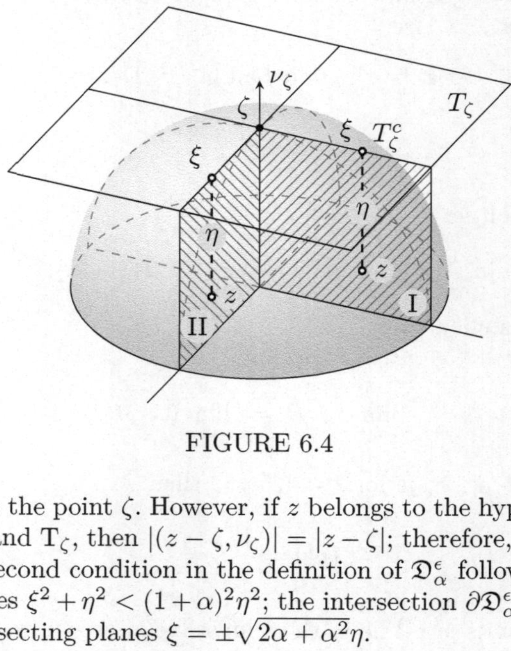

FIGURE 6.4

tangent to T^c_ζ at the point ζ. However, if z belongs to the hyperplane II, passing through ν_ζ and T_ζ, then $|(z-\zeta,\nu_\zeta)| = |z-\zeta|$; therefore, for z sufficiently close to ζ, the second condition in the definition of $\mathfrak{D}^\epsilon_\alpha$ follows from the first, and the first gives $\xi^2 + \eta^2 < (1+\alpha)^2\eta^2$; the intersection $\partial\mathfrak{D}^\epsilon_\alpha \cap$ II close to ζ is the pair of intersecting planes $\xi = \pm\sqrt{2\alpha+\alpha^2}\eta$.

Definition 6.3.20. If f is a complex valued function defined in Ω, we will say that f has an "*hypoadmissible limit*" at $\zeta \in \partial\Omega$ if

$$\lim_{\substack{z\to\zeta \\ z\in\mathfrak{D}^\epsilon_a(\zeta)}} f(z) \text{ exists for every } \alpha > 0$$

In other words, $z_j \to \xi$ hypoadmissibly means that the smallest approach region containing $\{z_j\}$ is nontangential in the complex normal direction and is asymptotically smaller than any admissible approach region in the complex tangential directions.

The first result in this direction is the extension of Lindelöf theorem to several complex variables to the effect that each normal function, which has a radial limit as $\zeta \to a \in \partial\mathbb{D}$, also has the same limit as $\zeta \to a$ along each path nontangential to $\partial\mathbb{D}$. For $n > 1$ it turns out that one can allow a tangential approach in certain directions but not others.

Now we have the following.

Theorem 6.3.21. *If the function f is holomorphic and normal in the unit ball $\mathbb{B} \subset \mathbb{C}^n$ and if it has a limit A as $z \to \zeta \in \partial\mathbb{B}$ in the direction of the real normal (i.e., the line $z = \zeta + t\nu_\zeta, t \in \mathbb{R}$), then it has the same limit as $z \to \zeta$ over points $z \in \mathfrak{D}_\alpha^\epsilon(\zeta)$.*

Proof. Without loss of generality, we assume that $\zeta = \mathbb{1} = (1,'0)$; since here $\varphi(z) = |z|^2 - 1$, then $\nabla\varphi = \bar{z}$ and $\nu_\zeta = \zeta$. The complex tangent plane to ∂B at the point ζ is denoted by T^c (for our choice of ζ it is defined by the equation $z_n = 1$), and by $T_\lambda^c = \left\{('z, 1-\lambda) \mid 'z \in \mathbb{C}^{n-1}\right\}$ a plane parallel to T^c.

We clarify the form of the intersection $\mathbb{B}_\lambda = T_\lambda^c \cap \mathbb{B}$. For this we pass a complex normal $N^c = \{(t, \ '0) \mid t \in \mathbb{C}\}$ to $\partial\mathbb{B}$ through ζ; its point of intersection with T_λ^c will be $c_\lambda = (1-\lambda,'0)$. The intersection $T_\lambda^c \cap \partial\mathbb{B}$ is defined by the conditions $|1-\lambda|^2 + |'z|^2 = 1$, $z_1 = 1 - \lambda$, the first of which is rewritten in the form $|'z|^2 = 2\operatorname{Re}\lambda - |\lambda|^2$. Thus, $\mathbb{B}_\lambda$ is a ball in the plane T_λ^c with center c_λ and radius $R_\lambda = \sqrt{2\operatorname{Re}\lambda - |\lambda|^2}$:

$$B_\lambda = \{z \in T_\lambda^c \mid |z - c_\lambda| < R_\lambda\} .$$

Now we clarify the form of the intersection $\mathbb{B}'_\lambda = T_\lambda^c \cap \mathfrak{D}_a^\epsilon(\mathbb{1})$. In the case of unit ball the set $\mathfrak{D}_a^\epsilon(\mathbb{1})$ is

$$\left\{z \in B \,\middle|\, |z_1 - 1|^2 + |'z|^2 < k\,|\operatorname{Re}(z_1 - 1)|^{1+\varepsilon}, \right.$$
$$\left. |z_1 - 1| < (1+\alpha)\,|\operatorname{Re}(z_1 - 1)| \right\}$$

Since $z_n - 1 = -\lambda$ on T_λ^c, then $\mathbb{B}'_\lambda$ is defined by the inequalities $|\lambda| < (1+\alpha)|\operatorname{Re}\lambda|$, $|'z|^2 < k|\operatorname{Re}\lambda|^{1+\varepsilon} - |\lambda|^2$, i.e., it is also a ball in the plane T_λ^c with the same center c_λ and radius $R'_\lambda = \sqrt{k|\operatorname{Re}\lambda|^{1+\varepsilon} - |\lambda|^2}$:

$$B'_\lambda = \{z \in T_\lambda^c \mid |z - c_\lambda| < R'_\lambda, \quad |\lambda| < (1+\alpha)|\operatorname{Re}\lambda|\} .$$

The ratio of the squares of the radii of the balls $\mathbb{B}'_\lambda$ and $\mathbb{B}_\lambda$ is

$$\left(\frac{R'_\lambda}{R_\lambda}\right)^2 = \frac{k|\operatorname{Re}\lambda|^{1+\varepsilon} - |\lambda|^2}{2\operatorname{Re}\lambda - |\lambda|^2} \le \frac{k|\operatorname{Re}\lambda|^\varepsilon}{2 - |\lambda||\lambda/\operatorname{Re}\lambda|}$$

and since we have $|\lambda/\operatorname{Re}\lambda| < 1 + \alpha$, then $R'_\lambda/R_\lambda \to 0$ as $\lambda \to 0$.

Notice that if f is normal and then it follows that

$$\operatorname{dist}_{\text{ph}}(f(c_\lambda), f(c_\lambda, c_\lambda +' z)) < M \cdot K_{B_\lambda}(c_\lambda, c_\lambda + 'z)$$

as we see from the definition of the integrated distance.

The Kobayashi metric for the unit ball is a constant multiple of the Bergman metric, for which there is an explicit formula

$$b_{\mathbb{B}}(0, z) = \sqrt{1+n} \ln \frac{1+|z|}{1-|z|} \qquad (z \in \mathbb{B}).$$

It is obtained if we assume (without loss of generality) that $z = (z_1, 0, \ldots, 0)$ and integrate

$$ds^2 = (n+1)\left\{\frac{|dz|^2}{1-|z|^2} + \sum_{\mu,\nu=1}^{n} \frac{\bar{z}_\mu z_\nu dz_\mu d\bar{z}_\nu}{(1-|z|^2)^2}\right\}$$

along a line segment.

Since $\ln \frac{1+x}{1-x}$ is an increasing function of x for $0 \leq x < 1$ we have

$$\operatorname{dist}_{\text{sph}}(f(c_\lambda), f(c_\lambda, c_\lambda + 'z)) < M \cdot \ln \frac{1 + \frac{R'_\lambda}{R_\lambda}}{1 - \frac{R'_\lambda}{R_\lambda}} \quad ('z \in B'_\lambda).$$

As $\lambda \to 0$ the right side of the previous inequality tends to 0. Since $f(c_\lambda) \to A$ it follows that f has admissible limit in $\mathbb{1}$. □

Corollary 6.3.22. *If the function f is a normal function in the domain $\Omega \subset \mathbb{C}^n$ with C^2 smooth boundary and has a limit as $z \to \xi \in \partial\Omega$ in the direction of the normal v, then it has the same limit as $z \to \xi$ through the points of $\mathfrak{D}^\epsilon_\alpha(\xi)$.*

Proof. Using the fact that the boundary Ω is a doubly smooth compact hypersurface, we have the number $r > 0$ possessing the following property: the inclusion

$$B_\xi(r) \subset \Omega$$

where $\mathbb{B}_\xi(r)$ is a ball of radius r, whose boundary contains the point ξ is valid for any polnt $\xi \in \partial\Omega$.

The proof is based on the remark that $\mathfrak{D}^\epsilon_\alpha(\xi)$ is a hypoadmissible domain of $\mathbb{B}_\xi(r)$ at the point ξ. □

Remark 6.3.23. It is a mistake, in our opinion, to define the admissible approach regions with respect to the boundary. Because, as Stein observed, there is trouble at non-convex points. Much better to define the admissible approach regions with respect to the normal vector.

It will be convenient to introduce a slightly modified family of approaches regions which also will be useful in the rest of the Chapter.

If $z = (z_1, \ldots, z_n) \in \mathbb{C}^n$, we write $z_k = x_k + iy_k$ with $x_k, y_k \in \mathbb{R}$. At times it will be convenient to use real variable notation by identifying $z = (x_1 + iy_1, \ldots, x_n + iy_n)$ with $(x_1, \xi) \in \mathbb{R}^{2n}$, where $\xi = (y_1, x_2, \ldots, x_n, y_n) \in \mathbb{R}^{2n-1}$.

After a unitary transformation of $\mathbb{C}^n$, if necessary, we may assume that the inner normal to $\partial\Omega$ at 0 points in the positive x_1 direction. Without loss of generality, there is a real valued C^∞ function ψ defined on $T_0(\partial\Omega) = \{(0, \xi) \mid \xi \in \mathbb{R}^{2n-1}\}$, so that $\partial\Omega = \{(\psi(\xi), \xi) \mid \xi \in \mathbb{R}^{2n-1}\}$ and $\Omega = \{(x_1, \xi) \mid x_1 > \psi(\xi)\}$. (This is certainly true in a neighborhood of 0 by the implicit function theorem, and our concerns are purely local here.) The fact that $T_0(\partial\Omega)$ is tangent to $\partial\Omega$ at 0 implies $\nabla\psi(0) = 0$.

For $z = (x_1, \xi) \in \Omega$, set

$$d(z) = \min\left(x_1, x_1 - \psi(\xi)\right)$$

and define an approach region $A_\alpha(\partial\Omega)$ by

$$A_\alpha(\partial\Omega) = \left\{z \in \mathbb{C}^n \mid |z|^2 < \alpha d(z), |y_1| < \alpha x_1\right\}$$

We shall write A_α instead of $A_\alpha(\partial\Omega)$. The following lemma shows that the regions A_α are "equivalent" to the admissible regions $\mathfrak{D}_\alpha(\xi)$ defined above. (For certain technical reasons, A_α will be easier to work with than $\mathfrak{D}_\alpha(0)$.)

Lemma 6.3.24. *Let* $\alpha > 0$.

(i) *There exists* $c = c(\alpha) > 0$ *so that if* $z = (x_1, \xi)$ *is sufficiently small and* $|z|^2 < \alpha d(z)$, *then* $d(z) \geq c x_1$.

(ii) *There exists* $\beta > 0$ *such that* $\mathfrak{D}_\alpha(0) \subset A_\beta$.

(iii) *There exists* $\gamma > 0$ *so that for sufficiently small* z, $z \in A_\alpha$ *implies* $z \in \mathfrak{D}_\alpha(0)$.

Proof.

(i) Since $\nabla\psi(0) = 0$, there exists $K < \infty$ so that $|\psi(\xi)| \leq K|\xi|^2$ for small $\xi \in \mathbb{R}^{2n-1}$. If $z = (x_1, \xi)$ satisfies $x_1 \geq 2K|\xi|^2$, then $x_1 - \psi(\xi) \geq x_1 - K|\xi|^2 \geq 1/2x_1$ when z is small. If z satisfies $x_1 \leq 2K|\xi|^2$ and $|z|^2 < \alpha\left(x_1 - \psi(\xi)\right)$, then $x_1 - \psi(\xi) \geq 1/2|\xi|^2 \geq (1/2K\alpha)x_1$. We may thus take $c = \min(1/2, 1/2K\alpha)$.

(ii) For $z \in \Omega$, let $r(z)$ denote the distance from z to $\partial\Omega$, and define

$$\Delta(z) = \min\left\{x_1, r(z)\right\}, z = (x_1, \xi)$$

($\Delta(z)$ is similar to the function $\delta_\zeta(z)$ defined above, except that it is negative when $x_1 < 0$.) In this setting,

$$\mathfrak{D}_\beta(0) = \left\{z \in \Omega \mid |z|^2 < \beta\Delta(z), |z_1| < (1+\beta)\Delta(z)\right\}.$$

If $z \in \mathfrak{D}_\beta(0)$, then $|z_1| < (1+\beta)x_1$, which implies $|y_1| < \sqrt{2\beta+\beta^2}x_1$. Also note that $\Delta(z) \le d(z)$ for every $z \in \Omega$. Thus if, $\sqrt{2\beta+\beta^2} \le \alpha$, we have $\mathfrak{D}_\beta(0) \subset A_\alpha$.

(iii) Observe that

$$\lim_{\substack{z\to 0\\ z\in\Omega}} \frac{r(z)}{x_1 - \psi(\xi)} = 1 \tag{6.25}$$

and hence

$$\lim_{\substack{z\to 0\\ z\in A_\alpha}} \frac{\Delta(z)}{d(z)} = 1 \tag{6.26}$$

((6.25) follows from the smoothness of ψ.) Thus, if $z \in A_\alpha$ and z is small, it follows from (6.26) that $|z|^2 < 2\alpha\Delta(z)$. By part **(i)** of the lemma, for small $z \in A_\alpha$ we have $d(z) \ge cx_1$, and hence $\Delta(z) \ge cx_1/2$ by (6.25). Thus $|z_1| \le x_1 + |y_1| \le (1+\alpha)x_1 \le 2(1+\alpha)\Delta(z)/c$ for sufficiently small $z \in A_\alpha$. Setting $\gamma = \max(2\alpha, 2(1+\alpha)/c)$, the proof of **(iii)** is complete.

□

We begin by proving the following proposition and discussing it.

Proposition 6.3.25. *Let f be a bounded, holomorphic function on the unit ball $\mathbb{B} \subseteq \mathbb{C}^2$. Let*

$$T = \left\{(s+i0, t+i0) \in \mathbb{C}^2 \mid s,t \in \mathbb{R}, 0 < s < 1, 0 < |t| < \sqrt{2-2s}\right\}$$

Let $\mathbb{1} = (1+i0, 0+i0)$. Suppose that

$$\lim_{T\ni z\to} f(z) \equiv \lambda \in \mathbb{C}$$

exists. Then, for any $\alpha > 1$,

$$\lim_{\mathcal{A}_\alpha(\mathbb{1})\ni z\to\mathbb{1}} f(z) = \lambda$$

Proof. The proof follows classical lines. We may assume that $\lambda = 0$. Define, for $j = 1, 2, \ldots$,

$$\Omega_j = \left\{(z_1,z_2) \in \mathbb{C}^2 \;\middle|\; \frac{1-2^{-j+1}}{2\alpha} \leqslant \operatorname{Re} z_1 < \frac{1-2^{-j-1}}{2\alpha}, \right.$$

$$\left. |\operatorname{Im} z_1| < 2^{-j+1}, |z_2| < \frac{\sqrt{2^{-j+1}}}{\sqrt{\alpha}} \right\}$$

Notice that

$$\bigcup_{j=1}^{\infty} \Omega_j \supseteq \mathcal{A}_\alpha(\mathbb{1})$$

Observe that, for $1 \leqslant j_0 \in \mathbb{Z}$, the map

$$\varphi_j(z_1, z_2) = \left(2^{j-j_0}(z_1 - 1) + 1, \sqrt{2^{j-j_0}}z_2\right)$$

sends Ω_j biholomorphically onto Ω_{j_0}. Also, following the paradigm of the classical Lindelöf principle on the disc, we may see that

$$\bigcup_{j=1}^{\infty} \Omega_j = \bigcup_{j=1}^{\infty} \varphi_j^{-1}(\Omega_{j_0}) \supseteq \mathcal{A}_\alpha(\mathbb{1})$$

Furthermore, we see that

$$T \cap \Omega_j = \left\{(s,t) \;\middle|\; \frac{1 - 2^{-j+1}}{2\alpha} \leqslant s < \frac{1 - 2^{-j-1}}{2\alpha}, 0 < t < \sqrt{2 - 2s}\right\}$$

Of course, φ_j maps $T \cap \Omega_j$ onto

$$T \cap \Omega_{j_0} = \left\{(s,t) \;\middle|\; \frac{1 - 2^{-j_0+1}}{2\alpha} \leqslant s < \frac{1 - 2^{-j_0-1}}{2\alpha}, 0 < t < \sqrt{2 - 2s}\right\}$$

Now consider the function

$$g_j \equiv f \circ \varphi_j^{-1} : \Omega_{j_0} \to \mathbb{C}$$

Being uniformly bounded, the g_j form a normal family. Let g_0 be a subsequential limit function.

It follows from our hypotheses that g_0 vanishes on $T \cap \Omega_{j_0}$. But $T \cap \Omega_{j_0}$, being two-dimensional and totally real, is a set of determinacy for holomorphic functions. Hence $g_0 \equiv 0$. Unraveling the logic, we find that if $K \subseteq \Omega_{j_0}$ is a compact set such that

$$\bigcup_{j=1}^{\infty} \varphi_j^{-1}(K) \supseteq \mathcal{A}_\alpha(\mathbb{1})$$

then $g_j \to 0$ uniformly on K. It follows that f itself has admissible limit 0 on K. □

The proof that we have just presented is misleadingly simple. For it is an artifact of the special geometry of T, and the way that it is embedded in space, that $\varphi_j^{-1}(T \cap \Omega_j) = T \cap \Omega_{j_0}$ for each j. For a very general sort of Lindelöf principle, we would like to replace the flat T with a rather arbitrary, two-dimensional, totally real surface. In that situation, the sets

$$\varphi_j^{-1}(T \cap \Omega_j)$$

could be pairwise disjoint. Thus, additional arguments will be required. It is worth noting that this problem arises even on the disc in the complex plane-in the situation where the hypothesis of the Lindelöf principle is the existence of a limit along a somewhat arbitrary curve rather than along a radius.

With these thoughts in mind, we now formulate a more sophisticated version of the Lindelöf principle on the ball:

Proposition 6.3.26. *Let f be a bounded, holomorphic function on the unit ball $\mathbb{B} \subseteq \mathbb{C}^2$. Let*

$$T = \left\{(s + i0, t + i0) \,\middle|\, s, t \in \mathbb{R}, 0 < s < 1, 0 < |t| < \sqrt{2 - 2s}\right\}$$

Suppose that $\rho : T \to \mathbb{R}^2$ is a C^2 function, with bounded first and second derivatives, such that (writing $\rho(s,t) = (\rho_1(s,t), \rho_2(s,t))$)

$$\mathcal{T} = \{(s + i\rho_1(s,t), t + i\rho_2(s,t)) \mid (s,t) \in T\}$$

is a two-dimensional, totally real manifold in $\mathbb{B} \subseteq \mathbb{C}^2$.

Let $\mathbb{1} = (1 + i0, 0 + i0)$. Suppose that

$$\lim_{\mathcal{T} \ni z \to \mathbb{1}} f(z) \equiv \lambda \in \mathbb{C}$$

exists. Then, for any $\alpha > 1$,

$$\lim_{\mathcal{A}_\alpha(\mathbb{1}) \ni z \to \mathbb{1}} f(z) = \lambda$$

This result is more nearly like the general Lindelöf principle that one may find in [148]. It hypothesizes the existence of a limit along a fairly "arbitrary" surface terminating at the boundary point $\mathbb{1} \in \partial\mathbb{B}$, and it concludes the existence of an admissible limit. It still leaves open the question of obtaining a result on more general domains, and also the question of treating normal functions. We save those two topics for subsequent sections.

Proof. Let Ω_j, ϕ_j, g_j be as in the proof of the preceding proposition. As before, we may assume that $\lambda = 0$ and we may obtain a subsequential limit function g_0. But it is important for us now to note that the sets

$$\mathcal{T}_j \cap \Omega_{j_0} \equiv \varphi_j\left(\mathcal{T} \cap \Omega_j\right)$$

are all graphs over $T \cap \Omega_{j_0}$ of functions τ_j, and each τ_j has, by design, bounded derivatives (with the bound uniform in j). Thus we may extract (using the Ascoli-Arzela theorem). A subsequence τ_{j_k} that converges uniformly, along with its first derivatives, on compacta to some τ_0. We pass to a corresponding subsequence of the g_j s, and continue to call the limit function g_0. Now let $\mathcal{T}_0$ be the graph of τ_0. It follows that $\mathcal{T}_0$ is a totally real, two-dimensional manifold. And certainly g_0 vanishes on $\mathcal{T}_0$. Thus $g_0 \equiv 0$. Arguing as in the proof of Proposition 6.3.26, we conclude that f has admissible limit 0 at $\mathbb{1}$. □

We continue, for the moment, to work on the unit ball $\mathbb{B} \subseteq \mathbb{C}^n$.

Now we have

Proposition 6.3.27. *Let $\mathbb{B} \subseteq \mathbb{C}^2$ be the unit ball. Let $f : \mathbb{B} \to \overline{\mathbb{C}}$ be holomorphic and normal. Let*

$$T = \left\{(s + i0, t + i0) \,\middle|\, s, t \in \mathbb{R}, 0 < s < 1, 0 < |t| < \sqrt{2 - 2s}\right\}$$

Suppose that $\rho : T \to \mathbb{R}^2$ *is a* C^2 *function, with bounded first and second derivatives, such that (writing* $\rho(s,t) = (\rho_1(s,t), \rho_2(s,t))$*)*

$$\mathcal{T} = \{(s + i\rho_1(s,t), t + i\rho_2(s,t)) \mid (s,t) \in T\}$$

is a two-dimensional, totally real manifold in $\mathbb{B} \subseteq \mathbb{C}^2$*.*

Let $\mathbb{1} = (1 + i0, 0 + i0)$*. Suppose that*

$$\lim_{\mathcal{T} \ni z \to \mathbb{1}} f(z) \equiv \lambda \in \mathbb{C}$$

exists. Then, for any $\alpha > 1$*,*

$$\lim_{\mathcal{A}_\alpha(\mathbb{1}) \ni z \to \mathbb{1}} f(z) = \lambda$$

Proof. The key here is to let

$$\Phi : \mathbb{B} \to \mathbb{B}$$

be the automorphism

$$\Phi(z_1, z_2) = \left(\frac{z_1 + 1/4}{1 + (1/4)z_1}, \frac{\sqrt{1 - 1/16}\, z_2}{1 + (1/4)z_1} \right)$$

Then we define Ω_{j_0} as before but now we let

$$\Omega_j = \Phi^{j - j_0}(\Omega_{j_0})$$

Here Φ^m is the mapping Φ composed with itself m times, $\ell > 0$.

Now the proof goes through just as that for Proposition 6.3.26. We merely must note that now we are examining the mappings $g_j \equiv f \circ \Phi^{-(j-j_0)}$. These are compositions of f with automorphisms. By the definition of "normal function," we may be sure that the g_j form a normal family. The proof is completed then as before. □

It should be stressed that it is misleading, indeed essentially incorrect, to think of normal functions on an arbitrary domain in terms of automorphisms of the domain. For most domains in $\mathbb{C}$ or $\mathbb{C}^n$ have only the identity as an automorphism (see [63] for a discussion of this notion). One of the main motivations for the development in [63] of normal functions using the Kobayashi metric was to address this difficulty. Thus, if we wish to prove a Lindelöf principle on general domains, we certainly cannot use the ideas in Section 6.3.3.

Instead we examine the invariant Kobayashi metric. Let us begin by looking at a strongly pseudoconvex domain Ω with C^2 boundary. Let $P \in \partial\Omega$. By normalizing coordinates, we may assume as usual that $P = \mathbb{1} = (1, 0)$, that $\operatorname{Re} z_1$ is the real normal direction, and that $\operatorname{Im} z_1$ is the complex normal direction. Thus $z_2, \ldots, z_n$ are the complex tangential directions at P. For simplicity, we assume as above that the dimension $n = 2$ (the statements and proofs in higher dimensions are analogous). We may assume that there is an internally tangent ball at $P \in \partial\Omega$. There is no loss of generality to assume that this ball is the unit ball. Thus we may define regions Ω_j just as on the unit ball above. And the maps φ_j are defined as before.

Theorem 6.3.28. *Let $\Omega \subseteq \mathbb{C}^2$ be a strongly pseudoconvex domain with C^2 boundary. Let $f : \Omega \to \overline{\mathbb{C}}$ be holomorphic and normal. Let*

$$T = \left\{(s + i0, t + i0) \;\middle|\; s, t \in \mathbb{R}, 0 < s < 1, 0 < |t| < \sqrt{2 - 2s}\right\}$$

Suppose that $\rho : T \to \mathbb{R}^2$ is a C^2 function, with bounded first and second derivatives, such that (writing $\rho(s,t) = (\rho_1(s,t), \rho_2(s,t))$)

$$\mathcal{T} = \{(s + i\rho_1(s,t), t + i\rho_2(s,t)) \mid (s,t) \in T\}$$

is a two-dimensional, totally real manifold in $\mathbb{B} \subseteq \mathbb{C}^2$.

Let $\mathbb{1} = (1 + i0, 0 + i0)$. Suppose that

$$\lim_{\mathcal{T} \ni z \to \mathbb{1}} f(z) \equiv \lambda \in \mathbb{C}$$

exists. Then, for any $\alpha > 1$,

$$\lim_{\mathcal{A}_\alpha(\mathbb{1}) \ni z \to \mathbb{1}} f(z) = \lambda$$

Proof. It is propitious to consider a holomorphic mapping $\psi : D \to \Omega_{j_0}$, where $D \subseteq \mathbb{C}$ is the unit disc. If $\psi(0) = p \in \Omega_{j_0}$, then we may take ψ to be an extremal function for the Kobayashi metric at the point p. Now look at $\mu_j : D \to \overline{\mathbb{C}}$ given by $f \circ \varphi_j^{-1} \circ \psi$.

Then we calculate that

$$\left|\mu_j'(0)\right| \leqslant \left|\nabla f\left(\varphi_j^{-1}(p)\right)\left[\left(\varphi_j^{-1} \circ \psi\right)'(0)\right]\right| \tag{6.27}$$

Of course, the second expression in brackets on the right is nothing other than the reciprocal of the Kobayashi metric for Ω_j at $\varphi_j^{-1}(p)$. The first expression, as we know from the definition of "normal function," is bounded from the Kobayashi metric on Ω (which is smaller than the Kobayashi metric on Ω_j) to the spherical metric on $\overline{\mathbb{C}}$. In sum, the expression (6.27) is bounded on compact subsets of D. And this bound is certainly independent of j, and also independent of the choice of p-as p ranges over a compact subset K of Ω_{j_0}.

We may obtain a similar estimate (just by composing with a Möbius transformation) for μ_j' at any point of a compact subset L of D. Of course, the estimate will depend on L.

So we may extract a normally convergent subsequence of μ_j'. Call the limit function μ_0. Arguing now as in the proof of Proposition 6.3.26, we see that there is a corresponding subsubsequence τ_{j_k} converging to τ_0 and a limiting totally real, two-dimensional manifold $\mathcal{I}_0$ in Ω_0 which is the graph of τ_0. Thus we find that μ_{j_k}' converges on $\mathcal{T}_0$. Putting together the convergence of the derivatives together with the convergence of the functions on $\mathcal{I}_0$, we see that the functions themselves converge uniformly on compact subsets of Ω_{j_0}. By the usual logic, we find that f has an admissible limit λ at P. □

6.3.6 Review of the Classical Theory of H^p Spaces on the Disc

Let $0 < p < \infty$. We define

$$H^p(\mathbb{D}) = \left\{ f \text{ holomorphic on } \mathbb{D} \;\middle|\; \sup_{0<r<1} \left[\frac{1}{2\pi} \int_0^{2\pi} \left| f\left(re^{i\theta}\right)\right|^p d\theta \right]^{1/p} \right\}$$
$$\equiv \|f\|_{H^p} < \infty$$

Also define

$$H^\infty(\mathbb{D}) = \left\{ f \text{ holomorphic on } \mathbb{D} \;\middle|\; \sup_{\mathbb{D}} |f| \equiv \|f\|_{H^p} < \infty \right\}$$

The fundamental result in the subject of H^p, or Hardy, spaces (and also the fundamental result of this section) is that if $f \in H^p(\mathbb{D})$ then the limit

$$\lim_{r \to 1^-} f\left(re^{i\theta}\right) \equiv \tilde{f}\left(e^{i\theta}\right)$$

exists for almost every $\theta \in [0, 2\pi)$. For $1 \leq p \leq \infty$, the function f can be recovered from $\tilde{f}$ by way of the Cauchy or Poisson integral formulas; for $p < 1$ this "recovery" process is more subtle and must proceed by way of distributions. Once this pointwise boundary limit result is established, then an enormous and rich mathematical structure unfolds (see Y. Katznelson [117], K. Hoffman [108], and J. Garnett [76]).

Recall that the Poisson kernel for the disc is

$$P_r\left(e^{i\theta}\right) = \frac{1}{2\pi} \frac{1 - r^2}{1 - 2r\cos\theta + r^2}$$

Let

$$\mathbf{h}^p(\mathbb{D}) = \left\{ f \text{ harmonic on } \mathbb{D} \;\middle|\; \sup_{0<r<1} \left[\frac{1}{2\pi} \int_0^{2\pi} \left| f\left(re^{i\theta}\right)\right|^p d\theta \right]^{1/p} \right\}$$
$$\equiv \|f\|_{\mathbf{h}^p} < \infty$$

and

$$\mathbf{h}^\infty(\mathbb{D}) = \left\{ f \text{ harmonic on } \mathbb{D} \;\middle|\; \sup_{\mathbb{D}} |f| \right\} \equiv \|f\|_{H^p} < \infty$$

Throughout this section, arithmetic and measure theory on $[0, 2\pi)$ (equivalently on $\partial\mathbb{D}$) is done by identifying $[0, 2\pi)$ with $\mathbb{R}/2\pi\mathbb{Z}$. See Y. Katznelson [117] for more on this identification procedure.

Proposition 6.3.29. *Let* $1 < p \leq \infty$ *and* $f \in \mathbf{h}^p(\mathbb{D})$. *Then there is an* $\tilde{f} \in L^p(\partial\mathbb{D})$ *such that*

$$f\left(re^{i\theta}\right) = \int_0^{2\pi} \tilde{f}\left(e^{i\psi}\right) P_r\left(e^{i(\theta-\psi)}\right) d\psi$$

Proof. Define $f_r\left(e^{i\theta}\right) = f\left(re^{i\theta}\right), 0 < r < 1$. Then $\{f_r\}_{0<r<1}$ is a bounded subset of $\left(L^{p'}(\partial\mathbb{D})\right)^*, p' = p/(p-1)$. By the Banach-Alaoglu theorem (see W. Rudin [189]), there is a subsequence f_{r_3} that converges weak-* to some $\tilde{f}$ in $L^p(\partial\mathbb{D})$. For any $0 < r < 1$, let $r < r_j < 1$. Then

$$f(re^{i\theta}) = f_{r_j}\left((r/r_j)\, e^{i\theta}\right) = \int_0^{2\pi} f_{r_j}\left(e^{i\psi}\right) P_{r/r_3}\left(e^{i(\theta-\psi)}\right) d\psi$$

because $f_{r_j} \in C(\bar{\mathbb{D}})$. Now $P_{r/r_j} \in C(\partial\mathbb{D}) \subseteq L^{p'}(\partial\mathbb{D})$. Thus, the right-hand side of the last equation is

$$\int_0^{2\pi} f_{r_j}\left(e^{i\psi}\right) P_r\left(e^{i(\theta-\psi)}\right) d\psi + \int_0^{2\pi} f_{r_j}\left(e^{i\psi}\right)\left[P_{r/r_j}\left(e^{i(\theta-\psi)}\right) - P_r\left(e^{i(\theta-\psi)}\right)\right] d\psi$$

As $j \to \infty$, the second integral vanishes (because the expression in brackets converges uniformly to 0) and the first integral tends to

$$\int_0^{2\pi} \tilde{f}\left(e^{i\psi}\right) P_r\left(e^{i(\theta-\psi)}\right) d\psi$$

This is the desired result. □

Remark: It is easy to see that the proof breaks down for $p = 1$ since L^1 is not the dual of any Banach space. This breakdown is not merely ostensible: The harmonic function

$$f\left(re^{i\theta}\right) = P_r\left(e^{i\theta}\right)$$

satisfies

$$\sup_{0<r<1} \frac{1}{2\pi} \int_0^{2\pi} \left|f\left(re^{i\theta}\right)\right| d\theta < \infty$$

but the Dirac δ mass is the only measure of which f is the Poisson integral.

Exercise for the Reader. If $f \in \mathbf{h}^1$, then there is a Borel measure μ_f on ∂D such that $f\left(re^{i\theta}\right) = P_r\left(\mu_f\right)\left(e^{i\theta}\right)$.

Proposition 6.3.30. *Let $f \in L^p(\partial\mathbb{D}), 1 \le p < \infty$. Then $\lim_{r\to 1^-} P_r f = f$ in the L^p norm.*

Remark: The result is false for $p = \infty$ if f is discontinuous. The correct analogue in the uniform case is that if $f \in C(\partial\mathbb{D})$ then $P_r f \to f$ uniformly. As an exercise, consider a Borel measure μ on $\partial\mathbb{D}$. Show that its Poisson integral converges in the weak-* topology to μ.

Proof of Proposition 6.3.30. If $f \in C(\partial\mathbb{D})$, then the result is clear by the solution of the Dirichlet problem. If $f \in L^p(\partial\mathbb{D})$ is arbitrary, let $\epsilon > 0$ and choose $g \in C(\partial\mathbb{D})$ such that $\|f - g\|_{L^p} < \epsilon$. Then

$$\begin{aligned}\|P_r f - f\|_{L^p} &\le \|P_r(f-g)\|_{L^p} + \|P_r g - g\|_{L^p} + \|g - f\|_{L^p}\\ &\le \|P_r\|_{L^1}\,\|f-g\|_{L^p} + \|P_r g - g\|_{L^p} + \epsilon\\ &\le \epsilon + o(1) + \epsilon\end{aligned}$$

as $r \to 1^-$. □

Proposition 6.3.31. *Let $K \subseteq \partial\mathbb{D}$ be compact, and let $\{I_\alpha\}_{\alpha\in A}$ be a covering of K by open intervals. Then there is a subcovering $\{I_{\alpha_j}\}_{j=1}^{M}$ such that every point of K is contained in at least one but not more than two of the I_{α_3}'s. (We call such a subcover a cover of valence two.)*

Proof. Since K is compact, $\{I_\alpha\}_{\alpha\in A}$ has a finite subcollection $\{I_{\alpha_j}\}_{j=1}^{J}$ that still covers K. Now, in sequence, discard any I_{α_J} that is contained in the union of the other intervals. □

Definition 6.3.32. If $f \in L^1(\partial\mathbb{D})$, let

$$Mf(\theta) = \sup_{R>0} \frac{1}{2R}\int_{-R}^{R} \left|f\left(e^{i(\theta-\psi)}\right)\right| d\psi$$

The function Mf is called the Hardy-Littlewood maximal function of f.

Definition 6.3.33. Let (X,μ) be a measure space and let $f : X \to \mathbb{C}$ be measurable. We say that f is of weak type $p, 0 < p < \infty$, if $\mu\{x \mid |f(x)| > \lambda\} \le C/\lambda^p$, all $0 < \lambda < \infty$. The space weak type ∞ is defined to be L^∞.

Lemma 6.3.34 (Chebycheff's Inequality)**.** *If $f \in L^p(X, d\mu)$, then f is weak type $p, 1 \le p < \infty$.*

Proof. Let $\lambda > 0$. Then

$$\mu\{x : |f(x)| > \lambda\} \le \int_{\{x \mid |f(x)|>\lambda\}} |f(x)|^p/\lambda^p d\mu(x) \le \lambda^{-p}\|f\|_{L^p}^p$$

□

Exercise for the Reader. There exist functions that are of weak type p but not in L^p, $1 \le p < \infty$.

Definition 6.3.35. An operator $T : L^p(X, d\mu) \to \{\text{measurable functions }\}$ is said to be of weak type (p,p), $0 < p < \infty$, if

$$\mu\{x \mid |Tf(x)| > \lambda\} \le C\|f\|_{L^p}^p/\lambda^p, \qquad \text{all } f \in L^p, \lambda > 0$$

Proposition 6.3.36. *The operator M is of weak type $(1,1)$.*

Proof. Let $\lambda > 0$. Set $S_\lambda = \{\theta \mid |Mf(e^{i\theta})| > \lambda\}$. Let $K \subseteq S_\lambda$ be a compact subset with $2m(K) \geq m(S_\lambda)$. For each $k \in K$, there is an interval $I_k \ni k$ with $|I_k|^{-1} \int_{I_k} |f(e^{i\psi})| \, d\psi > \lambda$. Then $\{I_k\}_{k\in K}$ is an open cover of K. By Proposition 6.3.31, there is a subcover $\{I_{k_j}\}_{j=1}^{M}$ of K of valence not exceeding 2. Then

$$\begin{aligned} m(S_\lambda) \leq 2m(K) \leq 2m\left(\bigcup_{j=1}^{M} I_{k_j}\right) &\leq 2\sum_{j=1}^{M} m(I_{k_j}) \\ &\leq \sum_{j=1}^{M} \frac{2}{\lambda} \int_{I_{k_j}} |f(e^{i\psi})| \, d\psi \\ &\leq \frac{4}{\lambda} \|f\|_{L^1} \end{aligned}$$

□

Proposition 6.3.37. *If $e^{i\theta} \in \partial\mathbb{D}, 1 < \alpha < \infty$, then there is a constant $C_\alpha > 0$ such that if $f \in L^1(\partial\mathbb{D})$, then*

$$\sup_{re^{2\phi} \in \Gamma_\alpha(e^{2\theta})} |P_r f(e^{i\phi})| \leq C_\alpha M f(e^{i\theta})$$

Proof. For $re^{i\phi} \in \Gamma_\alpha(e^{i\theta})$, we have

$$|\theta - \phi| \leq 2\alpha(1-r)$$

Therefore, for $1/\alpha \leq r < 1$, we obtain

$$\begin{aligned} &|P_r f(e^{i\phi})| \\ &\quad = \left| \frac{1}{2\pi} \int_0^{2\pi} f\left(e^{i(\phi-\psi)}\right) \frac{1-r^2}{1-2r\cos\psi + r^2} d\psi \right| \\ &\quad = \left| \frac{1}{2\pi} \int_0^{2\pi} f\left(e^{i(\phi-\psi)}\right) \frac{1-r^2}{(1-r)^2 + 2r(1-\cos\psi)} d\psi \right| \\ &\quad \leq \frac{4}{2\pi} \sum_{j=0}^{\log_2(\pi/\alpha(1-r))} \int_{S_j} \left| f\left(e^{i(\phi-\psi)}\right) \right| \frac{1-r^2}{(1-r)^2 + 2r\left(2^{j-1}\alpha(1-r)\right)^2} d\psi \\ &\qquad + \frac{1}{2\pi} \int_{|\psi| < \alpha(1-r)} \left| f\left(e^{i(\phi-\psi)}\right) \right| \frac{1-r^2}{(1-r)^2} d\psi \end{aligned}$$

where $S_j = \{\psi \mid 2^j\alpha(1-r) \leq |\psi| < 2^{j+1}\alpha(1-r)\}$. Now this is less than or equal to

$$
\begin{aligned}
&\frac{4\alpha}{4\pi\alpha^2}\sum_{j=0}^{\infty}\frac{1}{2^{2j-2}(1-r)}\int_{|\psi|<(2+2^{j+1})\alpha(1-r)}\left|f\left(e^{i(\theta-\psi)}\right)\right|d\psi \\
&\quad+\frac{2}{2\pi}\frac{1}{1-r}\int_{|\psi|<3\alpha(1-r)}\left|f\left(e^{i(\theta-\psi)}\right)\right|d\psi \\
&\leq\frac{32}{\pi}\sum_{j=0}^{\infty}2^{-j}\left[\frac{1}{2\alpha\left(2+2^{j+1}\right)(1-r)}\int_{|\psi|<(2+2^{j+1})\alpha(1-r)}\left|f\left(e^{i(\theta-\psi)}\right)\right|d\psi\right] \\
&\quad+\frac{6\alpha}{\pi}\frac{1}{2\cdot 3\alpha(1-r)}\int_{|\psi|<3\alpha(1-r)}\left|f\left(e^{i(\theta-\psi)}\right)\right|d\psi \\
&\leq\frac{32}{\pi}\cdot\sum_{j=0}^{\infty}2^{-j}Mf(\theta)+\frac{6\alpha}{\pi}Mf(\theta) \\
&\leq\frac{64}{\pi}Mf(\theta)+\frac{6\alpha}{\pi}Mf(\theta)
\end{aligned}
$$

If $0<r\leq 1/\alpha$, then

$$
\begin{aligned}
|P_r f(\phi)| &\leq \frac{1}{2\pi}\int_0^{2\pi}\left|f\left(e^{i(\phi-\psi)}\right)\right|(2\alpha/(\alpha-1))d\psi \\
&\leq \frac{2\alpha}{\alpha-1}Mf(\theta)
\end{aligned}
$$

□

Theorem 6.3.38. *Let $f\in\mathbf{h}^p(\mathbb{D})$ and $1<p\leq\infty$. Let $\tilde{f}$ be as in Proposition 6.3.29 and $1<\alpha<\infty$. Then*

$$
\lim_{\Gamma_\alpha(e^{e\theta})\ni z\to e^{2\theta}} f(z)=\tilde{f}\left(e^{i\theta}\right),\qquad a.e.\ e^{i\theta}\in\partial\mathbb{D}.
$$

Proof. It suffices to handle the case $p<\infty$ and f real-valued. If $\epsilon>0$ then choose $g\in C(\partial\mathbb{D})$ real-valued so that $\|\tilde{f}-g\|_{L^p(\partial D)}<\epsilon^2$. We know by the theory of the Dirichlet problem that

$$
\lim_{\Gamma_\alpha(e^{\ell\theta})\ni z\to e^{i\theta}} g(z)=g\left(e^{i\theta}\right),\qquad \text{all}\quad e^{i\theta}\in\partial\mathbb{D} \tag{6.28}
$$

where $g\left(re^{i\theta}\right)\equiv P_r g\left(e^{i\theta}\right)$. Therefore,

$$
\begin{aligned}
&m\left\{e^{i\theta}\;\middle|\;\limsup_{\Gamma_\alpha(e^{2\theta})\ni z\to e^{2\theta}}\left|f(z)-\tilde{f}\left(e^{i\theta}\right)\right|>\epsilon\right\} \\
&\leq m\left\{e^{i\theta}\;\middle|\;\limsup_{\Gamma_\alpha(e^{2\theta})\ni z\to e^{i\theta}}|f(z)-g(z)|>\frac{\epsilon}{3}\right\} \\
&\quad+m\left\{e^{i\theta}\;\middle|\;\lim_{\Gamma_\alpha(e^{2\theta})\ni z\to\to e^{i\theta}}\left|g(z)-g\left(e^{i\theta}\right)\right|>\frac{\epsilon}{3}\right\} \\
&\quad+m\left\{e^{i\theta}\;\middle|\;\left|g\left(e^{i\theta}\right)-\tilde{f}\left(e^{i\theta}\right)\right|>\frac{\epsilon}{3}\right\} \\
&\leq\left\{e^{i\theta}\;\middle|\;C_\alpha M(\tilde{f}-g)>\frac{\epsilon}{3}\right\}+0+\left(\frac{3\left\|g-\tilde{f}\right\|_{L^p}}{\epsilon}\right)^p
\end{aligned}
$$

In the last estimate, we used Lemma 6.3.34 and (6.28). Now the last line is majorized by

$$C'_\alpha \|\tilde{f} - g\|_{L^1}/(\epsilon/3) + 3^p\epsilon^p \leq C''_\alpha \|\tilde{f} - g\|_{L^p}/(\epsilon/3) + 3^p\epsilon^p$$
$$\leq C'''_\alpha \epsilon$$

It follows that

$$\lim_{\Gamma_\alpha(e^{i\theta})\ni z \to e^{i\theta}} f(z) = \tilde{f}\left(e^{i\theta}\right), \qquad \text{a.e.} \qquad e^{i\theta} \in \partial\mathbb{D}.$$

□

The informal statement of Theorem 6.3.38 is that f has nontangential boundary limits almost everywhere.

Theorem 6.3.38 contains essentially all that can be said about the boundary behavior of harmonic functions. Why are holomorphic functions better? The classical way of answering this question (which will, in the long run, be seen to be misleading) is to use Blaschke factorization.

Definition 6.3.39. If $a \in \mathbb{C}, |a| < 1$, then the Blaschke factor at a is

$$B_a(z) = \frac{z - a}{1 - \bar{a}z}$$

It is elementary to verify that $\mathbb{B}_a$ is holomorphic on a neighborhood of $\bar{\mathbb{D}}$ and that $\left|B_a\left(e^{i\theta}\right)\right| = 1$ for all θ.

Lemma 6.3.40. *If* $0 < r < 1$ *and* f *is holomorphic on a neighborhood of* $\bar{D}(0, r)$*, let* $p_1, \ldots, p_k$ *be the zeroes of* f *(listed with multiplicity) in* $D(0, r)$*. Assume that* $f(0) \neq 0$ *and that* $f\left(re^{it}\right) \neq 0$*, all* t*. Then*

$$\log|f(0)| + \log \prod_{j=1}^{k} r\,|p_J|^{-1} = \frac{1}{2\pi}\int_0^{2\pi} \log\left|f\left(re^{it}\right)\right| dt$$

Proof. The function

$$F(z) = \frac{f(z)}{\prod_{j=1}^{k} B_{p_j/r}(z/r)}$$

is holomorphic on a neighborhood of $\bar{\mathbb{D}}(0, r)$; hence $\log|F|$ is harmonic on a neighborhood of $\bar{\mathbb{D}}(0, r)$. Thus

$$\log|F(0)| = \frac{1}{2\pi}\int_0^{2\pi} \log\left|F\left(re^{it}\right)\right| dt$$
$$= \frac{1}{2\pi}\int_0^{2\pi} \log\left|f\left(re^{it}\right)\right| dt$$

or

$$\log|f(0)| + \log \prod_{j=1}^{k} \frac{r}{|p_j|} = \frac{1}{2\pi}\int_0^{2\pi} \log\left|f\left(re^{it}\right)\right| dt$$

□

Notice that, by the continuity of the integral, Lemma 6.3.40 holds even if f has zeros on $\{re^{\imath t}\}$.

Corollary 6.3.41. *If f is holomorphic in a neighborhood of $\bar{\mathbb{D}}(0,r)$, then*

$$\log|f(0)| \leq \frac{1}{2\pi}\int_0^{2\pi} \log\left|f\left(re^{\imath t}\right)\right| dt$$

Proof. The term $\log\prod_{j=1}^{k}|r/p_j|$ in Lemma 6.3.40 is positive. □

Corollary 6.3.42. *If f is holomorphic on $\mathbb{D}$, $f(0) \neq 0$ and $\{p_1, p_2, \ldots\}$ are the zeros of f counting multiplicities, then*

$$\log|f(0)| + \log\prod_{j=1}^{\infty}\frac{1}{|p_j|} \leq \sup_{0<r<1}\frac{1}{2\pi}\int_0^{2\pi}\log^+\left|f\left(re^{2t}\right)\right| dt$$

Proof. Apply Lemma 6.3.40, letting $r \to 1^-$. □

Corollary 6.3.43. *If $f \in H^p(\mathbb{D})$, $0 < p \leq \infty$, and $\{p_1, p_2, \ldots\}$ are the zeros of f counting multiplicities, then $\sum_{j=1}^{\infty}(1-|p_j|) < \infty$.*

Proof. Since f vanishes to finite order k at 0, we may replace f by $f(z)/z^k$ and assume that $f(0) \neq 0$. It follows from Corollary 6.3.42 that

$$\log\prod_{j=1}^{\infty}\left\{\frac{1}{|p_j|}\right\} < \infty$$

or $\prod(1/|p_j|)$ converges; hence $\prod|p_j|$ converges. So $\sum_j(1-|p_j|) < \infty$. □

Proposition 6.3.44. *If $\{p_1, p_2, \ldots\} \subseteq \mathbb{D}$ satisfy $\sum_j(1-|p_j|) < \infty, p_j \neq 0$ for all j, then*

$$\prod_{j=1}^{\infty}\frac{-\bar{p}_j}{|p_j|}B_{p_j}(z)$$

converges normally on $\mathbb{D}$.

Proof. Restrict attention to $|z| \leq r < 1$. Then the assertion that the infinite product converges uniformly on this disc is equivalent to the assertion that

$$\sum_j\left|1+\frac{\bar{p}_j}{|p_j|}B_{p_j}(z)\right|$$

converges uniformly. But

$$\begin{aligned}\left|1+\frac{\bar{p}_j}{|p_j|}B_{p_j}(z)\right| &= \left|\frac{|p_j| - |p_j|\bar{p}_j z + \bar{p}_j z - |p_j|^2}{|p_j|(1-z\bar{p}_j)}\right| \\ &= \left|\frac{(|p_j| + z\bar{p}_j)(1-|p_j|)}{|p_j|(1-z\bar{p}_j)}\right| \\ &\leq \frac{(1+r)(1-|p_j|)}{1-r}\end{aligned}$$

so the convergence is uniform. □

Definition 6.3.45. Let $0 < p \leq \infty$ and $f \in H^p(\mathbb{D})$. Let $\{p_1, p_2, \ldots\}$ be the zeros of f counted according to multiplicities. Let

$$B(z) = \prod_{j=1}^{\infty} \frac{-\bar{p}_j}{|p_j|} B_{p_j}(z)$$

(where each $p_j = 0$ is understood to give rise to a factor of z). Then $\mathbb{B}$ is a well-defined holomorphic function on $\mathbb{D}$ by Proposition 6.3.44 Let $F(z) = f(z)/B(z)$. By the Riemann removable singularities theorem, F is a well-defined, nonvanishing holomorphic function on $\mathbb{D}$. The representation $f = F \cdot B$ is called the canonical factorization of f.

Exercise for the Reader. All the assertions of Definition 6.3.45 hold for $f \in N(\mathbb{D})$, the Nevanlinna class.

Proposition 6.3.46. *Let $f \in H^p(\mathbb{D}), 0 < p \leq \infty$, and let $f = F \cdot B$ be its canonical factorization. Then $F \in H^p(\mathbb{D})$ and $\|F\|_{H^p(\mathbb{D})} = \|f\|_{H^p(\mathbb{D})}$.*

Proof. Trivially, $|F| = |f/B| \geq |f|$, so $\|F\|_{H^p} \geq \|f\|_{H^p}$. If $N = 1, 2, \ldots$, let

$$B_N(z) = \prod_{j=1}^{N} \frac{-\bar{p}_j}{|p_j|} B_{p_j}(z)$$

(where the factors corresponding to $p_j = 0$ are just z).

Let $F_N = f/B_N$. Since $\left|B_N\left(e^{it}\right)\right| = 1$, all t, it holds that $\|F_N\|_{H^p} = \|f\|_{H^p}$ (see [139, Lemma 2.1.17, p. 89] and the fact that $\mathbb{B}_N\left(re^{it}\right) \to B_N\left(e^{it}\right)$ uniformly in t as $r \to 1^-$.) If $0 < r < 1$, then

$$\begin{aligned}\left[\frac{1}{2\pi}\int_0^{2\pi} \left|F\left(re^{it}\right)\right|^p dt\right]^{1/p} &= \lim_{j\to\infty}\left[\frac{1}{2\pi}\int_0^{2\pi} \left|F_N\left(re^{it}\right)\right|^p dt\right]^{1/p} \\ &\leq \lim_{j\to\infty} \|F_N\|_{H^p} = \|f\|_{H^p}\end{aligned}$$

Therefore, $\|F\|_{H^p} \leq \|f\|_{H^p}$. □

Corollary 6.3.47. *If $\{p_1, p_2, \ldots\}$ is a sequence of points in $\mathbb{D}$ satisfying $\sum_j (1 - |p_j|) < \infty$ and if $B(z) = \prod_j (-\bar{p}_j/|p_j|) B_{p_j}(z)$ is the corresponding Blaschke product, then*

$$\lim_{r\to 1^-} B\left(re^{it}\right)$$

exists and has modulus 1 almost everywhere.

Proof. The conclusion that the limit exists follows from Theorem 6.3.38 and the fact that $\mathbb{B} \in H^\infty$. For the other assertion, note that the canonical factorization for $\mathbb{B}$ is $\mathbb{B} = 1 \cdot B$. Therefore, by Proposition 6.3.46,

$$\left[\frac{1}{2\pi}\int \left|\tilde{B}\left(e^{it}\right)\right|^2 dt\right]^{1/2} = \|B\|_{H^2} = \|1\|_{H^2} = 1$$

hence $\left|\tilde{B}\left(e^{it}\right)\right| = 1$ almost everywhere. □

Theorem 6.3.48. *If $f \in H^p(\mathbb{D}), 0 < p \leq \infty$, and $1 < \alpha < \infty$, then*

$$\lim_{\Gamma_\alpha(\mathbb{D})\ni z\to e^{i\theta}} f(z)$$

exists for almost every $e^{i\theta} \in \partial\mathbb{D}$ and equals $\tilde{f}\left(e^{i\theta}\right)$. Also, $\tilde{f} \in L^p(\partial\mathbb{D})$ and

$$\|\tilde{f}\|_{L^p} = \|f\|_{H^p} \equiv \sup_{0<r<1} \left[\frac{1}{2\pi}\int_0^{2\pi} \left|f\left(re^{i\theta}\right)\right|^p d\theta\right]^{1/p}$$

Proof. By Definition 6.3.45, write $f = B \cdot F$, where F has no zeros and $\mathbb{B}$ is a Blaschke product. Then $F^{p/2}$ is a well-defined H^2 function and thus has the appropriate boundary values almost everywhere. A fortiori, F has nontangential boundary limits almost everywhere. Since $\mathbb{B} \in H^\infty, B$ has nontangential boundary limits almost everywhere. It follows that f does as well. The final assertion follows from the corresponding fact for H^2 functions (exercise). □

6.3.7 Approach Regions and Balls in the Boundary

Lemma 6.3.49. *Let Ω be a bounded domain in $\mathbb{C}^n$ and let $\{z_j\}$, $\{z_j'\}$ be two sequences of points in Ω such that*

$$\lim_{j\to\infty} z_j = \xi, \quad \lim_{j\to\infty} K_\Omega(z_j, z_j') = 0.$$

If f is a normal function in Ω and $\lim_{j\to\infty} f(z_j) = f^*(\xi)$ then $\lim_{j\to\infty} f(z_j') = f^*(\xi)$.

Proof. Since f is by assumption normal in Ω from the definition of the distance K_Ω we deduce the inequality.

$$\frac{1}{\sqrt{M}} \inf \int_0^1 \frac{|g'(\ell)|}{1+|g(\ell)|^2} d\ell \leq K_\Omega(z_j, z_j'), \tag{6.29}$$

where M is some positive constant and the infimum is taken over all $g(\ell) = f(\gamma(\ell))$, $\ell \in [0,1]$, and γ is a smooth curve in Ω joining z_j with z_j'. The length of the curve $g(\ell)$, $g(0) = f(z_j)$, $g(1) = f(z_j')$, in the spherical metric $(1+|z|^2)^{-2} dz \wedge d\overline{z}$ is equal to

$$\int_0^1 \frac{|g'(\ell)|}{1+|g(\ell)|^2} d\ell\,.$$

By the hypothesis, $K_\Omega(z_j, z_j') \to 0$ as $j \to \infty$. The inequality (6.29) implies that the spherical distance between the points $f(z_j)$ and $f(z_j')$ tends to zero as $j \to \infty$ and therefore the assertion of the lemma follows. □

It follows from Lemma 6.3.49, Example 6.3.19 and Theorem 6.3.21 that the Kobayashy distance $k_B(z, \pi(z))$ is bounded on the admissible domain $\mathfrak{D}_\alpha(\zeta)$

by a constant depending only on α. Therefore, naturally arises the idea to define the approach regions by way of an invariant metric.

For convenience, we specialize now to the Kobayashi/Royden metric (although in many applications it is useful to take another metric). Call it ρ. Fix a domain $\Omega \subseteq \mathbb{C}^n$ with smooth (at least C^2) boundary. Let U be a tubular neighborhood of $\partial\Omega$ and take ϵ_0 to be one fourth of the distance of ${}^cU$ to $\partial\Omega$. As usual, let ν_P be the unit outward normal vector to $\partial\Omega$ at a boundary point $P \in \partial\Omega$. Fix a number $\beta > 0$. If $P \in \partial\Omega$, then we let $n_P = \{P - t\nu_P \mid 0 < t < \epsilon_0\}$.

Fix a number $\beta > 0$. If $P \in \partial\Omega$, then we let $n_P = \{P - t\nu_P \mid 0 < t < \epsilon_0\}$. We set

$$\mathcal{K}_\beta(P) = \{z \in \Omega \mid \rho(z, n_P) < \beta\}$$

Example 6.3.50. Let $\Omega = \mathbb{D}$, the unit disc in the complex plane. Fix a number $\beta > 0$. The set of points with Kobayashi/Royden distance less than β from the origin is $\mathbf{d} = \mathbb{D}\left(0, \left[e^{2\beta} - 1\right] / \left[e^{2\beta} + 1\right]\right)$. The approach region $\mathcal{K}_\beta$ is then the union of the discs with centers $\alpha + i0, 0 < \alpha < 1$, and Kobayashi/Royden radius β. Put in other terms, $\mathcal{K}_\beta$ is the union of the images $\phi_a(\mathbf{d}), -1 < a < 0$. Here the ϕ_a are the Möbius transformations.

To get an idea of what $\mathcal{K}_\beta(1 + i0)$ looks like, we calculate the trajectory of the extremal point $0 + i\left[e^{2\beta} - 1\right] / \left[e^{2\beta} + 1\right]$ on the boundary of $\mathbf{d}$ under the Möbius transformations just indicated. Let $r = \left[e^{2\beta} - 1\right] / \left[e^{2\beta} + 1\right]$. Then

$$\phi_a(ir) = \frac{-a\left(1 + r^2\right)}{1 + a^2r^2} + ir\frac{1 - a^2}{1 + a^2r^2}$$

We see that

$$\phi_a(ir) \cong -a + ir(1 - a)$$

Remembering that a is real (and negative), we find that the real part of $\phi_a(ir)$ has distance $1 - |a|$ from the boundary point $1 = 1 + i0$ and the imaginary part has a distance (a constant times) $1 - |a|$ from the real axis. This is the shape of the non-tangential approach.

Example 6.3.51. Let $\Omega \subseteq \mathbb{C}^n$ be a strongly pseudoconvex domain with C^2 boundary. Fix a point $P \in \partial\Omega$ and define $\mathcal{K}_\beta(P)$ as usual. For a strictly pseudoconvex domain in $\mathbb{C}^n$ Pinchuk and Khenkin independently have obtained a much more refined estimate, which takes into account the position of the vector v with respect to the complex structure of the boundary of the domain.

Theorem 6.3.52. *For each strictly pseudoconvex domain $\Omega \subset \mathbb{C}^n$ there exists positive constants k_1 and k_2 such that for all points $z \in \Omega$, sufficiently close to the boundary $\partial\Omega$, and all $v \in \mathbb{C}^n$*

$$k_1 \leqq C_\Omega(z, v)\left(\frac{|\pi_T(v)|}{\sqrt{\delta(z, \partial\Omega)}} + \frac{|\pi_N(v)|}{\delta(z, \partial\Omega)}\right)^{-1} \leqq k_2$$

where $\pi_T(v)$ and $\pi_N(v)$ are the projections of v onto the complex tangent plane and the complex normal to $\partial\Omega$ at the point $a = a(z)$ closest to z.

In the case of the ball $\mathbb{B}$ such a different behavior of $C_\Omega(z,w)$ near the boundary in d¡ff erent directions is evident. There is a refinement of Theorem 6.3.52, due to Graham [81].

Hence, we can conclude from this that there exist constants λ_1 and λ_2, depending on Ω and such that the ball in the Carathéodory metric c_Ω with center $z = \left(z_1^0, '0\right)$ and radius ε is enclosed between two ellipsoids

$$E\left(\lambda_j \varepsilon \sqrt{r}, \lambda_j \varepsilon r\right) = \left\{ \frac{\left|z_1 - z_1^0\right|^2}{\delta^2(z, \partial\Omega)} + \sum_{\nu=2}^{n} \frac{\left|z_\nu\right|^2}{\delta(z, \partial\Omega)} = \left(\lambda_j \varepsilon\right)^2 \right\}, \quad j = 1, 2$$

(all the semi-axes of these ellipsoids, except for the nth one, are equal to $\lambda_j \varepsilon \sqrt{r}$, while the nth semi-axis is equal to $\lambda_j \varepsilon r$).

From the shape of the ellipsoid $E\left(\lambda_j \varepsilon \sqrt{r}, \lambda_j \varepsilon r\right)$ we conclude that this approach region $\mathcal{K}_\beta$ is comparable to the approach regions defined by Stein in [198]. In particular, it is non-tangential in the complex normal directions and parabolic in the complex tangential directions.

It can be shown that if $\Omega' \subseteq \mathbb{C}^n$ is some smoothly bounded domain and $P \in \partial\Omega$ is not strongly pseudoconvex then the approach region $\mathcal{K}_\beta$ will always be strictly larger than the parabolic approach regions of Stein. Some of the examples below will explore what can happen in such a case.

Example 6.3.53. Let $\Omega \subseteq \mathbb{C}^2$ be a finite type domain with C^∞ boundary (see [139] for a detailed treatment of the concept of finite type.). It is important for this example that $n = 2$. Fix a point $P \in \partial\Omega$-say that P is a point of type m, which says in effect that P is flat to order m (in a complex-analytic sense) in complex tangential directions. Then the approach region $\mathcal{K}_\beta(P)$ will be comparable to the approach regions defined in [160]. [In fact [160] gives three equivalent definitions of these approach regions.] The details of this equivalences follow from the main results of [44].

In particular, the approach regions $\mathcal{K}_\beta$ in this setting are non-tangential in complex normal directions and have aperture of the shape $y = x^m$ in complex tangential directions. Of course, as previously indicated, the actual shape of the region will vary in a subtle semicontinuous way as P moves about the boundary.

Example 6.3.54. Let $\Omega \subseteq \mathbb{C}^n$ be a domain with C^2 boundary. Let $P \in \partial\Omega$ be a point of strong pseudoconcavity. The Kobayashi/Royden metric near such a point is well understood (see [138]), and the approach region $\mathcal{K}_\beta(P)$ may be seen to contain a full neighborhood $\mathbb{B}(P,r) \cap \Omega$. Thus one expects a Hardy space function to have an unrestricted limit at such a point.

Example 6.3.55. Let Ω be the domain

$$\left\{(z_1, z_2) \in \mathbb{C}^2 \mid |z| < 2, |z - (2,0)| > 1\right\}$$

Note that the boundary of this Ω is not smooth. However, we will only discuss boundary behavior near $(1,0) \in \Omega$, and the result of Theorem 6.3.60 is local.

We consider

$$U = \partial\Omega \cap \{z \in \mathbb{C} \mid |z - (1,0)| < 1/2\}$$

Let ρ be any of the Bergman, Caratheodory, or Kobayashi metrics on Ω and V the usual associated volume (see, for instance, [65] or [139]). Easy calculations with analytic discs as in the last example show that if $P \in U$ and $Z \in \Omega$ then the ρ-distance of Z to P is finite (see [138] for more on the behavior of the Kobayashi metric near such boundary points). In fact the integrated distance will be comparable to Euclidean distance. It follows that an approach region $\mathcal{L}_\alpha(P)$ will always contain a set of the form $\{z \in \Omega \mid |z - P| < c_0\}$. Thus approach at P will be unrestricted in all directions. Likewise the boundary balls $\beta(P, r)$ will be (comparable to) Euclidean balls $b(P, r)$. The compatibility condition thus follows near U because it holds for Euclidean balls.

Because the balls β are essentially Euclidean, we are forced to choose the gauge $\omega(r) = r^{2n-1}$. This gives a measure that is (mutually absolutely continuous with respect to) $(2n - 1)$ dimensional area measure on U. Thus Theorem 6.3.60 says that functions in H^p have boundary limits in the unrestricted sense at μ-almost every point of U. This is the Hartogs extension phenomenon or Kugelsatz in the category of H^p space theory: holomorphic functions on Ω will continue analytically past $(1, 0) \in \partial\Omega$ to the ball with center 0 and radius 2. The H^p space theory ought to reflect that fact, and because of Theorem 6.3.60 it does.

In fact, as J. Polking pointed out to us, one would like to have the H^p theory says that functions in H^p have unrestricted limits at every point of U. In Section 6.3.11, where we consider lower-dimensional phenomena, we are able to prove this fact also.

Fix $\alpha > 0$ once and for all (this is analogous to the fact that, in Stein's theory, the aperture of the approach regions is fixed once and for all). We now define balls in the boundary. In the classical theory, especially examples 1 and 2 in the chart, the singularity of the relevant kernel suggests the definition of balls in the boundary which in turn suggests the shape of the approach region. In our present more general setup, no detailed information about kernels is available; we begin with the approach regions $\mathcal{L}_\alpha$ and, in effect, define the balls in the boundary with reference to them. For $0 < r < \epsilon_0$ and $P \in \partial\Omega$ we define the ball $\beta(P, r) = \beta_\alpha(P, r) \subseteq \partial\Omega$ to be

$$\beta(P, r) = \pi\left(B\left(P - r\nu_P, \alpha\right)\right)$$

Observe that $\beta(P, r)$ is essentially the Euclidean normal projection to the boundary of that part of $\mathcal{L}_\alpha(P)$ that has Euclidean normal distance r from $\partial\Omega$. It will generally be useful to assume that the Euclidean diameter of $\beta(P, r)$ tends to 0 as $r \to 0$, but this is not always necessary.

It should be remarked that, ideally, one should use a geodesic ray terminating at P instead of the inward normal segment N_P at P. However, very little is known about these rays; in order to achieve maximum generality, we must use a Euclidean construct at this one step. It is straightforward to see that the definition of $\mathcal{L}_\alpha$ is asymptotically invariant under biholomorphic maps, which are C^1 to the boundary. This assertion holds in the following sense: The Jacobian matrix at a boundary point P of a biholomorphic mapping that is C^1 on the closure of the domain will of course vary continuously with P. Thus, this Jacobian will map the inward normal vector $-\nu_P$ at P to an image vector that is bounded (in angle) away from the (real) tangent plane at P. Easy estimates on the Kobayashi metric (see [137]) then show that there is a number $r > 0$ and a number $\alpha' < \infty$ such that, for any boundary point P and C^1 biholomorphism F on the closure of Ω, the image under F of $\{z \in \Omega \mid |z - P| < r\} \cap \mathcal{L}_\alpha(P)$ lies in $\mathcal{L}_{\alpha'}(F(P))$.

With (ρ, V) and α fixed as above, let g be any function that is defined on Ω. For $P \in \partial\Omega$, we recall the maximal function of Stein [198], which is defined by

$$m_\alpha g(P) \equiv \sup_{z \in \mathcal{L}_\alpha(P)} |g(z)|$$

We shall be interested instead in the maximal functions

$$\underline{m}_\alpha g(P) \equiv \liminf_{\mathcal{L}_\alpha(P) \ni z \to P} |g(z)|$$

and

$$\bar{m}_\alpha g(P) \equiv \limsup_{\mathcal{L}_\alpha(P) \ni z \to P} |g(z)|$$

Following Stein ([198]), our goal is to show that when f is a holomorphic function on Ω which is continuous on $\bar{\Omega}$ then the distribution function of $\bar{m}_\alpha f(P)$ is controlled by $\|f\|_{H^p}$.

6.3.8 Boundary Measures and Boundary Maximal Functions

We now construct measures on $\partial\Omega$ using the balls $\beta(P, r)$ and the volume V. Our construction is a variant of the Carathéodory measure construction as found in [68]. It will yield a family of measures that detect analytic objects of varying dimensions.

We think of the domain Ω, a compatible pair (ρ, V), and an $\alpha > 0$ as all being fixed. A gauge function ω is a function

$$\omega : \{x \in \mathbb{R} \mid x \geq 0\} \to \{x \in \mathbb{R} \mid x \geq 0\}$$

which is continuous and nondecreasing. Fix such an ω. Let $\delta > 0$. Let $E \subseteq \partial\Omega$ be a set. A countable collection of balls $\{\beta(P_i, r_i)\}_{i=1}^\infty$ is called a δ-admissible cover for E if $0 < r_i \leq \delta$ for all i and

$$\bigcup_i \beta(P_i, r_i) \supseteq E$$

Let $A_\delta(E)$ be the collection of all δ-admissible covers of E.

Define the outer measure

$$\mu_\delta(E) = \inf \left\{ \sum_i \omega(r_i) \cdot V\left(B\left(P_i - r_i\nu_{P_2}, \alpha\right)\right) \,\middle|\, \{\beta(P_i, r_i)\} \in A_\delta(E) \right\}$$

Observe that if $0 < \delta_1 < \delta_2$ then $\mu_{\delta_1}(E) \geq \mu_{\delta_2}(E)$. Therefore we may define

$$\mu(E) = \lim_{\delta \to 0+} \mu_\delta(E)$$

Then μ is the outer measure which we shall use to formulate the Fatou theorem in Section 6.3.10. The presence of the terms $V\left(B\left(P_i - r_i\nu_{P_2}, \alpha\right)\right)$ in the definition of μ has a practical role, as will be seen in the proof of the theorems, but it also has a philosophical role. The latter is that the condition of compatibility is invariant under scaling of V. Since we will estimate m_α in terms of boundary maximal functions involving the measure μ, and since the device for mediating between the two is the compatibility, it is necessary to build V into the definition of μ.

In practice, particularly in the examples in Section 6.3.7, we will choose $\omega(r)$ to be a power of r. The particular power chosen will be determined by

(1) the boundary geometry of the domain Ω,

(2) the dimension of the ambient space $\mathbb{C}^n$,

(3) the dimension of the analytic object in $\partial\Omega$ that is being studied.

We mandate that the balls $\beta(\zeta, r)$ not vary wildly. When this mild hypothesis is satisfied, we call the metric "*tame*." It is essentially automatic for the Kobayashi metric on a smoothly bounded domain. Further discussion of this point is relegated to Section 6.3.9.

We will assume in practice that if $\beta = \beta(\zeta, r)$ is a boundary ball and $\mathbb{B} = B\left(\zeta - r\nu_\zeta, \alpha\right)$ the metric ball in Ω of which it is the projection then

$$V(B) \cdot \omega(r) \leq C \cdot \mu(\beta) \tag{6.30}$$

We refer to this condition as property (6.30). In all practical applications in which the Carathéodory construction is successful (see [68]), this type of estimate holds. The purpose of the Carathéodory construction is to take a scalar-valued function on certain distinguished sets and turn it into a measure; however, one usually wants the measure of the distinguished sets to be comparable to the original scalar valued function on those sets.

We also will assume in practice that the measure $d\mu$ is absolutely continuous with respect to area measure $d\sigma$ on $\partial\Omega$. Note that this hypothesis, together with property (6.30), implies that the μ-measure of a boundary ball $\beta(\zeta, r)$ will always be positive and finite.

If μ is constructed as above and if $f \in L^1(d\mu)$ then we define

$$M_2 f(P) = \sup_{0 < r < \epsilon_0} \frac{1}{\mu(\beta(P, r))} \int_{\beta(P,r)} |f(\zeta)| d\mu(\zeta)$$

We reserve the notation $M_1 f(P)$ to denote the traditional Hardy-Littlewood maximal function defined using area measure $d\sigma$ and the isotropic balls b: We set

$$M_1 f(P) = \sup_{0<r<\epsilon_0} \frac{1}{\sigma(b(P,r))} \int_{b(P,r)} |f(\zeta)| d\sigma(\zeta)$$

whenever $f \in L^1(d\sigma)$.

6.3.9 Tame Metrics and Maximal Function Estimates

Fix a bounded domain Ω and a number $\lambda > 0$. Of course, $\alpha > 0$ is fixed as always. Fix attention on a relatively open subset U of $\partial\Omega$. Assume that a compatible pair (ρ, V) of a metric and a volume form have been fixed (for $U \subseteq \partial\Omega$) and assume that a gauge function ω has been selected so that the resulting measure μ (see Section 6.3.8) is absolutely continuous with respect to σ and has property (6.30) on U. Assume that

$$\operatorname{diam}(\beta(P,r)) \to 0 \quad \text{as} \quad r \to 0. \tag{6.31}$$

All of the constructs (balls, approach regions, etc.) from previous sections now apply.

We begin with a discussion of a covering lemma that deserves to be better known. Detailed material on this subject can be found in [68, pp. 145-150]. (See also [66] and [146] for a more leisurely treatment of these ideas.) A metric space is called "directionally limited" if, in effect, the number of points that can lie on a sphere of radius r and that are bounded from each other by $r > 0$ is a priori bounded above. The full technical definition is in [68, pp. 145]. Finite-dimensional normed vector spaces, Riemannian manifolds, and locally Euclidean spaces are the prime examples of directionally limited metric spaces.

We may think of $\partial\Omega$ as a metric space by defining the metric

$$\tau(a,b) = \min\{\inf\{r \mid a \in \beta(b,r)\}, \inf\{r \mid b \in \beta(a,r)\}\}$$

In case $\partial\Omega$ (or an open subset U of $\partial\Omega$) is a directionlly limited metric space when equipped with the metric τ, we call the original metric ρ on Ω tame. We note that in any of the examples described in Section 6.3.7 enough explicit information is available about the Kobayashi metric that one can verify the directionally limited hypothesis for τ directly.

In practice it is sometimes convenient to use, instead of the "directionally limited" hypothesis, a slightly weaker property called the "ball property." A metric σ satisfies the ball property if there is an a priori number K such that a metric ball $\mathbb{B}(P, 3r)$ can contain at most K balls $\mathbb{B}(Q,r)$. The ball property is often verified by having a volume estimate (as a power of r) on metric balls. It is proved in [146] that the ball property suffices in order for the conclusions of our covering lemma (Lemma 6.3.56 below) to hold.

Now the basic property of a directionally limited metric space is the following:

Lemma 6.3.56 ([68, p. 150]). *Let X be a directionally limited metric space. There is a number $N > 0$ (depending on the parameters in the definition of "directionally limited") such that if A is a subset of X and if $\mathcal{B}$ is a covering of A by balls then there are subfamilies $\mathcal{B}_1, \ldots, \mathcal{B}_N$ of $\mathcal{B}$ such that each $\mathcal{B}_j$ consists of balls that are pairwise disjoint and such that*

$$\bigcup_{j=1}^{N} \bigcup_{\beta \in \mathcal{B}_j} \beta \supseteq A$$

Thus there is such a number N for $\partial\Omega$ equipped with the balls β. We shall make use of this parameter in the proof of the next proposition.

Proposition 6.3.57. *Let f be an H^p function on Ω with radial boundary function f^*. Let $p > 0$. Fix $\lambda > 0$. Let*

$$E_\lambda = \{P \in U \mid \bar{m}_\alpha(|f|)(P) > \lambda\}$$

Then

$$\mu(E_\lambda) \leq \frac{C}{\lambda^p} \|M_1(|f^*|^p)\|_{L'(\partial\Omega, d\mu)}$$

Proof. If $P \in E_\lambda$ then, by definition of $\bar{m}_\alpha$, there is a sequence $\{z_j\} \subseteq \mathcal{L}_\alpha(P)$ which converges to P and such that $|f(z_j)| > \lambda$. For each j we have

$$\lambda^p \leq |f(z_j)|^p \leq \frac{C}{V(B(z_j, \alpha))} \int_{B(z_j,\alpha)} |f(\zeta)|^p dV(\zeta).$$

Here, of course, we use the compatibility of the pair (ρ, V) and the plurisubharmonicity of $|f|^p$. We rewrite the inequality as

$$\omega(\delta(z_j)) \cdot V(B(z_j)) \leq \frac{C}{\lambda^p} \int_{B(z_j,\alpha)} |f(\zeta)|^p \omega(\delta(z_j))\, dV(\zeta).$$

Now for each $P \in E_\lambda$ we may select a z_j such that $\delta(z_j) < \delta$. Call the selected point $z = z(P)$. Then the collection of balls $\beta(\pi(z(P)), \delta(z(P)))$ (recall that these are projections to the boundary of the metric balls $\mathbb{B}(z(P), \alpha)$) cover E_λ. Thus the lemma applies and we may find N subfamilies $\mathcal{B}_m$, each of which is pairwise disjoint, whose union covers E_λ. Now fix an m, $1 \leq m \leq N$, and sum the last inequality over the elements of $\mathcal{B}_m$. We find that

$$\sum_{\beta \in \mathcal{B}_m} \omega(r) \cdot V(B) \leq \frac{C}{\lambda^p} \sum_{\beta \in \mathcal{B}_m} \int_B |f|^p \omega(r) dV$$

Here we are using the by now familiar shorthand of $\beta = \beta(\zeta, r)$ for an element of $\mathcal{B}_m$ and $\mathbb{B} = B(\zeta - r\nu_\zeta, \alpha)$ for the metric ball in Ω of which it is the projection. Now we majorize each $|f(\zeta)|^p$ in the integrand by $M_1(|f^*(\pi(\zeta))|^p)$ as in Stein [199] and we invoke property (6.30).

We obtain

$$\sum_{\beta \in \mathcal{B}_m} \omega(r) \cdot V(B) \leq \frac{C}{\lambda^p} \sum_{\beta \in \mathcal{B}_m} \int_\beta M_1\left(|f^*|^p\right) d\mu + o(1)$$

as $\delta \to 0^+$. In fact an extra argument is required to justify this step: when the integrand is continuous, then for δ small, the integrand is essentially constant on each ball and the acceptable error is absorbed in $o(1)$ (because $\text{diam}(\beta(P,r)) \to 0$ as $r \to 0$); for a general integrand, we approximate f by translating f_ϵ (as explained in detail in [199] and [139]) and pass to the limit. Since the balls in $\mathcal{B}_m$ are pairwise disjoint, then we may majorize the sum on the right by the integral over all of $\partial\Omega$. Thus

$$\sum_{\beta \in \mathcal{B}_m} \omega(r) \cdot V(B) \leq \frac{C}{\lambda^p} \int_{\partial\Omega} M_1\left(|f^*|^p\right) d\mu + o(1)$$

Next we sum this inequality over m. Since the union of all the balls in the $\mathcal{B}_m$'s cover E_λ, the left-hand side then majorizes $\mu_\delta\left(E_\lambda\right)$ and we have

$$\mu_\delta\left(E_\lambda\right) \leq m \cdot \frac{C}{\lambda^p} \int_{\partial\Omega} M_1\left(|f^*|^p\right) d\mu + o(1)$$

Letting $\delta \to 0^+$ now gives the desired conclusion.

We note that a similar estimate for the maximal function $\underline{m}_\alpha$ also holds. □

The next maximal function estimate is included primarily because it is a standard part of H^p space theory; we will not use it explicitly in this paper but it will prove important in future papers on this material.

Proposition 6.3.58. *The Hardy-Littlewood maximal operator M_2 (see Section 6.3.8) is weak type $(1,1)$ and strong type (p,p) for $1 < p \leq \infty$.*

Proof. Since M_2 is obviously bounded on L^∞, it suffices (by the Marcinkiewicz Interpolation Theorem) to verify the weak type boundedness on L^1. However, this follows in an obvious way by the directionally limited property of $\partial\Omega$ with the balls β and the existence of the parameter N noted above: One simply replaces the usual covering lemma as found in [199] or [139] with the new covering lemma discussed here. The rest of the argument follows classical lines. □

Corollary 6.3.59. *The Lebesgue Differentiation Theorem holds for the measure μ and the balls β: if $f \in L^1(d\mu)$ then for μ-almost every $P \in \partial\Omega$ it holds that*

$$\limsup_{r \to 0} \sup_{\beta(Q,r) \ni P} \frac{1}{\mu(\beta(Q,r))} \int_{\beta(Q,r)} |f(\zeta) - f(P)| d\mu(\zeta) = 0$$

for almost every P.

Proof. This follows standard lines. See [200]. □

6.3.10 Statement and Proof of Principal Results

Let $\Omega \subseteq \mathbb{C}^n$ be a bounded domain with C^2 boundary. Let Ω be equipped with a metric ρ and a volume form V such that (ρ, V) is compatible with Ω. Assume that the metric ρ is tame (cf. Section 6.3.8). Fix $\alpha > 0$ and define approach regions $\mathcal{L}_\alpha$ as in Section 6.3.7. Fix a gauge function ω and define the corresponding measure $d\mu$. Let $U \subseteq \partial\Omega$ be open and assume that $d\mu$ is absolutely continuous with respect to $d\sigma$ and has property (6.30) on U.

Theorem 6.3.60. *If $f \in H^p(\Omega), 0 < p \leq \infty$, then for μ-almost every $P \in U$ it holds that*

$$\lim_{\mathcal{L}_\alpha(P)\ni z \to P} f(z) = f^*(P)$$

Before moving on to the proof, we shall make a few remarks about the significance of the hypothesis that $d\mu$ be absolutely continuous with respect to $d\sigma$. This hypothesis implies, in particular, that if h is the Poisson integral of a nonnegative function in $L^2(\partial\Omega, d\sigma)$ then h has nontangential boundary limits μ-almost everywhere. Now by a variant of the a Stein limits of sequences of operators theorem [197] (see Sawyer's theorem [192] and Nikishin's theorem in [79]) we may conclude that the classical Hardy-Littlewood maximal operator M_1 is of weak type $(2, 2)$ with respect to the measure $d\mu$. In particular, M_1 maps L^2 to L^s for every $s < 2$. It is this last result that we require in the proof of Theorem 6.3.60.

Proof of Theorem 6.3.60. Fix an $f \in H^p(\Omega)$. As in [198] or [141], we may restrict attention to a subdomain Ω_1 in Ω with the property that $\partial\Omega_1 \cap \partial\Omega$ is the closure of a relatively open set U in $\partial\Omega$. Shrinking U if necessary, we may suppose that the measure $d\mu$ is absolutely continuous with respect to $d\sigma$ on U and has property (6.30). We also assume that there is a unit vector ν, normal to $\partial\Omega$ at some point P of U and such that $\{z - \epsilon\nu \mid z \in \Omega_1\}$ is relatively compact in Ω for ϵ small. Define $f_\epsilon(z) = f(z - \epsilon\nu)$. This function is continuous on the closure of Ω_1 and holomorphic on Ω_1. A calculation shows that $f \in H^p(\Omega_1)$ and that the H^p norm of f_ϵ on Ω_1 is not greater than a constant times the H^p norm of f on Ω, independent of ϵ.

Now fix $\delta > 0$ and set

$$S_\delta = \left\{ P \in \partial\Omega \,\middle|\, \left| \lim_{\mathcal{L}_\alpha \ni z \to P} \sup \mathrm{Re}(z) - \liminf_{\mathcal{L}_\alpha \ni z \to P} \mathrm{Re}\, f(z) \right| > \delta \right\}$$

We wish to estimate $\mu(S_\delta)$. Of course, the passage to estimates on $\mathrm{Re}\, f$ (and a similar argument will apply to $\mathrm{Im}\, f$) is merely a device to allow us to use limsup and liminf. It has no significance for the argument.

We majorize $\mu(S_\delta)$ in an obvious way by

$$\mu\left(\left\{z \,\middle|\, \limsup_{\mathcal{L}_\alpha \ni z \to P} |f(z) - f_\epsilon(z)| > \delta/3\right\}\right)$$
$$+ \mu\left(\left\{z \,\middle|\, \left|\limsup_{\mathcal{L}_\alpha \ni z \to P} \operatorname{Re} f_\epsilon(z) - \liminf_{\mathcal{L}_\alpha \ni z \to P} \operatorname{Re} f_\epsilon(z)\right| > \delta/3\right\}\right)$$
$$+ \mu\left(\left\{z \,\middle|\, \liminf_{\mathcal{L}_\alpha \ni z \to P} |f_\epsilon(z) - f(z)| > \delta/3\right\}\right).$$

For fixed ϵ, the second term in this sum of three is zero (since f_ϵ is continuous up to the boundary). It remains to estimate the first and third terms. Since they are estimated in the same way, we treat the first. Fix $p/2 < s < p$. By Proposition 6.3.57 applied to the plurisubharmonic function $|f - f_\epsilon|^{s/2}$ it is majorized by

$$\frac{C}{\delta} \cdot \left\|M_1\left(|(f - f_\epsilon)^*|^s\right)\right\|_{L'(d\mu)} \leq \frac{C}{\delta} \cdot \left\|M_1\left(|(f - f_\epsilon)^*|^{p/2}\right)\right\|_{L^{2(s/p)(P)}(d\mu)}^{p/(2s)} \tag{6.32}$$

However, $f_\epsilon \to f^*$ in $L^p(d\mu)$ of $\partial\Omega \cap \partial\Omega_1$ (by a standard argument using the maximal function estimates of Proposition 6.3.57 — see [199] or [141, pp. 312–313]). Thus $\left\| |(f - f_\epsilon)^*|^{p/2} \right\|_{L^2(d\mu)} \to 0$ and therefore, by the boundedness of M_1 from $L^2(d\mu)$ to $L^{2s/p}(d\mu)$,

$$\left\|M_1\left(|(f - f_\epsilon)^*|^{p/2}\right)\right\|_{L^{2s/p}(d\mu)} \to 0$$

Thus we may choose f_ϵ such that the right side of (6.32) is less than δ. We see that

$$\mu(S_\delta) < 2\delta$$

It follows that the limsup and the liminf agree μ-almost everywhere on U. Since a similar argument applies to $\operatorname{Im} f$, the theorem is proved. □

6.3.11 Lower-Dimensional Phenomena

In this section we give a brief discussion of how our measure-theoretic construction distinguishes certain lower-dimensional sets in $\partial\Omega$ and we discuss boundary behavior of holomorphic functions on these sets.

For the first example, we restrict attention (for convenience only) to the unit ball $\mathbb{B}$ in $\mathbb{C}^2$ equipped with the Kobayashi metric and corresponding volume V as discussed in Section 6.3.9. We use the gauge function $\omega(r) = r$ to construct the measure μ. The result is a measure analogous to a one-dimensional Hausdorff measure but which assigns Euclidean one-dimensional length to complex normal curves and zero length to complex tangential curves. The reason for this is as follows: Consider a boundary curve $\gamma : [0, a] \to \partial\Omega$ of

Euclidean length λ. If the curve is complex normal then, for $\delta > 0$, the curve can be covered by approximately λ/δ balls $\beta(P, \delta)$ (since said balls have radius comparable to δ in the complex normal direction). Since each such ball has ω-gauge δ, the approximation to the μ-measure of γ is essentially λ. However, if the curve γ is complex tangential, then the curve can be covered by $\ll \lambda/r$ balls (one would like to say by $\lambda/\sqrt{r}$ balls, but that is not quite accurate — see the discussion below), and the approximations to the μ-measure converge to zero.

Now that this measure has been constructed, we may imitate the proof of Proposition 6.3.57 and Theorem 6.3.60 on a fixed complex normal curve γ using the one-dimensional measure μ. Observe that we do not attempt to do analysis on the entire boundary at once, but rather on the one-dimensional curve γ. Everything goes through as before with a few important caveats:

(1) One cannot expect the boundary limit of an $H^p(\Omega)$ function along a one-dimensional curve to be in L^p of that curve. More specifically, if that curve is the boundary of a transversal analytic disc — say the disc $\mathfrak{d} = \{(\zeta, 0) \mid |\zeta| < 1\} \subseteq B\mathbb{C}^2$ — then one cannot expect the restriction of an $f \in H^p(\Omega)$ to lie in $H^p(\mathfrak{d})$. Thus we can prove nothing about boundary limits of $H^p(\Omega)$ functions along boundary curves. According to the theory that has already been established by Nagel/Rudin [159], Ramey [174], Cima/Krantz [52], and others, this was to be expected.

(2) We may expect a result for $H^\infty(\Omega)$ because the restriction of a bounded function to a lower-dimensional set is still bounded. We may use the maximal function estimates to show that the functions f_ϵ, defined as in the proof of Theorem 6.3.60, are Cauchy in $L^p(\gamma)$ for $p < \infty$. Then a pointwise convergence result follows just as in the proof of Theorem 6.3.60.

(3) The hypothesis that $d\mu$ be absolutely continuous with respect to $d\sigma$ and have property (6.30) is replaced by the fact (not hypothesis) that in this one-dimensional setting $d\mu$ is absolutely continuous with respect to one-dimensional Hausdorff measure on the curve γ. The argument as to why this implies that the classical Hardy-Littlewood operator is weak type (2,2) with respect to the measure $d\mu$ on the curve then proceeds via a Poisson operator for a second-order uniformly elliptic partial differential operator on a smooth, transverse two-dimensional surface with γ in its boundary.

(4) It was discovered by Nagel/Rudin [159] and independently by the present author (unpublished) that in fact the pointwise convergence that one obtains for H^∞ functions along complex normal curves — indeed even for the Lindelöf principle at a single point — is not the full admissible convergence (in the strongly pseudoconvex case) or $\mathcal{L}$-convergence (in more general cases). In fact the convergence takes place

only through so-called hypoadmissible regions which are asymptotically smaller than the admissible or $\mathcal{L}$-regions. A precise rendition of the concept of "asymptotically smaller" is given in [206] or [139]. Until now, there has been no satisfactory explanation for why the convergence on lower-dimensional sets is worse than on sets of positive measure. For instance, in the case of the Lindelöf Principle in one complex variable the hypothesis of radial limit for a bounded analytic function at a point gives rise to a nontangential limit at that point — not a limit through an asymptotically smaller region. Why are matters worse in several variables? The geometric method presented here gives an answer to this question: Still restricting attention to Ω being the unit ball $\mathbb{B}$ in $\mathbb{C}^2$, let $\gamma(t) = \left(e^{it}, 0\right)$. An H^∞ function f on $\mathbb{B}$ will have limits almost everywhere along γ in the hypoadmissible sense just described, but not in the full admissible sense.

In particular, the points $a_\delta = (1-\delta, \sqrt{\delta})$ which approach $\mathbb{1} = (1,0)$ as $\delta \to 0^+$ have the property that $\lim_{\delta \to 0^+} f\left(e^{i\theta} a_\delta\right)$ will in general not agree with $\lim_{\delta \to 0^+} f\left(e^{i\theta}(1-\delta)\right)$ on a set of positive one-dimensional measure in γ (take $f(z_1, z_2) = (z_2)^2 / (1-z_1)$ to see how the failure takes place at a point). The reason for this, from the point of view of our one-dimensional measure μ constructed above, is as follows. Let $\mathcal{L}_\alpha(\mathbb{1})$ be an approach region with base point $\mathbb{1}$ and with α chosen so that the tangential radius of the approach region at Euclidean distance δ from the boundary is essentially $\sqrt{\delta}$. In particular, the approach region should just contain the points a_δ. Then a metric ball $\beta(a_\delta, \alpha)$ will certainly intersect the curve γ, but only barely. In particular, the intersection will be a segment of length much smaller than the anticipated length δ (because the ball $\beta(a_\delta, \alpha)$ tapers at its ends). Therefore the control that $\bar{m}_2$ provides for points near the edge of the approach region (for these lower-dimensional phenomena) is less than the control provided for points near the center of the approach region. Thus hypoadmissible approach regions (see [139] for details of this terminology) are required in the study of lower-dimensional phenomena.

(5) One might be tempted to construct a μ using $\omega(r) = \sqrt{r}$ in order to detect complex tangential curves. However, such a measure would assign infinite length to complex normal curves; also the covering properties of the balls β in the tangential directions are poor. So such a measure would be useless.

Next we take another look at the Hartogs extension phenomenon, or Kugelsatz. Consider the domain $\mathbb{B}(0,2)\backslash\bar{\mathbb{B}}(2,1)$ discussed in Example 6.3.55 of Section 6.3.7 and focus attention on a neighborhood of the boundary point $(1,0)$. Define the gauge function $\omega(r) = 1$. Then the only submanifolds of $\partial\Omega$ on which μ is well-behaved are the discrete sets, in particular the singletons. A maximal function inequality is derived just as in the proof of Proposition 6.3.57. This inequality may then be used to prove, for instance, that at any point P in $\partial\Omega$ near $\mathbb{1}$ at which a holomorphic function has a radial limit also has an $\mathcal{L}$-limit. But in the present situation, an $\mathcal{L}$-limit (as discussed in Example 6.3.55 of Section 6.3.7) is an unrestricted limit. Thus we conclude that f extends continuously to P. This is a version of the Hartogs extension phenomenon in the context of Hardy space theory.

7

Sokhotskiii–Plemelj Theorem

The is a theory in complex analysis which helps in evaluating certain integrals. The real-line version of it (see below) is often used in physics, although rarely referred to by name (see, for example [6, 94]). The main theorem of this theory is named after Julian Sokhotskii, who proved it in 1868, and Josip Plemelj, who rediscovered it (see [171, Chapter 14, 136]) as a main ingredient of his solution of the Riemann—Hilbert problem in 1908.

7.1 The Improper Integral

Occasionally an isolated pole will be directly on the contour of an integration, causing the integral to diverge.

In accordance with the original definition as the limit of a sum, it has a meaning only for bounded functions. If the integrand is infinite at a point c of the integration range, $a \leqq c \leqq b$, then the integral of such a function is given a meaning by means of a second passing to the limit. Thus, a neighborhood of the point c is cut out, the integral is taken over the remaining part, and passing to the limit is carried out, where the length of the segment cut out tends to zero. Let c be an interior point of the interval. The limit

$$\lim_{\substack{\varepsilon_1 \to 0 \\ \varepsilon_2 \to 0}} \left[\int_a^{c-\varepsilon_1} f(x)dx + \int_{c+\epsilon_2}^{b} f(x)dx \right] ,$$

if it exists, is called the *improper integral* of the unbounded function $f(x)$ in the interval (a, b).

In this definition, it is significant that the neighborhood cut out is entirely arbitrary, and only its length tends to zero. Consequently, ε_1 and ε_2 tend to zero according to an arbitrary law, independently of each other.

Consider the integral.

$$\int_b^a \frac{dx}{x-c} , \quad (a < c < b)$$

DOI: 10.1201/9781003619581-7

Computing it as an improper integral, we have

$$\begin{aligned}\int_a^b \frac{dx}{x-c} &= \lim_{\substack{\varepsilon_1\to 0\\ \varepsilon_2\to 0}} \left[-\int_a^{c-\varepsilon_1} \frac{dx}{c-x} + \int_{c+\varepsilon_2}^b \frac{dx}{x-c} \right] \\ &= \ln\frac{b-c}{c-a} + \lim_{\substack{\varepsilon_2\to 0\\ \varepsilon_2\to 0}} \ln\frac{\varepsilon_1}{\varepsilon_2}.\end{aligned} \tag{7.1}$$

It is evident that the limit of the last expression depends on the manner in which ε_1 and ε_2 tend to zero. Consequently, the integral regarded as improper does not exist. It is called a *singular integral.*

Cauchy's Principal Value Method assigns values to a divergent integral by making the assumption that each limit converges at the same rate.

However, it can be given a meaning if a dependence between ε_1 and ε_2 can be established. Assuming that the interval cut-out is situated symmetrically with respect to the point c, i.e.,

$$\varepsilon_1 = \varepsilon_2 = \varepsilon\,, \tag{7.2}$$

we are led to the concept of the Cauchy principal value of the improper integral.

Definition 7.1.1. The *Cauchy principal value* of the singular integral

$$\int_a^b \frac{dx}{x-c}\,, \quad (a<c<b)$$

is the expression

$$\lim_{\varepsilon\to 0}\left[\int_a^{c-\epsilon} \frac{dx}{x-c} + \int_{c+\varepsilon}^b \frac{dx}{x-c}\right].$$

Taking into account formula (7.1) and condition (7.2) we have

$$\int_a^b \frac{dx}{x-c} = \ln\frac{b-c}{c-a} \tag{7.3}$$

Consider now the more general integral

$$\int_a^b \frac{\varphi(x)}{x-c}dx$$

where $\varphi(x)$ is a function that obeys in the interval (a, b) the Hölder condition. This case can easily be reduced to the preceding simpler one by representing the integral in the form

$$\int_a^b \frac{\varphi(x)}{x-c}dx = \int_a^b \frac{\varphi(x)-\varphi(c)}{x-c}dx + \varphi(c)\int_a^b \frac{dx}{x-c}.$$

In view of the Hölder condition

$$\left|\frac{\varphi(x)-\varphi(c)}{x-c}\right| < \frac{A}{|x-c|^{1-\lambda}},$$

the first integral exists as an improper integral, and the second is identical with (7.3). Consequently, we obtain the following result: the singular integral

$$\int_a^b \frac{\varphi(x)}{x-c}dx$$

where $\varphi(x)$ satisfies the Hölder condition, exists in the sense of the Cauchy principal value, and it is equal to

$$\int_a^b \frac{\varphi(x)}{x-c}dx = \int_a^b \frac{\varphi(x)-\varphi(c)}{x-c}dx + \varphi(c)\ln\frac{b-c}{c-a}.$$

To denote the singular integral, some authors use special symbols, $⨍$ (valeur principle).

7.2 The Cauchy Principal Value

We introduce the topic of this section by considering the integral

$$I = \int_0^\infty \frac{\sin x}{x}dx = \int_0^\infty \frac{e^{ix}-e^{-ix}}{2ix}dx \tag{7.4}$$

We would like to separate this into two individual integrals, but if we do so, each integral will diverge at $x = 0$ because the integrand behaves there like $1/x$. However, the integral as written in Eq. (7.4) does not diverge, and we can without changing its value, write it as

$$I = \lim_{\varepsilon\to 0+}\int_\varepsilon^\infty \frac{e^{ix}-e^{-ix}}{2ix}dx = \lim_{\varepsilon\to 0+}\left[\int_\varepsilon^\infty \frac{e^{ix}}{2ix}dx - \int_\varepsilon^\infty \frac{e^{-ix}}{2ix}dx\right]$$

The notation 0+ indicates that ε is restricted to approach zero through positive values, so that both integrals will be individually convergent for nonzero ε. If we next make a change of variable from x to $-x$ in the second of the above integrals, we get

$$I = \lim_{\varepsilon\to 0+}\left[\int_{-\infty}^{-\varepsilon} \frac{e^{ix}}{2ix}dx + \int_\varepsilon^\infty \frac{e^{ix}}{2ix}dx\right]. \tag{7.5}$$

This is still a convergence result.

Expressions like that of Eq. (7.5) arise often enough that they have been given a name and notation. The right-hand side of that expression for I is called a *Cauchy principal value* integral, and in this book such integrals are written with a line through the integral sign, indicating that the integration range is cut open at the point where the integrand is singular. Thus, we write

$$I = ⨍_{-\infty}^{\infty} \frac{e^{ix}}{2ix} dx \tag{7.6}$$

It is usually left to the reader to identify the singular point at which the integration range is cut.

It is important to realize that the notation of Eq. (7.6) does not cause the convergence of a divergent integral, in the present example

$$I' = \int_{-\infty}^{\infty} \frac{e^{ix}}{2ix} dx, \qquad \text{(divergent)}$$

but instead defines a useful related quantity that does converge to a definite value. Convergence is achieved because the two partial ranges end at the same distance from the singularity; a plot of $1/x$ near $x = 0$, Fig. 7.1, illustrates how the infinite contributions near the singularity cancel.

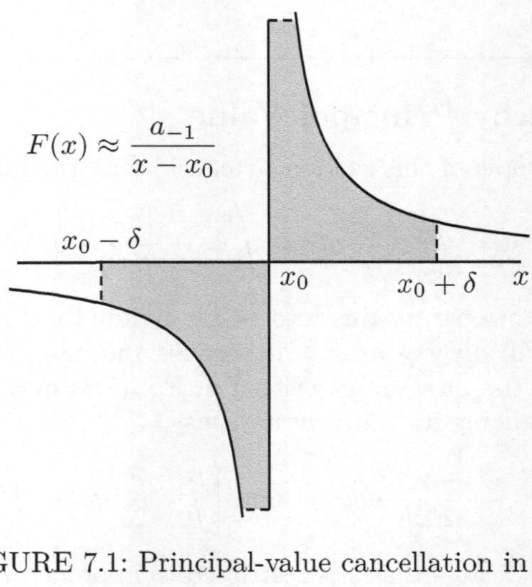

FIGURE 7.1: Principal-value cancellation in $1/x$.

We can evaluate the integral I by considering the contour integral

$$J = \int_C \frac{e^{iz}}{2iz} dz$$

where C is the contour shown in Fig. 7.2.

The two pieces of the straight line along the real axis together evaluate to I; note that to have the integrand analytic everywhere on the contour we cannot connect these lines through the singular point $z = 0$, but must go

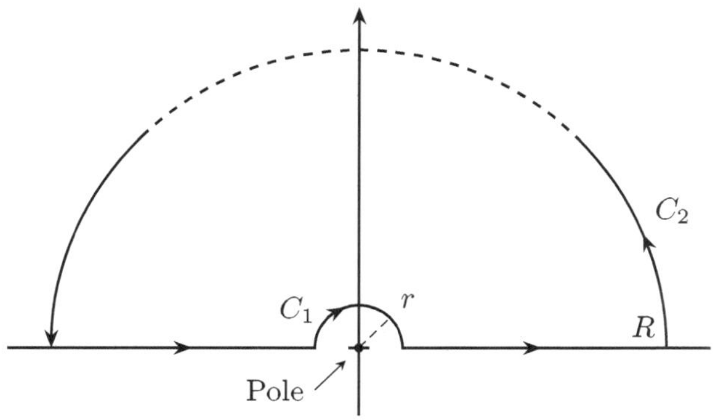

FIGURE 7.2: A contour including a Cauchy principal value integral.

around it. Invoking Jordan's lemma, the large arc C_2 closing the contour in the upper half-plane does not contribute to J. However, the small arc C_1 near $z = 0$ does make a contribution to J and must be analyzed further.

Noting that there is a simple pole at $z = 0$, we see that the semicircular arc about that point constitutes half the circuit about that pole, and therefore, since we are integrating [(residue) $id\theta$], we get $\pi i \times (residue)$ instead of $2\pi i \times (residue)$. This is a general result obtained when we go halfway around a first-order pole. We point out, however, that the same result is not obtained for the half-circuit of a higher-order pole.

In the present problem, the contour of Fig. 7.2 is a half-circuit in the negative direction (and it leaves the pole outside the contour). There are therefore no singularities encircled by our contour and $J = 0$. Taking these observations into consideration and noting that the residue of the integrand is $1/2i$, we find that

$$J = I - \pi i \left(\frac{1}{2i}\right) = 0, \qquad \text{so} \qquad I = \frac{\pi}{2}$$

This is an important integral, so we write it more explicitly:

$$I = \int_0^\infty \frac{\sin x}{x} dx = \frac{\pi}{2} \tag{7.7}$$

We could have drawn the contour for J to make the small half-circle go below $z = 0$, with the result that the pole is half-enclosed in the positive direction, and that it is now within the contour. With these changes, J is now nonzero but we get the same result for I.

To develop this topic further, consider a Cauchy principal value integral

$$I = ⨍_{-\infty}^{\infty} \frac{f(x)}{x-a} dx \tag{7.8}$$

and a related contour integral of the form

$$J = \int_C \frac{f(z)}{z-a} dz$$

where C is a contour similar to Fig. 7.2, $f(z)$ (with $z = x + iy$) is analytic in the upper half-plane, and $f(z)$ approaches zero for large y sufficiently rapidly that the large arc closing the contour makes no contribution to the integral. We also note that the integrand is analytic in the entire region enclosed by the contour, so $J = 0$. Examining the various pieces of the contour, we see that the small clockwise arc makes a contribution corresponding to $(-1/2)$ times the residue at $z = a$ (compare with the earlier discussion leading to Eq. (7.7)), namely $-\pi i f(a)$. Moreover, the straight-line portions of the integral correspond to the Cauchy principal value integral I of Eq. (7.8). We therefore get

$$J = I - \pi i f(a) = 0$$

which leads to the formula needed here:

$$⨍_{-\infty}^{\infty} \frac{f(x)}{x-a} dx = \pi i f(a) \tag{7.9}$$

If we set $f(x) = u(x) + iv(x)$ and change the name of the quantity a to x_0, Eq. (7.9) can be decomposed into the two real equations

$$u(x_0) = \frac{1}{\pi} ⨍_{-\infty}^{\infty} \frac{v(x)}{x-x_0} dx, \quad v(x_0) = -\frac{1}{\pi} ⨍_{-\infty}^{\infty} \frac{u(x)}{x-x_0} dx \tag{7.10}$$

These are the formulas known as *dispersion relations.* As is obvious from their form, they enable determination of the real part of $f(x)$ from its imaginary part, and vice versa. Equations (7.10) are also known as those defining *Hilbert transforms*; if the first of these equations is interpreted as giving the transform of v, the second provides a formula for the inverse transform of u.

The Cauchy principal value has implications for complex variable theory. Suppose now that, instead of having a break in the integration path from $x_0 - \delta$ to $x_0 + \delta$, we connect the two parts of the path by a circular arc passing, in the complex plane, either above or below the singularity at x_0. Let's continue the discussion in conventional complex-variable notation, denoting the singular point as z_0, so our arc will be a half circle (of radius δ) passing either counterclockwise below the singularity at z_0 or clockwise above z_0. We restrict further analysis to singularities no stronger than $1/(z - z_0)$, so we are dealing with a simple pole. Looking at the Laurent expansion of the function $f(z)$ to be integrated, it will have initial terms

$$\frac{a_{-1}}{z-z_0} + a_0 + \cdots$$

and the integration over a semicircle of radius δ will take (in the limit as $\delta \to 0^+$) one of the two forms (in the polar representation $z - z_0 = re^{i\theta}$, with

$dz = ire^{i\theta}d\theta$ and $r = \delta$):

$$I_{\text{over}} = \int_{\pi}^{0} d\theta i\delta e^{i\theta}\left[\frac{a_{-1}}{\delta e^{i\theta}} + a_0 + \cdots\right]$$
$$= \int_{\pi}^{0}\left(ia_{-1} + i\delta e^{i\theta}a_0 + \cdots\right)d\theta \to -i\pi a_{-1} \quad (7.11)$$

$$I_{\text{under}} = \int_{\pi}^{2\pi} d\theta i\delta e^{i\theta}\left[\frac{a_{-1}}{\delta e^{i\theta}} + a_0 + \cdots\right]$$
$$= \int_{\pi}^{2\pi}\left(ia_{-1} + i\delta e^{i\theta}a_0 + \cdots\right)d\theta \to i\pi a_{-1} \quad (7.12)$$

Note that all but the first term of each of Eqs. (7.11) and (7.12) vanish in the limit as $\delta \to 0^+$, and that each of these equations yields a result that is in magnitude half the value that would have been obtained by a full circuit around the pole. The signs associated with the semicircles correspond as expected to the direction of travel, and the two semicircular integrals average to zero.

We occasionally will want to evaluate a contour integral of a function $f(z)$ on a closed path that includes the two pieces of a Cauchy principal value integral $-\!\!\!\!\!\int f(z)dz$ with a simple pole at z_0, a semicircular arc connecting them at the singularity, and whatever other curve C is needed to close the contour (see Fig. 7.2).

These contributions combine as follows, noting that in the figure the contour passes over the point z_0:

$$-\!\!\!\!\!\!\int f(z)dz + I_{\text{over}} + \int_{C_2} f(z)dz = 2\pi i \sum \text{residues (other than at } z_0)$$

which rearranges to give

$$-\!\!\!\!\!\!\int f(z)dz = -\,I_{\text{over}} - \int_{C_2} f(z)dz + 2\pi i \sum \text{residues (other than at } z_0). \quad (7.13)$$

On the other hand, we could have chosen the contour to pass under z_0, in which case, instead of Eq. (7.13) we would get

$$-\!\!\!\!\!\!\int f(z)dz = -\,I_{\text{under}} - \int_{C_2} f(z)dz + 2\pi i \sum \text{residues (other than at } z_0) + 2\pi i a_{-1}\,, \quad (7.14)$$

where the residue denoted a_{-1} is from the pole at z_0. Equations (7.13) and (7.14) are in agreement because $2\pi i a_{-1} - I_{\text{under}} = -I_{\text{over}}$, so for the purpose of evaluating the Cauchy principal value integral, it makes no difference whether we go below or above the singularity on the original integration path.

7.3 The Pole Expansion of Meromorphic Functions

Analytic functions $f(z)$ that have only isolated poles as singularities are called meromorphic. Mittag-Leffler showed that, instead of making an expansion about a single regular point (a Taylor expansion) or about an isolated singular point (a Laurent expansion), it was also possible to make an expansion each of whose terms arise from a different pole of $f(z)$. Mittag-Leffler's theorem assumes that $f(z)$ is analytic at $z = 0$ and at all other points (excluding infinity) with the exception of discrete simple poles at points $z_1, z_2, \ldots$, with respective residues $b_1, b_2, \ldots$ We choose to order the poles in a way such that $0 < |z_1| \leq |z_2| \leq \cdots$, and we assume that in the limit of large z, $|f(z)/z| \to 0$. Then Mittag-Leffler's theorem states that

$$f(z) = f(0) + \sum_{k=1}^{\infty} b_k \left(\frac{1}{z - z_k} + \frac{1}{z_k} \right) . \tag{7.15}$$

To prove the theorem, we make the preliminary observation that the quantity being summed in Eq. (7.15) can be written as

$$\frac{z b_k}{z_k \left(z_k - z\right)} ,$$

suggesting that it might be useful to consider a contour integral of the form

$$I_N = \int_{C_N} \frac{f(w) dw}{w(w - z)}$$

where w is another complex variable and C_N is a circle enclosing the first N poles of $f(z)$. Since C_N, which has a radius we denote R_N, has total arc length $2\pi R_N$, and the absolute value of the integrand asymptotically approaches $|f(R_N)| / R_N^2$, the large z behavior of $f(z)$ guarantees that $\lim_{R_N \to \infty} I_N = 0$.

We now obtain an alternative expression for I_N using the residue theorem. Recognizing that C_N encircles simple poles at $w = 0, w = z$, and $w = z_k$, $k = 1 \ldots N$, that $f(w)$ is nonsingular at $w = 0$ and $w = z$, and that the residue of $f(z)/w(w - z)$ at z_k is just $b_k / z_k \left(z_k - z\right)$, we have

$$I_N = 2\pi i \frac{f(0)}{-z} + 2\pi i \frac{f(z)}{z} + \sum_{k=1}^{N} \frac{2\pi i b_k}{z_k \left(z_k - z\right)} .$$

Taking the large N limit, in which $I_N = 0$, we recover Mittag-Leffler's theorem, Eq. (7.15). The pole expansion converges when the condition $\lim_{z \to \infty} |f(z)/z| = 0$ is satisfied.

7.4 On the Limit Values of Cauchy Integrals

For the sake of simplicity, we consider a region which is bounded by a single, sufficiently smooth curve C. This curve is to have everywhere a tangent the direction of which varies in a continuous manner.

Let $\varphi(z)$ be a holomorphic function on C, i.e., holomorphic in every point of C. Then the Cauchy-type integral

$$\frac{1}{2\pi i}\int_C \frac{\varphi(\zeta)d\zeta}{\zeta - z}$$

cannot be evaluated for any z on the curve C. However, this integral renders two analytic functions, $F(z)$ regular inside and $F_1(z)$ regular outside the line of integration.

A contour integral of a complex-valued function with a pole c on the contour strictly does not exist. However, it may have the Cauchy principal value if this concept is extended for this case.

We will prove that the function

$$F(z) = \frac{1}{2\pi i}\int_C \frac{\varphi(\zeta)d\zeta}{\zeta - z} \tag{7.16}$$

(or $F_1(z) = \frac{1}{2\pi i}\int_C \frac{\varphi(\zeta)d\zeta}{\zeta - z}$) approaches a certain finite limit as the point z approaches, an arbitrary point z_0 of the contour C while remaining inside C (or outside C).

These limit values of the Cauchy type integral (7.16) therefore form a function $\varphi_i(z_0)$ defined at all points z_0 of the contour C (respectively, a function $\varphi_e(z_0)$ if the point z approaches the point z_0 while remaining outside C). As we shall show, this last will be analytic at all points z_0 and is closely related to the boundary function $\varphi(z_0)$.

In the Cauchy-type integral

$$F(z) = \frac{1}{2\pi i}\int_C \frac{\varphi(\zeta)d\zeta}{\zeta - z},$$

by z we mean a point lying inside the contour C.

The question naturally arises whether it is possible to consider the value of the Cauchy integral on the contour C, i.e., what is the meaning of the expression

$$\frac{1}{2\pi i}\int_C \frac{f(\zeta)d\zeta}{\zeta - z_0} \tag{7.17}$$

when $z_0 \in C$.

With the usual understanding of the integration process, formula (7.17) is meaningless, since the function $\frac{\varphi(\zeta)}{\zeta - z_0}$ is not integrable, generally speaking, along C when the point z_0 is on the line C.

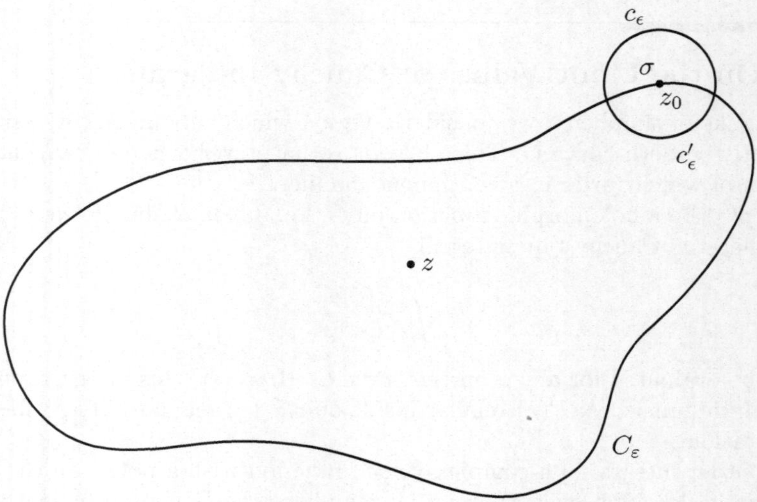

FIGURE 7.3

What do we mean here? We must first determine the meaning of the expression (7.17).

For this purpose, we describe from the point z_0, as the center, a circle of arbitrarily small radius ε and denote by σ the smallest arc of line C cut off by this circle, containing the point z_0 (Fig. 7.3).

By removing the arc σ from the line C, we denote the remaining part of the line C by C_ϵ. The expression

$$\frac{1}{2\pi i}\int_{C_\varepsilon}\frac{\varphi(\zeta)d\zeta}{\zeta - z_0} \tag{7.18}$$

makes sense for any $\varepsilon > 0$, since the function $\varphi(\zeta)/(\zeta - z_0)$ is continuous along C_ε. By forcing ε to tend to zero, we show that expression (7.18) tends to a certain finite limit, which we will take as the value of integral (7.17).

Indeed, we close the line C_ε using the arc c_ε of the constructed circle, lying outside C (Fig. 7.3). Thus, we obtain a piecewise smooth closed contour $\Gamma_\varepsilon = C_\varepsilon + c_\varepsilon$, for which the point z_0 will be internal. Let us now consider the Cauchy integral:

$$F(z) = \frac{1}{2\pi i}\int_C\frac{\varphi(\zeta)d\zeta}{\zeta - z}$$

where z is a point inside the contour C.

Since the function $\varphi(C)/(\zeta - z)$ is analytic on the contours C and $\Gamma_\varepsilon = C_\varepsilon + c_\varepsilon$, as well as between them, then by the Cauchy theorem, the integral we are dealing with is independent of the path, we have:

$$F(z) = \frac{1}{2\pi i}\int_C \frac{\varphi(\zeta)d\zeta}{\zeta - z} = \frac{1}{2\pi i}\int_{C_\varepsilon} \frac{\varphi(\zeta)d\zeta}{\zeta - z} + \frac{1}{2\pi i}\int_{c_\varepsilon} \frac{\varphi(\zeta)d\zeta}{\zeta - z}.$$

Passing to the limit in this equality at $z \to z_0$, we obtain:

$$\varphi_i(z_0) = \lim_{z\to z_0} F(z) = \frac{1}{2\pi i}\int_{C_\varepsilon} \frac{\varphi(\zeta)d\zeta}{\zeta - z_0} + \frac{1}{2\pi i}\int_{c_\varepsilon} \frac{\varphi(\zeta)d\zeta}{\zeta - z_0} \tag{7.19}$$

This formula shows that the limit values $\varphi_i(z_0)$ of the integral of the type Cauchy functions exist at each point of the contour C and form an analytic function on the contour C.

To find the dependence of the function $\varphi_i(z_0)$ on the boundary function $\varphi(z_0)$, we must pass to the limit in formula (7.19), forcing ε to tend to zero. Let us first calculate the limit to which the integral

$$\frac{1}{2\pi i}\int_{c_\varepsilon} \frac{\varphi(\zeta)d\zeta}{\zeta - z_0}$$

tends.

Setting in the last integral

$$\zeta - z_0 = \varepsilon e^{i\theta}$$

along the arc c of the circle, we have: $d\zeta = \varepsilon i e^{i\theta} d\theta$. In addition, setting

$$\varphi(\zeta) = \varphi(z_0) + \eta,$$

we see that on c_ε the modulus η tends to zero together with ε. So we get:

$$\frac{1}{2\pi i}\int_{c_\varepsilon} \frac{\varphi(\zeta)d\zeta}{\zeta - z_0} = \frac{1}{2\pi i}\int_{c_\varepsilon} \frac{\varphi(z_0)+\eta}{\varepsilon e^{i\theta}} i\varepsilon e^{i\theta} d\theta = \frac{\varphi(z_0)}{2\pi}\int_{c_\varepsilon} d\theta + \frac{1}{2\pi}\int_{c_\varepsilon} \eta d\theta .$$

Since

$$\lim_{\varepsilon\to 0}\int_{c_\varepsilon} d\theta = \pi, \quad \lim_{\varepsilon\to 0}\int_{c_\varepsilon} \eta d\theta = 0,$$

we have

$$\lim_{\varepsilon\to 0}\frac{1}{2\pi i}\int_{c_\varepsilon} \frac{\varphi(\zeta)d\zeta}{\zeta - z_0} = \frac{1}{2}\varphi(z_0). \tag{7.20}$$

Passing to the limit at $\varepsilon \to 0$, from equalities (7.19) we see that the expression

$$\frac{1}{2\pi i}\int_{C_\varepsilon} \frac{\varphi(\zeta)d\zeta}{\zeta - z_0}$$

tends to a certain finite limit, which we have agreed to denote by

$$\frac{1}{2\pi i}\int_{C} \frac{\varphi(\zeta)d\zeta}{\zeta - z_0}$$

and call the value of the Cauchy-type integral at the point z_0 of the contour C. Thus, we see that the integral

$$\frac{1}{2\pi i}\int_C \frac{\varphi(\zeta)d\zeta}{\zeta - z_0}$$

has a well-defined meaning, and formula (7.19), after passing to the limit at $\varepsilon \to 0$, becomes the following:

$$\varphi_i\left(z_0\right) = \frac{1}{2\pi i}\int_C \frac{\varphi(\zeta)d\xi}{\zeta - z_0} + \frac{1}{2}\varphi\left(z_0\right). \tag{7.21}$$

Formula (7.21) gives an expression for the limiting values $\varphi_i(z_0)$ of the Cauchy type integral from within the contour C in terms of the values of the integral itself on the contour and the given boundary function $\varphi_i(z_0)$.

To obtain a similar formula for $\varphi_e\left(z_0\right)$, consider the contour $\Gamma'_\varepsilon = C_\varepsilon + c'_\varepsilon$ obtained after closing the line C_ε by means of the arc c'_ε of a small circle lying inside C and traversed in the negative direction (Fig. 7.3). Since the function $\varphi(\zeta)/(\zeta - z_0)$, where the point z is outside C, is analytic on the contours C and Γ'_e, as well as between them, then by the Cauchy theorem we have:

$$F_1(z) = \frac{1}{2\pi i}\int_C \frac{\varphi(\zeta)d\zeta}{\zeta - z} = \frac{1}{2\pi i}\int_{C_\varepsilon} \frac{\varphi(\zeta)d\zeta}{\zeta - z} - \frac{1}{2\pi i}\int_{c'_\varepsilon} \frac{\varphi(\zeta)d\zeta}{\zeta - z}.$$

Passing to the limit in this equality as $z \to z_0$, we obtain:

$$\varphi_e\left(z_0\right) = \lim_{z\to z_0} F_1(z) = \frac{1}{2\pi i}\int_{C_\varepsilon} \frac{\varphi(\zeta)d\zeta}{\zeta - z_0} - \frac{1}{2\pi i}\int_{c'_\varepsilon} \frac{\varphi(\zeta)d\zeta}{\zeta - z_0}. \tag{7.22}$$

From the last equality we see that the limit values $\varphi_e\left(z_0\right)$, like $\varphi_i\left(z_0\right)$, form an analytic function on the contour. Noting that by the Cauchy formula it will be:

$$\frac{1}{2\pi i}\int_{c_\varepsilon} \frac{\varphi(\zeta)d\zeta}{\zeta - z_0} + \frac{1}{2\pi i}\int_{c'_\varepsilon} \frac{\varphi(\zeta)d\zeta}{\zeta - z_0} = \varphi\left(z_0\right),$$

we will get from here by passing to the limit at $\varepsilon \to 0$ due to (7.20) the following result:

$$\lim_{\varepsilon\to 0}\frac{1}{2\pi i}\int_{c'_e} \frac{\varphi(\zeta)d\zeta}{\zeta - z_0} = \frac{1}{2}\varphi\left(z_0\right).$$

Therefore, forcing ε to tend to zero, from equality (7.22) we obtain:

$$\varphi_e\left(z_0\right) = \frac{1}{2\pi i}\int_C \frac{\varphi(\zeta)d\zeta}{\zeta - z_0} - \frac{1}{2}\varphi\left(z_0\right). \tag{7.23}$$

Formula (7.23) expresses the outer limit values $\varphi_e\left(z_0\right)$ of the Cauchy type integral in terms of the boundary function $\varphi(z_0)$ and the values of integral on the contour.

Adding the formulas (7.21) and (7.23) and dividing the resulting equality in half, we find:

$$\frac{1}{2\pi i}\int_C \frac{\varphi(\zeta)d\zeta}{\zeta - z_0} = \frac{\varphi_i(z_0) + \varphi_e(z_0)}{2} \tag{7.24}$$

i.e., the value of the Cauchy-type integral at any point of the integration contour is equal to the arithmetic mean of its limit values. Subtracting formulas (I) and (II), we obtain:

$$\varphi(z_0) = \varphi_i(z_0) - \varphi_e(z_0)\,, \tag{7.25}$$

i.e., the value of the boundary function at any point of the contour is equal to the difference of the limit values at this point of the Cauchy type integral.

These formulas (7.24) and (7.25) are equivalent to the previous formulas (7.21) and (7.23).

In particular, if the function $\varphi(z)$ is analytic everywhere inside the integration contour C, as well as on C, then the Cauchy type integral becomes a Cauchy integral; its limit values $\varphi_e(z_0)$ and $\varphi_e(z_0)$ in this case will be $\varphi(z_0)$ and 0, respectively [cf. formula 7.25], and the value of the Cauchy integral itself at the point z_0 of the integration contour C according to formula (7.24) will be equal to $1/2\varphi(z_0)$.

Formulas (7.21) and (7.23) of this section were first obtained by the Russian mathematician Yu. V. Sokhotsky in 1873 under more general assumptions regarding the function $\varphi(\xi)$ (see Subection 7.5 of this Chapter). Josip Plemelj, rediscovered these formulas (see [171, Chapter 14, 136]) and use them as a main ingredient of his solution to the Riemann—Hilbert problem in 1908.

Now formulas (7.21) and (7.23) are called *Sochocki-Plemelj formulas.* The Sochocki—Plemelj formulas relate the limiting boundary values of these two analytic functions at a point z on C and the Cauchy principal value of the integral.

Subsequent generalizations relax the smoothness requirements on curve C and the function φ.

7.5 The Boundary Behavior of Analytic Functions

We have investigated the limit values of the Cauchy-type integral, assuming that the boundary function $\varphi(\zeta)$ is analytic at all points of the closed integration contour C, while the Cauchy-type integral makes sense if $\varphi(\zeta)$ is an arbitrary continuous function along the integration contour C (closed or not) and even discontinuous under some condition of integrability along C.

A version of Sochocki—Plemelj formulas will now be established. The proof which we shall give is due to Plemelj [171].

With a view to later needs, we now study the boundary values of analytic functions. We require the necessary and sufficient condition which must be satisfied by a continuous function defined on a closed curve, in order that it may represent the boundary values of an analytic function, regular everywhere inside the closed curve.

Again, we consider a region which is bounded by a single, sufficiently smooth curve C. This curve is to have everywhere a tangent the direction of which varies in a continuous manner.

We are now given a function $f(\zeta)$ which varies continuously as ζ moves along the curve C and form the integral

$$F(z) = \frac{1}{2\pi i}\int \frac{f(\zeta)\mathrm{d}\zeta}{\zeta - z} \tag{7.26}$$

taken over the closed boundary curve. At every point which does not lie on this curve, the function $F(z)$ is determined uniquely. We can expand $F(z)$ at every such point z_0 in a series of powers of $z - z_0$. Hence $F(z)$ renders two analytic functions, regular inside and outside the line of integration respectively. By Cauchy's theorem, we now know that this integral gives inside the curve of integration the function $f\left(z^+\right)$, provided $f(\zeta)$ corresponds to the inner boundary value of an analytic function, regular inside the curve C. First, if $f(\zeta) \equiv 1$, we have

$$\frac{1}{2\pi i}\int \frac{d\zeta}{\zeta - \zeta'} = \begin{cases} 1 & \text{for } \zeta' \text{ inside } C\,, \\ 0 & \text{for } \zeta' \text{ outside } C\,. \end{cases}$$

Thus, if $z \in C$, then we can rewrite (7.26) in the form

$$F\left(\zeta'\right) = \begin{cases} \frac{1}{2\pi i}\int \frac{f(\zeta)-f(z)}{\zeta-\zeta'} d\zeta + f(z) & \text{for } \zeta' \text{ inside } C\,, \\ \frac{1}{2\pi i}\int \frac{f(\zeta)-f(z)}{\zeta-\zeta'} d\zeta & \text{for } \zeta' \text{ outside } C\,. \end{cases}$$

In order now to find the boundary values of $F(z)$, we let ζ' tend to z from inside and outside C respectively. Let

$$\begin{aligned} \zeta' &= z + \varepsilon i \text{ be an internal point,} \\ \zeta' &= z - \varepsilon i \text{ be an external point,} \end{aligned}$$

and cause ζ' to approach z by making $\varepsilon \to 0$, so that ζ' tends to z along the normal to C at z.

Then the above equation can be written in the form

$$\begin{aligned} F(z + \varepsilon i) &= \frac{1}{2\pi i}\int \frac{f(\zeta) - f(z)}{\zeta - z - \varepsilon i} d\zeta + f(z)\,, \\ F(z - \varepsilon i) &= \frac{1}{2\pi i}\int \frac{f(\zeta) - f(z)}{\zeta - z + \varepsilon i} d\zeta\,. \end{aligned} \tag{7.27}$$

Set

$$\lim_{\varepsilon\to 0} F(z + \varepsilon i) = F\left(z^+\right), \quad \lim_{\varepsilon\to 0} F(z - \varepsilon i) = F\left(z^-\right)$$

if these limiting values exist.

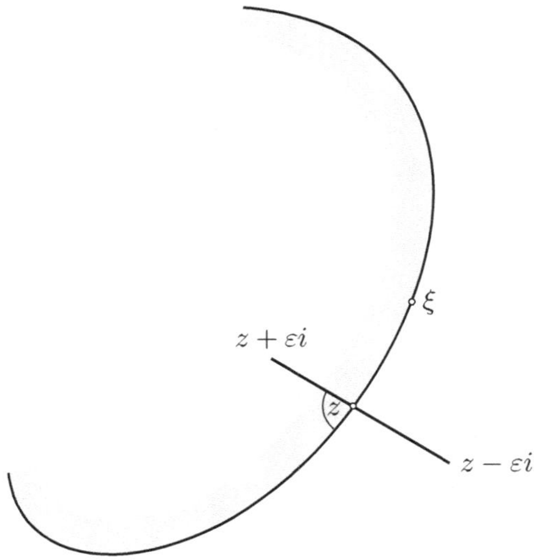

FIGURE 7.4

Remark that both the integrals in (7.27) can only exist simultaneously as $\varepsilon \to 0$. The difference of the integrals which we can write in the form

$$\frac{\varepsilon}{\pi} \int \frac{f(\zeta) - f(z)}{(\zeta - z)^2 + \varepsilon^2} d\zeta$$

then gives us the value zero.

This can be verified by dividing the line of integration into two parts by separating from it a small arc around the point z. The integral over the remainder of the curve of integration gives in the limit $\varepsilon \to 0$ the value zero. We shall now find an estimate for the integral over the small arc. Since $f(\zeta)$ was assumed to be continuous, we can make $f(\zeta) - f(z)$ as small as we please, say, smaller than g for every ζ on the arc ζz, provided it is small enough. Hence we find that the integral is smaller than

$$\frac{g}{\pi} \int \frac{|d(\zeta - z)/\varepsilon|}{|1 + (\zeta - z)/\varepsilon^2|}$$

As $\varepsilon \to 0$ the closer this integral approaches the value π, the smaller is the arc, since eventually it is approximated by the tangent. As a consequence of the continuity of $f(\zeta)$, we can also make g as small as we please. Hence follows the above statement and

$$\lim_{\varepsilon \to 0} [F(z + \varepsilon i) - F(z - \varepsilon i)] = f(z) \, .$$

Thus, the two limiting values for $\epsilon \to 0$ can only exist simultaneously and they must be equal. If the limiting value exists, we can write

$$\begin{aligned} \lim_{\varepsilon\to 0} \int \frac{f(\zeta)-f(z)}{\zeta - z - \varepsilon i} d\zeta &= \lim_{\varepsilon\to 0} \int \frac{f(\zeta)-f(z)}{\zeta - z + \varepsilon i} d\zeta \\ &= \int \frac{f(\zeta)-f(z)}{\zeta - z} d\zeta \,, \end{aligned} \tag{7.28}$$

where the integral on the right-hand side represents the principal value defined by this formula. If the function $f(\zeta)$ is continuous in such a manner that

$$|f(\zeta_2) - f(\zeta_1)| < \text{ const. } |\zeta_2 - \zeta_1|^{\mu} \,, \quad \mu = |\mu| > 0 \,,$$

we can readily verify that the limit (7.28) exists. It is given by the integral

$$\int \frac{f(\zeta)-f(z)}{\zeta - z} d\zeta$$

taken over the entire curve, from which a small arc around z has been excluded which is then made to tend to zero.

Assuming $f(\zeta)$ to have this property, we can write for the boundary values $F(z^+)$ and $F(z^-)$ the formulae

$$\begin{aligned} F(z^+) &= \frac{1}{2\pi i} \int \frac{f(\zeta)-f(z)}{\zeta - z} d\zeta + f(z) \,, \\ F(z^-) &= \frac{1}{2\pi i} \int \frac{f(\zeta)-f(z)}{\zeta - z} d\zeta \,. \end{aligned} \tag{7.29}$$

These formulae determine completely the boundary values of the integral. If the function $f(\zeta)$ is given, we find in this way the inner and outer boundary values of the two analytic functions represented by the integral (7.26) in the internal and external regions.

Now, let $f^+(\zeta)$ be some analytic function, regular inside C. If we take for z a point outside C, the function $f^+(\zeta)/(\zeta - z)$ will be regular for values of ζ inside C, because z is outside. By Cauchy's theorem, the integral over C vanishes, i.e., in (7.26) $F(z) = 0$ for all external points; hence also $F(z^+) = f^+(z)$ as we know already from Cauchy's integral representation. Thus the condition $F(z^-) = 0$ in (7.29) is necessary as well as sufficient.

In a similar manner, we also find the condition for $f^-(z)$ to be the boundary value of an analytic function which is regular outside C and has the constant value c at infinity.

If z is an internal point, then the function $[f(\zeta) - c]/(\zeta - z)$ is analytic regular for every ζ outside C and vanishes at infinity at least like $1/\zeta^2$. In order to be able to apply Cauchy's theorem at infinity, we will bound this region still by a very large circle and consider the annular region thus obtained. The integral over the large circle will be independent of its radius and equal to zero, which is readily verified by letting the radius grow beyond all bounds.

In the integral for $F(z)$, we have taken this time $f^-(\zeta) - c$ instead of $f(\zeta)$ and have found for every internal point z the value zero, i.e., also $F(z^+) = 0$. Taking this into consideration in (7.29), we find the condition

$$0 = \frac{1}{2\pi i} \int \frac{f^-(\zeta) - f^-(z)}{\zeta - z} d\zeta + f^-(z) - c\,.$$

The second equation (7.29) says now that $F(z^-) = -f^-(z) + c$, so that $f^-(z) = -F(z^-) + c$ really is the boundary value of a function $-F(z) + c$, regular outside C. Therefore here again the condition $F(z^+) = 0$ is necessary and sufficient.

We can summarize the above results in the following theorems:

The necessary and sufficient conditions for $f^+(z)$ or $f^-(z)$ to be the boundary values of analytic functions, regular inside or outside a closed curve with the value c at infinity, are

$$\frac{1}{2\pi i} \int \frac{f^+(\zeta) - f^+(z)}{\zeta - z} d\zeta = 0, \tag{7.30}$$

or

$$\frac{1}{2\pi i} \int \frac{f^-(\zeta) - f^-(z)}{\zeta - z} d\zeta + f^-(z) - c \equiv 0 \tag{7.31}$$

respectively.

However, if for any given continuous boundary functions $f^+(z)$ or $f^-(z)$ the expressions (7.31) or (7.30) are not identically equal to zero, then (7.31) represents the boundary value of a function which is regular inside the curve C, and (7.30) that of a function which is regular outside and vanishes at infinity, where the integrals in the expressions (7.30) and (7.31) are defined by (7.28). These theorems are certainly valid if $f(\zeta)$ is assumed to be continuous so that for any two boundary points ζ_1 and ζ_2

$$|f(\zeta_2) - f(\zeta_1)| < |\zeta_2 - \zeta_1|^\mu \ \text{const.},$$

where μ is a positive number. In that case not only do the integrals in (7.28) exist, but they lead even to functions which have the same type of continuity along the line of integration as the function $f(\zeta)$.

We consider now the existence of the integral (7.28). We remove from the path of integration a small arc $z_1 z z_2$ surrounding z and extend the integrals

$$\int \frac{f(\zeta) - f(z)}{\zeta - z - \varepsilon i} d\zeta \qquad \text{and} \qquad \int \frac{f(\zeta) - f(z)}{\zeta - z + \varepsilon i} d\zeta$$

over the remainder of the closed curve. The difference of these two integrals

$$2\varepsilon i \int \frac{f(\zeta) - f(z)}{(\zeta - z)^2 + \varepsilon^2} d\zeta$$

will tend to zero as $\varepsilon \to 0$. For the type of continuity prescribed above, these integrals, taken over the arc $\widehat{z_1 z z_2}$, will tend to zero as the length of this arc becomes infinitesimally small. Since we have assumed that the tangent at the boundary point z of the curve exists, we can compare in our estimate $|\zeta - z|$ with the arc length $\widehat{\zeta z}$ between ζ and z and replace $|d\zeta|$ by ds. Although the integrand s^{u-1} in the integral $\int s^{u-1} ds$ is not finite at $s = 0$, the limit of this integral exists for the lower limit $s = 0$. The absolute values of the integrals for both arcs $\widehat{z_1 z}$ and $\widehat{z z_2}$ can be compared with $(s_1^\mu + s_2^\mu)$ const., where s_1 and s_2 are the arcs $\widehat{z_1 z}$ and $\widehat{z z_2}$ respectively. This estimate gives us for the integral the order of magnitude of $(s_1 + s_2)^\mu$, and we see that the integrals tend to zero together with the length of the arc $\widehat{z_1 z z_2}$. In this manner we recognize that we arrive at the improper integral defined by (7.28) if we take the integral

$$I(z) = \int \frac{f(\zeta) - f(z)}{\zeta - z} d\zeta$$

over the boundary excluding the small arc $\widehat{z_1 z z_2}$ which contains the point z and which is then made to tend to zero. Next, we study the continuity of the integral (7.28). We prove that for the above type of continuity of $f(\zeta)$ the difference $I(z_2) - I(z_1)$ has the same order of magnitude as $|z_2 - z_1|^\mu$, so that $I(z)$ has the same type of continuity along the boundary as $f(z)$. For this purpose we take on the boundary outside the arc $\widehat{z_1 z_2}$ on each side points ζ_1 and ζ_2, so that the distances $|\zeta_1 - z_1|$ and $|\zeta_2 - z_2|$ have the same order of magnitude as $|z_2 - z_1|$. If we now take in the difference $I(z_2) - I(z_1)$, the integrals over that part of the boundary which remains after the removal of the arc $\zeta_1 \zeta_2$, we can express this difference in the form

$$\begin{aligned} &\int \frac{f(\zeta) - f(z_2)}{z_2} d\zeta - \int \frac{f(\zeta) - f(z_1)}{\zeta - z_1} d\zeta \\ &= (z_2 - z_1) \int \frac{f(\zeta) - f(z_1)}{(\zeta - z_1)(\zeta - z_2)} d\zeta + [f(z_1) - f(z_2)] \int \frac{d\zeta}{\zeta - z_2}. \end{aligned}$$

The quotient $(\zeta - z_1) / (\zeta - z_2)$ in the third integral is bounded for all ζ in the range of integration; as a result of the small magnitude of $|z_2 - z_1|$, it is in a large part of it approximately equal to unity. If we take into consideration that one has for smaller values of $|\zeta - z_1|$ the estimate $|f(\zeta) - f(z_1)| < |\zeta - z_1|^\mu$ const., we find that, as long as ζ is near z_1, the integral is of order $s^{\mu-1}$; as a result of the factor $z_2 - z_1$, it is therefore again of order s^μ. The second integral on the right-hand side has a value which is equal to the angle $\angle \zeta_2 z_2 \zeta_1$, i.e., approximately equal to π, because the arc $\widehat{\zeta_1 \zeta_2}$ is small (Fig. 7.5). As regards the difference $I(z_2) - I(z_1)$, we found so far a contribution of the order of $|z_2 - z_1|^\mu$. If we still now extend the integral for the difference $I(z_2) - I(z_1)$ to the arc $\widehat{\zeta_1 z_1 z_2 \zeta_2}$, we can obtain estimates for each individual integral; we find then for each an order of magnitude $|\zeta_2 - \zeta_1|^\mu$, i.e., of $|z_2 - z_1|^\mu$ with $\mu > 0$.

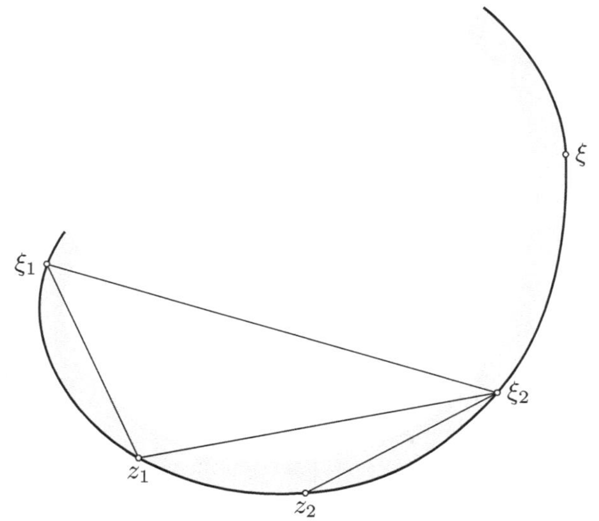

FIGURE 7.5

We should still mention that it is not necessary to assume such a degree of regularity of the boundary curve. It can be shown that all the theorems maintain their validity if we admit that the tangent has discontinuous changes in direction at a finite number of points. At least this is true as long as we do not have cusps.

8

The Dirichlet Problem

In the case of the unit disc the problem of constructing a harmonic function with given boundary values always has a solution. Since, for the latter, the Dirichlet problem is solved by the Poisson integral (see, for example, [85, Theorem 7.3.4 p. 216]), this disposes of the simply connected case. This is because, by the Riemann mapping theorem, all simply connected domains with more than one boundary point are conformally equivalent to the unit disc.

The classical Dirichlet problem of determining a harmonic function from its values on the boundary of a given multiply connected domain occurs in various fields of pure and applied mathematics.

It is known that the Dirichlet problem can be solved for any domain Ω such that each boundary point is the end point of a line segment whose other points are exterior to Ω (see [3, Theorem 9, p. 251]). The hypothesis is fulfilled if Ω and its complement have a common boundary consisting of a finite number of simple closed curves with a tangent at each point.

8.1 Some Properties of Harmonic Functions

If U is open in $\mathbb{C}$ and $u \in \mathcal{C}(U)$, its Laplacian is defined to be

$$\Delta u = \frac{\partial^2 u}{\partial x^2} + \frac{\partial^2 u}{\partial y^2}.$$

In terms of $\partial/\partial z$ and $\partial/\partial \overline{z}$, this can be rewritten in the form

$$\Delta u = \frac{\partial^2 u}{\partial \overline{z} \partial z}.$$

A harmonic function $u(x, y)$ is a C^2-solution of the Laplace equation

$$\Delta u = 0.$$

Let U be an open subset of $\mathbb{C}$, $f \in \mathcal{O}(U)$. Then $f, \operatorname{Re} f$, and $\operatorname{Im} f$ and harmonic functions.

DOI: 10.1201/9781003619581-8

Whenever u is harmonic in a simply connected domain U, an analytic function f can always be determined so that $u = \operatorname{Re} f$ in U.

In fact, if u is a harmonic function,

$$\frac{\partial^2 u}{\partial \bar{z} \partial z} = 0,$$

which shows that $\frac{\partial u}{\partial z}$ is a holomorphic function in U. The differential form $\omega := 2\frac{\partial u}{\partial z} dz$ is closed since

$$d\omega = \bar{\partial}\left(2\frac{\partial u}{\partial z} dz\right) = 2\frac{\partial^2 u}{\partial \bar{z} \partial z} d\bar{z} \wedge dz = 0\,.$$

Thus there is an $h \in \mathcal{E}(U)$ such that $\omega = dh$. We have, therefore,

$$\frac{\partial h}{\partial z} = 2\frac{\partial u}{\partial z} \quad \text{and} \quad \frac{\partial h}{\partial \bar{z}} = 0\,.$$

Hence h is a holomorphic function. Furthermore, since u is real valued we have

$$d\bar{h} = 2\frac{\partial u}{\partial \bar{z}} d\bar{z}$$

and

$$d\left(\frac{h + \bar{h}}{2}\right) = \frac{\partial u}{\partial z} dz + \frac{\partial u}{\partial \bar{z}} d\bar{z} = du\,.$$

It follows that for some real constant k, we have $u = \operatorname{Re} h + k$. Letting $f = h + k$ we have obtained a holomorphic function in U such that $u = \operatorname{Re} f$. The function f is uniquely determined up to the addition of a purely imaginary constant.

Note that a harmonic function is locally the real part of some analytic function, but it need not be the real part of an analytic function on all of U. An example of such a function is

$$u(x, y) = \frac{1}{2} \log\left(x^2 + y^2\right)\,.$$

The function u is harmonic on the annulus, $0 \leqslant r < x^2 + y^2 < R$, but, because locally $u(x, y)$ is the real part of $\log z$, there is no analytic function on all of the annulus whose real part equals u.

A function v such that $u = \operatorname{Re} f$ and $v = \operatorname{Im} f$ for the same holomorphic function f is called a *harmonic conjugate* of u. This concept is invariant under holomorphic transformations.

If $f(z) = u(x, y) + iv(x, y)$ is analytic, then $u = u(x, y)$ is harmonic and $v = v(x, y)$ is the harmonic conjugate of u, with u and v satisfying the Cauchy-Riemann equations

$$\frac{\partial u}{\partial x} = \frac{\partial v}{\partial y}, \quad \frac{\partial u}{\partial y} = -\frac{\partial v}{\partial x}\,.$$

Given a continuous function φ on the circle $|z| = R$, the Poisson formula

$$u(z) = \frac{1}{2\pi}\int_0^{2\pi} \frac{R^2 - |z|^2}{|\zeta - z|^2}\varphi\left(Re^{i\phi}\right) d\phi, \quad \zeta = Re^{i\phi} \tag{8.1}$$

represents a harmonic function in $|z| < R$ with $\lim_{z\to\zeta} u(z) = \varphi(\zeta)$. Moreover, if u is harmonic in $|z| < R$ and continuous on $|z| \leq R$, then u has the Poisson representation (8.1) in terms of its boundary values. When the point z is at the center of the disk, (8.1) reduces to the mean value property

$$u(0) = \frac{1}{2\pi}\int_0^{2\pi} u\left(Re^{i\phi}\right) d\phi\,.$$

This can also be expressed in terms of a real mean value

$$u(0) = \frac{1}{\pi R^2}\int_0^{2\pi}\int_0^R u(r,\phi) r dr d\phi\,.$$

A continuous function u on a domain U which satisfies the mean value property for all sufficiently small disks contained in U is necessarily harmonic (cf., e.g., Helms [102, Corollary 2.9]).

We may conclude from the preceding that if $\{u_n\}$ is a sequence of harmonic functions converging uniformly on compact subsets of a domain U to a function u, then u is harmonic in U.

Another consequence of the formula (8.1) is the well-known Harnack's inequality:

Theorem 8.1.1. *If h is a positive harmonic function on $U \subset \mathbb{C}$ and K is a compact subset of U, then there exists a constant $m = m(U, K)$ such that*

$$h(z) \leq m \cdot h(\zeta)$$

for all $z, \zeta \in K$.

Observe that the constant m is independent of the harmonic function u.

Definition 8.1.2. A family $\mathcal{H}$ of harmonic functions is normal in a domain U if every sequence $\{u_n\} \subseteq \mathcal{H}$ contains a subsequence which converges uniformly on compact subsets of U to either a harmonic function or to $\pm\infty$.

Unfortunately, the real parts of a normal family of analytic functions do not necessarily constitute a normal family of harmonic functions. For instance, if $z \in U$, define

$$\begin{aligned} f_j(z) &= jx + i\left(jy + j^2\right), \quad j = 1, 2, 3, \ldots \\ &= u_j(x, y) + iv_j(x, y) \end{aligned}$$

i.e.,

$$f_j(z) = jz + ij^2.$$

Then $f_j(z)$ is analytic and $f_j \to \infty$ uniformly in U, so that $\{f_j\}$ is normal in U. However, this is certainly not the case for the family of harmonic functions $\{u_j(x,y) = jx \mid j = 1,2,3,\ldots\}$.

Nevertheless, it is evident that: if $\mathcal{F}$ is a normal family of analytic functions which do not admit ∞ as a limit, then $\mathcal{H} = \operatorname{Re}\mathcal{F}$ is a normal family of harmonic functions.

On the other hand, if $\mathcal{H} = \{u\}$ is a normal family of harmonic functions in a simply connected domain U, and $\mathcal{F}$ is the corresponding family of analytic functions $\{f\}$ in U such that $\operatorname{Re} f = u$, then $\mathcal{F}$ is normal. In fact, take $\{f_j\} \subseteq \mathcal{F}$, where $\operatorname{Re} f_j = u_j \in \mathcal{H}$.

First, suppose that $\{u_j\}$ has a convergent subsequence, again denoted by $\{u_j\}$, converging normally to $\pm\infty$. Since $|f_j| \geq |u_j|$, the sequence $\{f_j\}$ converges normally to ∞.

Next, suppose that the subsequence $\{u_j\}$ converges normally in U to a harmonic function u. Then $\{u_j\}$ is locally bounded, and hence so is $\{f_j'\}$, since

$$f_j' = \frac{\partial u_j}{\partial x} - i\frac{\partial u_j}{\partial y}.$$

Consequently $\{f_j\}$ has a normally convergent subsequence by Marty's theorem, proving the result.

In common with families of analytic functions, there is the harmonic analogue of Montel's theorem:

Theorem 8.1.3. *A locally bounded family $\mathcal{H}$ of harmonic functions on a domain U is normal.*

Proof. If suffices to show that the family $\mathcal{H}$ is normal at each point of U since normality in a domain is equivalent to pointwise normality just as in the analytic case.

Let $\{h_j\}$ be a sequence in $\mathcal{H}$ and z_0 an arbitrary point of U, with $D = D(z_0; r) \subseteq U$ a disk in which $|h_j(z)| \leq M$ for $j = 1,2,3,\ldots$. Then there are analytic functions f_j in D such that $h_j = \operatorname{Re} f_j$, so that the analytic function $F_j = e^{f_j}$ satisfies

$$e^{-M} \leq |F_j| = e^{h_j} \leq e^M$$

By virtue of Montel's theorem, there is a subsequence $\{F_{j_k}\}$ satisfying $F_{j_k} \to F$ normally in D and F is analytic there. Since $e^{-M} \leq |F| \leq e^M$, the uniform continuity of the log function on $\left[e^{-M}, e^M\right]$ implies

$$h_{n_k} = \log|F_{j_k}| \to \log|F| = h$$

normally in D, where h is harmonic. This completes the proof.

The family $\mathcal{H}$ is clearly compact as well. □

From the normality criterion of boundedness, the weaker criterion of being bounded above or bounded below can be derived. Firstly,

Theorem 8.1.4. *The family $\mathcal{H}^+$ of positive harmonic functions on a domain U is normal.*

Proof. Again it suffices to show that $\mathcal{H}^+$ is normal in a neighborhood of an arbitrary point $z_0 \in U$. Given a sequence $\{u_j\} \subseteq \mathcal{H}^+$, choose a subsequence, again denoted by $\{u_j\}$, for which $u_j(z_0)$ converges to a limit $L \leq \infty$. Then in some closed disk $K(z_0; r) \subseteq U$, an application of the Harnack Inequality with $u_j = u$ gives

$$\frac{1}{\kappa} u_j(z_0) \leq u_j(z) \leq \kappa u_j(z_0), \quad z \in K(z_0; r)$$

If $L < \infty$, then $\{u_j(z)\}$ is bounded, hence normal in a neighborhood of z_0. If $L = \infty$, then $u_j \to \infty$ uniformly on $K(z_0; r)$, and normality at z_0 is again achieved. □

Corollary 8.1.5. *If $\mathcal{H}^+$ is a family of positive harmonic functions in U such that $u(z_0) = a_0 \in \mathbb{R}$ for all $u \in \mathcal{H}^+$, then $\mathcal{H}^+$ is normal and compact.*

Furthermore, it is evident from the theorem that a family of harmonic functions which is bounded above by M or bounded below by m, is normal by considering the functions $M - u$ or $u - m$, respectively. Another immediate consequence is

Theorem 8.1.6 (Harnack's Principle). *Let $\{u_j\}$ be a sequence of harmonic functions in a domain U which satisfy $u_{j+1} \geq u_j, j = 1, 2, 3 \ldots.$ Then either $u_j \to +\infty$ uniformly on compact subsets of U or $\{u_j\}$ converges to a harmonic function, uniformly on compact subsets.*

For, the functions $u_j - u_1, j = 1, 2, 3, \ldots$, are nonnegative harmonic and as such form a normal family. The sequence $\{u_j - u_1\}$ is also monotone and either converges or diverges to $+\infty$ at each point of U. By the normality, this sequence either converges normally to a harmonic function or to $+\infty$ in U, and likewise for the sequence $\{u_j\}$.

Corollary 8.1.7. *A family $\mathcal{H}$ of harmonic functions in a domain U which omit one specific real value α is normal.*

In fact, for each $u \in \mathcal{H}$, either $u > \alpha$ or $u < \alpha$ in U. This is so because if $u(z_1) > \alpha$ and $u(z_2) < \alpha, z_1, z_2 \in U$, then u would assume the value α at some point z on each curve in U joining z_1 and z_2. Thus $\mathcal{H}$ can be partitioned into two subfamilies

$$\mathcal{H}_1 = \{u \in \mathcal{H} \mid u > \alpha\}, \quad \mathcal{H}_2 = \{u \in \mathcal{H} \mid u < \alpha\}$$

The normality of $\mathcal{H}_1$ and $\mathcal{H}_2$ is then deduced, and so $\mathcal{H}$ is normal.

8.2 Some Properties of Subharmonic Functions

Definition 8.2.1. Suppose that $v(z)$ is a continuous, real-valued function inside a domain U, and V is any domain inside U. If for any harmonic function $u(z)$ inside $V, v(z) - u(z)$ satisfies the maximum principle inside V, then $v(z)$ is said to be a *subharmonic function* inside the domain U.

Theorem 8.2.2.

(1) Suppose $v_1(z)$ and $v_2(z)$ are subharmonic functions inside a domain U. Then $v(z) = \max[v_1(z), v_2(z)]$ is also subharmonic inside U.

(2) Suppose $v(z)$ is a subharmonic function inside the domain U, and $\mathbb{D}(a, r)$ is a disc inside U. If we use $H(z)$ to denote the function determined by the Poisson formula, then the function

$$V(z) = \begin{cases} H(z), & z \in \overline{\Delta} \\ v(z), & z \in D - \Delta \end{cases}$$

is also subharmonic in U.

Proof.

(1) The continuity of $v(z) = \max[v_1(z), v_2(z)]$ inside U is obvious. Next, we use the definition of subharmonic function to prove that $v(z)$ is subharmonic inside U. Suppose V is a domain inside U, and $u(z)$ is a harmonic function inside V with $v(z) - u(z)$ taking a maximum value at a point $a \in G$. We may safely assume that $v(a) = v_1(a)$, so that inside V we then have

$$\begin{aligned} v_1(z) - u(z) \le v(z) - u(z) &\le v(a) - u(a) \\ &= v_1(a) - u(a) \end{aligned}$$

Thus, by Definition 8.2.1, $v_1(z) - u(z)$ is constant inside V, and by the above inequality, we know that $v(z) - u(z)$ is also constant inside V, thereby showing that $v(z)$ is subharmonic inside U.

(2) It is easy to see that $V(z)$ is a subharmonic function inside both Δ and $U - \overline{\Delta}$. Suppose z_1 is a boundary point of Δ, and choose any harmonic function $u(z)$ in a neighborhood of z_1. If $V(z) - u(z)$ attains a maximum at the point z_1, then $v(z) - u(z)$ also attains a maximum at z_1. By Definition 8.2.1, $v(z) - u(z)$ is a constant, and we have

$$\begin{aligned} v(z) - u(z) \le V(z) - u(z) &\le V(z_1) - u(z_1) \\ &= v(z_1) - u(z_1) \end{aligned}$$

hence $V(z) - u(z)$ is a constant. This then proves that $V(z)$ is subharmonic inside U.

□

8.3 The Dirichlet Problem for Analytic and Harmonic Functions

Suppose that U is a bounded domain in the z-plane $\mathbb{C}$, whose boundary consists of $N+1$ smooth closed curves,[1] $\gamma_0, \gamma_1, \ldots, \gamma_N$, and $\gamma_1, \ldots, \gamma_N$ lie inside the bounded domain enclosed by γ_0. There is no harm in assuming that $z = 0 \in U$.

Let $g(t)$ denote a continuous real-valued function on γ, and let $M = \max_{t\in\gamma} |g(t)|$.

Definition 8.3.1. The *Dirichlet boundary value problem*, or first boundary value problem, for harmonic functions on a domain U is the problem of finding a harmonic function $u(z)$ inside U which is continuous over the closed domain $\overline{U} = U + \gamma$, and satisfies the boundary condition:

$$u(t) = g(t) \quad \text{for } t \in \gamma\,.$$

In order to use the methods of subharmonic functions to prove the solvability of Dirichlet boundary value problems, we introduce a family of functions $\mathcal{G} = \{v(z)\}$ on U, such that $\mathcal{G}$ consists of all the functions $v(z)$ subharmonic in U, such that the upper limit of each $v(z)$ at every point t of the boundary γ does not exceed $g(t)$; in other words,

$$\overline{\lim_{z\to t}}\, v(z) \le g(t), \quad t \in \gamma\,.$$

We begin by proving an important property of this family $\mathcal{G}$.

Theorem 8.3.2. *Let $u(z)$ represent the function which is the least upper bound (abbreviated to "l.u.b.") of all the functions in $\mathcal{G}$, i.e., let*

$$u(z) = \ l.u.b.\ v(z), \quad v(z) \in \mathcal{G}\,.$$

Then $u(z)$ is harmonic inside U.

Proof. Since $g(t) \le M$ for $t \in \gamma$ and $v(z) \in \mathcal{G}$, we have $u(z) \le M$ for $z \in U$.

Choose arbitrary $a \in U$. Suppose that $\overline{\mathbb{D}(a,r)} \subset U$. By the hypothesis of the theorem, we know that there exists a sequence of functions $v_j(z) \in \mathcal{G}, n = 1, 2, \ldots$, such that

$$\lim_{n\to\infty} v_j\,(a) = u\,(a)\,.$$

[1] Suppose L is a curve, whose parametric representation is $z = z(t), a \le t \le b$. If $z(t)$ has a nowhere zero continuous derivative $z'(t)$ over the interval $[a,b]$ (meaning a one-sided derivative at the endpoints a and b), then L is said to be a *smooth curve*. Furthermore, if $z(a) = z(b)$ and $z'(a) = z'(b)$, then L is said to be a *smooth closed curve*. The condition of smoothness is equivalent to having continuously varying tangent lines.

Set $V_j(z) = \max\{v_1(z), \ldots, v_j(z)\}$. By Theorem 8.2.2 (1), $V_j(z)$ is subharmonic inside U, and $\{V_j(z)\}$ is a nondecreasing sequence of functions in $\mathcal{G}$. Let $U_j(z)$ denote a harmonic function inside the disc $\mathbb{D}(a,r)$, which is continuous on $\overline{\mathbb{D}(a,r)}$, and such that $U_j(z) = V_j(z)$ for $|z-a| = r$. Define

$$W_j(z) = \begin{cases} U_j(z) & \text{for } z \in \mathbb{D}(a,r) \\ V_j(z) & \text{for } z \in U \setminus \mathbb{D}(a,r) \end{cases}$$

According to Theorem 8.2.2 (2), $W_j(z) \in \mathcal{G}$, and $\{W_j(z)\}$ is a nondecreasing sequence of functions inside U. Moreover, since at the point a,

$$v_j\,(a) \le V_j\,(a) \le W_j\,(a) \le u\,(a) \le M$$

we have

$$\lim_{n\to\infty} W_j\,(a) = u\,(a)$$

Then Theorem 8.1.6 gives us the fact that the nondecreasing sequence of harmonic functions $\{W_j(z)\}$ converges inside $\mathbb{D}(a,r)$ to a harmonic function $U(z)$ such that $U(z) \le u(z)$ for $z \in \mathbb{D}(a,r)$, and $U\,(a) = u\,(a)$.

Next, we choose any point $z_1 \in \mathbb{D}(a,r)$, and a sequence of subharmonic functions $\{v_j^*(z)\}$ from the family of functions $\mathcal{G}$ such that

$$\lim_{j\to\infty} v_j^*\,(z_1) = u\,(z_1)$$

Let

$$V_j^*(z) = \max\left[v_1(z), v_1^*(z), \ldots, v_j(z), v_j^*(z)\right]$$

and let $U_j^*(z)$ denote the harmonic function inside the disc $\mathbb{D}(a,r)$ such that it is continuous on $\overline{\mathbb{D}(a,r)}$, and $U_j^*(z) = V_j^*(z)$ for $|z-a| = r$. Put

$$W_j^*(z) = \begin{cases} U_j^*(z), & z \in \mathbb{D}(a,r) \\ V_j^*(z), & z \in U \setminus \mathbb{D}(a,r) \end{cases}$$

As before, we can prove that the nondecreasing sequence of harmonic functions $\{W_j^*(z)\}$ converges inside $\mathbb{D}(a,r)$ to a harmonic function $U^*(z)$ such that $U(z) \le U^*(z)$ for $z \in \mathbb{D}(a,r)$, and $U^*\,(z_1) = u\,(z_1)$. From this it follows that the harmonic function $U(z) - U^*(z)$ attains a maximum value of 0 at the point a in $\mathbb{D}(a,r)$, whence $U(z) \equiv U^*(z)$ for $z \in \mathbb{D}(a,r)$, and $u\,(z_1) = U\,(z_1)$ for any point z_1 in $\mathbb{D}(a,r)$. This proves that $u(z)$ is harmonic inside the disc $\mathbb{D}(a,r)$, and due to the arbitrary choice of $a \in D$, $u(z)$ is also harmonic inside the domain U. □

Theorem 8.3.3. *There exists a unique solution to the Dirichlet boundary value problem.*

Proof. It is simple to prove the uniqueness of the solution by using the maximum-minimum principle, so let us prove the existence.

Suppose $u(z)$ is the harmonic function as described in Theorem 8.3.2; we want to prove that

$$\lim_{z\to t}(z) = \Phi(z), \quad t \in \gamma$$

We only need to prove that for any positive number $\varepsilon > 0$, we have

$$\begin{cases} \overline{\lim}_{z\to t_0} u(z) \le \Phi(t_0) + \varepsilon, \\ \underline{\lim}_{z\to t_0} u(z) \ge \Phi(t_0) - \varepsilon, \end{cases} \quad t_0 \in \gamma \tag{8.2}$$

For this we need to construct a barrier function,

$$\omega(z) = \operatorname{Im}\left[e^{-i\alpha}\sqrt{\frac{z-t_0}{z-t_1}}\right]$$

where t_1 is a point outside of $\overline{U}$ such that with the exception of the point t_0, the line segment $\overline{t_0 t_1}$ is completely contained in the complement of $\overline{U}$, where we fix a single-valued analytic branch of $\sqrt{(z-t_0)/(z-t_1)}$ inside U, and where the constant α is so chosen as to satisfy both $\omega(t_0) = 0$ and $\omega(z) > 0$ for every z in $\overline{U} - \{t_0\}$. In particular, $\omega(t)$ is a positive harmonic function inside U.

Given any positive number $\varepsilon > 0$, we can choose a suitably small positive number $r > 0$, such that for $t \in \gamma \cap \mathbb{D}(a, r)$ we have

$$|g(t) - \Phi(t_0)| < \varepsilon$$

Let $\omega_0 = \min_{z\in\overline{U}-\mathbb{D}(a,r)} \omega(z) > 0$, and consider the harmonic function

$$V(z) = \Phi(t_0) + \varepsilon + \frac{\omega(z)}{\omega_0}[M - \Phi(t_0)]$$

where $M = \max_{t\in\gamma} |g(t)|$. It is easy to see that for $t \in \gamma \cap \mathbb{D}(a, r)$ we have

$$V(t) \ge \Phi(t_0) + \varepsilon > g(t)$$

and for $t \in \gamma \setminus \mathbb{D}(a, r)$ we have

$$V(t) \ge M + \varepsilon > g(t)$$

By the maximum principle for subharmonic functions, we know that $v(z) < V(z)$ for every function $v(z)$ in the family $\mathcal{G}$. Hence, we have $u(z) \le V(z)$ for $z \in U$, for the harmonic function $u(z)$ in Theorem dp-thm1 In particular,

$$\overline{\lim_{z\to t_0}}\, u(z) \le V(t_0) = \Phi(t_0) + \varepsilon$$

which is just the first inequality in (8.2). In order to prove the second inequality in (8.2), we consider the function

$$W(z) = \Phi(t_0) - \varepsilon - \frac{\omega(z)}{\omega_0}[M + \Phi(t_0)]$$

Since

$$W(t) \leq \Phi(t_0) - \varepsilon < g(t)$$

for $t \in \gamma \cap \mathbb{D}(a,r)$, and

$$W(t) \leq -M - \varepsilon < g(t)$$

for $t \in \gamma - \mathbb{D}(a,r)$, we see that the harmonic function $W(z) \in B$. Hence, $u(z) \geq W(z)$ for $z \in U$, so that

$$\underline{\lim}_{z \to t_0} u(z) \geq W(t_0) = \Phi(t_0) - \varepsilon$$

Theorem 8.3.3 follows. □

Remark 8.3.4. It is clear from the preceding proof that we can weaken the conditions on the boundary γ of the domain U, i.e., it suffices to assume that every point t on γ is the endpoint of a segment which, except for the point t, lies on the outside of the closed domain $\overline{U}$.

Remark 8.3.5. It follows from the uniqueness property that $U(\zeta)$[2] remains invariant under transformations of the group G. In fact, the functions $U(S(\zeta))$ and $U(\zeta)$ are both solutions of the problem that we have just examined and hence they are identical in $|\zeta| < 1$. If we now return to the z-plane, the function $U(\zeta)$ becomes the function $u(z)$, which is single-valued and bounded in B and which satisfies the Dirichlet problem posed above. It also follows from what was said above that this problem has a unique solution. We note that if certain of the curves constituting the boundary Γ degenerated into points, the Dirichlet problem for the domain obtained would not always have a solution. To see this, note that the function $u(z)$ which provides a solution to the Dirichlet problem for such a domain is, by virtue of what was said above, a harmonic function at all isolated boundary points. Therefore, it is uniquely determined by its values on the remaining boundary curves. This shows that the values of the function $u(z)$ at isolated boundary points cannot be chosen arbitrarily.

Remark 8.3.6. Thus, we have proved the existence and uniqueness of the solution to the Dirichlet problem for an arbitrary finitely connected domain U bounded by Jordan curves and for arbitrary given values on the boundary Γ that constitute either a continuous function or a function with a finite or countable set of points of discontinuity. To find the solution of an arbitrary Dirichlet problem for a given domain U, it is sufficient to know the solution of any one particular Dirichlet problem. Here, it is a question of finding the Green's function.

[2] In the present case, with the possible exception of a countable set of points on $|\zeta| = 1$.

8.4 The Green's Function

We now give the definition of the Green's function and then some examples that illustrate the Green's function and its use.

Definition 8.4.1. Suppose that Ω is a bounded domain whose boundary is a finite number of disjoint smooth simple closed curves (Fig. 8.1).

Let p be a point in Ω. The Green's function for Ω with singularity at p is the function $G(z; p)$ with these three properties:

(a) $G(z; p)$ is harmonic for z in Ω, $z \neq p$,

(b) $G(z; p) + \log |z - p|$ is harmonic near p,

(c) $G(z; p) = 0$ if z is in the boundary of Ω.

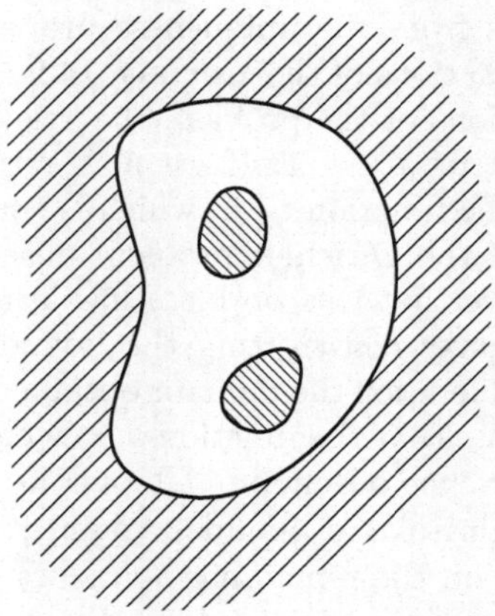

FIGURE 8.1: A domain bounded by three disjoint piecewise smooth simple closed curves

The Green's function for an unbounded domain is defined exactly as in **(a)**, **(b)**, and **(c)** above if p is not the point at ∞. If ∞ is a point in the domain Ω, then the Green's function for Ω with singularity at ∞ is the function $G(z; \infty)$ satisfying

(a)' $G(z; \infty)$ is harmonic on $\Omega \backslash \{\infty\}$,

(b)' $G(z; \infty) - \log |z|$ is harmonic for $|z|$ large,

(c)' $G(z; \infty) = 0$ if z is in the boundary of Ω.

Such a function $G(z;p)$ can be found in the following way. Let v be the harmonic function on Ω with $v(z) = \log|z-p|$ if z is in the boundary of Ω; then set

$$G(z;p) = v(z) - \log|z-p|.$$

In accordance with the maximum principle, $g(z,\zeta)$ must be positive everywhere in U. If a Green' s function $g(z,\zeta)$ exists, then, for finite ζ the difference $g(z,\zeta) - \log(z-\zeta)$ when assigned the appropriate value at $z=\zeta$, at the point $z=\zeta$ is harmonic in U and continuous in $\bar{U}$ and assumes on the boundary Γ the values $-\log(1/|z-\zeta|)$; that is, this difference is a solution of a particular Dirichlet problem for the domain U.

Conversely, if $u(z)$ is a solution of this Dirichlet problem, then the function $-\log(z-\zeta) + u(z)$ is obviously a Green's function for the domain U. An analogous situation holds for $\zeta = \infty$.

Under a single sheeted mapping of the domain U by means of a function $z' = f(z)$ onto another domain U' bounded by Jordan curves, a Green's function $g(z,\zeta)$ is transformed into the function $g\left(f^{-1}(z'), f^{-1}(\zeta')\right)$, which, obviously, is a Green's function for the domain U'.

This shows that, if we know a Green's function for one domain, we can, by means of univalent mappings, obtain Green's functions for various other domains of the same order of connectedness.

In the case of a simply connected domain U bounded by a closed Jordan curve, an analytic function $h(z,\zeta)$ the real part of which is a Green's function $g(z,\zeta)$ for a domain U is obviously continuous on $\bar{B}$ except at the point $z=\zeta$. If, in addition, the boundary of the domain U is an analytic curve, then $p(z,\zeta)$ is holomorphic on that curve. It turns out that these properties also hold in the case of finitely connected domains.

Let us consider first a finitely connected domain U bounded by arbitrary closed Jordan curves. Let $g(z,\zeta)$ denote a Green's function of U and let $h(z,\zeta)$ denote its harmonic conjugate. As we shall see, the last function is non-single-valued in U. Consider the function $F(z) = g - ih$. Let us take an arbitrary points z_0 on one of the boundary curves of the domain U and let us describe a circle γ about it sufficiently small that it intersects the curve and there are no other boundary curves either inside the circle γ or on it. The part of the domain U lying inside γ consists of simply connected domains. We denote by B_0 whichever one of these domains has the point z_0 on its boundary. Let us map B_0 onto the halfplane $\mathcal{F}(t) > 0$ in such a way that $z = z_0$ is mapped into $t = 0$. If we denote the mapping function by $z = \phi(t)$, the function $F(\phi(t))$ is holomorphic in the domain $\mathcal{F}(t) > 0$ and its real part $u(t)$ is also continuous in $\mathcal{F}(t) \geq 0$ and is nonnegative. Noting that $u(t) = 0$ at points on the real axis sufficiently close to $t = 0$, we describe a circle $|t| = \rho$ sufficiently small that $u(t) = 0$ on the diameter lying on the real axis.

Let us now define a function $H(t)$ on $|t| = \rho$ as follows: $H(t)$ coincides with the function $u(t)$ on the upper semicircle and it assumes at points in the lower semicircle the same values as the function $u(t)$ assumes at the conjugate points in the upper halfplane with opposite sign. The function $H(t)$ is obviously continuous on the circle $|\zeta| = \rho$. Consider the integral

$$H(t) = \frac{1}{2\pi}\int_0^{2\pi} H\left(pe^{i\theta}\right)\frac{p^2 - r^2}{p^2 - 2pr\cos(\theta - \varphi) + r^2}d\theta, t = re^{i\varphi}$$

It defines a function $H(t)$ that is harmonic in $|t| < \rho$ and continuous on $|t| \leq \rho$.

It is obviously equal to 0 at points on the real axis and it coincides with $u(t)$ on the upper boundary semicircle. Consequently, $H(t)$ coincides with $u(t)$ on the entire boundary of the upper half-disk and therefore throughout this entire halfdisk. If $\Phi(t)$ is an analytic function in $|t| < \rho$ and its real part is the function $H(t)$, then the difference $\Phi(t) - F(\phi(t))$ has real part equal to zero in the upper half-disk. Therefore, this difference must be a constant in $|t| < \rho$. This last fact shows that the function $F(\phi(t))$ can be extended in a neighborhood of $t = 0$ beyond the real axis and, in particular, it is a continuous function at the point $b = 0$. Returning to the z-plane, we conclude that the function $F(z)$, considered in $\bar{B}$, is continuous at the point z_0. Since z_0 is an arbitrary point on the boundary of the domain U, this means that $F(z)$ is continuous on $\bar{B}$ except, of course, at the point ζ. Furthermore, since $\Phi'(0)$ is obviously nonzero, it follows that $f'(z) = \Phi'(t)/\phi'(t) \neq 0$ in a part of a sufficiently small neighborhood of the point z_0 contained in U. This means that the zeros of the function $F'(z)$ cannot cluster on the boundary of the domain U. Consequently, $F'(z)$ and hence $h_z'(z, \zeta)$ bave only finitely many zeros. Finally, since the function $w = \Phi(t)$ maps the part of the neighborhood of the point $t = 0$ lying above the real axis into the part of the neighborhood of the point $w = \Phi(0)$ that lies to the right of a vertical line passing through the point $w = \Phi(0)$, it follows that, as we move along the boundary of U in a neighborhood of z_0, the function $h(z, \zeta)$ always changes in the same direction. More precisely, as z describes the boundary of the domain U in the positive direction, $h(z, \zeta)$ increases. Now, if z_0 is an interior point of an analytic arc on the boundary of U, then the function $\phi(t)$ is holomorphic at the point $t = 0$. But then, the function $F(z)$ and hence the function $p(z, \zeta)$ is holomorphic at the point z_0.

8.5 The Green's Function and the Riemann Mapping Function

In the case of a simply connected region there is a simple connection between Green's function and the Riemann mapping function.

Theorem 8.5.1. *Let $f_a(z)$ be the Rieman map from the simply connected domain U onto the unit disc $\mathbb{D}$ for fixed $a \in U$, satisfying $f_a(a) = 0$, $f_a'(a) > 0$. The function*

$$g(z, a) := -\log|f_a(z)|, \quad z \in U$$

has the following properties:

(i) *$g(z,a)$ is harmonic in $z \in U \setminus \{a\}$,*

(ii) *$\log|z-a| + g(z,a)$ is harmonic in the neighborhood of a,*

(iii) *$\lim_{z\to\partial U} g(z,a) = 0$.*

Proof.

(i) In $U \setminus \{a\}$ we have

$$\begin{aligned}\Delta g(z,a) &= -4\frac{\partial^2}{\partial z\partial\overline{z}}\frac{1}{2}\log|f_a(z)|^2\\ &= -2\frac{\partial^2}{\partial z\partial\overline{z}}\left(\log f_a(z) + \log\overline{f_a(z)}\right) = 0\end{aligned}$$

(ii)

$$\log|z-a| + g(z,a) = \log\left|\frac{z-a}{f_a(z)}\right| = \operatorname{Re}\log\frac{z-a}{f_a(z)}$$

Because $\frac{f_a(z)}{z-a}$ is analytic and non-vanishing in the neighborhood of a the function $\log\left|\frac{z-a}{f_a(z)}\right|$ is harmonic there.

(iii) Let $\{z_n\}$ be a sequence in U with all its accumulation points on the boundary ∂U. Then all accumulation points of the image sequence $\{w_n\}$, $w_n = f_a(z_n) \in \mathbb{D}$, are lying on $\partial\mathbb{D}$. For if otherwise $\hat{w}$, $|\hat{w}| < 1$ would be an accumulation point of $\{w_n\}$ there would exist a subsequence $\{w_{n_k}\}$ of $\{w_n\}$ with limit $\hat{w} \in \mathbb{D}$. Let $\hat{z} \in U$ be the preimage of $\hat{w}$, $\hat{w} = f_a(\hat{z})$ and z_{n_k} those of w_{n_k}, $w_{n_k} = f_a(z_{n_k})$. The inverse mapping of $w = f_a(z)$ which will shortly be denoted by $z = z(w)$ is analytic in the neighborhood of $\hat{z}$, too. By continuity we have

$$\hat{z} = z(\hat{w}) = \lim_{k\to+\infty} z(w_{n_k}) = \lim_{k\to+\infty} z_{n_k} \in \partial U.$$

This contradicts $\hat{z} \in U$. Therefore all accumulation points of $\{w_n\}$ are on $\partial\mathbb{D}$. This proves property **(iii)**.

□

Theorem 8.5.2. *The function $g(z,a)$ has the additional properties*

1. *$0 < g(z,a)$,*
2. *$g(z,a) = g(a,z)$,*
3. *It is uniquely given by properties **(i)** to **(iii)**,*

4. *If ϕ is any schlicht mapping from U onto U, then the Green's function is*

$$g(z,a) = -\log\left|\frac{\phi(z)-\phi(a)}{1-\overline{\phi(a)}\phi(z)}\right|.$$

Proof.

1. This property of g follows from the maximum principle for harmonic functions applied to $g(\cdot, a)$ in the domain $U\setminus\{z \mid |z-a|<\varepsilon\}$ for small enough positive ε.

2. The function

$$f(\zeta) := \frac{f_a(\zeta)-f_a(z)}{1-\overline{f_a(z)}f_a(\zeta)}$$

maps U onto $\mathbb{D}$ for any fixed $z, a \in U$ with $z \neq a$. Moreover, $f(z)=0$ and the function $\frac{f_z(\zeta)}{f(\zeta)}$ has a removable singularity at $\zeta = z$ with

$$\lim_{\zeta\to z}\frac{f_z(\zeta)}{f(\zeta)} = \frac{f_z'(z)}{f'(z)} = 1-|f_a(z)|^2 \leq 1$$

where for $z \neq a$ we have a strong inequality. Applying the maximum principle by observing

$$\lim_{\zeta\to\partial U}\left|\frac{f_z(\zeta)}{f(\zeta)}\right| = 1$$

we find

$$|f_z(\zeta)| \leq |f(\zeta)|, \quad \zeta \in U$$

especially for $\zeta = a$

$$|f_z(a)| \leq |f_a(z)|$$

By interchanging the roles of z and a the inverse inequality can be proved in the same way. Hence, equality holds in the last relation giving

$$g(z,a) = g(a,z)$$

3. Let g_1, g_2 be two functions with properties (i) to (iii), then $g_1 - g_2$ is a harmonic function in the entire domain U with vanishing boundary values of ∂U. Thus $g_1 - g_2 = 0$ in U by the maximum principle.

4. Together with ϕ the linear transformation $\frac{\phi-\phi(a)}{1-\overline{\phi(a)}\phi}$ maps U onto U. Obviously,

$$-\log\frac{|\phi-\phi(a)|}{\left|1-\overline{\phi(a)}\phi\right|}$$

has the properties (i) to (iii), so it is the Green's function of U.

□

Remark 8.5.3. The existence of the Green's function for a given domain U can be proved if the Dirichlet problem for harmonic functions can be solved for U. The Dirichlet problem demands us to find a harmonic function in U attaining prescribed boundary values on the boundary ∂U. In the case of continuous boundary values this problem can be shown to be (uniquely) solvable for a wide class of domains by the method of Perron-Rado-Riesz. If then $u(z,a)$ is the harmonic function in U satisfying

$$\lim_{z \to \zeta} u(z,a) = \log |\zeta - a| , \quad \zeta \in \partial U$$

for $a \in U$ fixed then

$$g(z,a) := \log \frac{1}{|z-a|} + u(z,a)$$

is the Green's function for U. See [205, p. 4], [62, p. 263].

There are domains without a Green's function. If e.g., U has an isolated boundary point a Green's function should vanish at this point but would be positive in a neighborhood which is impossible because of the minimum principle, see [205, p. 2]. Another class of domains having no Green's function are domains whose boundary has vanishing capacity, see [62, p. 281], [205, p. 54].

Remark 8.5.4. The result of Theorem 8.5.1 can be inverted. If $g(z,a)$ is the Green's function of a simply connected domain U and $h(z,a)$ its harmonic conjugate[3], then $f_a(z) := \exp[-(g+ih)(z,a)]$ is the Riemann mapping function for U, satisfying $\lim_{z \to \zeta} |f_a(z)| = 1$ for $\zeta \in \partial U$ and $f_a(a) = 0$, $f_a'(a) \neq 0$.

8.6 Harmonic Measure

We now define the concept of a harmonic measure inside an $(N+1)$-connected domain U.

Definition 8.6.1. Suppose $\omega_j(z) \quad (0 \leq j \leq N)$ is a harmonic function inside the domain U, which is continuous on the closed domain $\overline{U}$. If on the boundary γ we have

$$\omega_j(t) = \begin{cases} 1 & \text{for } t \in \gamma_j \\ 0 & \text{for } t \in \gamma_k, k \neq j , \end{cases}$$

then $\omega_j(z)$ is said to be a *harmonic measure* on U.

[3]Solving the Cauchy-Riemann equations with a given harmonic function, the conjugate harmonic function can locally be calculated. This conjugate harmonic functionin general is not a single-valued function. But it is when the domain U under consideration is simply connected.

On an $(N+1)$-connected domain U, we have $N+1$ different harmonic measures $\omega_j(z)$ $(j = 0, 1, \ldots, N)$, the existence of which we know from Theorem 8.3.3. The harmonic function $\sum_{j=0}^{N} \omega_j(z)$ has the value 1 on the boundary γ, and according to the maximum principle for harmonic functions, we have

$$\sum_{j=0}^{N} \omega_j(z) = 1 \quad \text{for } z \in U\,.$$

However, $0 < \omega_j(z) < 1$ for $z \in U, j = 0, 1, \ldots, N$.

Let us suppose for the multi-connected domain U, that $u(z)$ is the solution to Dirichlet boundary value problem inside U. Then the conjugate harmonic function $v(z)$ determined by the integral

$$v(z) = \int_a^z -\frac{\partial u}{\partial y}dx + \frac{\partial u}{\partial x}dy + v\,(a)\,, \quad a = x_0 + iy_0 \in U$$

is, in general, not single-valued in U. Nevertheless, the multi-valued analytic function $f(z) = u(z) + iv(z)$ is then the solution to the Dirichlet problem for (multi-valued) analytic functions on the domain U.

In the case of a simply-connected domain we have frequently used the fact that the Dirichlet problem is uniquely soluble. The same is true in the case of the above problem for a multiply-connected domain, if only the unknown function be subject to no additional conditions. But we need that the unknown function is not simply harmonic but, moreover, it constitutes the real part of a single-valued analytic function. (Generally speaking it is the real part of a many-valued analytic function.) Under this additional assumption the Dirichlet problem is in the general case insoluble. To make it soluble it is necessary to introduce certain arbitrary parameters into the prescribed value on the contour; this leads to the so-called modified Dirichlet problem. We now present the precise formulation.

Definition 8.6.2. The so-called *modified Dirichlet boundary value problem* for analytic functions on an $(N+1)$-connected domain U is one of finding a (single-valued) analytic function $F(z)$ inside U, such that it is continuous up to the boundary γ of U and satisfies the boundary conditions:

$$\operatorname{Re} F(t) = g(t) + h(t), \quad t \in \gamma, \quad \operatorname{Im} F(a) = b \tag{8.3}$$

where $g(t)$ is a real-valued function on γ possessing Hölder continuous partial derivatives, and

$$h(t) = \begin{cases} 0, & t \in \gamma_0 \\ h_j, & t \in \gamma_j, j = 1, \ldots, N \end{cases}$$

where the $h_j (j = 1, \ldots, N)$ are N indeterminate real constants, a is a point on γ_0, and b is a real constant.

Theorem 8.6.3. *The solution to the modified Dirichlet boundary value problem on an $(N+1)$-connected domain U exists and is unique.*

Proof. We first prove the uniqueness of the solution. Suppose $F_1(z)$ and $F_2(z)$ and two solutions to modified Dirichlet boundary value problem, and let $F(z) = F_1(z) - F_2(z)$. Then $F(z)$ is analytic inside U, and satisfies the boundary conditions:

$$\operatorname{Re} F(t) = h(t), \quad t \in \gamma, \quad \operatorname{Im} F(a) = 0$$

Letting $u(z) = \operatorname{Re} F(z)$ and $v(z) = \operatorname{Im} F(z)$, we have by Green's formula,

$$\begin{aligned}
&\iint_U \left[(u_x)^2 + (u_y)^2\right] dxdy \\
&= \iint_U \left[(uu_x)_x + (uu_y)_y\right] dxdy = \int_\gamma uu_x dy - uu_y dx \\
&= \int_\gamma u\frac{\partial u}{\partial n} ds = \int_\gamma u\frac{\partial v}{\partial s} ds = \sum_{j=1}^N h_j \int_{\gamma_i} dv = 0
\end{aligned}$$

where $u_x = \frac{\partial u}{\partial x}$ and $u_y = \frac{\partial u}{\partial y}$. From the preceding equation we know that $u_x = u_y = 0$ inside U, and that

$$\begin{aligned}
u(z) &= \int_a^z u_x dx + u_y dy = 0, \quad z \in U \\
v(z) &= \int_a^z -u_y dx + u_x dy = 0
\end{aligned}$$

Thus, $F(z) = u(z) + iv(z) \equiv 0$, i.e., $F_1(z) \equiv F_2(z)$ for $z \in U$.

To prove the existence of a solution, we use Theorem 8.3.3 to first find a solution $u(z)$ to Dirichlet boundary value problem satisfying the boundary conditions:

$$u(t) = g(t), \quad t \in \gamma$$

and we write

$$\beta_j = -\frac{1}{2\pi}\int_{\gamma_j} \frac{\partial u}{\partial n} ds, \quad j = 1, 2, \ldots, N \tag{8.4}$$

If $\beta_j = 0, j = 1, \ldots, N$, we use $v(z)$ to denote the conjugate harmonic function of $u(z)$, so that since

$$\int_{\gamma_j} \frac{\partial u}{\partial n} ds = \int_{\gamma_j} \frac{\partial v}{\partial s} ds = \int_{\gamma_j} dv = 0, \quad j = 1, \ldots, N$$

$v(z)$ is single-valued in U. Thus, $F(z) = u(z) + iv(z)$ is a single-valued analytic function in U which is a solution to the Dirichlet boundary value problem with $h_j = 0, j = 1, \ldots, N$, in (8.3). If in (8.4) the $\beta_j (j = 1, \ldots, N)$ are not all 0, we write

$$r_{jk} = \frac{1}{2\pi}\int_{\gamma_j} \frac{\partial \omega_k}{\partial n} ds, \quad j, k = 1, \ldots, N$$

where $\omega_k(z)(k = 1, \ldots, N)$ are all harmonic measures. We now prove:

$$J = \begin{vmatrix} r_{11} & \cdots & r_{1N} \\ \cdots & \cdots & \cdots \\ r_{N1} & \cdots & r_{NN} \end{vmatrix} \neq 0 \tag{8.5}$$

Suppose $J = 0$, then there exist real constants $c_1, \ldots, c_N$ not all equal to 0, such that the function

$$f(z) = \sum_{k=1}^{N} c_k \omega_k(z)$$

satisfies the condition

$$\frac{1}{2\pi} \int_{\gamma_j} \frac{\partial f}{\partial n} ds = 0, \quad j = 1, \ldots, N$$

Using $g(z)$ to denote the conjugate harmonic function of $f(z)$, then since

$$\frac{1}{2\pi} \int_{\gamma_j} \frac{\partial f}{\partial n} ds = \frac{1}{2\pi} \int_{\gamma_j} \frac{\partial g}{\partial s} ds = 0, \quad j = 1, \ldots, N$$

we know $g(z)$ is single-valued in U and therefore, $f(z)+ig(z)$ is a singlevalued analytic function in U satisfying the boundary conditions of the modified Dirichlet boundary value problem:

$$\text{Re}[f(t) + ig(t)] = f(t) = c_j, \quad t \in \gamma_j, j = 0, \ldots, N$$

where $c_0 = 0$. If we let $g_0 = g(a)$, then by the uniqueness of the solution to modified Dirichlet boundary value problem, we know that

$$f(z) + ig(z) = ig_0$$

whence

$$f(z) = \sum_{k=1}^{N} c_k \omega_k(z) \equiv 0$$

which is impossible. (8.5) therefore holds. Hence, there exist real constants $h_k(k = 1, \ldots, N)$ such that

$$\sum_{k=1}^{N} h_k \left(\frac{1}{2\pi} \int_{\gamma_j} \frac{\partial \omega_k}{\partial n} ds \right) = \beta_j, \quad j = 1, \ldots, N$$

The harmonic function determined by these $h_k(k = 1, \ldots, N)$,

$$H(z) = u(z) + \sum_{k=1}^{N} h_k \omega_k(z)$$

now satisfies

$$\begin{aligned}\frac{1}{2\pi}\int_{\gamma_j}\frac{\partial U}{\partial n}ds &= -\beta_j + \sum_{k=1}^{N}\frac{h_k}{2\pi}\int_{\gamma_j}\frac{\partial \omega_k}{\partial n}ds \\ &= 0, \quad j = 1, \ldots, N\end{aligned}$$

Thus, the conjugate harmonic function $V(z)$ in U of $H(z)$ satisfies the conditions:

$$\int_{\gamma_j}\frac{\partial V}{\partial s}ds = 0, \quad j = 1, \ldots, N$$

that is, $V(z)$ is single-valued in U, and $V(a) = b$. Hence

$$F(z) = H(z) + iV(z), \quad z \in D,^{3}$$

is the solution to modified Dirichlet boundary value problem satisfying (8.3). □

It is easily seen that by simply adding a complex constant to the solution of the modified Dirichlet boundary value problem in the preceding theorem, we obtain the following result.

Corollary 8.6.4. *There exists a unique analytic function $F(z)$ in the domain U which is continuous on $\overline{U}$, and satisfies the boundary conditions:*

$$\operatorname{Re} F(t) = g(t) + h_j, \quad t \in \gamma_j, \quad j = 0, \ldots, N$$

where the $h_j (j = 0, 1, \ldots, N)$ are suitably chosen real constants, and $F(0) = c$, c a complex constant.

We can use either Theorem 8.6.3 or Corollary 8.6.4 to prove the existence of a univalent analytic function which conformally maps a bounded simply connected domain onto the unit disc, and also to prove the existence of a univalent analytic function which transforms a multi-connected domain into several canonical domains.

9

The Ahlfors Map

The idea of the Ahlfors map is to come up with a substitute for the Riemann mapping theorem in the case of domains that are multiply connected. The elegant answer to this question is due to Ahlfors, and it builds on the proof of the Riemann mapping theorem in a very nice way. The informed reader will want to compare the Ahlfors map to the map that comes from the uniformization theorem.

9.1 An Extremum for Bounded Functions in Multiply Connected Domains

The Ahlfors function was first investigated, but not named, by Ahlfors [2] and Ahlfors and Beurling [1, p. 101].

To explain Ahlfors's result we begin with the classical Riemann mapping theorem for simply connected domains. It is today, of course, included in many textbooks on complex analysis; cf. [87, Theorem 6.4.2, p. 190, Theorem 6.6.3, p. 197].

The domain U in $\overline{\mathbb{C}}$ is called *simply connected* if the closed set $\mathbb{C} \setminus U$ is connected.

Theorem 9.1.1 (Riemann mapping theorem). *Let $U \subset \mathbb{C}$ be a simply connected, proper, open subset. Then U is biholomorphic to the unit disc $\mathbb{D}$. Moreover, if one fixes $a \in U$ and $\varphi \in [0, 2\pi)$, then there is a unique such map f_a with $f_a(a) = 0$ and the argument of the derivative of f_a at a, $\arg f_a'(a) = \varphi$.*

The unique function specified by Theorem 9.1.1 will be called the *Riemann map* associated to a.

Riemann is justly credited with the theorem that bears his name, despite the fact that, as was pointed out by Weierstrass, the proposed proof contains a serious flaw. A long sequence of modified proofs were then developed during the following half-century, but all of them imposed assumptions concerning the smoothness of the boundary. Finally, H. Poincaré succeeded in mapping a region conformally onto a unit disc, without regard for boundary behavior (thus *avoiding completely any assumption about the boundary*). Since then

DOI: 10.1201/9781003619581-9

numerous contributors have provided alternative proofs and have investigated delicate problems concerning the behavior of the mapping function near the boundary. Since 1851, when Riemann presented his flawed proof, many major analysts have participated in what has become a substantial part of complex analysis. A very partial listing of names with which the reader should become familiar is the following: C. Neumann, H.A. Schwarz, D. Hilbert, R. Courant, P. Koebe, E. Lindelöf, P. Painleve, and C. Carathéodory (and, of course, Riemann, Weierstrass, and Poincaré).

The Riemann mapping theorem is one of the most important results of complex analysis. It implies e.g., that any simply connected domain U in $\mathbb{C}$ is homeomorphic to $\mathbb{C}$ and thus that any closed curve in U is homotopic to a point.

Because conformal equivalence is an equivalence relation, it is symmetric and transitive, and so the Riemann mapping theorem implies that any two simply connected domains are conformally equivalent, for we can use the unit disc as an intermediate stage.

On the other hand, two domains belonging to a different classes cannot be conformally equivalent. This can be seen quickly as follows: Without loss of generality, we may confine attention to the sphere, $\overline{\mathbb{C}}$, and the unit disc $\mathbb{D}$. The sphere is compact, while neither $\mathbb{C}$ nor $\mathbb{D}$ is compact; since the continuous (a fortiori the conformal) image of a compact set is compact, we are left only with the question whether $\mathbb{C}$ and $\mathbb{D}$ are conformally equivalent; by the Liouville Theorem, they are not conformally equivalent, and so the above assertion is confirmed.

Recall that a simply connected domain is one with no holes or punctures; every closed curve lying in a simply connected domain U must have its interior contained entirely in U also. It is not hard to see that if U can be mapped one-to-one onto the disc by a continuous mapping f (analyticity not required), then U must be simply connected. And if f is analytic as well, then U cannot equal $\mathbb{C}$ because of Liouville's Theorem. The converse, however, the existence of the conformal mapping f, is not at all obvious.

Further Comments

1. The Riemann Mapping Theorem is an existence theorem. It does not give an explicit method of writing down the mapping function f. In the paper for every simply connected region in $\mathbb{C}$ we give a series of increasingly better approximations of the Riemann map in the Riemann mapping theorem. The idea is to approximate a domain by a polygon using a Whitney decomposition, then use the Schwarz—Christoffel formula to compute the map onto the polygonal approximation, and finally use the Carathéodory convergence theorem to obtain the Riemann map in the limit.

2. In particular, the theorem does not assert that the analytic mapping can be accomplished by a linear fractional transformation in all cases. We know this is false.

3. You may have found the Riemann Mapping Theorem surprising: a topological property (simple connectedness) guaranteeing analytic equivalence. For any two annuli such as U_1 and U_2 surely are topologically equivalent surely there exists a one-to-one continuous mapping of U_1 onto U_2 with continuous inverse, yet now the mapping cannot be analytic unless $R_1/r_1 = R_2/r_2$. And what if the domain U has more than one hole?

What happens if we drop the condition that the competing functions are injective? That is, what happens if we maximize over the larger class of functions that are simply analytic and bounded by 1 in U? The answer is the same—that the solution to this new (relaxed?) extremal problem is again a conformal mapping of U onto $\mathbb{D}$ (unique if normalized as above).

This is not obvious—certainly Koebe's proof cannot be easily modified to eliminate the assumption of injectivity. That an apparently less restrictive extremal problem has the same solution as the original problem emerges as a special case of Ahlfors's theorem:

> The true equivalent of Schwarz's lemma for single-valued functions in a multiply-connected domain.

The quote is from Ahlfors's 1947 paper "Bounded analytic functions." He considers precisely the extremal problem above in the setting of multiply-connected domains. Here's a partial statement of his result (we will give a more complete version later).

Theorem 9.1.2 (Ahlfors mapping theorem, first version). *Let U be a domain of connectivity $m \geq 1$, none of whose boundary components reduce to a point. Fix a point $a \in U$ and consider the extremal problem*

$$\max |f'(a)|$$

where f is analytic (and single-valued) in U with $|f| < 1$. The solution of this problem is an $m:1$ map onto $\mathbb{D}$.

9.2 A General Discussion of Conformal Mappings between Multiply Connected Domains

Let U and U' be regions in the z-plane and the w-plane, respectively. Suppose there exists an analytic function $w = f(z)$ that is a bijection of the region U onto the region U'; then, because $f(z)$ is univalent in U, $f'(z)$ never vanishes in U, and hence the mapping preserves angles. We say the regions U and

U' are conformally equivalent. (It is simple to verify that this is actually an equivalence relation.)

Clearly, a multiply connected region cannot be conformally equivalent to a simply connected region.

Riemann's mapping theorem permits us to conclude that any two simply connected domains, with the exception of the whole plane, can be mapped conformally onto each other, or that they are conformally equivalent.

If we try to introduce the concept of the standard domain, or, as we shall also say, the *canonical domain*, into the theory of conformal mapping of multiply-connected domains, we meet two initial obstacles.

The first, and less serious, one is due to the fact that a conformal mapping is continuous and thus preserves the order of connectivity of a domain. In order to stay on an elementary level, we will limit ourselves to the general case of conformal mappings of multiply-connected domains of finite connectivity, including the case of connectivity greater than 2. Oded Schramm has found a way to treat domains of infinite connectivity.

First we prove that two domains with different connectivity numbers cannot be conformally mapped onto each other.

Theorem 9.2.1. *Suppose U_1 is an n-connected domain on the z-plane, and $w = f(z)$ is a univalent meromorphic function in U_1 which maps U_1 onto the domain U_2 in the w-plane. Then U_2 is also an n-connected domain.*

Proof. We may assume that U_1 contains the point at ∞ and that the function $w = f(z)$ maps $z = \infty$ into $w = \infty$, for otherwise the desired assumption can be obtained through a linear fractional transformation. Now suppose $\Gamma_j (j = 1, 2, \ldots)$ are the n components of the boundary of the domain U_1 and we use n polygons $P_j (j = 1, 2, \ldots)$ to enclose each of the $\Gamma_j (j = 1, 2, \ldots)$, respectively, such that each polygon is in the exterior of the other polygons. The function $w = f(z)$ maps these n polygons $P_j (j = 1, 2, \ldots, n)$ onto n sectionally smooth Jordan closed curves $L_j (j = 1, 2, \ldots, n)$, where each curve is in the exterior of those others remaining.

Assuming the connectivity number of the domain U_2 to be $m < n$, then there is at least one L_j among the n closed curves $L_j (j = 1, 2, \ldots, n)$, whose interior belongs to U_2. Consider the inverse function $z = g(w)$ of $w = f(z)$ which univalently maps L_j onto P_j. By Theorem 13.4.3 of Appendix, $z = g(w)$ conformally maps the interior region of L_j onto that of P_j, which in turn belongs to U_1. This contradiction proves the connectivity number of U_2 to be $m \geq n$. By interchanging U_1 and U_2 and considering the inverse function $z = g(w)$ of $w = f(z)$, we can prove by the same argument that the connectivity number of U_2 is $m \leq n$. Hence, the connectivity number of U_2 is exactly m. □

The preceding theorem shows the necessity for two multiply connected domains to have the same connectivity number if one can conformally map one onto the other. It therefore becomes necessary to introduce distinct canonical domains for each order of connectivity.

The second difficulty is caused by the fact that no exact equivalent of the Riemann mapping theorem holds in the multiply-connected case. It is not true that any two domains of the same order of connectivity can be conformally mapped onto each other.

Let A be the *annulus* in $\mathbb{C}$ defined by

$$A = \{z \in \mathbb{C} \mid 0 < r < |z| < R\}.$$

The number $M = \log(R/r)$ is called the *modulus of* A. Two annuli with the same modulus are clearly biholomorphic to each other under a suitable homothetic transformation.

Theorem 9.2.2. *Any doubly connected domain can be conformally mapped onto an annulus.*

Proof. Let D denote a doubly connected domain in the $\mathbb{C}$. If D has an isolated boundary point z_0, then consider a simple connected domain D_* such that $D_* \ni z_0$. By the Riemann mapping theorem, we can conformally map D_* onto $\mathbb{D}$, or $\mathbb{C}$, such that z_0 is mapped into the point $w = 0$. The domain D then is conformally mapped onto the annulus:

$$\mathbb{D}_* \text{ or } \mathbb{C}_*.$$

Suppose now Γ_1 and Γ_2 are the two components of the boundary of D, each of which contains at least two points, and suppose one of them, say Γ_1, is bounded. The complement of Γ_2 in $\mathbb{C}$ is necessarily an open set, and while this open set may be a simply connected domain or an open set consisting of several nonintersecting simply connected domains, one domain among these must be a simply connected domain D_1 containing D. We can thus conformally map D_1 onto the disc $|z'| < 1$, and under this mapping Γ_2 is mapped onto a connected closed set Γ_2' in the disc $|z'| > 1$, and D is mapped onto a domain D' whose boundary is just the union of Γ_2' and $|z'| = 1$. Suppose the simply connected component of the complement of Γ_2' in the z'-plane which contains D' is denoted by D_2. We can then conformally map D_2 onto the domain $|z''| > 1$, such that the point $z' = \infty$ is mapped into the point $z'' = \infty$. Under this mapping, the circle $|z'| = 1$ is mapped into an analytic Jordan curve in the domain $|z''| > 1$ and D' is mapped onto D''. The boundary of D'' is the union of this Jordan curve and the circle $|z''| = 1$. Combining the above two conformal mappings, we have established a conformal mapping from the domain D onto D''.

Next we prove that the domain D'' can be conformally mapped onto a certain annulus: $1 < |w| < h$. We divide the proof into several steps: We first map the domain $|z''| > 1$ onto the left half-plane $\operatorname{Re} t > 0$ through the function $t = \ln z''$. We should point out that this transformation is not one-to-one, for any point z'' in $|z''| > 1$ corresponds to an infinite point set

$$t + 2k\pi i, \quad k = 0, \pm 1, \pm 2, \ldots,$$

of $\mathrm{Re}\, t > 0$. Furthermore, this transformation also maps the domain D'' onto a vertical strip domain H in the t-plane. H is a simply connected domain, one part of whose boundary is the imaginary axis and the other part of which is an infinitely analytic curve. If a point $t \in H$, then $t + 2k\pi i \in H$ for $k = 0, \pm 1, \pm 2, \ldots$.

Next, we conformally map the domain H onto the strip $G \equiv \{0 < \mathrm{Re}\,\zeta < h\}$, such that the boundary points $t = -\infty i, 0, \infty i$ of H map into the boundary points $\zeta = -\infty i, 0, \infty i$ of G. Under this mapping, the point $t = 2\pi i$ is mapped into a point a_0 on the positive imaginary axis. We may assume that $a_0 = 2\pi i$, for otherwise the desired result could be achieved by a similarity transformation. Letting $\zeta = f(t)$ represent the aforementioned conformal mapping, and $t = g(\zeta)$ its inverse function, we can prove that

$$f(t + 2k\pi i) = f(t) + 2k\pi i \text{ and } g(\zeta + 2k\pi i) = g(\zeta) + 2k\pi i, \quad k = \pm 1, \pm 2, \ldots \tag{9.1}$$

In fact, $f(t + 2\pi i)$ and $f(t) + 2\pi i$ both conformally map the domain H onto the domain G, and the points $t = -\infty i, 0, \infty i$ into the points $\zeta = -\infty i$, $2\pi i, \infty i$, respectively. By the uniqueness of the mapping function, we know that the first equality in (9.1) holds for $k = 1$; the remaining cases are also easily proven.

Finally, we use the function $w = e^{\zeta}$ to map the domain G onto the annulus: $1 < |w| < e^h$. This transformation is also one-to-one, for if a point $\zeta \in G$, then $\zeta + 2k\pi i \quad (k = 0, \pm 1, \pm 2, \ldots)$ also corresponds to the same point $w = e^{\zeta}$.

Combining the above three mappings yields $w = e^{f(\ln z'')}$, which maps the domain D'' onto the annulus $1 < |w| < e^h$. Since from (9.1), $w = e^{f(\ln z'')}$ and $z'' = e^{g(\ln w)}$ establish a one-to-one correspondence between the domain D'' and the annulus, the function $w = e^{f(\ln z'')}$ is univalent inside D''. Furthermore, this function is analytic inside D'' and its inverse function is also analytic inside the annulus $1 < |w| < e^h$. □

Let A_1 and A_2 be two annuli:

$$A_k = \{z \in \mathbb{C} \mid 1 < |z| < r_k\}, \quad k = 1, 2\,.$$

Suppose $f : A_1 \to A_2$ be a holomorphic mapping, where $r_1 \neq r_2$, and consider the harmonic function

$$h(z) = \log r_2 \log |f(z)| - \log r_2 \log |z|.$$

Clearly, $h(z)$ is harmonic in A_1 and has vanishing boundary values. $h(z)$ is therefore identically zero, and the analytic function

$$g(z) = \log r_1 \log f(z) - \log r_2 \log z\,,$$

of which $h(z)$ is the real part, reduces to an imaginary constant, i.e.,

$$g(z) = \log r_1 \log f(z) - \log r_2 \log z = i\alpha \tag{9.2}$$

If z describes the circle $|z| = r_1$ in the positive sense, $w = f(z)$ describes $|w| = r_2$ in the positive sense, and both $\arg\{z\}$ and $\arg\{f(z)\}$ grow by 2π. Since $\arg\{z\} = \mathrm{Im}\{\log z\}$, it follows from (9.2) that the imaginary part of $g(z)$ grows by the amount $2\pi(\log r_1 - \log r_2)$ if z describes $|z| = r_1$. But $\mathrm{Im}\{g(z)\} = \alpha = \text{const.}$, and we must therefore have $\log r_1 = \log r_2$, that is, $r_1 = r_2$, contrary to our assumption that $r_1 \neq r_2$.

Hence, we may say that two annuli are of the same conformal type if, and only if, their moduli are the same.

If we pass now to doubly-connected domains G we have a new situation: Two such domains are conformally equivalent if and only if they are conformally equivalent to an annulus with the same conformal module. (This annulus can degenerate.)

Therefore we must try to find a system of canonical domains with the property that each multiply connected domain is conformally equivalent to one and only one canonical domain. Theorem 9.2.1 shows that becomes necessary to introduce distinct canonical domains for each order of connectivity. The choice of canonical domains is to a certain extent arbitrary, and there are several types with equally simple properties.

Hence, we cannot expect to find a canonical domain onto which all domains of a given order of connectivity can be mapped. This difficulty is overcome by giving a more flexible definition of a canonical domain which determines the general geometric character of the domain without fixing its moduli. The more important of these domains are designed in Fig. 9.1.

9.3 Conformal Mappings For Canonical Domains

By a domain with parallel slits, we mean a domain obtained from the whole plane by removing several mutually parallel line segments (cf. Fig. 9.1 (a)). In this section we want to prove an existence and uniqueness theorem for conformal mappings from a multiply connected domain onto such a domain.

The precise statement is given by the following theorem.

Theorem 9.3.1. *Every multiply connected domain G in $\mathbb{C}$ can be mapped by a univalent function onto a parallel slit domain, with slits of prescribed angle θ. Moreover, if ζ is a given point of G which is mapped to ∞, then the expansion of the mapping about $z = \zeta$ has the form*

$$w = \frac{1}{z-\zeta} + a_1(z-\zeta) + \dots \qquad \textit{or} \qquad w = z + \frac{a_1}{z} + \dots \tag{9.3}$$

according to whether ζ is finite or ∞.

There exist today many proofs of Theorem 9.3.1, but the simplest proof uses the extremal property with respect to $\mathrm{Re}\, a_1$ in the development in (9.3); cf. [80, p. 179], [91, p. 82], [163, p. 346] and many other books [147].

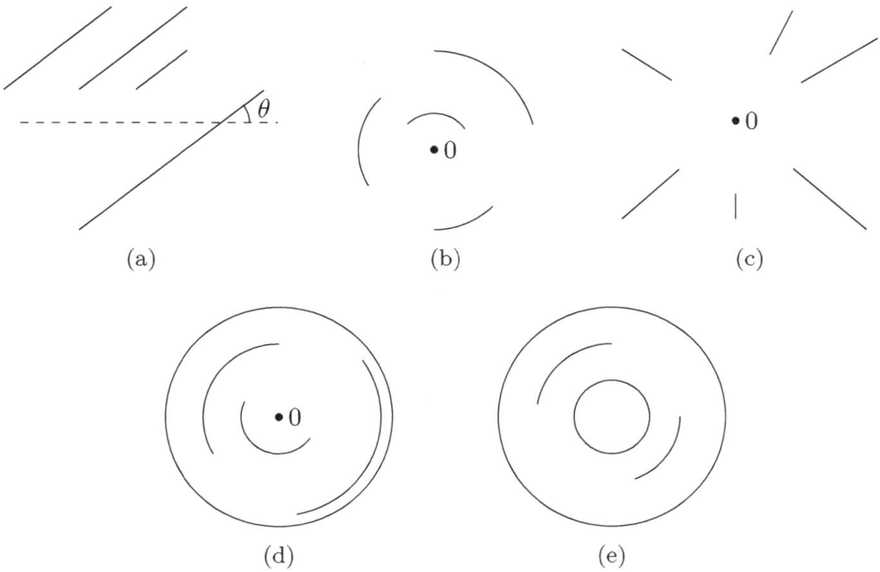

FIGURE 9.1: Examples of (a) a parallel slit domain (b) a circular slit domain (c) a radial slit domain (d) a circle with concentric circular slits (e) a circular ring with concentric circular slits.

The proof will use a technical result which is a direct corollary of Green's formula (see, for example [187, 10.45 Green's theorem, p.282]). The classical Green's formula states that if U is a domain bounded by a finite number of positively oriented curves and functions $P(x, y)$ and $Q(x, y)$ are continuously differentiable in the closure of U, then

$$\int_U [\partial_x P(x,y) - \partial_y Q(x,y)]\, dxdy = \int_{\partial U} Q(x,y)dx + P(x,y)dy$$

This formula can be rewritten in a complex form. Let $F(z, \bar{z}) = F(x, y)$ be a (real) continuously differentiable function; then

$$\begin{aligned}\int_U \partial_{\bar{z}} F(z,\bar{z}) dxdy &= \frac{1}{2}\int_U [\partial_x F(x,y) + i\partial_y F(x,y)]\, dxdy \\ &= \frac{1}{2i}\int_{\partial U} F(x,y)dz.\end{aligned}$$

Lemma 9.3.2. *Let Ω be a multiply connected domain bounded by analytic curves, and let f and g be functions analytic in Ω with the possible exception of a finite number of poles in Ω. Assume that they are regular up to the boundary* *and that $f - g$ is analytic everywhere in Ω. Then*

$$\int_\Omega |f' - g'|^2 \, dxdy = \mathrm{Re}\left(\frac{1}{i}\int_{\partial\Omega}(\bar{f} - \bar{g})f'dz\right) + \frac{1}{2i}\int_{\partial\Omega}\bar{g}g'dz + \overline{\frac{1}{2i}\int_{\partial\Omega}\bar{f}f'dz}. \tag{9.4}$$

Proof. Using the classical Green's formula rewriten in complex form and the fact that $\partial_{\bar{z}}\bar{f} = \bar{f}'$, we can write

$$\int_\Omega |f' - g'|^2 = \int_\Omega (f' - g')\overline{(f' - g')} = \frac{1}{2i}\int_{\partial\Omega}(f' - g')\overline{(f - g)}dz$$
$$= \frac{1}{2i}\int_{\partial\Omega}\overline{(f - g)}f'dz + \frac{1}{2i}\int_{\partial\Omega}\bar{g}g'dz - \frac{1}{2i}\int_{\partial\Omega}\bar{f}g'dz$$

We integrate the last integral by parts (the integrated term vanishes since $\partial\Omega$ is a union of closed curves)

$$\int_{\partial\Omega}\bar{f}g'dz = \int_{\partial\Omega}\bar{f}dg = -\int_{\partial\Omega}gd\bar{f} = -\overline{\int_{\partial\Omega}\bar{g}f'dz}$$

Finally, we add and subtract the conjugate of the first term to obtain

$$\mathrm{Re}\left(\frac{1}{i}\int_{\partial\Omega}(\bar{f} - \bar{g})f'dz\right) + \frac{1}{2i}\int_{\partial\Omega}\bar{g}g'dz + \overline{\frac{1}{2i}\int_{\partial\Omega}\bar{g}f'dz} - \overline{\frac{1}{2i}\int_{\partial\Omega}\overline{(f - g)}f'dz}$$

The last two terms simplify to

$$\overline{\frac{1}{2i}\int_{\partial\Omega}\bar{f}f'dz}$$

and we obtain (9.4). □

Proof of Theorem 9.3.1. To prove this theorem, we consider $S_\theta = S_{z_0,\theta}$ the class of all functions univalent in Ω that have expansion as in (9.3). We state that these functions are admissible.

The main idea of the proof is to show that there is an admissible function Let us assume for a while that a slit domain mapping f_θ does exist. Let f be any other admissible function. We can apply Lemma 9.3.2 to f_θ and f to obtain

$$\int_\Omega |f'_\theta - f'|^2 = \mathrm{Re}\left(\frac{1}{i}\int_{\partial\Omega}\left(\bar{f}_\theta - \bar{f}\right)f'_\theta\right) + \frac{1}{2i}\int_{\partial\Omega}\bar{f}f' - \overline{\frac{1}{2i}\int_{\partial\Omega}\bar{f}_\theta f'_\theta} \tag{9.5}$$

Since f_θ maps onto a slit domain, $e^{-i\theta} f'_\theta dz$ is real everywhere on the boundary. This allows one to write

$$f'_\theta(z)dz = e^{2i\theta}\overline{f'_\theta(z)dz}, \quad z \in \partial\Omega. \tag{9.6}$$

This identity can be used to simplify some integrals in the above formula. We start with the first one. Using (9.6) and the residue calculus,

$$\frac{1}{i}\int_{\partial\Omega} \left(\bar{f}_\theta - \bar{f}\right) f'_\theta dz = \frac{e^{2i\theta}}{i}\overline{\int_{\partial\Omega} \left(f_\theta - f\right) f'_\theta dz} = \overline{\left(2\pi e^{-2i\theta}\left(a_\theta - a_1\right)\right)},$$

where a_θ and a_1 are the coefficients of f_θ and f in front of $(z - z_0)$ for finite z_0 and in front of z^{-1} if $z_0 = \infty$. Thus, the first term in (9.5) simplifies to

$$2\pi \operatorname{Re}\left(e^{-2i\theta}\left(a_\theta - a_1\right)\right)$$

Using Green's formula once again, we obtain that the second term in (9.5) is $-A$ where $A = \text{Area}(\mathbb{C}\backslash \text{f}(\Omega))$ is the area of the domains inside the images of boundary components of $f(\Omega)$. The minus sign comes from the fact that the boundary is positively oriented with respect to $f(\Omega)$ instead of its complement. Using (9.6) one more time, we simplify the last term in (9.5)

$$\int_{\partial\Omega} \bar{f}_\theta f'_\theta dz = e^{2i\theta}\overline{\int_{\partial\Omega} f_\theta f'_\theta dz} = \frac{e^{2i\theta}}{2}\overline{\int_{\partial\Omega} d\left(f_\theta^2\right)} = 0.$$

Combining all these simplifications together, we arrive at

$$\int_\Omega \left|f'_\theta - f'\right|^2 = 2\pi \operatorname{Re}\left(e^{-2i\theta}\left(a_\theta - a_1\right)\right) - A$$

Since $A \geq 0$, we have that $\operatorname{Re}\left(e^{-2i\theta} a_\theta\right) \geq \operatorname{Re}\left(e^{-2i\theta} a_1\right)$. Moreover, the equality happens if and only if

$$\int_\Omega \left|f'_\theta - f'\right|^2 = 0$$

which happens if and only if f and f_θ differ by a constant. In particular, this implies that if a normalized map onto a slit domain exists, then it must be unique.

Applying this argument to $f(z) = 1/\left(z - z_0\right)$ for finite z_0 or $f(z) = z$ if $z_0 = \infty$, we can see that $\operatorname{Re}\left(e^{-2i\theta} a_\theta\right) \geq 0$. Moreover, equality happens if and only if f and f_θ differ by a constant.

Let us consider the problem of maximizing $\operatorname{Re}\left(e^{-2i\theta} a_1\right)$ over all functions in $S_{z_0,\theta}$. For simplicity, we assume that $z_0 = \infty$. The proof of the other case is almost identical.

First, this class is not empty since $f(z) = z$ belongs to this class. By the Cauchy formula, the functional $f \mapsto a_1$ is continuous with respect to the local uniform convergence in $\widehat{\mathbb{C}}$. Since the family S_θ is normal, the normalization near z_0 implies that the limit of any convergent sequence is univalent. This follows from essentially the same argument as in the simply connected case. This gives us that there is a function $F(z) = z + A_1/z + \cdots$ which maximizes our functional over all functions from S_θ. It remains to be shown that this F maps onto a slit domain.

Assume that this is not true and there is a boundary component γ such that its image is not a slit with the slope θ. Let Ω' be the domain exterior to $F(\gamma)$. This domain is simply connected. By the simply connected Riemann mapping theorem, there is a conformal map from this domain onto the complement of the unit disc which maps infinity to itself. Composing this map with an appropriate Möbius transformation, a rotation of the Joukowsky map, we can find a map ϕ from Ω' onto a domain which is a complement of a slit with the slope θ. Moreover, we can choose this map to have expansion at infinity of the form $\phi(z) = z + a_\theta z^{-1} + \cdots$. We already know that $\operatorname{Re}\left(e^{-2i\theta} a_\theta\right) > 0$ since $F(\gamma)$ is not a straight slit. The composition $\phi(F)$ is also a univalent map on Ω. Expanding the composition at infinity, we get

$$\phi(F(z)) = z + \frac{a_\theta + A_1}{z} + \cdots$$

This implies that $\phi(F) \in S_\theta$ and it has a larger value of the functional than F. This contradiction implies that F maps onto a slit domain. □

Parallel slit regions may be thought of as canonical regions, but they are not all conformally inequivalent, even if it is required that the point at ∞ should correspond to itself—cf., for example, [3, p. 260].

Another important type of canonical mapping consists of the *Circular Slit* (resp. *Radial Slit*) *mapping* of Ω.

In both cases, we can normalize a map in such a way that two given points z_1 and z_2 from U are mapped to the origin and infinity. Let us consider a family of functions f that are univalent in U, $f(z_1) = 0$, and there is a simple pole of residue 1 at z_2. The function that maximizes $|f'(z_1)|$ maps the domain onto a circular slit domain and the function that minimizes $|f'(z_1)|$ maps the domain onto a radial slit domain. We will not prove these results here. Interested readers can find the proofs and further discussion of the uniformization of multiply connected domains in Nehari [163, Chapter VII].

9.4 Harmonic Measures

In the following U denotes a plane region of connectivity $m > 1$. The components of the complement are denoted by $C_1, C_2, \ldots, C_m$, and we take C_m to be the unbounded component. Without loss of generality we can and will assume that no C_k reduces to a point, for it is clear that a point component is a removable singularity of any mapping function, and consequently the mappings remain the same if this isolated boundary point is added to the region.

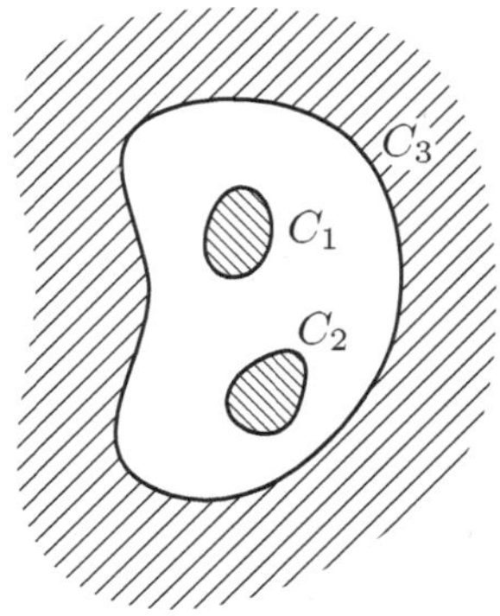

FIGURE 9.2: A domain of commectivity 3

The complement of C_m is a simply connected region U'. By Riemann's theorem, U' can be mapped conformally onto the unit disk $\mathbb{D}$; under this mapping U is transformed into a new region, and the images of $C_1, \ldots, C_{m-1}$ are the bounded components of its complement. For the sake of convenience we agree to use the same notations as before the mapping; in particular, C_m is now the set $\mathbb{C} \setminus \bar{\mathbb{D}}$. The unit circle $\partial\mathbb{D}$, traced in the positive direction, will be denoted by γ_m and is called the outer contour of the new region U.

Consider now the complement of C_1 with respect to the extended plane. This is again a simply connected region, and we map it onto the outside of the unit circle with ∞ corresponding to itself. The image of γ_m is a directed closed analytic curve which we continue to denote by γ_m, just as we keep all the other notations. In addition we define the inner contour γ_1 to be the unit circle in the new plane, traced in the negative direction.

The process can evidently be repeated until we end up with a region U bounded by an outer contour γ_m and $m-1$ inner contour $\gamma_1, \ldots, \gamma_{m-1}$ (Fig. 9.3). It is important to note that the index of a contour with respect to an arbitrary point in the plane can be readily computed. For instance, at the stage where $\gamma_k, k < n$, is the unit circle, the index of γ_k is -1 with respect to interior points of C_k and 0 with respect to all other points not on γ_k. The subsequent mappings will not change this state of affairs. The fact is clear, and a formal proof based on proof based on the argument principle can easily be given. One shows in the same way that the outer contour γ_m has the index 0 with respect to interior points of C_m and the index 1 with respect to all other points not on γ_m. It follows that the cycle $C = \gamma_1 + \gamma_2 + \cdots + \gamma_m$ bounds U in the sense of Chap. 3, Sec. 3.2.2, Definition 3.2.18 The distinction between outer and inner contours is coincidental, for evidently an inversion with respect to an interior point of C_k will make γ_k the outer contour.

It is clear that Theorem 8.3.3 applies to U (see Remarck 8.3.4). As a matter of fact the existence of a barrier is completely obvious since any contour can be transformed into a circle.

Suppose now that we solve the Dirichlet problem in U with the boundary values 1 on γ_k and 0 on the other contours. The solution is denoted by $\omega_k(z)$,

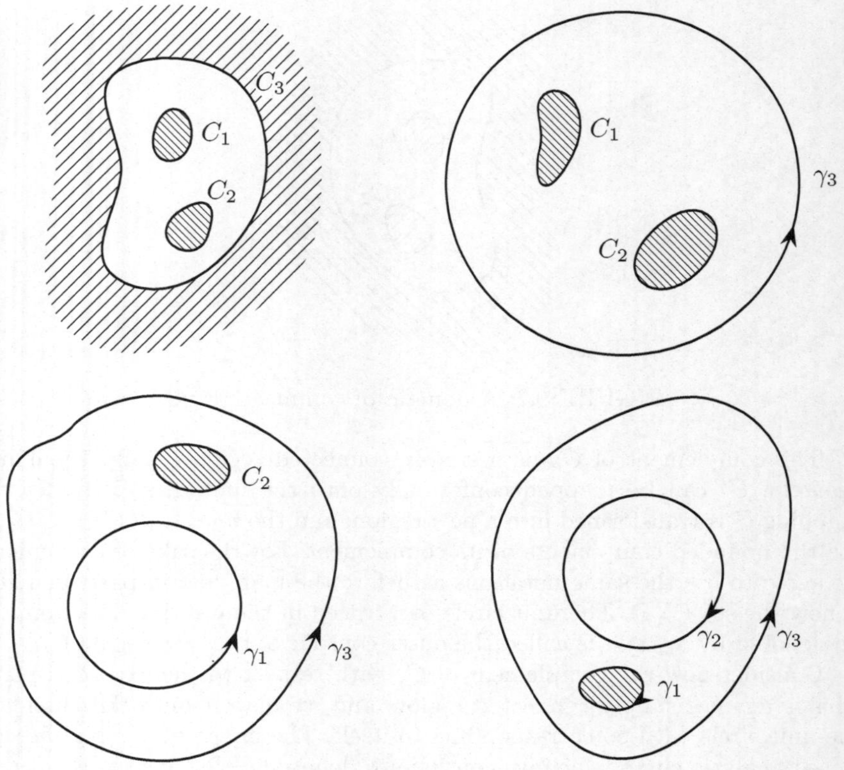

FIGURE 9.3: Transformations of a multiply connected region.

and it is called *harmonic measure of* γ_k with respect to the region U. We have clearly $0 < \omega_k(z) < 1$ in U and

$$\omega_1(z) + \omega_2(z) + \cdots + \omega_n(z) \equiv 1$$

If we map U so that γ_i becomes a circle, then ω_k can be continued across γ_i according to the reflection principle. We conclude that ω_k is harmonic in the closed region U in the sense that it can be extended to a larger region.

The contours $\gamma_1, \ldots, \gamma_{m-1}$ form a homology basis for the cycles in U, homology being understood with respect to an unspecified larger region. The conjugate harmonic differential of ω_k has periods

$$\alpha_{kj} = \int_{\gamma_i} \frac{\partial \omega_k}{\partial n} ds = \int_{\gamma_i}^{*} d\omega_k$$

along γ_j. We assert that no linear combination

$$\lambda_1\omega_1(z) + \lambda_2\omega_2(z) + \cdots + \lambda_{m-1}\omega_{m-1}(z)$$

with constant coefficients can have a single-valued conjugate function unless all the λ_i are zero. To see this, suppose that this expression were the real part of an analytic function $f(z)$. By the reflection principle, $f(z)$ would have an analytic extension to the closure of U. The real part of $f(z)$ would be constantly equal to λ_i on $\gamma_i, i = 1, \ldots, m-1$, and zero on γ_n. Consequently, each contour would be mapped onto a vertical line segment. If w_0 does not lie on any of these segments, a single-valued branch of $\arg(f(z) - w_0)$ can be defined on each contour. It follows the argument principle that $f(z)$ cannot take the value w_0 in U. But then $f(z)$ must reduce to a constant, for otherwise the image of U would certainly contain points not on the line segments. We conclude that the real part of $f(z)$ is identically zero, and hence the boundary values λ_i must all vanish.

What we have proved is that the homogeneous system of linear equations

$$\lambda_1\alpha_{1j} + \lambda_2\alpha_{2j} + \cdots + \lambda_{m-1}\alpha_{m-1,j} = 0 \quad (j = 1, \ldots, m-1) \tag{9.7}$$

has only the trivial solution $\lambda_i = 0$, for these are the conditions under which $\lambda_1\omega_1 + \cdots + \lambda_{m-1}\omega_{m-1}$ has a single-valued conjugate. By the theory of linear equations any inhomogeneous system of equations with the same coefficients as (9.7) must have a solution. In particular, we conclude that it is possible to solve the system

$$\begin{cases} \lambda_1\alpha_{11} + \lambda_2\alpha_{21} + \cdots + \lambda_{m-1}\alpha_{m-1,1} = 2\pi \\ \lambda_1\alpha_{12} + \lambda_2\alpha_{22} + \cdots + \lambda_{m-1}\alpha_{m-1,2} = 0 \\ \cdots\cdots + \cdots \\ \lambda_1\alpha_{1,m-1} + \lambda_2\alpha_{2,m-1} + \cdots + \lambda_{m-1}\alpha_{m-1,m-1} = 0 \\ \lambda_1\alpha_{1n} + \lambda_2\alpha_{2n} + \cdots\cdots + \lambda_{m-1}\alpha_{m-1,n} = -2\pi \end{cases}$$

where the last equation is a consequence of the $m-1$ first (because $\alpha_{k1} + \alpha_{k2} + \cdots + \alpha_{kn} = 0$). In other words, we can find a multiple-valued integral $f(z)$ with periods $\pm 2\pi i$ along γ_1 and γ_n and all other periods equal to zero, the real part being constantly equal to λ_k on γ_k (we set $\lambda_n = 0$).

The function $F(z) = e^{f(z)}$ is then single-valued. We prove:

Theorem 9.4.1. *The function $F(z)$ effects a one-to-one conformal mapping of U onto the annulus $1 < |w| < e^{\lambda_1}$ minus $n-2$ concentric arcs situated on the circles $|w| = e^{\lambda_i}, i = 2, \ldots, m-1$.*

The mapping is illustrated in Fig. 9.4. The contours γ_1 and γ_n are in one-to-one correspondence with the full circles, while the other are flattened into circular slits. It should be imagined that each slit has two edges which together with the endpoints form a closed contour.

Proof of Theorem 9.4.1. The proof is by use of the argument principle. We know that $F(z)$ is analytic with a constant modulus on each contour. The number of roots of the equation $F(z) = w_0$ is given by

$$\frac{1}{2\pi i}\int_{\gamma_1}\frac{F'(z)dz}{F(z)-w_0}+\frac{1}{2\pi i}\int_{\gamma_2}\frac{F'(z)dz}{F(z)-w_0}+\cdots+\frac{1}{2\pi i}\int_{\gamma_n}\frac{F'(z)dz}{F(z)-w_0} \tag{9.8}$$

at any rate if w_0 is not taken on the boundary. For $w_0 = 0$ the terms in (9.8) are known, being equal to $1, 0, \ldots, 0, -1$, respectively. The integral over γ_1

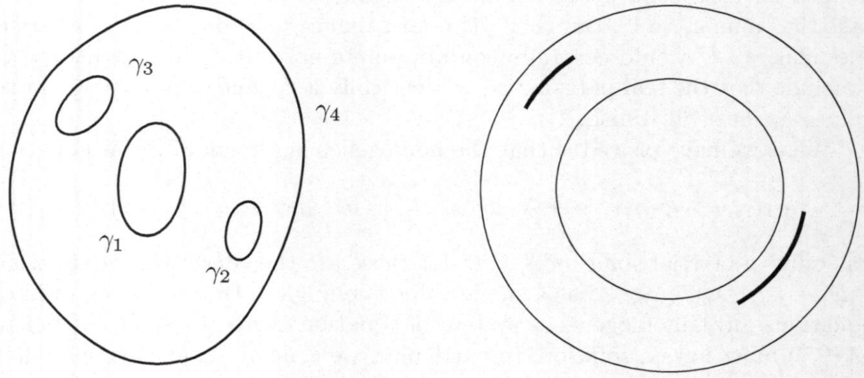

FIGURE 9.4: Concentric slit region

remains constantly equal to 1 for $|w_0| < e^{\lambda_1}$, and it vanishes for $|w_0| > e^{\lambda_1}$; similarly, the last integral is -1 for $|w_0| < 1$ and 0 for $|w_0| > 1$. The integrals over $\gamma_k, 1 < k < n$, vanish for all w_0 with $|w_0| \neq e^{\lambda_k}$. Suppose now that the value w_0 is actually taken by $F(z)$; inasmuch as U must be mapped onto an open set, we can choose $|w_0| \neq$ all e^{λ_i}. For this w_0 the expression (9.8) must be positive. But that is possible only if $1 < |w_0| < e^{\lambda_1}$. Thus $\lambda_1 > 0$ and, by continuity, $0 \leqq \lambda_i \leqq \lambda_1$.

From here on the proof could be completed by means of a purely topological argument. It is more instructive, however, and in fact simpler, to draw the conclusion from the argument principle. When there are simple poles on the boundary, the residue theorem continues to hold provided that the contour integral is replaced by its Cauchy principal value, and provided that the sum of the residues includes one-half of the residues on the boundary. In the present situation the second convention means that a value taken on the boundary is counted with half its multiplicity. The computation of the principal values causes no difficulty. If $|w_0| = e^{\lambda_k}$, we find that

$$⨍_{\gamma_k}\frac{F'(z)dz}{F(z)-w_0} = \frac{1}{2}\int_{\gamma_k}\frac{F'(z)dz}{F(z)}$$

for by elementary geometry (or direct computation)

$$d\arg\left(F(z)-w_0\right) = \frac{1}{2}d\arg F(z)$$

Consequently, the principal values in (9.8) are $\frac{1}{2}$ for $k = 1, 0$ for $2 \leqq k \leqq n-1, -\frac{1}{2}$ for $k = n$.

We conclude now that each value on the circle $|w_0| = 1$ or $|w_0| = e^{\lambda_1}$ is taken one-half time, that is to say once on the boundary; this proves that γ_1 and γ_n are mapped in a one-to-one manner and that $0 < \lambda_i < \lambda_1$, $i \neq 1, n$. Next, if $1 < |w_0| < e^{\lambda_1}$, it follows that w_0 is taken either once in the interior, twice on the boundary, or once on the boundary with the multiplicity 2. On each contour $\gamma_2, \ldots, \gamma_{n-1}$ a single-valued branch of $\arg F(z)$ can be defined, and the values of multiplicity 2 correspond to relative maxima and minima of $\arg F(z)$. There is at least one maximum and one minimum, and there cannot be more or else $F(z)$ would pass more than twice through the same values. Furthermore, the difference between the maximum and the minimum must be $< 2\pi$, which shows that each contour is mapped onto a proper arc. Finally, the arcs which correspond to different contours must be disjoint. □

We have proved the complete Theorem 9.4.1, and in addition we have been able to describe the correspondence of the boundaries. The significance of the theorem is that we can map U onto a canonical region bounded by two circles and $n-2$ concentric circular slits; by way of normalization, the radius of the inner circle is chosen equal to 1. For a given choice of γ_1 and γ_n the canonical mapping is uniquely determined up to a rotation. This follows from the fact that the system (9.8) has only one solution.

The shape of a canonical region of connectivity n depends on $3n-6$ real constants. In fact, the position and size of each slit are determined by three numbers, a total of $3n-6$; the thickness of the annulus gives one additional parameter, but another parameter must be discounted to allow for the arbitrary rotation.

9.5 Existence of the Ahlfors Map

One typical application of Montel's theorem is to guarantee the existence of extremal functions for "compact" extremal problems. To illustrate the principle, we consider the extremal problem of maximizing the derivative at a prescribed point over a compact family of analytic functions.

Let U be a connected, bounded open subset of the complex plane. Then $H^\infty(U)$ is the space of bounded analytic functions in the domain U with the norm

$$\|f\|_\infty = \sup_{z \in U} |f(z)|, \quad f \in H^\infty(U).$$

Let $w \in U$ and put

$$\gamma = \sup\{|h'(w)| \mid h \in H^\infty(U), \|h\|_\infty \leqslant 1\}.$$

A function $h \in H^\infty(U)$, $\|h\|_\infty = 1$, for which $h'(w) = \gamma$ will be called *extremal.*

The extremal function has a number of interesting properties, some of which are developed in the theorems following this useful elementary result.

Lemma 9.5.1. *At least one extremal exists.*

Proof. Assume that $\gamma \not\equiv 0$. Otherwise, any constant $e^{i\theta}$, $\theta \in [0, 2\pi]$, would be an extremal. Since γ is defined as a supremum at a given point a, there exists a sequence $\{h_j\}$ in $H^\infty(U)$ with $\|h_j\|_\infty \leq 1$ such that $\lim_{j\to\infty} |h_j'(a)| = \gamma$. Note that $\{h_j\}$ forms a normal family because this sequence is uniformly bounded by 1 in U. Then, by Montel's theorem, there is a subsequence $\{h_{j_k}(z)\}$ which tends to $h(z)$ uniformly convergent on compact subsets of U. By Weierstrass' theorem $h(z)$ is holomorphic on U and the sequence $\{h'_{j_k}\}$ tends to h', again locally uniformly on U. Hence,

$$|h'(a)| = \lim_{j\to\infty} |h'_{j_k}(a)| = \gamma. \tag{9.9}$$

Since we assumed $\gamma \neq 0, h$ is not constant. We must now show that $\|h\|_\infty = 1$. The pointwise convergence of the subsequence $\{h_{j_k}\}$ implies that the limit function h is bounded by 1. Now assume that

$$\|h\|_\infty = \sup_{z\in U} |h(z)| = m < 1$$

Clear that $m > 0$. Otherwise, h must be constant. Set $g = h/m$. Then $g \in H^\infty(U)$ and

$$\|g\|_\infty = \|h/m\|_\infty \leq 1.$$

It is clear that

$$|g'(a)| = |h'(a)/m| > |h'(a)|,$$

a new function which has a bigger derivative than h and is therefore h is not the extremal. This contradiction shows that $\|h\|_\infty = 1$. □

Theorem 9.5.2 provides an elementary proof that there is only one extremal function.

Theorem 9.5.2. *There is precisely one extremal and it vanishes at w.*

Proof. Let F be an extremal. We first show that $F(w) = 0$. For if F takes values in $\mathbb{D}$ then so does

$$F_1(z) = \frac{F(z) - F(a)}{1 - \overline{F(a)}F(z)},$$

for which

$$F_1'(z) = \frac{F'(z)}{1 - \overline{F(a)}F(z)}$$

Thus if $F(a) \neq 0$ then

$$|F_1'(a)| = \frac{|F'(a)|}{1 - |F(a)|^2} > |F'(a)|$$

We also note that it's possible to modify the extremal problem in ways that, depending on the approach, can make it easier to work with. Obviously, all rotations $e^{i\theta}F(z)$ of competing functions are again competing functions, so we can first assume the problem is to maximize $F'(a)$ among functions with $F'(a) > 0$. Also, if $F(z)$ has values in $\mathbb{D}$ then $g(z) = \frac{4}{\pi}\tan^{-1} F(z)$ has values in $-1 < \operatorname{Re} z < 1$. If $F(a) = 0$ and $F'(a) > 0$ then

$$g'(a) = \frac{4}{\pi}F'(a) > 0$$

Hence the original extremal problem is equivalent to maximizing $\operatorname{Re} g'(a)$ over the class of analytic functions in U satisfying $|\operatorname{Re} g(z)| < 1$. Next, if both F_1 and F_2 are extremals, then so is $\frac{1}{2}(F_1 + F_2)$. The proof will thus be complete if we show that any extremal F is an extreme point of the unit ball of H^∞. Suppose, then, that $g \in H^\infty(U)$ and $\|F \pm g\|_\infty \leqslant 1$. Consequently,

$$|F|^2 \pm 2\operatorname{Re} F\overline{g} + |g|^2 \leqslant 1$$

so that

$$|F|^2 + |g|^2 \leqslant 1$$

Hence

$$|g|^2 \leqslant 1 - |F|^2 = (1 + |F|)(1 - |F|) \leqslant 2(1 - |F|)\,.$$

Define $h = \frac{1}{2}(g^2)$; then $h \in H^\infty$ and

$$|F| + |h| \leqslant 1 \qquad \text{on } U. \tag{9.10}$$

We shall show that (9.10) implies that h vanishes identically; hence, so does g, and F will be proved to be an extreme point of the unit ball of $H^\infty(U)$. First note that $h(w) = 0$. For otherwise there is a choice of λ, $|\lambda| = 1$, which gives the function $F + \lambda F h$ a larger derivative at p than F, a contradiction since this function is bounded by 1 in U:

$$|F + \lambda F h| \leqslant |F| + |F||h| \leqslant |F| + |h| \leqslant 1\,.$$

If h is not identically zero, let r be the order of its zero at $p, r \geqslant 1$. Let ε be a small complex number and consider

$$F(z) + \varepsilon h(z)(z - p)^{-r+1} = G(z).$$

Then $|G'(w)| = \left|\gamma + \varepsilon h^{(r)}(w)/r!\right| > \gamma$, for an appropriate choice of the argument of ε. Furthermore, once out of a neighborhood of p the quantity $|\varepsilon||z - p|^{1-r}$ is less than 1 so that $|G| \leqslant 1$ by the maximum principle. This contradiction establishes that $h \equiv 0$ and the theorem follows. □

The unique function specified by Theorem 9.5.2 will be called the *Ahlfors map* for p and U. The extremal value $A = |F'(w)|$ can be regarded as the best constant for which the Schwarz lemma (infinitesimal version) holds with respect to $w \in D$. The Ahlfors function was first investigated, but not named, by Ahlfors [2] and Ahlfors and Beurling [1, p. 101]. For the subject of the Ahlfors map and for details the reader is referred to the nice book of Fisher [71]; the presentation here is very much like his.

9.5.1 The Ahlfors and Riemann Mappings

The classical Schwarz lemma is part of the grist of every complex analysis text. It gives a way to relate the derivative of a holomorphic function to its more global behavior at a lower level. A version of the result is as follows.

Lemma 9.5.3 (The Schwarz Lemma). *Let $f : \mathbb{D} \to \mathbb{D}$ be holomorphic with $f(0) = 0$. Then,*

(a) $|f(z)| \leq |z|$ for all $z \in \mathbb{D}$

(b) If for some $z_0 \neq 0$ we have $|f(z_0)| = |z_0|$, then f is a rotation.

(c) $|f'(0)| \leq 1$, and if equality holds then f is a rotation.

Proof. The classical argument is to consider $g(z) = f(z)/z$. On a circle $|z| = 1 - \epsilon$, we see that $|g(z)| \leq 1/(1-\epsilon)$. Thus $|f(z)| \leq |z|/(1-\epsilon)$. Since this inequality holds for all $\epsilon > 0$, part (a) follows. The Cauchy estimates show that $|f(0)| \leq 1$

For the uniqueness, if $|f(z)| = |z|$ for some $z \neq 0$, then $|g(z)| = 1$. The maximum modulus principle then forces $|f(z)| = |z|$ for all z, and hence f is a rotation. If, instead, $|f(0)| = 1$, then $|g(0)| = 1$ and again the maximum modulus principle yields that f is a rotation. □

To explain Ahlfors's result let us recall that by the Riemann mapping theorem (Theorem), if $U \neq \mathbb{C}$ is simply connected and $a \in U$, there exists a unique conformal map ψ from U onto $\mathbb{D}$ with $\psi(a) = 0$, $\psi'(a) > 0$. Moreover, ψ appears as the solution of an extremal problem which is very similar to the one defining the Ahlfors map. We shall call ψ the *Riemann map* corresponding to a.

Theorem 9.5.4. *For a simply connected domain $U \neq \mathbb{C}$, the Ahlfors map corresponding to $a \in U$ equals the Riemann map corresponding to a.*

Proof. Let $a \in U$ and let F be the corresponding Ahlfors function. The idea is to compose the Ahlfors function with a Riemann map from the disc onto U. Let ψ be the Riemann map corresponding to a, and let $\phi = \psi^{-1}$, $\psi : U \to \mathbb{D}$. Then $\phi(0) = a$ and since $\psi(\phi(z)) = z$, $z \in \mathbb{D}$,

$$\phi'(0) = \frac{1}{\psi'(a)}.$$

The composition $F \circ \phi$ maps $\mathbb{D}$ into itself and satisfies $F \circ \phi(0) = 0$. Then by the Schwarz lemma (Lemma 9.5.3) we have that

$$|(F \circ \phi)'(0)| = \left| \frac{F'(a)}{\psi'(p)} \right| \leq 1$$

On the other hand, ψ is in the unit ball of $H^\infty(U)$, hence $0 < \psi'(p) \leq F'(a)$. Thus $\psi'(p) = F'(a)$ and by the uniqueness of the Ahlfors function it follows that $F = \psi$. □

Corollary 9.5.5. *Let $U = \mathbb{D}$ and $a \in \mathbb{D}$. Then the Ahlfors function corresponding to a is $F(z) = \frac{z-a}{1-\bar{a}z}$.*

Proof. This follows directly from Theorem 9.5.4 since the automorphism in the statement is the Riemann map from $\mathbb{D}$ onto itself, corresponding to the point a. □

9.6 Ahlfors Functions in Multiply Connected Domains

We close with a general theorem of Ahlfors which refers to domains with real-analytic boundary and can be seen as a generalization of the Riemann mapping theorem.

Theorem 9.6.1. *Let U be bounded by m disjoint analytic Jordan simple closed curves $\gamma_1, \ldots, \gamma_m$ and let $\mathcal{F}$ be the Ahlfors function for U and $a \in U$. Then $\mathcal{F}$ extends analytically over each γ_j and maps each γ_j homomorphically on $\partial\mathbb{D}$:*

$$F' \text{ is not zero on any } \gamma_j\,.$$

$$F \text{ maps } U \text{ onto } \mathbb{D} \text{ exactly } m \text{ times}\,.$$

Proof. The complete boundary will be denoted by Γ. For Green's function with a finite positive pole ζ we shall use the notation $g(z, \zeta)$. We also introduce the harmonic measure $\omega_k(z)$ of the contour γ_k; this is a bounded harmonic function which is 1 on γ_k and 0 on all other contours. The sum $\omega_1(z) + \cdots + \omega_m(z)$ is identically equal to 1.

That the extremal problem posed in the theorem has a solution is obvious. Let $f_0(z)$ denote one of the extremal functions. Since $|f_0'(a)| > 0$, the point $z = a$ is a simple zero of the function $f_0(z)$.

Suppose that the extremal function $f_0(z)$ has, in addition to a, zeros at the points $a_1, \ldots, a_N$ in the domain U. These need not be all the zeros; in fact, $f(z)$ may a priori have infinitely many zeros.

Let these zeros be moved to new positions a'_k, and consider the function $f_1(z)$ defined by

$$\log|f_1(z)| = \log|f_0(z)| + \sum_{k=1}^{N} \left(g\left(z, a'_k\right) - g\left(z, a_k\right)\right)$$

This function is again of absolute value ≤ 1, but it is single-valued only if

$$\int_{\gamma_i} \frac{\partial}{\partial n} \sum_{k=1}^{N} \left(g\left(z, a'_k\right) - g\left(z, a_k\right)\right) ds = \nu_i \cdot 2\pi$$

where the ν_i are integers. (All normal derivatives are taken in the direction of the outer normal.)

By a simple application of Green's theorem

$$\int_{\gamma_i} \frac{\partial}{\partial n} g\left(z, a_i\right) ds = \int_{\gamma} \omega_i \frac{\partial}{\partial n} g\left(z, a_i\right) ds = -2\pi\omega_i\left(a_i\right)$$

Hence, if we assume all the $\omega_i = 0$, the condition will read

$$\omega_i\left(a'_1\right) + \cdots + \omega_i\left(a'_N\right) = \omega_i\left(a_1\right) + \cdots + \omega_i\left(a_N\right) \quad (i = 1, \ldots, m) \tag{9.11}$$

By the extremal property of $f(z)$ these conditions must imply

$$\lim_{z \to \infty} \left(\log|f_1(z)| - \log|f(z)|\right) = \sum_{i=1}^{N} \left(g\left(a'_i\right) - g\left(a_i\right)\right) \leq 0$$

In other words, under condition (9.11) the function

$$F\left(a'_1, \ldots, a'_k\right) = \sum_{i=1}^{N} g\left(a'_i\right)$$

will attain its maximum for $a'_j = a_j$.

We now use the ordinary Lagrangean necessary condition for an extremum. For the sake of rigor we state and prove the precise theorem referred to:

Lemma 9.6.2. *Let $F_0\left(x_1, \ldots, x_m\right), F_1\left(x_1, \ldots, x_m\right), \ldots, F_n\left(x_1, \ldots, x_m\right)$ be of class C^1 in a neighborhood of $\left(x_1^0, \ldots, x_m^0\right)$. If F_0 has a relative extremum at $\left(x_1^0, \ldots, x_m^0\right)$ under the conditions $F_1 = \cdots = F_n = 0$, the rank of the matrix*

$$\left\| \frac{\partial F_i}{\partial x_i} \right\|$$

at $\left(x_1^0, \ldots, x_m^0\right)$ cannot exceed n. The result implies that there exist multipliers λ_i, not all $= 0$, such that

$$\lambda_0 \frac{\partial F_0}{\partial x_i} + \lambda_1 \frac{\partial F_1}{\partial x_i} + \cdots + \lambda_n \frac{\partial F_n}{\partial x_i} = 0 \quad (j = 1, \ldots, m)$$

Proof. The proof is immediate. If the rank were $n+1$ the functions $y_i = F_i(x_1, \ldots, x_m)$ would map a neighborhood of $(x_1^0, \ldots, x_m^0)$ onto an open set. This set would cut the line $y_1 = \cdots = y_n = 0$ along an open segment, and y_0 could not have an extremum. Note that in this formulation we need not know that there exist values $x_i \neq x_i^0$ which fulfill the conditions, but for the same reason we cannot assert that $\lambda_0 \neq 0$. □

Using this lemma we infer the existence of multipliers λ_i, such that the partial derivatives of

$$h(z) = -\lambda_0 g(z) + \lambda_1 \omega_1(z) + \cdots + \lambda_m \omega_m(z) \tag{9.12}$$

vanish at $a_1, \ldots, a_N$. The case $\lambda_0 = 0, \lambda_1 = \lambda_2 = \cdots = \lambda_m$ can be excluded, for the m-th condition being a consequence of the first $m-1$ we can always choose $\lambda_m = 0$. It follows that $h(z)$ is not constant.

But a function of the form (9.12) can have at most $m-1$ critical points if $\lambda_0 \neq 0$, and at most $m-2$ if $\lambda_0 = 0$. This is seen by application of the argument principle to $w'(z)$, where $w(z)$ is the analytic function whose real part is $h(z)$. Assuming that no critical points lie on the boundary we obtain

$$\int_{\Gamma_i} d\arg \frac{dw}{dz} = -\int_{\Gamma_i} d\arg dz = 2\pi.$$

At infinity, $w = -\lambda_0 \log z + c_0 + c_p/z^\nu + \cdots$ and $w'(z) = -\lambda_0/z + \cdots$ for $\lambda_0 \neq 0$, $w'(z) = -pc_p/z^{p+1} + \cdots$ for $\lambda_0 = 0$. The variation of the argument on a large circle is thus -2π or $-(p+1)2\pi$. Hence the number of finite critical points, counted with their multiplicities, turns out to be $n-1$ or $n-1-p$. The result remains true even if there are critical points on the boundary, provided they are counted with half their multiplicity. We may therefore state this preliminary result:

The zeros of the extremal function $f(z)$ are critical points of a function of form (9.12). Their number is at most $m-1$ and if this maximal number is attained, $\lambda_0 \neq 0$.

Now suppose that $a_1, \ldots, a_N$, $N \leq m-1$, constitute all the zeros of the function $f_0(z)$ inside U. Then, the function

$$H(z) = \log|f_0(z)| + g(z,a) + \sum_{k=1}^{m} g(z, a_k) \tag{9.13}$$

is harmonic in U and $H(z) \leq 0$ in U.

Since this last inequality is satisfied in the part of the domain U contained in a small neighborhood of an arbitrary point of the boundary Γ, if we shift to the disk by means of a univalent mapping and use the boundary properties of nonnegative harmonic functions in a disk (see [54, Theorem 2.4, p. 21]), we conclude that $H(z)$ has definite limiting values almost everywhere on Γ along nontangential paths and these values define a summable function on Γ.

Now, let $f_2(z)$ denote the varied function obtained from the relationship

$$\log |f_2(z)| = \log |f_0(z)| + u(z) \tag{9.14}$$

where $u(z)$ is a harmonic function in U.

To make $f_2(z)$ single-valued in U, we require that the conjugate $v(z)$ to $u(z)$ be single-valued in U. For $|f_2(z)|$ not to exceed 1 in U, it will be sufficient to require that $u(z)$ be bounded in U and that $H(z) + u(z)$ be nonpositive almost everywhere on Γ.

This is true because, in the first place, by virtue of (9.13), and (9.14)

$$|f_2(z)| = e^{H(z)+u(z)} \cdot e^{-g(z,a)-\sum_{k=1}^{N} g(z,a_k)} |f_0(z)| \tag{9.15}$$

in U. Furthermore, under the assumptions made, $u(z)$ also has definite limiting values almost everywhere on Γ.

From Green's formula

$$H(z) + u(z) = \frac{1}{2\pi} \int_{\Gamma_\epsilon} (H(z') + u(z')) \frac{dg(z', z)}{dn} ds$$

where Γ_ϵ is the set of curves $g(z', z) = \epsilon$ (where ϵ is a sufficiently small positive number), we obtain Green's formula for the domain U

$$H(z) + u(z) = \frac{1}{2\pi} \int_{\Gamma} (H(z') + u(z')) \frac{dg(z', z)}{dn} ds$$

by taking the limit as $\epsilon \to 0$. (As always, the normal n is directed inside the domain U.) It follows that $H(z) = u(z) \leq 0$ in U. Therefore, in accordance with (9.15), $|f_2(z)| \leq 1$ in U. Thus, $f_2(z) \in \mathcal{F}$.

If we rewrite (9.14) in the form

$$\log \left| \frac{f_2(z)}{z-a} \right| = \log \left| \frac{f_0(z)}{z-a} \right| + u(z)$$

and let z approach a, we obtain

$$\log |f_2'(a)| = \log |f_0'(a)| + u(a)$$

Then, by virtue of the extremality of the function $f_0(z)$, we have $u(a) \leq 0$. By using Green's formula (as derived above), we see that this leads to the inequality

$$u(a) = \frac{1}{2\pi} \int_{\Gamma} u(z') \frac{dg(z', a)}{dn} ds \leqslant 0 \tag{9.16}$$

Furthermore, if we denote by $\Gamma_{\epsilon,l,l}$ a curve in U that is close to Γ_l and defined by the equation $\omega_l(z) = 1 - \epsilon$, where ϵ is a sufficiently small positive number, and if we denote by $\Gamma_{\epsilon,l,k}$, where $k \neq l$, a curve in U close to Γ_k and defined by the equation $\omega_l(z) = \epsilon$ with the same ϵ, then the single-valuedness of $v(z)$ in U can be expressed in the form of $m-1$ equations (for sufficiently small ϵ):

$$\begin{aligned}
(v(z))_{\Gamma_\epsilon,l,l} &= \int_{\Gamma_{\epsilon,l,l}} dv(z) \\
&= -\int_{\Gamma_{\epsilon,l,l}} \frac{du}{dn} ds \\
&= -\int_{\Gamma_{\epsilon,l,l}} (\omega_l(z)+\epsilon)\frac{du}{dn} ds \\
&= \int_{\bigcup_{k=1}^{n} \Gamma_{\epsilon,l,k}} (\omega_l(z)-\epsilon)\frac{du}{dn} ds + 2\epsilon \int_{\Gamma_{\epsilon,l,l}} \frac{du}{dn} ds \\
&= \int_{\bigcup_{k=1}^{m} \Gamma_{\epsilon,l,k}} u(z)\frac{d\omega_l(z)}{dn} ds + 2\epsilon \int_{\Gamma_{\epsilon,l,l}} \frac{du}{dn} ds \\
&= 0
\end{aligned}$$

where $l = 1, \ldots, m-1$.

Here, ϵ_0 is a fixed small positive number. This is equivalent to satisfaction of the $m-1$ limiting (as $\epsilon \to 0$) equations

$$\int_\Gamma u(z)\frac{d\omega_l(z)}{dn} ds = 0, \quad l = 1, \ldots, m-1 \tag{9.17}$$

Turning now to the function (9.13), let us suppose that, for some $q > 0$, there exists a subset E of Γ of positive measure such that $H(z) \leq -q$ for $z \in E$. Then, for $u(z)$ we may take the function defined by the boundary values

$$u(z) = \begin{cases} 0 \text{ on } \Gamma \text{ outside } E \\ \varepsilon\left(-\lambda_0 \frac{dg(z,a)}{dn} + \sum_{k=1}^{m-1} \lambda_m \frac{d\omega_k}{dn}\right) \text{ on } E, \end{cases}$$

where the λ_k, for $k = 1, \ldots, m-1$, are constants and ϵ is a constant of either sign but sufficiently small absolute value. (To construct $u(z)$, we need only shift from the domain U to the disk $|\zeta| < 1$, then use Poisson's formula to construct a harmonic function on that disk from the new boundary values, and finally return to the domain U with it.) For this function $u(z)$, conditions (9.17) take the forms

$$\left.\begin{array}{c} -\lambda_0 \int_\Gamma \frac{dg}{dn}\frac{d\omega_l}{dn} ds + \sum_{k=1}^{m-1} \lambda_k \int_\Gamma \frac{d\omega_k}{dn} \cdot \frac{d\omega_l}{dn} ds = 0 \\ l = 1, \ldots, m-1 \end{array}\right\} \tag{9.18}$$

and, by virtue of the choice of the λ_k, for $k = 0, 1, \ldots, m-1$, they must be satisfied. But conditions (12) constitute a system of $m-1$ homogeneous linear equations with n unknowns λ_k, $k = 0, 1, \ldots, m-1$. Therefore, this system has nonzero solutions.

Let us now denote by λ_k, $k = 0, \ldots, m-1$, any nonzero system of solutions and let us construct from them the function $u(z)$ in (9.16). We then obtain an inequality in which the left-hand member includes the factor ϵ, which may be of either sign. This leads to the condition

$$-\lambda_0 \int_E \left(\frac{dg(z,a)}{dn}\right)^2 ds + \sum_{k=1}^{m-1} \lambda_k \int_E \frac{d\omega_k}{dn} \cdot \frac{dg(z,a)}{dn} ds = 0 \tag{9.19}$$

If we multiply equation (9.19) by $-\lambda_0$ and equations (9.18) by λ_l (for each $l = 1, \ldots, m-1$) and then add all the resulting equations, we obtain

$$\int_E \left(-\lambda_0 \frac{dg(z,a)}{dn} + \lambda_1 \frac{d\omega_1}{dn} + \cdots + \lambda_{m-1} \frac{d\omega_{m-1}}{dn}\right)^2 ds = 0.$$

It follows that, almost everywhere on E,

$$-\lambda_0 \frac{dg}{dn} + \sum_{k=1}^{m-1} \lambda_k \frac{d\omega_k}{dn} = 0$$

Since, in addition, everywhere on Γ we have

$$-\lambda_0 \frac{dg}{ds} + \sum_{k=1}^{m-1} \lambda_k \frac{d\omega_k}{ds} = 0$$

we conclude that the function

$$w'(z) = -\lambda_0 p_z'(z,a) + \sum_{k=1}^{m} \lambda_k w_k'(z)$$

which is regular on Γ, vanishes on E. But then $w'(z) \equiv 0$ in U and hence the function

$$-\lambda_0 g(z,a) + \sum_{k=1}^{m} \lambda_k \omega_k(z)$$

is constant in U.

However, this cannot be the case because of the choice of the λ_k for $k = 0, 1, \ldots, m$. This contradiction shows that $H(z) = 0$ almost everywhere on Γ. But then, if we apply Privalov's uniqueness theorem (see Theorem 6.2.1) to a suitable subset of the domain U, we conclude chat $H(z) \equiv 0$ in U; that is, in U we have

$$\log |f_0(z)| = -g(z,a) - \sum_{k=1}^{m} g\left(z, a_k\right)$$

and

$$f_0(z) = e^{-p(z,a) - \sum_{k=1}^{m} p(a_1 a_k)}$$

Now, the conditions under which $f_0(z)$ is single-valued in U take the form

$$\frac{1}{2\pi} \left(\arg f_0(z)\right)_{\Gamma_l} = \omega_l(a) + \sum_{k=1}^{m} \omega_l\left(a_k\right) = v_l, \quad l = 1, \ldots, n \tag{9.20}$$

where the ω_l are integers. Obviously, $\nu_l \geq 1$, for $l = 1, \ldots, n$. If we sum (9.20) over $l = 1, \ldots, n$, we obtain

$$m + 1 = \sum_{v=1}^{n} v_l \geqslant n$$

But $m \leq n - 1$. Consequently, $m = n - 1$ and $\nu_l = 1$ for $l = 1, \ldots, n$.

Summarizing what we have said, we conclude that the function $f_0(z)$ has exactly n zeros in the domain U and that it is equal to unity in modulus on Γ. Then, by evaluating $(\arg(f_0(z) - w_0))_\Gamma$, we prove that the function $f_0(z) - w_0$ has, for each w_0 in the disk $|w_0| < 1$ exactly n zeros (counting each one according to its multiplicity) in U, and this means that the function $w = f_0(z)$ maps the domain U onto the n-sheeted disk $\mathbb{D}$. This completes the proof of the theorem. □

We see that the Riemann mapping theorem corresponds to $m = 1$, and, for that case, that Koebe's extremal problem can be relaxed to the family of functions that are analytic but need not be one-to-one. There is a slight catch here. Ahlfors needs a certain amount of boundary smoothness to form integrals over ∂U, and he uses some preliminary conformal mapping to assume that ∂U consists of analytic Jordan curves (a standard technique when working with finitely connected domains). This requires the Riemann mapping theorem! [The general methods in Bell's book [20] assume C^∞ boundaries.] To what extent these assumptions can be weakened or worked around we do not know.

10

Reproducing Kernels in Complex Analysis

This chapter discusses the role of reproducing kernels in complex analysis, both in one and several complex variables. We give applications to the study of the boundary smoothness of biholomorphic mappings and to other parts of geometric analysis. We develop the theme of constructible kernels versus non-constructible kernels, and compare and contrast the two.

Most readers will be familiar with the Cauchy kernel, which comes from the Cauchy integral formula. But there are many other important reproducing kernels in this subject. Among these are the Bergman kernel, the Szegő kernel, and the Poisson-Szegő kernel. And that list only scratches the surface of the matter.

10.1 Introduction

In the complex analysis of one variable, the Cauchy integral formula plays a central role. Indeed, virtually all of the fundamental results in the subject are derived from that simple formula.

In the realm of harmonic functions, the Poisson kernel and integral formula play a central role. Fundamental results, such as the Fatou theorem about nontangential boundary behavior of harmonic and holomorphic functions, depend critically on the nature of the Poisson kernel, and particularly on the shape of its singularity.

In the complex analysis of several variables—where the nature of the analysis depends on the geometry of the domain in question—the nature of kernels plays an even more critical role. On the one hand, it is generally very difficult to calculate the relevant kernels. On the other hand, knowledge of the kernels is a valuable tool.

DOI: 10.1201/9781003619581-10

10.2 Harmonic Functions and the Poisson Kernel

If we take it on faith that the Dirichlet problem on the disc always has a solution (see, for instance, [84]), then we may discover a formula for that solution in the following fashion.

Let f be an L^2 function on ∂D, the boundary of the unit disc. Then we may expand f, in the L^2 topology, as a Fourier series:

$$\begin{aligned} f(e^{i\theta}) &= \sum_{j=-\infty}^{\infty} \widehat{f}(j) e^{ij\theta} \\ &= \sum_{j=-\infty}^{\infty} \frac{1}{2\pi} \int_0^{2\pi} f(t) e^{-itj}\, dt \cdot e^{ij\theta} \\ &= \int_0^{2\pi} f(t) \left(\frac{1}{2\pi} \sum_{j=-\infty}^{\infty} e^{ij(\theta - t)} \right) dt\,. \end{aligned}$$

Now each character e^{ijt} has a harmonic extension to the disc given by $\varphi_j(re^{it}) \equiv r^{|j|} e^{ijt}$. So it is natural to consider that the harmonic extension of f to the disc D is given by

$$f(re^{i\theta}) = \int_0^{2\pi} f(t) \left(\frac{1}{2\pi} \sum_{j=-\infty}^{\infty} r^{|j|} e^{ij(\theta-t)} \right) dt$$

It behooves us then to calculate the sum

$$P_r(e^{i(\theta-t)}) = \frac{1}{2\pi} \sum_{j=-\infty}^{\infty} r^{|j|} e^{ij(\theta-t)}\,.$$

Now we have

$$P_r(e^{i(\theta-t)}) = \frac{1}{2\pi} \sum_{-\infty}^{-1} r^{-j} e^{ij(\theta-t)} + \frac{1}{2\pi} \sum_{1}^{\infty} r^j e^{ij(\theta-t)} + \frac{1}{2\pi} \equiv I + II + III\,.$$

We see that

$$\begin{aligned} I &= \frac{1}{2\pi} \sum_1^{\infty} r^j e^{-ij(\theta-t)} \\ &= \frac{1}{2\pi} \sum_0^{\infty} r^{j+1} e^{-i(j+1)(\theta-t)} \\ &= \frac{1}{2\pi} re^{-i(\theta-t)} \sum_{j=0}^{\infty} \left(re^{-i(\theta-t)} \right)^j \\ &= \frac{1}{2\pi} re^{-i(\theta-t)} \cdot \frac{1}{1 - re^{-i(\theta-t)}} \\ &= \frac{1}{2\pi} \frac{r}{e^{i(\theta-t)} - r}\,. \end{aligned}$$

A similar calculation shows that

$$II = \frac{1}{2\pi}\frac{r}{e^{-i(\theta-t)} - r}.$$

As a result,

$$\begin{aligned} P_r(e^{i(\theta-t)}) &= \frac{1}{2\pi}\cdot\frac{r}{e^{i(\theta-t)} - r} + \frac{1}{2\pi}\cdot\frac{r}{e^{-i(\theta-t)} - r} + \frac{1}{2\pi} \\ &= \frac{1}{2\pi}\cdot\frac{1-r^2}{1-2r\cos(\theta-t)+r^2}. \end{aligned}$$

This is the *Poisson kernel.*

The basic result about this kernel is as follows.

Theorem 10.2.1. *Let f be a continuous function on the unit circle $\mathbb{T} = \partial D$. Then the function*

$$F(re^{i\theta}) = P_r * f(e^{i\theta}).$$

is continuous on $\overline{D}$ and harmonic on D. Note that $F(e^{i\theta}) = f(e^{i\theta})$.

10.3 Introduction to the Bergman Theory

Stefan Bergman introduced his reproducing kernel in 1921. It was anticipated somewhat by the thesis of S. Bochner which was later published in [31]. Since Hilbert space was a very new idea at the time, and Bergman's approach made quite an impact.

In this section we work only in $\mathbb{C}$, the complex plane. Let $\Omega \subset \mathbb{C}$ be a domain. Define the *Bergman space*

$$A^2(\Omega) = \left\{ f \text{ holomorphic on } \Omega \;\middle|\; \int_\Omega |f(z)|^2 \, dA(z)^{1/2} \right\} \equiv \|f\|_{A^2(\Omega)} < \infty.$$

Observe that $A^2(\Omega)$ is a closed subspace of $L^2(\Omega)$. The orthogonal subspace B is also of interest. Of course, $L^2(\Omega) = A^2(\Omega) \oplus B$.

Lemma 10.3.1. *Let $K \subset \Omega$ be compact. There is a constant $C_K > 0$, depending on K, such that*

$$\sup_{z\in K} |f(z)| \le C_K \|f\|_{A^2(\Omega)}, \quad \text{all } f \in A^2(\Omega).$$

Proof. In what follows, $D(z,r)$ is the open disc with center z and radius r. Also dA denotes 2-dimensional area measure.

Since K is compact, there is an $r(K) = r > 0$ so that, for any $z \in K, D(z,r) \subset \Omega$. Therefore, for each $z \in K$ and $f \in A^2(\Omega)$, the mean value theorem tells us that

$$\begin{aligned} |f(z)| &= \frac{1}{A(D(z,r))}\left|\int_{D(z,r)} f(t)\,dA(t)\right| \\ &\le (A(D(z,r)))^{-1/2}\|f\|_{L^2(D(z,r))} \\ &\le C\cdot r^{-1}\|f\|_{A^2(\Omega)} \\ &\equiv C_K\|f\|_{A^2(\Omega)}. \end{aligned}$$

□

Lemma 10.3.2. *The space $A^2(\Omega)$ is a Hilbert space with the inner product $\langle f,g\rangle \equiv \int_\Omega f(z)\overline{g(z)}\,dA(z)$.*

Proof. Everything is clear except for completeness. Let $\{f_j\} \subset A^2$ be a sequence that is Cauchy in norm. Since L^2 is complete there is an L^2 limit function f. We need to see that f is holomorphic. But Lemma 10.3.1 yields that norm convergence implies normal (uniform on compact sets) convergence. And holomorphic functions are closed under normal limits. Therefore f is holomorphic and $A^2(\Omega)$ is complete. □

Lemma 10.3.3. *For each fixed $z \in \Omega$, the functional*

$$\Phi_z : f \mapsto f(z), \quad f \in A^2(\Omega)$$

is a continuous linear functional on $A^2(\Omega)$.

Proof. This is immediate from Lemma 10.3.1 if we take K to be the singleton $\{z\}$. □

We may now apply the Riesz representation theorem to see that there is an element $k_z \in A^2(\Omega)$ such that the linear functional Φ_z is represented by inner product with k_z: if $f \in A^2(\Omega)$, then for all $z \in \Omega$ we have

$$f(z) = \Phi_z(f) = \langle f, k_z\rangle.$$

Definition 10.3.4. The Bergman kernel is the function $K(z,\zeta) = \overline{k_z(\zeta)}$, $z,\zeta \in \Omega$. It has the reproducing property

$$f(z) = \int K(z,\zeta)f(\zeta)\,dA(\zeta), \quad \forall f \in A^2(\Omega).$$

Proposition 10.3.5. *The Bergman kernel $K(z,\zeta)$ is conjugate symmetric: $K(z,\zeta) = \overline{K(\zeta,z)}$.*

Proof. By its very definition, $\overline{K(\zeta,\cdot)} \in A^2(\Omega)$ for each fixed ζ. Therefore, the reproducing property of the Bergman kernel gives

$$\int_\Omega K(z,t)\overline{K(\zeta,t)}\,dA(t) = \overline{K(\zeta,z)}.$$

On the other hand,

$$\begin{aligned}\int_\Omega K(z,t)\overline{K(\zeta,t)}\,dA(t) &= \overline{\int K(\zeta,t)\overline{K(z,t)}\,dA(t)}\\ &= \overline{\overline{K(z,\zeta)}}\\ &= K(z,\zeta).\end{aligned}$$

□

Proposition 10.3.6. *The Bergman kernel is uniquely determined by the properties that it is an element of $A^2(\Omega)$ in z, is conjugate symmetric, and reproduces $A^2(\Omega)$.*

Proof. Let $K'(z,\zeta)$ be another such kernel. Then

$$\begin{aligned}K(z,\zeta) = \overline{K(\zeta,z)} &= \int K'(z,t)\overline{K(\zeta,t)}\,dA(t)\\ &= \overline{\int K(\zeta,t)\overline{K'(z,t)}\,dA(t)}\\ &= \overline{\overline{K'(z,\zeta)}} = K'(z,\zeta)\,.\end{aligned}$$

□

Since $L^2(\Omega)$ is a separable Hilbert space then so is its subspace $A^2(\Omega)$. Thus there is a complete orthonormal basis $\{\phi_j\}_{j=1}^\infty$ for $A^2(\Omega)$.

Proposition 10.3.7. *Let K be a compact subset of Ω. Then the series*

$$\sum_{j=1}^\infty \phi_j(z)\overline{\phi_j(\zeta)}$$

sums uniformly on $K \times K$ to the Bergman kernel $K(z,\zeta)$.

Proof. Let $\{\phi_j\}$ be a complete orthonormal basis for the Bergman space. By the Riesz-Fischer and Riesz representation theorems, we obtain

$$\begin{aligned}
\sup_{z\in K}\sqrt{\left(\sum_{j=1}^{\infty}|\phi_j(z)|^2\right)} &= \sup_{z\in K}\left\|\{\phi_j(z)\}_{j=1}^{\infty}\right\|_{l^2} \\
&= \sup_{\substack{\|\{a_j\}\|_{l^2}=1 \\ z\in K}}\left|\sum_{j=1}^{\infty}a_j\phi_j(z)\right| \\
&= \sup_{\substack{\|f\|_{A^2}=1 \\ z\in K}}|f(z)| \\
&\le C_K. \qquad (10.1)
\end{aligned}$$

In the last lines we have used Lemma 10.3.1 and the Hahn-Banach theorem. Therefore

$$\sum_{j=1}^{\infty}\left|\phi_j(z)\overline{\phi_j(\zeta)}\right| \le \left(\sum_{j=1}^{\infty}|\phi_j(z)|^2\right)^{1/2}\left(\sum_{j=1}^{\infty}|\phi_j(\zeta)|^2\right)^{1/2}$$

and the convergence is uniform over $z,\zeta \in K$. For fixed $z \in \Omega$, (10.1) shows that $\{\phi_j(z)\}_{j=1}^{\infty} \in l^2$. Hence we have that $\sum \phi_j(z)\overline{\phi_j(\zeta)} \in \overline{A^2(\Omega)}$ as a function of ζ. Let the sum of the series be denoted by $K'(z,\zeta)$. Notice that K' is conjugate symmetric by its very definition. Also, for $f \in A^2(\Omega)$, we have

$$\int K'(\cdot,\zeta)f(\zeta)\,dA(\zeta) = \sum \hat{f}(j)\phi_j(\cdot) = f(\cdot),$$

where convergence is in the Hilbert space topology. [Here f_j is the jth Fourier coefficient of f with respect to the basis $\{\phi_j\}$.] But Hilbert space convergence dominates pointwise convergence (Lemma 10.3.1) so

$$f(z) = \int K'(z,\zeta)f(\zeta)\,dA(\zeta), \quad \text{all } f \in A^2(\Omega).$$

Therefore K' is the Bergman kernel. □

Remark 10.3.8. It is worth noting explicitly that the proof of Proposition 10.3.7 shows that

$$\sum \phi_j(z)\overline{\phi_j(\zeta)}$$

equals the Bergman kernel $K(z,\zeta)$ *no matter what the choice* of complete orthonormal basis $\{\phi_j\}$ for $A^2(\Omega)$.

Proposition 10.3.9. *If Ω is a bounded domain in $\mathbb{C}$ then the mapping*

$$P : f \mapsto \int_\Omega K(\cdot, \zeta) f(\zeta)\, dA(\zeta)$$

is the Hilbert space orthogonal projection of $L^2(\Omega, dA)$ onto $A^2(\Omega)$.

Proof. Notice that, by Fubini's theorem, P is idempotent and self-adjoint and that $A^2(\Omega)$ is precisely the set of elements of L^2 that are fixed by P. □

Given a holomorphic mapping $\Phi : \Omega_1 \to \Omega_2$, we may ask to compare the complex derivative Φ' with the real Jacobian determinant $J_{\mathbb{R}}(\Phi)$. The fact of the matter, and this is easy to check just using the Cauchy-Riemann equations, is that

$$\det J_{\mathbb{R}}(\Phi) = |\Phi'|^2 .$$

A holomorphic mapping $f : \Omega_1 \to \Omega_2$ of domains $\Omega_1 \subset \mathbb{C}, \Omega_2 \subset \mathbb{C}$ is said to be *biholomorphic* (or *conformal*) if it is one-to-one, onto, and $f'(z) \neq 0$ for every $z \in \Omega_1$.

In what follows we denote the Bergman kernel for a given domain Ω by K_Ω.

Proposition 10.3.10. *Let Ω_1, Ω_2 be domains in $\mathbb{C}$. Let $f : \Omega_1 \to \Omega_2$ be conformal. Then*

$$f'(z) K_{\Omega_2}(f(z), f(\zeta)) \overline{f'(\zeta)} = K_{\Omega_1}(z, \zeta).$$

Proof. Let $\phi \in A^2(\Omega_1)$. Then, by change of variable,

$$\begin{aligned}
&\int_{\Omega_1} f'(z) K_{\Omega_2}(f(z), f(\zeta)) \overline{f}'(\zeta) \phi(\zeta)\, dA(\zeta) \\
&= \int_{\Omega_2} f'(z) K_{\Omega_2}(f(z), \widetilde{\zeta}) \overline{f}'(f^{-1}(\widetilde{\zeta})) \phi(f^{-1}(\widetilde{\zeta})) \\
&\quad \times \det J_{\mathbb{R}} f^{-1}(\widetilde{\zeta})\, dA(\widetilde{\zeta}).
\end{aligned}$$

By the remark following Proposition 10.3.9, this simplifies to

$$f'(z) \int_{\Omega_2} K_{\Omega_2}(f(z), \widetilde{\zeta}) \left\{ \left(f'(f^{-1}(\widetilde{\zeta})) \right)^{-1} \phi\left(f^{-1}(\widetilde{\zeta}) \right) \right\} dA(\widetilde{\zeta}).$$

By change of variables, the expression in braces $\{\ \}$ is an element of $A^2(\Omega_2)$. So the reproducing property of K_{Ω_2} applies and the last line equals

$$= f'(z) \left(f'(z) \right)^{-1} \phi\Big(f^{-1}(f(z)) \Big) = \phi(z) .$$

By the uniqueness of the Bergman kernel, the proposition follows. □

Proposition 10.3.11. *For $z \in \Omega$ relatively compact in $\mathbb{C}$ it holds that $K_\Omega(z,z) > 0$.*

Proof. Now

$$K_\Omega(z,z) = \sum_{j=1}^{\infty} |\phi_j(z)|^2 \geq 0.$$

If in fact $K(z,z) = 0$ for some z then $\phi_j(z) = 0$ for all j hence $f(z) = 0$ for every $f \in A^2(\Omega)$. By Proposition 10.3.7 this is absurd. □

Definition 10.3.12. For any $\Omega \subset \mathbb{C}$ we define a Hermitian metric on Ω by

$$g(z) = \frac{\partial^2}{\partial z \partial \bar{z}} \log K(z,z), \quad z \in \Omega.$$

This means that the square of the length of a tangent vector ξ at a point $z \in \Omega$ is given by

$$|\xi|^2_{B,z} = g(z)\xi\bar{\xi}\,.$$

The metric that we have just defined is called the *Bergman metric.*

In a Hermitian metric g, the length of a C^1 curve $\gamma : [0,1] \to \Omega$ is given by

$$l(\gamma) = \int_0^1 \left(g(\gamma(t))\, \gamma'(t)\, \overline{\gamma}'(t) \right)^{1/2} dt.$$

If P, Q are points of Ω then their distance $d_\Omega(P,Q)$ in the metric is defined to be the infimum of the lengths of all piecewise C^1 curves connecting the two points.

It is not *a priori* obvious that the Bergman metric for a bounded domain Ω is given by a positive definite matrix at each point.

Proposition 10.3.13. *Let $\Omega_1, \Omega_2 \subset \mathbb{C}$ be domains and let $f : \Omega_1 \to \Omega_2$ be a biholomorphic mapping. Then f induces an isometry of Bergman metrics:*

$$|\xi|_{B,z} = |f'(z)\xi|_{B,f(z)}$$

for all $z \in \Omega_1, \xi \in \mathbb{C}$. Equivalently, f induces an isometry of Bergman distances in the sense that

$$d_{\Omega_2}(f(P), f(Q)) = d_{\Omega_1}(P,Q)$$

for all $P, Q \in \Omega_1$.

Proof. This is a formal exercise but we include it for completeness:

From the definitions, it suffices to check that

$$g^{\Omega_2}(f(z))\,(f'(z)w)\left(\overline{f'(z)w}\right) = g^{\Omega_1}(z) w\overline{w} \qquad (10.2)$$

for all $z \in \Omega, w \in \mathbb{C}$. But, by Proposition 10.3.10,

$$\begin{aligned} g^{\Omega_1}(z) &= \frac{\partial^2}{\partial z \partial \bar{z}} \log K_{\Omega_1}(z,z) \\ &= \frac{\partial^2}{\partial z \partial \bar{z}} \log \left\{ |f'(z)|^2 K_{\Omega_2}(f(z), f(z)) \right\} \\ &= \frac{\partial^2}{\partial z \partial \bar{z}} \log K_{\Omega_2}(f(z), f(z)) \qquad (10.3) \end{aligned}$$

since $\log |f'(z)|^2$ is locally

$$\log (f') + \log \left(\overline{f'}\right) + C$$

hence is annihilated by the mixed second derivative. But line (10.3) is nothing other than

$$g^{\Omega_2}(f(z)) \frac{\partial f(z)}{\partial z} \frac{\partial \overline{f(z)}}{\partial \bar{z}}$$

and (10.2) follows. □

Proposition 10.3.14. *Let $\Omega \subset\subset \mathbb{C}$ be a domain. Let $z \in \Omega$. Then*

$$K(z,z) = \sup_{f \in A^2(\Omega)} \frac{|f(z)|^2}{\|f\|^2_{A^2}} = \sup_{\|f\|_{A^2(\Omega)}=1} |f(z)|^2 .$$

Proof. Now

$$\begin{aligned} K(z,z) &= \sum |\phi_j(z)|^2 \\ &= \left(\sup_{\|\{a_j\}\|_{l^2}=1} \left| \sum \phi_j(z) a_j \right| \right)^2 \\ &= \sup_{\|f\|_{A^2}=1} |f(z)|^2 , \end{aligned}$$

by the Riesz-Fischer theorem,

$$= \sup_{f \in A^2} \frac{|f(z)|^2}{\|f\|^2_{A^2}} .$$

□

10.3.1 Calculating the Bergman Kernel

We present three methods for calculating the Bergman kernel on the unit disc $D \subset \mathbb{C}$. For the first we use Proposition 10.3.7.

Construction of the Bergman Kernel with an Orthonormal Basis

Now we first notice that $\{\zeta^j\}_{j=0}^{\infty}$ forms an orthogonal basis for $A^2(D)$. This follows just from circular symmetry. In order to use Proposition 10.3.7, we need an *orthonormal* basis. So we calculate that

$$\int_D |\zeta^j|^2 \, dA(\zeta) = \int_0^{2\pi} \int_0^1 r^{2j} \cdot r \, dr d\theta = \frac{2\pi}{2j+2} .$$

Hence a complete orthonormal system for $A^2(D)$ is

$$\left\{ \frac{\sqrt{2j+2} \cdot \zeta^j}{\sqrt{2\pi}} \right\}_{j=0}^{\infty} .$$

We then may calculate that

$$K_D(z,\zeta) = \sum_{j=0}^{\infty} \frac{z^j \overline{\zeta}^j (2j+2)}{2\pi} .$$

Set $\alpha = z\overline{\zeta}$. Then we have

$$\begin{aligned}
K_D(z,\zeta) &= \frac{1}{2\pi} \sum_{j=0}^{\infty} (2j+2)\alpha^j \\
&= \frac{1}{\pi} \sum_{j=0}^{\infty} (j+1)\alpha^j \\
&= \left(\frac{\alpha}{\pi} \sum_{j=0}^{\infty} \alpha^j \right)' \\
&= \left(\frac{\alpha}{\pi} \cdot \frac{1}{1-\alpha} \right)' \\
&= \frac{1}{\pi} \cdot \frac{1}{(1-\alpha)^2} \\
&= \frac{1}{\pi} \cdot \frac{1}{(1-z\overline{\zeta})^2} .
\end{aligned}$$

This is our first calculation of the Bergman kernel for D.

Construction of the Bergman Kernel by Way of Differential Equations

It is actually possible to obtain the Bergman kernel of a domain in the plane from the Green's function for that domain (see [142, Section 1.3.3]). Let us now summarize the key ideas. Unlike the first Bergman kernel construction, the present one will work for *any* domain with C^2 boundary. Thanks to work of Garabedian [75], one can say rather precisely what the Green's function of any planar domain is (see also [113]).

First, the fundamental solution for the Laplacian in the plane is the function

$$\Gamma(\zeta, z) = \frac{1}{2\pi} \log |\zeta - z| .$$

This means that $\Delta_\zeta \Gamma(\zeta, z) = \delta_z$ in the sense of distributions. [Observe that δ_z denotes the Dirac "delta mass" at z and Δ_ζ is the Laplacian in the ζ variable.] In more prosaic terms, the condition is that

$$\int \Gamma(\zeta, z) \cdot \Delta\varphi(\zeta) \, d\xi d\eta = \varphi(z)$$

for any C^∞ function φ with compact support. We write, as usual, $\zeta = \xi + i\eta$. [This topic is treated in detail in [141, Ch. 0, 1].]

Given a domain $\Omega \subseteq \mathbb{C}$, the *Green's function* is posited to be a function $G(\zeta, z)$ on $\Omega \times \Omega$ that satisfies

$$G(\zeta, z) = \Gamma(\zeta, z) - F_z(\zeta) ,$$

where $F_z(\zeta) = F(\zeta, z)$ is a particular harmonic function in the ζ variable (to be specified momentarily). Moreover, it is mandated that $G(\,\cdot\,, z)$ vanish on the boundary of Ω. One constructs the function $F(\,\cdot\,, z)$, for each fixed z, by solving a Dirichlet problem with boundary data $\Gamma(\,\cdot\,, z)$. Again, the reference [141, p. 40] has all the particulars. It is worth noting, and this point is not completely obvious but is discussed in [141, Ch. 1], that the Green's function is a symmetric function of its arguments.

The next proposition establishes a striking connection between the Bergman kernel and the classical Green's function.

Proposition 10.3.15. *Let $\Omega \subseteq \mathbb{C}$ be a bounded domain with C^2 boundary. Let $G(\zeta, z)$ be the Green's function for Ω and let $K(z, \zeta)$ be the Bergman kernel for Ω. Then*

$$K(z, \zeta) = 4 \cdot \overline{\frac{\partial^2}{\partial\zeta \partial\overline{z}} G(\zeta, z)} . \tag{10.4}$$

Proof. Our proof will use a version of Stokes's theorem written in the notation of complex variables. This matter was discussed earlier in the book. It says that, if $u \in C^1(\overline{\Omega})$, then

$$\oint_{\partial U} u(\zeta) \, d\zeta = 2i \cdot \iint_U \frac{\partial u}{\partial\overline{\zeta}} \, d\xi \, d\eta ,$$

where again $\zeta = \xi + i\eta$. The reader is invited to convert this formula to an expression in ξ and η and to confirm that the result coincides with the standard real-variable version of Stokes's theorem that can be found in any calculus book (see, e.g., [101, 25]).

Now we already know that

$$G(\zeta, z) = \frac{1}{4\pi} \log(\zeta - z) + \frac{1}{4\pi} \log \overline{(\zeta - z)} + F(\zeta, z) . \tag{10.5}$$

Here we think of the logarithm as a multivalued holomorphic function; after we take a derivative, the ambiguity (which comes from an additive multiple of $2\pi i$) goes away.

Differentiating with respect to z (and using subscripts to denote derivatives), we find that

$$G_z(\zeta, z) = \frac{1}{4\pi}\frac{-1}{\zeta - z} + F_z(\zeta, z)\,.$$

We may rearrange this formula to read

$$\frac{1}{\zeta - z} = -4\pi \cdot G_z(\zeta, z) + 4\pi F_z(\zeta, z)\,.$$

We know that G, as a function of ζ, vanishes on $\partial\Omega$. Hence so does G_z. Let $f \in C^2(\overline{\Omega})$ be holomorphic on Ω. It follows that the Cauchy formula

$$f(z) = \frac{1}{2\pi i}\oint_{\partial\Omega} \frac{f(\zeta)}{\zeta - z}\,d\zeta$$

can be rewritten as

$$f(z) = -2i \oint_{\partial\Omega} f(\zeta) F_z(\zeta, z)\,d\zeta\,.$$

Now we apply Stokes's theorem (in the complex form) to rewrite this last as

$$f(z) = 4 \cdot \iint_\Omega (f(\zeta)F_z)_{\overline{\zeta}}(\zeta, z)\,d\xi\,d\eta\,,$$

where $\zeta = \xi + i\eta$. Since f is holomorphic and F is real-valued, we may conveniently write this last formula as

$$f(z) = 4 \cdot \iint_\Omega f(\zeta)\overline{F_{\zeta\overline{z}}}(\zeta, z)\,d\xi\,d\eta\,.$$

Now formula (10.5) tells us that $F_{\zeta\overline{z}} = G_{\zeta\overline{z}}$. Therefore we have

$$f(z) = \iint_\Omega f(\zeta) 4\overline{G_{\zeta\overline{z}}}(\zeta, z)\,d\xi\,d\eta\,. \tag{10.6}$$

With a suitable limiting argument, we may extend this formula from functions f that are holomorphic and in $C^2(\overline{\Omega})$ to functions in $A^2(\Omega)$.

It is straightforward now to verify that $4\overline{G_{\zeta\overline{z}}}$ satisfies the first three characterizing properties of the Bergman kernel, just by examining our construction. The crucial reproducing property is, of course, formula (10.6). Then it follows that

$$K(z, \zeta) = 4 \cdot \overline{\frac{\partial^2}{\partial\zeta\partial\overline{z}} G(\zeta, z)}\,.$$

That is the desired result. □

It is worth noting that the proposition we have just established gives a practical method for confirming the existence of the Bergman kernel—by relating it to the Green's function, whose existence is elementary. See [99] for a version of these techniques in the several complex variables context.

Now let us calculate. Of course, the Green's function of the unit disc D is

$$G(\zeta, z) = \frac{1}{2\pi} \log |\zeta - z| - \frac{1}{2\pi} \log |1 - \zeta\overline{z}| \,,$$

as a glance at any classical complex analysis text will tell us (see, for example, [3] or [107]). Verify the defining properties of the Green's function for yourself.

With formula (10.4) in mind, we can make life a bit easier by writing

$$\begin{aligned} G(\zeta, z) = {} & \frac{1}{4\pi} \log(\zeta - z) + \frac{1}{4\pi} \log(\overline{\zeta - z}) \\ & - \frac{1}{4\pi} \log(1 - \zeta\overline{z}) - \frac{1}{4\pi} \log\left(\overline{1 - \zeta\overline{z}}\right) . \end{aligned}$$

Here we think of the expression on the right as the concatenation of four multivalued functions, in view of the ambiguity of the logarithm function. This ambiguity is irrelevant for us because the derivative of the Green's function is still well defined (i.e., the derivative annihilates additive constants).

Now we readily calculate that

$$\frac{\partial G}{\partial \overline{z}} = \frac{1}{4\pi} \cdot \frac{-1}{\overline{\zeta} - \overline{z}} + \frac{1}{4\pi} \cdot \frac{\zeta}{1 - \zeta\overline{z}}$$

and

$$\frac{\partial^2 G}{\partial \zeta \partial \overline{z}} = \frac{1}{4\pi} \cdot \frac{1}{(1 - \zeta\overline{z})^2} .$$

In conclusion, we may apply Proposition 10.3.15 to see that

$$K(z, \zeta) = \frac{1}{\pi} \cdot \frac{1}{(1 - z \cdot \overline{\zeta})^2} .$$

This result is consistent with that obtained in the other two calculations (see above and below).

Construction of the Bergman Kernel by Way of Conformal Invariance

Let $D \subseteq \mathbb{C}$ be the unit disc. First we notice that, if either $f \in A^2(D)$ or $\overline{f} \in A^2(D)$, then

$$f(0) = \frac{1}{\pi} \iint_D f(\zeta) \, dA(\zeta) . \tag{10.7}$$

This is the standard, two-dimensional area form of the mean-value property for holomorphic or harmonic functions.

Of course, the constant function $u(z) \equiv 1$ is in $A^2(D)$, so it is reproduced by integration against the Bergman kernel. Hence, for any $w \in D$,

$$1 = u(w) = \iint_D K(w,\zeta)u(\zeta)\, dA(\zeta) = \iint_D K(w,\zeta)\, dA(\zeta)\,,$$

or

$$\frac{1}{\pi} = \frac{1}{\pi} \iint_D K(w,\zeta)\, dA(\zeta)\,.$$

By (10.7), we may conclude that

$$\frac{1}{\pi} = K(w,0)$$

for any $w \in D$.

Now, for $a \in D$ fixed, consider the Möbius transformation

$$h(z) = \frac{z-a}{1-\overline{a}z}\,.$$

We know that

$$h'(z) = \frac{1-|a|^2}{(1-\overline{a}z)^2}\,.$$

We may thus apply Proposition 10.3.10 with $\phi = h$ to find that

$$\begin{aligned} K(w,a) &= h'(w) \cdot K(h(w),h(a)) \cdot \overline{h'(a)} \\ &= \frac{1-|a|^2}{(1-\overline{a}w)^2} \cdot K(h(w),0) \cdot \frac{1}{1-|a|^2} \\ &= \frac{1}{(1-\overline{a}w)^2} \cdot \frac{1}{\pi} \\ &= \frac{1}{\pi} \cdot \frac{1}{(1-w\overline{a})^2}\,. \end{aligned}$$

This is our formula for the Bergman kernel.

10.3.2 The Poincaré-Bergman Metric on the Disc

If $D \subset \mathbb{C}$ is the unit disc, $z \in D$, then Definition 10.3.12 shows that

$$|w|_{B,z} = \left\{ \frac{2|w|^2}{(1-|z|^2)^2} \right\}^{1/2} = \frac{\sqrt{2}|w|}{1-|z|^2},$$

where the subscript B indicates that we are working in the Bergman metric. We now use this formula to derive an explicit expression for the Poincaré distance from $0 \in D$ to $r + i0 \in D, 0 < r < 1$. Call this distance $d(0, r)$. Then

$$d(0,r) = \inf \left\{ \int_0^1 |\gamma'(t)|_{B,\gamma(t)} dt \;\middle|\; \gamma \text{ is a curve in } D, \right. \\ \left. \gamma(0) = 0, \gamma(1) = r + i0 \right\}.$$

Elementary comparisons show that, among curves of the form

$$\psi(t) = t + iw(t), \quad 0 \le t \le 1,$$

the curve $\gamma(t) = tr + i0$ is the shortest in the Poincaré metric. Further elementary arguments show that a general curve of the form $\psi(t) = v(t) + iw(t)$ is always longer than some corresponding curve of the form $t + i\widetilde{w}(t)$. We leave the details of these assertions to the reader. Thus

$$\begin{aligned} d(0,r) &= \int_0^1 \frac{\sqrt{2}r}{(1-(rt)^2)} dt \\ &= \sqrt{2} \int_0^r \frac{1}{1-t^2} dt \\ &= \frac{1}{\sqrt{2}} \log \left(\frac{1+r}{1-r} \right). \end{aligned}$$

Since rotations are conformal maps of the disc we may next conclude that

$$d(0, re^{i\theta}) = \frac{1}{\sqrt{2}} \log \left(\frac{1+r}{1-r} \right).$$

Finally, if w_1, w_2 are arbitrary, then the Möbius transformation

$$\phi : z \mapsto \frac{z - w_1}{1 - \overline{w}_1 z}$$

satisfies $\phi(w_1) = 0, \phi(w_2) = (w_2 - w_1)/(1 - \overline{w}_1 w_2)$. Then Proposition 10.3.15 yields that

$$\begin{aligned} d(w_1, w_2) &= d\left(0, \frac{w_2 - w_1}{1 - \overline{w}_1 w_2}\right) \\ &= \frac{1}{\sqrt{2}} \left(\frac{1 + \left|\frac{w_2 - w_1}{1 - \overline{w}_1 w_2}\right|}{1 - \left|\frac{w_2 - w_1}{1 - \overline{w}_1 w_2}\right|} \right). \end{aligned}$$

We note in passing that the expression $\rho(w_1, w_2) \equiv |(w_1 - w_2)/(1 - \overline{w}_1 w_2)|$ is called the *pseudohyperbolic distance*. It is also conformaly invariant, but it does *not* arise from integrating an infinitesimal metric (i.e., lengths of tangent vectors at a point). A fuller discussion of both the Poincaré metric and the pseudo hyperbolic metric on the disc may be found in [77].

10.4 Preliminary Facts About Boundary Smoothness

The Riemann mapping theorem is a powerful result. It allows us to transfer the function theory of one proper, simply connected domain in the plane to another. But for many applications, we need to know something about the boundary regularity of a conformal mapping. Does the mapping extend univalent and continuously to the boundary? Does it extend smoothly to the boundary?

Curiously, the second of these questions was answered first—by Paul Painlevé in his thesis [170]. About 35 years later, Constantin Carath'eodory [37] answered the first question (and it is the first question that is typically treated in textbooks, not the second—although [86] is one of the exceptions).

For many purposes, continuity to the boundary is insufficient. In [5], studies were made of the Bergman kernel on a variety of domains with different boundary geometries. The behavior of the kernel on a domain Ω was studied by considering a conformal mapping $\varphi : \Omega \to D$ and the corresponding transformation formula

$$K_\Omega(z, \zeta) = K_D(\varphi(z), \varphi(\zeta))\varphi'(z)\overline{\varphi'(\zeta)}$$

(see Proposition 10.3.10). In order to pass back and forth, it was essential to know that the expression $|\varphi'(z)|$ is bounded above and below:

$$c_1 \leq |\varphi'(z)| \leq c_2 \,. \tag{10.8}$$

Note also that a set of inequalities like (10.8) tells us that the amount of stretching that occurs under the map φ is bounded above and below; this fact was also crucial in the results of [5].

A result that implies the kind of estimates we have been discussing is the following

Theorem 10.4.2. *Let $\Omega \subset \mathbb{C}$ be a bounded domain in $\mathbb{C}$ with sufficiently smooth boundary. Then any conformal mapping $\varphi : \Omega \to D$ will extend univalentl and continuously differentiable to $\overline{\Omega}$, and the inverse mapping φ^{-1} will extend univalently and continuously differentiable to $\overline{D}$.*

With this theorem in hand, we know that $|\varphi'|$ extends continuously to the compact set $\overline{\Omega}$. Thus it is bounded. In other words,

$$|\varphi'(z)| \leq C \quad \text{for all } z \in \Omega \,.$$

Likewise, $|(\varphi^{-1})'|$ extends continuously to the compact set $\overline{D}$. So it is bounded. But then

$$\left| \frac{1}{\varphi'(\varphi^{-1})(z)} \right| = |(\varphi^{-1})'(z)| \leq C \,,$$

and hence[1]

$$\left|\varphi'(\varphi^{-1}(z))\right| \geq \frac{1}{C}.$$

In the statement of Theorem 10.4.2, we have intentionally been imprecise about what "sufficiently smooth" means, since that is one of the main points of the rest of this section. For the moment, we shall content ourselves with proving the next result (see [22]). First some definitions.

We shall require a brief discussion of the concept of boundary smoothness. We will offer two approaches to the question.

Boundary Smoothness by Way of Calculus: We will only consider domains with finitely many boundary components. Each boundary component will be a simple closed curve—see Figure 10.1.

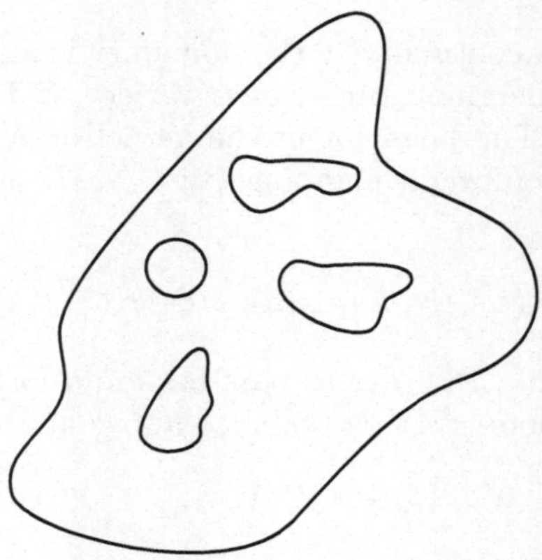

FIGURE 10.1: A domain with finitely many boundary components.

If S^1 is the unit circle in the plane, parametrized by $t \mapsto e^{it}$, then we may think of each boundary curve γ_j as given by

$$\gamma_j : S^1 \to \mathbb{C}.$$

Let $k \in \{0, 1, 2, \ldots\}$. We say that $\partial\Omega$ is C^k if γ_j' is never zero, each j, and the first k partial derivatives of the function γ_j exist, each j, and each derivative is continuous.

[1]Implicit in this discussion is an important point that ought to be explicitly enunciated. Namely, if Ω_1 and Ω_2 are domains with C^2 boundary and $\varphi : \Omega_1 \to \Omega_2$ is a conformal map, then there are constants $c, C > 0$ such that $c \cdot \text{dist}(\varphi(z), \partial\Omega_2) \leq \text{dist}(z, \partial\Omega_1) \leq C \cdot \text{dist}(\varphi(z), \partial\Omega_2)$. In other words, φ preserves—qualitatively speaking—the distance to the boundary. This assertion follows from the Hopf lemma, a result that we treat below.

Boundary Smoothness by Way of a Defining Function: In geometric analysis it is frequently useful to think of a domain Ω in space as the sublevel set of a function $\rho = \rho_\Omega$. For example, the unit disc is given by

$$D = \left\{ z \in \mathbb{C} \mid \rho(z) \equiv |z|^2 - 1 < 0 \right\} . \tag{10.9}$$

It is a nice exercise with the implicit function theorem to see that any domain whose boundary consists of C^1 curves can be written as the sublevel set of some function ρ, as in equation (10.9). We usually demand that $\nabla\rho \neq 0$ on $\partial\Omega$ in order to prevent degeneracies and, more particularly, so that $\nabla\rho(P)$ gives a well-defined outward normal vector at each boundary point P. We say that $\Omega = \{z \in \mathbb{C} \mid \rho(z) < 0\}$ has C^k boundary, $k \in \{0, 1, 2, \ldots\}$, if there is a defining function ρ that is C^k.

We say that a bounded domain $\Omega \subseteq \mathbb{C}$ has C^∞ (or *smooth*) boundary if there is a function $\rho : \mathbb{C} \to \mathbb{R}$ such that ρ is C^∞,

$$\Omega = \{z \in \mathbb{C} \mid \rho(z) < 0\} ,$$

and $\nabla\rho$ is nowhere zero on $\partial\Omega$.

It is worth noting that possession of a defining function says nothing about the topology of the domain. The domain could be simply connected or not.

Example 10.4.4. Let

$$\Omega = \left\{ z \in \mathbb{C} \;\middle|\; \rho(z) = \left(\frac{z + \overline{z}}{2}\right)^2 + \left(\frac{z - \overline{z}}{8i}\right)^2 - 1 < 0 \right\} .$$

Then Ω is the region bounded by an ellipse. Certainly Ω has C^∞ boundary.

Theorem 10.4.5. *Let $k \in \{0, 1, 2, \ldots\}$. Then there is an integer $N = N(k)$ such that, if $\Omega \subset \mathbb{C}$ is a bounded domain with C^N boundary, then any conformal mapping $\varphi : \Omega \to D$ will extend to be univalent and C^k from $\overline{\Omega}$ to $\overline{D}$. Likewise, the inverse mapping φ^{-1} will extend to be univalent and C^k from $\overline{D}$ to $\overline{\Omega}$. It may be shown—although it requires considerable extra effort—that N may be taken to be $k + 1$, or even $k + \epsilon$.*

This theorem is properly attributed to P. Painlevé, who proved it in his thesis [170]. Over the years, it has been sharpened through work of Kellogg, Warschawski, Pommerenke, and others. We shall indicate some of these results in what follows.

Let $\Omega \subseteq \mathbb{C}$ be a domain. We say that Ω satisfies *Bell's Condition R* if, for each $k \in \{0, 1, 2, \ldots\}$, there is an $l \in \{0, 1, 2, \ldots\}$ such that, whenever $f : \Omega \to \mathbb{C}$ has bounded derivative up to order l, then $P_\Omega f$ (the Bergman projection of f) has bounded derivatives up to Condition A R order k. Condition R has proved to be of historical importance in the study of holomorphic mappings. In this section we prove that $D = D(0, 1)$ satisfies a version of Condition R (see Theorem 10.4.13). This will be the key to the proof below of Theorem 10.4.5.

If Ω is a domain and $\Phi : \Omega \to \mathbb{C}$, then we say that Φ is k times boundedly continuously differentiable, and we write $\Phi \in C_b^k(\Omega)$, if all partial derivatives $(\partial/\partial x)^s(\partial/\partial y)^t\Phi$ of order $s+t \le k$ on Ω (the interior of Ω, not the boundary) exist, are continuous, and are bounded. We let $C_b^\infty(\Omega) = \cap_k C_b^k(\Omega)$. If $\Phi \in C_b^k(\Omega)$, then we set

$$\|\Phi\|_{C_b^k(\Omega)} = \sum_{s+t\le k} \sup_\Omega \left\| \left(\frac{\partial}{\partial x}\right)^s \left(\frac{\partial}{\partial y}\right)^t \Phi(x+iy) \right\|.$$

We also need an idea of the continuity of derivatives at the boundary: A function $\Phi : \overline{\Omega} \to \mathbb{C}$ is called C^k if all partial derivatives of Φ up to and including order k extend continuously to $\overline{\Omega}$. As before, $C^\infty(\overline{\Omega})$ is defined to be $\cap_k C^k(\overline{\Omega})$.

Definition 10.4.6. If $f, g \in C^k(\overline{\Omega})$, then we say that f and g *agree up to order* k on $\partial\Omega$ if

$$\left(\frac{\partial}{\partial x}\right)^s \left(\frac{\partial}{\partial y}\right)^t (f(z) - g(z))\bigg|_{\partial\Omega} = 0$$

for all $s + t \le k$.

Lemma 10.4.7. *For every $k = 0, 1, 2, \ldots$ there is a function $\lambda_k \in C^k(\mathbb{R})$ such that*

1. $\lambda_k(x) = 0$ *if* $x \le \frac{1}{3}$,
2. $\lambda_k(x) = 1$ *if* $x \ge \frac{2}{3}$,
3. $0 \le \lambda_k(x) \le 1$ *for all* $x \in \mathbb{R}$.

Proof. Let

$$\psi(x) = \begin{cases} x^{k+1} & \text{if } x \ge 0, \\ 0 & \text{if } x < 0. \end{cases}$$

Obviously $\psi \in C^k(\mathbb{R})$. Define

$$\phi(x) = \psi\left(x - \frac{1}{3}\right) \cdot \psi\left(-x + \frac{2}{3}\right).$$

Then $\phi \in C^k$ and $\phi(x) \ne 0$ only if $1/3 < x < 2/3$. Observe that $\phi(x) \ge 0$ for all x. Let

$$u(x) = \int_{-\infty}^x \phi(t)\, dt.$$

Then $u \in C^k(\mathbb{R}), u(x) = 0$ for $x \le 1/3$, and $u \equiv c$ a positive constant for $x \ge 2/3$. Define

$$\lambda_k(x) = \frac{1}{c} u(x).$$

This function has all the required properties. □

It is actually possible to find a C^∞ function with properties **1**, **2**, and **3** of Lemma 10.4.7.

Lemma 10.4.8. *If $\Omega \subseteq \mathbb{C}$ is a domain with C^k boundary, $\Omega = \{z \in \mathbb{C} \mid \rho(z) < 0\}$, and if U is a neighborhood of $\partial\Omega$ on which $|\nabla\rho| \geq c > 0$, then there is a C^k function α_k on $\overline{\Omega}$ such that*

1. $\alpha_k = 0$ *on* $\Omega \setminus U$,

2. $\alpha_k = 1$ *in a neighborhood of* $\partial\Omega$.

Proof. Choose a number $\epsilon > 0$ such that, if $-\epsilon < \rho(z) < 0$, then $z \in U \cap \Omega$. Let λ_k be as in the last lemma. Define

$$\alpha_k(z) = \lambda_k\left(1 + \frac{\rho(z)}{\epsilon}\right).$$

Then α_k has all the desired properties. □

Lemma 10.4.9. *Let $\Omega = \{z \in \mathbb{C} \mid \rho(z) < 0\}$ be a bounded domain with C^∞ boundary. Given $g \in C^\infty(\overline{\Omega})$, choose $f(z) = \rho(z)g(z)$. If $h \in A^2(\Omega)$ (the Bergman space—see Section 10.3), then*

$$\int_\Omega \overline{h(\zeta)}\frac{\partial}{\partial\zeta} f(\zeta)\, d\xi\, d\eta = 0.$$

Proof. We want to integrate by parts, but h is not defined on $\partial\Omega$ so we need a limiting argument.

Let $\epsilon > 0$ be small and let $\Omega_\epsilon = \{z \mid \rho(z) < -\epsilon\}$. Then define $f_\epsilon(z) = (\rho(z) + \epsilon)g(z)$. Notice that $f_\epsilon \in C^\infty(\overline{\Omega}_\epsilon)$ and $f_\epsilon\big|_{\partial\Omega_\epsilon} = 0$. Also $h \in C^\infty(\overline{\Omega}_\epsilon)$. So, by Green's theorem (or just the one-variable fundamental theorem of calculus, applied one variable at a time),

$$\int_{\Omega_\epsilon} \overline{h(\zeta)}\frac{\partial}{\partial\zeta} f_\epsilon(\zeta)\, d\xi\, d\eta = -\int_{\Omega_\epsilon}\left(\frac{\partial}{\partial\zeta}\overline{h(\zeta)}\right)\cdot f_\epsilon(\zeta)\, d\xi\, d\eta. \tag{10.10}$$

There is no boundary term since $h \cdot f_\epsilon\big|_{\partial\Omega_\epsilon} \equiv 0$. But $h \in A^2(\Omega)$ so $(\partial/\partial\zeta)\overline{h} \equiv 0$ and the last expression is 0.

Finally, let

$$\chi_\epsilon(\zeta) = \begin{cases} 1 & \text{if} \quad \zeta \in \Omega_\epsilon\,, \\ 0 & \text{if} \quad \zeta \notin \Omega_\epsilon. \end{cases}$$

Then, using (10.10),

$$\begin{aligned}
&\left|\int_\Omega \overline{h(\zeta)}\frac{\partial}{\partial\zeta} f(\zeta)\, d\xi\, d\eta\right| \\
&= \left|\int_\Omega \overline{h(\zeta)}\frac{\partial}{\partial\zeta} f(\zeta)\, d\xi\, d\eta - \int_\Omega \overline{h(\zeta)}\chi_\epsilon(\zeta)\frac{\partial}{\partial\zeta} f_\epsilon(\zeta)\, d\xi\, d\eta\right| \\
&\leq \int_\Omega \left|\overline{h(\zeta)}\right| \left|\frac{\partial f}{\partial\zeta}(\zeta)\right| |1 - \chi_\epsilon(\zeta)|\, d\xi\, d\eta \\
&\quad + \int_\Omega |\overline{h(\zeta)}||\chi_\epsilon(\zeta)| \cdot \left|\frac{\partial f}{\partial\zeta}(\zeta) - \frac{\partial f_\epsilon}{\partial\zeta}(\zeta)\right|\, d\xi\, d\eta\,.
\end{aligned}$$

By the Cauchy–Schwarz inequality, this is

$$\leq \left(\int_\Omega |\overline{h(\zeta)}|^2 \left| \frac{\partial f}{\partial \zeta}(\zeta) \right|^2 d\xi\, d\eta \right)^{1/2} \cdot \left(\int_\Omega |1 - \chi_\epsilon(\zeta)|^2 \, d\xi\, d\eta \right)^{1/2}$$
$$+ \left(\int_\Omega |\overline{h(\zeta)}|^2 \, d\xi\, d\eta \right)^{1/2} \cdot \left(\int_\Omega \left| \frac{\partial f}{\partial \zeta}(\zeta) - \frac{\partial f_\epsilon}{\partial \zeta}(\zeta) \right|^2 d\xi\, d\eta \right)^{1/2} .$$

Now $\int_\Omega |1 - \chi_\epsilon(\zeta)|^2 \, d\xi\, d\eta$ clearly tends to zero as $\epsilon \to 0^+$. Also we see that $\partial f_\epsilon / \partial \zeta \to \partial f / \partial \zeta$ uniformly. So, letting $\epsilon \to 0$, we obtain

$$\int_\Omega \overline{h(\zeta)} \frac{\partial f}{\partial \zeta}(\zeta) \, d\xi\, d\eta = 0.$$

That is the desired conclusion. □

If $\Phi : \Omega_1 \to \Omega_2$ is conformal, then we might expect the Bergman kernel for Ω_1 to be related to that for Ω_2. This is indeed the case, as our Proposition 10.3.10 showed. We review it now:

Theorem 10.4.10. *If $\Phi : \Omega_1 \to \Omega_2$ is conformal then*

$$K_{\Omega_1}(z, w) = \Phi'(z) K_{\Omega_2}(\Phi(z), \Phi(w)) \overline{\Phi'(w)}.$$

We conclude this section with Bell's projection formula, which is based on Theorem 10.4.10.

If $\Omega \subseteq \mathbb{C}$ is a domain and K its Bergman kernel, then for any square integrable f on Ω we define

$$P_\Omega f(z) = \int_\Omega f(\zeta) K_\Omega(z, \zeta) \, d\xi\, d\eta.$$

This is the *Bergman projection.*

Recall that, if Ω is a bounded domain in $\mathbb{C}$, then the mapping

$$P : f \mapsto \int_\Omega K(\cdot, \zeta) f(\zeta) \, dA(\zeta)$$

is the Hilbert space orthogonal projection of $L^2(\Omega, dA)$ onto $A^2(\Omega)$.

Next is Bell's formula:

Proposition 10.4.11. *Let Ω_1, Ω_2 be domains and let*

$$\Phi : \Omega_1 \to \Omega_2$$

be a conformal map. Let $f \in L^2(\Omega_1)$. Then

$$P_{\Omega_2}((\Phi^{-1})' \cdot (f \circ \Phi^{-1})) = (\Phi^{-1})' \cdot ((P_{\Omega_1} f) \circ \Phi^{-1}).$$

Proof. We sketch the proof. For any $f \in L^2(\Omega_1)$ and $g \in L^2(\Omega_2)$,

$$\begin{aligned}
\langle P_{\Omega_2}((\Phi^{-1})' \cdot (f \circ \Phi^{-1})), g\rangle_2 &= \langle (\Phi^{-1})' \cdot (f \circ \Phi^{-1}), P_{\Omega_2} g\rangle_2 \\
&= \langle f, ((P_{\Omega_2} g) \circ \Phi) \cdot \Phi' \rangle_1 \\
&= \langle P_{\Omega_1} f, ((P_{\Omega_2} g) \circ \Phi) \cdot \Phi' \rangle_1 \\
&= \langle ((P_{\Omega_1} f) \circ \Phi^{-1}) \cdot (\Phi^{-1})', P_{\Omega_2} g\rangle_2 \\
&= \langle ((P_{\Omega_1} f) \circ \Phi^{-1}) \cdot (\Phi^{-1})', g\rangle_2 .
\end{aligned}$$

Since $g \in L^2(\Omega_2)$ was arbitrary, we may conclude that Bell's identity is proved. □

The next lemma, due to Bell, is central to the theory.

Proposition 10.4.12. *If $\Omega = \{z \in \mathbb{C} \mid \rho(z) < 0\}$ is a bounded domain with C^∞ boundary, if $u \in C^{k+1}(\overline{\Omega})$, and if $k \geq 1$, then there is a $g \in C^k(\overline{\Omega})$ that agrees with u up to order k on $\partial\Omega$ and such that $P_\Omega g = 0$.*

Proof. Let α_k be as in Lemma 10.4.8. We define g by induction. For the C^1 case, let

$$v_1(z) = \frac{\partial}{\partial z} w_1(z) ,$$

where

$$w_1(z) = \frac{\alpha_1(z) \cdot u(z) \cdot \rho(z)}{(\partial\rho/\partial z)(z)} .$$

Then

$$\begin{aligned}
v_1(z) &= \alpha_1(z) \cdot u(z) + \rho(z) \cdot \frac{\partial}{\partial z}\left(\frac{\alpha_1(z) \cdot u(z)}{(\partial\rho/\partial z)(z)}\right) \\
&\equiv \alpha_1(z) \cdot u(z) + \rho(z) \cdot \eta_1(z) ,
\end{aligned}$$

where η_1 is defined by this identity and is continuous on $\overline{\Omega}$. Then

$$v_1 - u = \rho(z) \cdot \eta_1(z) \quad \text{near} \ \ \partial\Omega.$$

(So v_1 and u agree to order *zero* on $\partial\Omega$.) In particular, $v_1 - u\big|_{\partial\Omega} = 0$. Also

$$P_\Omega v_1(z) = \int_\Omega K(z,\zeta) \frac{\partial}{\partial \zeta} w_1(\zeta)\, d\xi\, d\eta = 0$$

by Lemma 10.4.9.

Suppose inductively that we have constructed w_{l-1} and $v_{l-1} = \frac{\partial}{\partial z} w_{l-1}$ such that v_{l-1} agrees to order $(l-1) - 1$ with u on $\partial\Omega$ and $P_\Omega v_{l-1} = 0$. We shall now construct a function w_l of the form

$$w_l = w_{l-1} + \theta_l \cdot \rho^l \tag{10.11}$$

such that $v_l = \frac{\partial}{\partial z} w_l$ agrees with u up to order $l-1$ on $\partial\Omega$ and $P_\Omega v_l = 0$.

Let α_l be as in Lemma 10.4.8 and define a differential operator $\mathcal{D}$ on $\overline{\Omega}$ by

$$\mathcal{D}(\phi) = \frac{\alpha_l}{|\partial\rho/\partial z|^2} \operatorname{Re}\left(\frac{\partial \rho}{\partial z}\frac{\partial \phi}{\partial \overline{z}}\right).$$

Notice that

$$\mathcal{D}\rho(z) = 1 \tag{10.12}$$

when $z \in \partial\Omega$. We define

$$\theta_l = \frac{\alpha_l \mathcal{D}^{l-1}(u - v_{l-1})}{l!\partial\rho/\partial z}. \tag{10.13}$$

Then, with w_l defined as in (refkernel-eq5121) and $v_l = \frac{\partial}{\partial z} w_l$, we have

$$\begin{aligned}\mathcal{D}^{l-1}(u - v_l) &= \mathcal{D}^{l-1}u - \mathcal{D}^{l-1}\frac{\partial}{\partial z}(w_{l-1} + \theta_l \cdot \rho^l) \\ &= \mathcal{D}^{l-1}(u - v_{l-1}) - \theta_l \frac{\partial \rho}{\partial z} \cdot (\mathcal{D}\rho)^{l-1} \cdot l! \\ &\quad + (\text{terms that involve a factor of } \rho)\,.\end{aligned}$$

If $z \in \partial\Omega$, then (using (10.12), (10.13))

$$\mathcal{D}^{l-1}(u - v_{l-1}) - \mathcal{D}^{l-1}(u - v_{l-1}) + 0 = 0. \tag{10.14}$$

Since any directional derivative at $P \in \partial\Omega$ is a linear combination of

$$\mathcal{D} = a(P)\frac{\partial}{\partial x} + b(P)\frac{\partial}{\partial y} \quad \text{and} \quad \tau = -b(P)\frac{\partial}{\partial x} + a(P)\frac{\partial}{\partial y},$$

we may re-express our task as follows: We need to see that

$$(\tau)^s \mathcal{D}^t(u - v_l)\Big|_{\partial\Omega} = 0$$

for all $s+t = l-1$ (notice that the case $s+t < l-1$ follows from the inductive hypothesis and the explicit form of w_l in (10.11)). The case $s = 0$, $t = l-1$ was treated in (10.14). If $s \geq 1$, then we write

$$\tau^s \mathcal{D}^t(u - v_l) = \tau(\tau^{s-1}\mathcal{D}^t(u - v_l)). \tag{10.15}$$

Since $(s-1) + t = l-2$, the expression in parentheses is 0 on $\partial\Omega$ by the inductive hypothesis. But τ is a directional derivative *tangent* to $\partial\Omega$ (because $\mathcal{D}$ is normal); hence (10.15) is 0.

The induction is now complete, and v_l has been constructed. We set $v_l = g$. Then

$$P_\Omega g(z) = \int_\Omega K_\Omega(z,\zeta) g(\zeta) d\xi\, d\eta = \int_\Omega K_\Omega(z,\zeta)\frac{\partial}{\partial \zeta} w_k(\zeta)\, d\xi\, d\eta = 0$$

by Lemma 10.4.9. □

We shall use Bell's lemma (Proposition 10.3.10) twice. Our first use right now is on the disc. First note the following two simple facts:

(a) If K_D is the Bergman kernel for the disc, then

$$\left|\left(\frac{\partial}{\partial z}\right)^k K(z,w)\right| = \left|\frac{(k+1)!\overline{w}^k}{\pi(1-z\cdot\overline{w})^{k+2}}\right| \le \frac{(k+1)!}{(1-|w|)^{k+2}}.$$

(b) If $u \in C^k(\overline{D})$ and if u *vanishes to* order k at ∂D (i.e., u agrees with the zero function to order k at ∂D), then there is a $C > 0$ such that $|u(z)| \le C \cdot (1-|z|)^k$.

Theorem 10.4.13 (Condition R for the disc). *If $k \ge 1$ and $u \in C^{k+2}(\overline{D})$, then $\|P_D u\|_{C^{k-1}(D)} < \infty$.*

Proof. Use Proposition 10.4.12 to find a function $v \in C^{k+1}(\overline{D})$ that agrees with u to order $k+1$ on ∂D and such that $Pv = 0$. Then $P_D(u-v) = P_D u$, and $u - v$ vanishes to order $k+1$ on ∂D. In particular, by observation **(b)** above, $|u(\zeta) - v(\zeta)| \le C \cdot (1-|\zeta|)^{k+1}$. Then, for $j \le k-1$, we have

$$\begin{aligned}\left|\left(\frac{\partial}{\partial z}\right)^j P_D u(z)\right| &= \left|\left(\frac{\partial}{\partial z}\right)^j P_D(u-v)(z)\right| \\ &= \left|\int_D \left(\frac{\partial}{\partial z}\right)^j K_D(z,\zeta)(u-v)(\zeta)\, d\xi\, d\eta\right| \\ &\le \int_D (j+1)!(1-|\zeta|)^{-j-1} C \cdot (1-|\zeta|)^{k+1}\, d\xi\, d\eta\,,\end{aligned}$$

where we have used observation **(a)** above. This last integral is clearly bounded, independent of z. □

Remark 10.4.14. Item **(b)** above actually holds on any bounded domain Ω with C^k boundary: If $u \in C^k(\overline{\Omega})$ vanishes to order k on $\partial\Omega$, then there is a $C > 0$ such that

$$|u(z)| \le C \cdot \delta_\Omega(z)^k.$$

Here, for $z \in \Omega$,

$$\delta_\Omega(z) = \inf_{w \notin \Omega} |z - w|.$$

Of course, δ_Ω could also be replaced here by any C^k-smooth defining function. This is immediate from the definitions and the Taylor expansion in the normal direction (in terms of powers of the defining function ρ).

10.5 Smoothness to the Boundary of Conformal Mappings

Let $\Omega = \{z \in \mathbb{C} \mid \rho(z) < 0\}$ be a bounded and simply connected domain with C^∞ boundary. Let $F : D \to \Omega$ be a conformal mapping. We wish to show that the one-to-one continuation of F to $\overline{D}$ (provided by Theorem 10.4.2) is actually in $C^\infty(\overline{D})$. For this we need a few lemmas. The first is a classical result from the theory of partial differential equations due to Hopf.

Lemma 10.5.1 (Hopf's lemma)**.** *Let $U \subseteq \mathbb{C}$ be smoothly bounded. Let u be a harmonic function on U, continuous on the closure $\overline{U}$. Suppose that u assumes a local maximum value at $P \in \partial U$. Let ν be the unit outward normal vector to ∂U at P. Then the one-sided lower derivative $\partial u/\partial \nu$, defined to be*

$$\frac{\partial u}{\partial \nu} = \liminf_{t \to 0^+} \frac{u(P) - u(P - t\nu)}{t},$$

is positive.

Proof. It is convenient to make the following normalizations: Assume that u assumes the value 0 at P and is negative nearby and inside U; finally take the negative of our function so that u has a local *minimum* at P.

Now, since U has smooth boundary, there is an internally tangent disc at P. After scaling, we may as well suppose that it is the unit disc and that $P = 1 + i0$. Thus we may restrict our positive function u, with the minimum value 0 at $P = 1$, to the closed unit disc. Note in particular that $u(0) > 0$. Set $C = u(0) > 0$.

The Harnack inequality shows that $u(r) \geq [(1-r)/(1+r)]u(0)$, hence

$$\frac{u(1) - u(r)}{1 - r} = \frac{-u(r)}{1 - r} \leq -\frac{u(0)}{1 + r} \equiv \frac{-C}{1 + r} \leq -\frac{C}{2}.$$

The desired inequality for the normal derivative of u now follows. ☐

Remark 10.5.2. It is worth noting that the definition of the derivative and the fact that P is a local maximum guarantee—just from first principles—that the indicated one-sided lower normal derivative will be nonnegative. The Hopf lemma asserts that this derivative is actually positive.

Lemma 10.5.3. *If Ω is a bounded, simply connected domain with C^∞ boundary and if $F : D \to \Omega$ is a biholomorphic mapping, then there is a constant $C > 0$ such that*

$$\delta_\Omega(F(z)) \leq C(1 - |z|), \quad \text{all } z \in \Omega.$$

Here $\delta_\Omega(z) = \inf_{w \notin \Omega} |z - w|$.

Proof. The issue has to do only with points z near the boundary of D, and thus with points where $F(z)$ is near the boundary of Ω. Consider the function $w \mapsto \log|F^{-1}(w)|$. This function is defined for all w sufficiently near $\partial\Omega$ and indeed on $\Omega \setminus \{F(0)\}$. And it is harmonic there. Moreover, it is continuous on $(\Omega \setminus \{F(0)\}) \cup \partial\Omega$, with value 0 on $\partial\Omega$. In particular, it attains a (global) maximum at every point of $\partial\Omega$, since $|F^{-1}(w)| < 1$ if $w \in \Omega \setminus \{F(0)\}$. So the Hopf lemma applies. The logarithm function has nonzero derivative at all points. The conclusion of the lemma follows from combining this fact with the "normal derivative" conclusion of the Hopf lemma. That is, we show that

$$\frac{\partial}{\partial\nu}|F^{-1}|\bigg|_P \geq c > 0$$

at each $P \in \partial D$. [Note that here, as in the Hopf lemma, no differentiability at boundary points is assumed: The derivative estimates are on the "lower derivative" only, which, as a lim inf, always exists and has value in the extended reals.] □

Lemma 10.5.4. *With F, Ω as above and $k \in \{0, 1, 2, \ldots\}$ it holds that*

$$\left|\left(\frac{\partial}{\partial z}\right)^k F(z)\right| \leq C_k(1 - |z|)^{-k}.$$

Proof. Since F takes values in Ω, it follows that F is bounded. Now apply the Cauchy estimates on $D(z, (1 - |z|))$. □

Lemma 10.5.5. *If $\psi \in C^{2k+2}(\overline{\Omega})$ vanishes to order $2k + 1$ on $\partial\Omega$, then $F' \cdot (\psi \circ F) \in C_b^k(D)$. That is, $F' \cdot (\psi \circ F)$ has bounded derivatives up to and including order k.*

Proof. For $j \leq k$ we have

$$\left|\left(\frac{\partial}{\partial z}\right)^j (F' \cdot (\psi \circ F))\right| = \left|\sum_{l=0}^{j} \binom{j}{l} \left(\frac{\partial}{\partial z}\right)^l (F') \cdot \left(\frac{\partial}{\partial z}\right)^{j-l} (\psi \circ F)\right|.$$

But

$$\left(\frac{\partial}{\partial z}\right)^{j-l} (\psi \circ F) \tag{10.16}$$

is a linear combination, with complex coefficients, of terms of the form

$$\left[\left(\left(\frac{\partial}{\partial z}\right)^m \psi\right)(F(z))\right] \cdot \left(\frac{\partial}{\partial z}\right)^{n_1} F(z) \cdots \left(\frac{\partial}{\partial z}\right)^{n_k} F(z)$$

where $m \leq j - l$ and $n_1 + \cdots + n_k \leq j - l$. So (10.16) is dominated by

$$\begin{aligned} &C \cdot \delta_\Omega(F(z))^{2k+1-(j-l)}(1 - |z|)^{-n_1} \cdots (1 - |z|)^{-n_k} \\ &\qquad\qquad \leq C \cdot \delta_\Omega(F(z))^{k+1} \cdot (1 - |z|)^{-k-1}. \end{aligned}$$

By Lemma 10.5.3, this is

$$\leq C \cdot (1-|z|)^{k+1} \cdot (1-|z|)^{-k-1} \leq C\,.$$

□

Remark 10.5.6. There is a remarkable formula of Faà di Bruno that formalizes the expansion for the higher derivatives of a composition. While the identity was first discovered in the eighteenth century, it is still being studied today (in the higher-dimensional version).

Lemma 10.5.7. *Let $G : D \to \mathbb{C}$ be holomorphic and have the property that*

$$\left| \left(\frac{\partial}{\partial z}\right)^j G(z) \right| \leq C_j < \infty$$

for $j = 0, \ldots, k+1$. Then each $(\partial/\partial z)^j G$ extends continuously to $\overline{D}$ for $j = 1, \ldots, k$.

Proof. It is enough to treat the case $k = 0$. The general case follows inductively.

If $P \in \partial D$, then we define

$$G(P) = \int_0^1 G'(tP) \cdot P dt + G(0).$$

It is clear that this defines a continuous extension of G to ∂D. □

Theorem 10.5.8 (Painlevé). *If $\Omega = \{z \in \mathbb{C} \mid \rho(z) < 0\}$ is a bounded, simply connected domain with C^∞ boundary and $F : D(0,1) \to \Omega$ is a conformal mapping, then $F \in C^\infty(\overline{D})$ and $F^{-1} \in C^\infty(\overline{\Omega})$.*

Proof. Let $k \in \mathbb{N}$. Proposition 10.4.12 applied to the function $u = 1$ on Ω, there is a function $v \in C^{2k+8}(\overline{\Omega})$ such that v agrees with u up to order $2k+8$ on $\partial\Omega$ and $P_\Omega v = 0$. Then $\phi \equiv 1 - v$ satisfies $P_\Omega(\phi) = P_\Omega 1 - P_\Omega v = 1$ and ϕ vanishes to order $2k+8$ on $\partial\Omega$. By Lemma 10.5.5, $F' \cdot (\phi \circ F) \in C_b^{k+3}(D)$. By Lemma 10.5.7, $F' \cdot (\phi \circ F) \in C^{k+2}(\overline{D})$. By Theorem 10.4.13, $P_D(F' \cdot (\phi \circ F)) \in C^{k-1}(D)$ and has $k-1$ bounded derivatives. But the transformation law (Proposition 10.4.11) tells us that

$$P_D(F' \cdot (\phi \circ F)) = F' \cdot ((P_\Omega \phi) \circ F) = F' \cdot 1 = F'.$$

Thus F' has bounded derivatives up to order $k-1$. By Lemma 10.5.7, all derivatives of F up to order $k-2$ extend continuously to $\overline{D}$. Since k was arbitrary, we may conclude that $F \in C^\infty(\overline{D})$.

To show that $F^{-1} \in C^\infty(\overline{\Omega})$, it is enough to show that the Jacobian determinant of F as a real mapping on $\overline{D}$ does not vanish at any boundary point of D (we already know it is everywhere nonzero on D). Since F is holomorphic, F continues to satisfy the Cauchy—Riemann equations on $\overline{D}$. So it is enough to check that, at each point of $\overline{D} \setminus D$, some first derivative of F is nonzero. This assertion follows from a Hopf lemma argument analogous to the proof of Lemma 10.5.1. Details are left as an exercise. □

Classically, Theorem 10.5.8 was proved by studying Green's potentials. The result dates back to P. Painlevé's thesis. All the ideas in the proof we have presented here are due to S. Bell and E. Ligocka [17]. An account of Bell's approach, in a more general context, can be found in [22].

10.6 The Bochner-Martinelli Formula in Several Complex Variables

One variable is special because one has the Cauchy integral formula. In this formula the kernel is the same for all domains, and it is holomorphic in the free variable z. In several variables there is the Bochner-Martinelli formula. In this one the kernel is the same for all domains, but it is NOT holomorphic in the free variable z. Of course, both situations have the Szego and the Bergman kernels. They do have kernels which are holomorphic in the free variable z, but the kernels are difficult or impossible to compute.

If ω is a differential form on $\Omega \subset \mathbb{R}^N$ then, in coordinates, ω can be written as a finite sum of terms of the form $\omega_\alpha dx^\alpha$, where α is a multi-index and ω_α is a smooth function. Differential forms on $\mathbb{C}^n$ may be written in this fashion also since $\mathbb{C}^n$ is canonically identified with $\mathbb{R}^{2n}$. However, it is much more convenient to use complex notation. Thus if $\Omega \subset \mathbb{C}^n$ and ω is a differential form on Ω, then ω is a sum of terms of the form $\omega_{\alpha\beta} dz^\alpha d\overline{z}^\beta$, where α, β are multi-indices with $|\alpha| \leq n, |\beta| \leq n$. If $0 \leq p, q \leq n$ and

$$\omega = \sum_{|\alpha|=p, |\beta|=q} \omega_{\alpha\beta} dz^\alpha d\overline{z}^\beta,$$

then ω is said to be a differential form of *type* (or *bidegree*) (p, q).

In classical advanced calculus only a differential form of total degree m may be integrated on a space or surface or manifold of (real) dimension m. Likewise, in our new notation, only forms of type (p, q) with $p + q = m$ may be integrated on a space or surface or manifold of (real) dimension m.

If $\omega = \sum_{\alpha,\beta} \omega_{\alpha\beta} dz^\alpha d\overline{z}^\beta$, then we define

$$\partial\omega = \sum_{j=1}^{n} \sum_{\alpha,\beta} \frac{\partial \omega_{\alpha\beta}}{\partial z_j} dz_j \wedge dz^\alpha \wedge d\overline{z}^\beta \,,$$

$$\overline{\partial}\omega = \sum_{j=1}^{n} \sum_{\alpha,\beta} \frac{\partial \omega_{\alpha\beta}}{\partial \overline{z}_j} d\overline{z}_j \wedge dz^\alpha \wedge d\overline{z}^\beta \,.$$

Letting d denote the usual exterior differential operator on forms (see [188, 68]), we see by a straightforward calculation that $d = \partial + \overline{\partial}$.

Notice that when f is a C^1 *function* (or $(0,0)$ form) then

$$\overline{\partial} f = \sum_j \frac{\partial f}{\partial \overline{z}_j} d\overline{z}_j \, .$$

Since the differentials $d\overline{z}_j$ are linearly independent, we conclude that $\overline{\partial} f \equiv 0$ on Ω if and only if $\partial f / \partial \overline{z}_j \equiv 0$ for $j = 1, \ldots, n$. That is, $\overline{\partial} f \equiv 0$ on Ω if and only if f is holomorphic in each variable separately.

The language of differential forms is needed in order to formulate Stokes's theorem:

Theorem 10.6.1 (Stokes). *Let $\Omega \subset \mathbb{C}^n$ be a bounded open set with C^1 boundary. Let ω be a differential form of bidegree (p,q) with coefficients in $C^1(\overline{\Omega})$. Then*

$$\int_{\partial\Omega} \omega = \int_\Omega d\omega = \int_\Omega \partial\omega + \overline{\partial}\omega \, .$$

Standard references for Stokes's theorem are W. Rudin [187] and G. de Rham [59]. The Stokes's theorem that we have recorded here is the standard one simply expressed in complex notation.

The *full Cauchy integral formula* in $\mathbb{C}$ is a formula not just about holomorphic functions but about all continuously differentiable functions. We now derive this more general result for all $\mathbb{C}^n, n \geq 1$, and learn what consequences it has for the function theory of both one and several variables.

Definition 10.6.2. On $\mathbb{C}^n$ we let

$$\omega(z) \equiv \qquad dz_1 \wedge dz_2 \wedge \ldots \wedge dz_n$$

$$\eta(z) \equiv \qquad \sum_{j=1}^n (-1)^{j+1} z_j dz_1 \wedge \ldots \wedge dz_{j-1} \wedge dz_{j+1} \wedge \ldots \wedge dz_n \, .$$

The form η is sometimes called the *Leray form.* We will often write $\omega(\overline{z})$ to mean $d\overline{z}_1 \wedge \ldots \wedge d\overline{z}_n$ and likewise $\eta(\overline{z})$ to mean $\sum_{j=1}^n (-1)^{j+1} \overline{z}_j d\overline{z}_1 \wedge \ldots \wedge d\overline{z}_{j-1} \wedge d\overline{z}_{j+1} \wedge \ldots \wedge d\overline{z}_n$.

The genesis of the Leray form is explained by the following lemma.

Lemma 10.6.3. *For any $z_0 \in \mathbb{C}^n$, any $\epsilon > 0$, we have*

$$\int_{\partial B(z^0,\epsilon)} \eta(\overline{z}) \wedge \omega(z) = n \int_{B(z^0,\epsilon)} \omega(\overline{z}) \wedge \omega(z) \, .$$

Proof. Notice that $d\eta(\overline{z}) = \overline{\partial}\eta(\overline{z}) = n\omega(\overline{z})$. Therefore, by Stokes's theorem,

$$\int_{\partial B(z^0,\epsilon)} \eta(\overline{z}) \wedge \omega(z) = \int_{B(z^0,\epsilon)} d[\eta(\overline{z}) \wedge \omega(z)] \, .$$

Of course, the expression in [ ] is saturated in dz's so, in the decomposition $d = \partial + \overline{\partial}$, only the term $\overline{\partial}$ will not die. Thus the last line equals

$$\int_{B(z^0,\epsilon)} [\overline{\partial}(\eta(\overline{z}))] \wedge \omega(z) = n \int_{B(z^0,\epsilon)} \omega(\overline{z}) \wedge \omega(z) .$$

□

Remark 10.6.4. Notice that, by change of variables,

$$\begin{aligned}\int_{B(z^0,\epsilon)} \omega(\overline{z}) \wedge \omega(z) &= \int_{B(0,\epsilon)} \omega(\overline{z}) \wedge \omega(z) \\ &= \epsilon^{2n} \int_{B(0,1)} \omega(\overline{z}) \wedge \omega(z) .\end{aligned}$$

A straightforward calculation shows that

$$\begin{aligned}&\int_{B(0,1)} \omega(\overline{z}) \wedge \omega(z) \\ &\qquad = (-1)^{q(n)} \cdot (2i)^n \cdot (\text{volume of the unit ball in } \mathbb{C}^n \approx \mathbb{R}^{2n}) ,\end{aligned}$$

where $q(n) = [n(n-1)]/2$. We denote the value of this integral by $W(n)$.

Theorem 10.6.5 (Bochner–Martinelli). *Let $\Omega \subset \mathbb{C}^n$ be a bounded domain with C^1 boundary. Let $f \in C^1(\overline{\Omega})$. Then, for any $z \in \Omega$, we have*

$$\begin{aligned}f(z) = &\frac{1}{nW(n)} \int_{\partial\Omega} \frac{f(\zeta)\eta(\overline{\zeta} - \overline{z}) \wedge \omega(\zeta)}{|\zeta - z|^{2n}} \\ &- \frac{1}{nW(n)} \int_{\Omega} \frac{\overline{\partial} f(\zeta)}{|\zeta - z|^{2n}} \wedge \eta(\overline{\zeta} - \overline{z}) \wedge \omega(\zeta).\end{aligned}$$

Proof. Fix $z \in \Omega$. We apply Stokes's theorem to the form

$$L_z(\zeta) \equiv \frac{f(\zeta)\eta(\overline{\zeta} - \overline{z}) \wedge \omega(\zeta)}{|\zeta - z|^{2n}}$$

on the domain $\Omega_{z,\epsilon} \equiv \Omega \setminus \overline{B}(z,\epsilon)$, where $\epsilon > 0$ is chosen so small that $\overline{B}(z,\epsilon) \subset \Omega$. Note that Stokes's theorem does not apply to forms that have a singularity; thus we may not apply the theorem to L_z on any domain that contains the point z in either its interior or its boundary. This observation helps to dictate the form of the domain $\Omega_{z,\epsilon}$. As the proof develops, we shall see that it also helps to determine the outcome of our calculation.

Notice that

$$\partial(\Omega_{z,\epsilon}) = \partial\Omega \cup \partial B(z,\epsilon)$$

but that the two pieces are equipped with opposite orientations.

Thus, by Stokes,

$$\int_{\partial\Omega} L_z(\zeta) - \int_{\partial B(z,\epsilon)} L_z(\zeta) = \int_{\partial\Omega_{z,\epsilon}} L_z(\zeta)$$
$$= \int_{\Omega_{z,\epsilon}} d_\zeta(L_z(\zeta)) . \tag{10.17}$$

Notice that we consider z to be fixed and ζ to be the variable. Now

$$\begin{aligned} d_\zeta L_z(\zeta) &= \overline{\partial}_\zeta L_z(\zeta) \\ &= \frac{\overline{\partial} f(\zeta) \wedge \eta(\overline{\zeta} - \overline{z}) \wedge \omega(\zeta)}{|\zeta - z|^{2n}} \\ &\quad + f(\zeta) \cdot \left[\sum_{j=1}^{n} \frac{\partial}{\partial \overline{\zeta}_j} \left(\frac{\overline{\zeta}_j - \overline{z}_j}{|\zeta - z|^{2n}} \right) \right] \omega(\overline{\zeta}) \wedge \omega(\zeta). \end{aligned} \tag{10.18}$$

Observing that

$$\frac{\partial}{\partial \overline{\zeta}_j} \left(\frac{\overline{\zeta}_j - \overline{z}_j}{|\zeta - z|^{2n}} \right) = \frac{1}{|\zeta - z|^{2n}} - n \frac{|\overline{\zeta}_j - \overline{z}_j|^2}{|\zeta - z|^{2n+2}},$$

we find that the second term on the far right of (10.18) dies and we have

$$d_\zeta L_z(\zeta) = \frac{\overline{\partial} f(\zeta) \wedge \eta(\overline{\zeta} - \overline{z}) \wedge \omega(\zeta)}{|\zeta - z|^{2n}}.$$

Substituting this identity into (10.17) yields

$$\int_{\partial\Omega} L_z(\zeta) - \int_{\partial B(z,\epsilon)} L_z(\zeta) = \int_{\Omega_{z,\epsilon}} \frac{\overline{\partial} f(\zeta) \wedge \eta(\overline{\zeta} - \overline{z}) \wedge \omega(\zeta)}{|\zeta - z|^{2n}} . \tag{10.19}$$

Next we remark that

$$\begin{aligned} \int_{\partial B(z,\epsilon)} L_z(\zeta) &= f(z) \int_{\partial B(z,\epsilon)} \frac{\eta(\overline{\zeta} - \overline{z}) \wedge \omega(\zeta)}{|\zeta - z|^{2n}} \\ &\quad + \int_{\partial B(z,\epsilon)} \frac{(f(\zeta) - f(z))\, \eta(\overline{\zeta} - \overline{z}) \wedge \omega(\zeta)}{|\zeta - z|^{2n}} \\ &\equiv T_1 + T_2 . \end{aligned} \tag{10.20}$$

Since $|f(\zeta) - f(z)| \leq C|\zeta - z|$ (and since each term of $\eta(\overline{\zeta} - \overline{z})$ has a factor of some $\overline{\zeta}_j - \overline{z}_j$) it follows that the integrand of T_2 is of size $O(|\zeta - z|)^{-2n+2} \approx \epsilon^{-2n+2}$. Since the surface over which the integration is performed has area $\approx \epsilon^{2n-1}$, it follows that $T_2 \to 0$ as $\epsilon \to 0^+$.

By Lemma 10.6.3, we also have

$$\begin{aligned} T_1 &= \epsilon^{-2n} f(z) \int_{\partial B(z,\epsilon)} \eta(\overline{\zeta} - \overline{z}) \wedge \omega(\zeta) \\ &= n\epsilon^{-2n} f(z) \int_{B(0,\epsilon)} \omega(\overline{\zeta}) \wedge \omega(\zeta) \\ &= nW(n) f(z) \end{aligned} \tag{10.21}$$

Finally, (10.19–10.21) yield that

$$\left(\int_{\partial\Omega} L_z(\zeta) \right) - nW(n) f(z) + o(1) = \int_{\Omega_{z,\epsilon}} \overline{\partial} f(\zeta) \wedge \left[\frac{\eta(\overline{\zeta} - \overline{z})}{|\zeta - z|^{2n}} \right] \wedge \omega(\zeta) .$$

Since

$$\left| \frac{\eta(\overline{\zeta} - \overline{z})}{|\zeta - z|^{2n}} \right| = O(|\zeta - z|^{-2n+1}) ,$$

the last integral is absolutely convergent as $\epsilon \to 0^+$ (remember that $\overline{\partial} f$ is bounded). Thus we finally have

$$f(z) = \frac{1}{nW(n)} \int_{\partial\Omega} L_z(\zeta) - \frac{1}{nW(n)} \int_{\Omega} \overline{\partial} f(\zeta) \wedge \frac{\eta(\overline{\zeta} - \overline{z})}{|\zeta - z|^{2n}} \wedge \omega(\zeta) .$$

This is the Bochner–Martinelli formula. □

Corollary 10.6.6. *If $\Omega \subset \mathbb{C}$ is a bounded domain with C^1 boundary and if $f \in C^1(\overline{\Omega})$ then, for any $z \in \Omega$,*

$$f(z) = \frac{1}{2\pi i} \int_{\partial\Omega} \frac{f(\zeta)}{\zeta - z} d\zeta - \frac{1}{2\pi i} \int_{\Omega} \frac{\left(\partial f(\zeta) / \partial \overline{\zeta} \right)}{\zeta - z} d\overline{\zeta} \wedge d\zeta .$$

Proof. It is necessary only to note that, when $n = 1$,

$$\omega(\zeta) = d\zeta, \qquad \eta(\overline{\zeta} - \overline{z}) = \overline{\zeta} - \overline{z}, \qquad \text{and} \quad nW(n) = 2\pi i .$$

□

Corollary 10.6.7. *With hypotheses as in Corollary 10.6.6, and the additional assumption that $\overline{\partial} f = 0$ on Ω, we have*

$$f(z) = \frac{1}{2\pi i} \int_{\partial\Omega} \frac{f(\zeta)}{\zeta - z} d\zeta .$$

10.7 The Bergman Kernel in Several Complex Variables

The basic theory of the Bergman kernel in several complex variables is exactly like that in one complex variable. The proofs are all soft analysis, and proceed step by step in just the same way.

All these ideas have been gathered under the rubric of "Hilbert space with reproducing kernel," an idea first formulated by Aronszajn in [7]. All that is really needed to make the theory run is a version of our Lemma 10.3.1. For the Bergman space, the proof of this lemma in one variable works just as well in several variables. For the Szegő space, an application of the Bochner—Martinelli formula does the job.

10.8 The Szegő and Poisson-Szeő Kernels

The basic theory of the Szegő kernel is similar to that for the Bergman kernel—they are both special cases of a general theory of "Hilbert spaces with reproducing kernel" (see [7]). Thus we only outline the basic steps here, leaving details to the reader.

Let $\Omega \subset \mathbb{C}^n$ be a bounded domain with C^2 boundary. Let $A(\Omega)$ be those functions continuous on $\overline{\Omega}$ that are holomorphic on Ω. Let $H^2(\partial\Omega)$ be the space consisting of the closure in the $L^2(\partial\Omega, d\sigma)$ topology of the restrictions to $\partial\Omega$ of elements of $A(\Omega)$ (here $d\sigma$ is $(2n-1)$-dimensional area measure on the boundary). Then $H^2(\partial\Omega)$ is a proper Hilbert subspace of $L^2(\partial\Omega)$. Each element $f \in H^2(\partial\Omega)$ has a natural holomorphic extension to Ω given by its Poisson integral Pf. It is the case (see [141, Ch. 8]) that, for σ-almost every $\zeta \in \partial\Omega$, it holds that

$$\lim_{\epsilon\to 0+} f(\zeta - \epsilon\nu_\zeta) = f(\zeta).$$

Here, as usual, ν_ζ is the unit outward normal vector to $\partial\Omega$ at the point ζ.

For each fixed $z \in \Omega$ the functional

$$\psi_z : H^2(\partial\Omega) \ni f \mapsto Pf(z)$$

is continuous (why?). Let $k_z(\zeta)$ be the Hilbert space representative for the functional ψ_z. Define the Szegő kernel $S(z,\zeta)$ by the formula

$$S(z,\zeta) = \overline{k_z(\zeta)}, z \in \Omega, \quad \zeta \in \partial\Omega.$$

If $f \in H^2(\partial\Omega)$ then

$$Pf(z) = \int_{\partial\Omega} S(z,\zeta) f(\zeta) d\sigma(\zeta)$$

for all $z \in \Omega$. We shall not explicitly formulate and verify the various uniqueness and extremal properties for the Szegő kernel. The reader is invited to consider these topics.

Let $\{\phi_j\}_{j=1}^{\infty}$ be an orthonormal basis for $H^2(\partial\Omega)$. Define

$$S'(z,\zeta) = \sum_{j=1}^{\infty} \phi_j(z)\overline{\phi_j(\zeta)}, \quad z,\zeta \in \Omega.$$

For convenience we tacitly identify here each function with its Poisson extension to the interior of the domain. Then, for $K \subset \Omega$ compact, the series defining S' converges uniformly on $K \times K$. By a Riesz-Fischer argument, $S'(\cdot,\zeta)$ is the Poisson integral of an element of $H^2(\partial\Omega)$ and $S'(z,\cdot)$ is the conjugate of the Poisson integral of an element of $H^2(\partial\Omega)$. So S' extends to $(\overline{\Omega} \times \Omega) \cup (\Omega \times \overline{\Omega})$, where it is understood that all functions on the boundary are defined only almost everywhere. The kernel S' is conjugate symmetric. Also, by Riesz-Fischer theory, S' reproduces $H^2(\partial\Omega)$. Since the Szegő kernel is unique, it follows that $S = S'$.

The Szegő kernel may be thought of as representing a map

$$S : f \mapsto \int_{\partial\Omega} f(\zeta)S(\cdot,\zeta)d\sigma(\zeta)$$

from $L^2(\partial\Omega)$ to $H^2(\partial\Omega)$. Since S is self-adjoint and idempotent, it is the Hilbert space projection of $L^2(\partial\Omega)$ to $H^2(\partial\Omega)$.

The Poisson-Szegő kernel is obtained by a formal procedure from the Szegő kernel: this procedure manufactures a *positive* reproducing kernel from one that is not necessarily positive. Note in passing that, just as we argued for the Bergman kernel in the last section, $S(z,z)$ is never 0 when $z \in \Omega$.

Proposition 10.8.1. *Define*

$$\mathcal{P}(z,\zeta) = \frac{|S(z,\zeta)|^2}{S(z,z)}, \quad z \in \Omega,\ \zeta \in \partial\Omega.$$

Then for any $f \in A(\Omega)$ and $z \in \Omega$ it holds that

$$f(z) = \int_{\partial\Omega} f(\zeta)\mathcal{P}(z,\zeta)d\sigma(\zeta).$$

Proof. Fix $z \in \Omega$ and $f \in A(\Omega)$ and define

$$u(\zeta) = f(\zeta)\frac{\overline{S(z,\zeta)}}{S(z,z)}, \quad \zeta \in \partial\Omega.$$

Then $u \in H^2(\partial\Omega)$ hence

$$\begin{aligned} f(z) = u(z) &= \int_{\partial\Omega} S(z,\zeta)u(\zeta)d\sigma(\zeta) \\ &= \int_{\partial\Omega} \mathcal{P}(z,\zeta)f(\zeta)d\sigma(\zeta). \end{aligned}$$

This is the desired formula. □

Remark 10.8.2. In passing to the Poisson-Szegő kernel we gain the advantage of positivity of the kernel (for more on this circle of ideas, see [141, Ch. 1, 8] and also [116, Ch. 1]). However, we lose something in that $\mathcal{P}(z,\zeta)$ is no longer holomorphic in the z variable nor conjugate holomorphic in the ζ variable. The literature on this kernel is rather sparse and there are many unresolved questions.

As an exercise, use the paradigm of Proposition 10.8.1 to construct a positive kernel from the Cauchy kernel on the disc (be sure to first change the notation in the usual Cauchy formula so that it is written in terms of arc length measure on the boundary). What familiar kernel results?

Like the Bergman kernel, the Szegő and Poisson-Szegő kernels can almost never be explicitly computed. They can be calculated asymptotically in a number of important instances, however (see [69, 35]). We will give explicit formulas for these kernels on the ball.

Lemma 10.8.3. *The functions $\{z^\alpha\}$, where α ranges over multi-indices, are pairwise orthogonal and span $H^2(\partial B)$.*

Proof. The orthogonality follows from symmetry considerations. For the completeness, notice that it suffices to see that the span of $\{z^\alpha\}$ is dense in $A(B)$ in the uniform topology on the boundary. By the Stone-Weierstrass theorem, the closed algebra generated by $\{z^\alpha\}$ and $\{\overline{z}^\alpha\}$ is all of $C(\partial B)$. But the monomials $\overline{z}^\alpha, \alpha \neq 0$, are orthogonal to $A(B)$ (use the power series expansion about the origin to see this). The claimed density follows. □

Lemma 10.8.4. *Let* $\mathbf{1} = (1, 0, \ldots, 0)$. *Then*

$$S(z, \mathbf{1}) = \frac{(n-1)!}{2\pi^n} \frac{1}{(1-z_1)^n}.$$

Proof. We have that

$$S(z,\mathbf{1}) = \sum_\alpha \frac{z^\alpha \cdot 1^\alpha}{\|z_1^\alpha\|^2_{L^2(\partial B)}} = \sum_{k=0}^\infty \frac{z_1^k}{\eta(k)} = \frac{1}{2\pi^n} \sum_{k=0}^\infty \frac{z_1^k (k+n-1)!}{k!} \tag{10.22}$$

$$= \frac{(n-1)!}{2\pi^n} \sum_{k=0}^\infty \binom{k+n-1}{n-1} z_1^k \tag{10.23}$$

$$= \frac{(n-1)!}{2\pi^n} \frac{1}{(1-z_1)^n}.$$

□

Lemma 10.8.5. *Let ρ be a unitary rotation on $\mathbb{C}^n$. For any $z \in \overline{B}, \zeta \in \partial B$, we have that $S(z,\zeta) = S(\rho z, \rho\zeta)$.*

Proof. This is a standard change of variables argument and we omit it. □

Theorem 10.8.6. *The Szegő kernel for the ball is*

$$S(z,\zeta) = \frac{(n-1)!}{2\pi^n}\frac{1}{(1-z\cdot\overline{\zeta})^n}.$$

Proof. Let $z \in B$ be arbitrary. Let ρ be the unique unitary rotation such that ρz is a multiple of 1. Then, by Lemma 10.8.4,

$$S(z,\zeta) = S(\rho^{-1}1,\zeta) = S(1,\rho\zeta) = \overline{S(\rho\zeta,1)} = \frac{(n-1)!}{2\pi^n}\frac{1}{\left(1-(\overline{\rho\zeta})\cdot 1\right)^n}$$
$$= \frac{(n-1)!}{2\pi^n}\frac{1}{\left(1-\overline{\zeta}\cdot(\rho^{-1}1)\right)^n} = \frac{(n-1)!}{2\pi^n}\frac{1}{(1-z\cdot\overline{\zeta})^n}.$$

□

Corollary 10.8.7. *The Poisson-Szegő kernel for the ball is*

$$\mathcal{P}(z,\zeta) = \frac{(n-1)!}{2\pi^n}\frac{(1-|z|^2)^n}{|1-z\cdot\overline{\zeta}|^{2n}}.$$

10.9 Constructible versus Non-Constructible Kernels

A kernel is *constructible* if it can be written down explicitly. A kernel is *non-constructible* (or *canonical*) if it is produced abstractly by some functional-analytic argument.

In the theory of one complex variable, all canonical kernels (the Cauchy kernel, the Poisson kernel) are constructible, and all constructible kernels are canonical.

Not so in several complex variables. Certainly, in $\mathbb{C}^n$, we may produce the Bergman and Szegő kernels abstractly. But, in general (except on domains with a great deal of symmetry like the ball or the polydisc or the bounded symmetric domains of Cartan) these kernels are quite difficult to construct. Fefferman [69] produced an asymptotic expansion for the Bergman kernel on a strongly pseudoconvex domain (and Boutet de Monvel/Sjöstrand [35] did something similar for the Szegő kernel), but these arguments are quite deep and difficult.

It non-constructible kernels in $\mathbb{C}^n$ was considered to be quite exciting around 1970 when, independently, G.M. Henkin [103], H. Grauert and I. Lieb [82], and E. Ramirez [175] were able to actually *construct* reproducing kernels on strongly pseudoconvex domains. These constructions were highly nontrivial, but they produced a formula quite analogous to the Cauchy kernel of one complex variable. In particular, these new reproducing kernels $K(z,\zeta)$ had the important property that *they were holomorphic in the free z variable.*

This meant that the new kernels $K(z, \zeta)$ would not only *reproduce* holomorphic functions; they would also *create* holomorphic functions. That is to say, if $d\mu$ were any finite measure, then

$$f(z) \equiv \int K(z, \zeta)\, d\mu(\zeta)$$

will be a holomorphic function in z.

The new kernels proved to be particularly useful. For instance, they could be used to produce explicit solutions to the equation $\overline{\partial} u = f$ (see [104], for instance, or [141, Ch. 5]). They could also be used to attack various problems in the theory of function algebras.

Here we sketch the Henkin construction for a domain Ω which is strongly convex (all boundary curvatures positive). For the full details of the matter, consult [141, Ch. 5].

We begin with an idea called the Cauchy–Fantappié formula or formalism. This is a device for creating reproducing formulas in complex analysis. Good references for this notion are [178] and [133].

Recall the notation $\omega(z)$ and $\eta(z)$ from our discussion of the Bochner—Martinelli formula in Definition 10.6.2, ff. Further recall that a straightforward calculation shows that

$$\int_{B(0,1)} \omega(\overline{z}) \wedge \omega(z)$$
$$= (-1)^{q(n)} \cdot (2i)^n \cdot (\text{volume of the unit ball in } \mathbb{C}^n \approx \mathbb{R}^{2n}),$$

where $q(n) = [n(n-1)]/2$. We denote the value of this integral by $W(n)$.

Theorem 10.9.1. *Let $\Omega \subset \mathbb{C}^n$ be a domain with C^1 boundary. Let $w(z,\zeta) = (w_1(z,\zeta), \ldots, w_n(z,\zeta))$ be a C^1, vector-valued function on $\overline{\Omega} \times \overline{\Omega}\{diagonal\}$ that satisfies*

$$\sum_{j=1}^{n} w_j(z,\zeta)(\zeta_j - z_j) \equiv 1.$$

Then we have, for any $f \in C^1(\overline{\Omega}) \cap \{holomorphic\ functions\ on\ \Omega\}$ and any $z \in \Omega$, the formula

$$f(z) = \frac{1}{nW(n)} \int_{\partial\Omega} f(\zeta)\eta(w) \wedge \omega(\zeta).$$

Of course, it should be understood here that ζ is the variable of integration and z is a free variable.

Sketch of Proof: For convenience we restrict attention to complex dimension 2. We may assume that $z = 0 \in \Omega$.

(a) If $\alpha^1 = (a_1^1, a_2^1), \alpha^2 = (a_1^2, a_2^2)$ are 2-tuples of C^1 functions on $\overline{\Omega}$ that satisfy $\sum_{j=1}^{2} a_j^k(\zeta) \cdot (\zeta_j - z_j) = 1$, let

$$B(\alpha^1, \alpha^2) = \sum_{\sigma \in S_2} \epsilon(\sigma) a^1_{\sigma(1)} \wedge \overline{\partial}(a^2_{\sigma(2)}),$$

where S_2 is the symmetric group on 2 letters (which of course has just two elements) and $\epsilon(\sigma)$ is the signature of the permutation σ. Prove that B is independent of α^1.

(b) It follows that $\overline{\partial}B = 0$ on $\overline{\Omega}\setminus\{0\}$ (indeed $\overline{\partial}B$ is an expression like B with the expression $a^1_{\sigma(1)}$ replaced by $\overline{\partial}a^1_{\sigma(1)}$).

(c) Use **(b)**, especially the paranthetical remark, to prove inductively that if $\beta^1 = (b_1^1, b_2^1), \beta^2 = (b_1^2, b_2^2)$, then there is a form γ on $\Omega\setminus\{0\}$ such that

$$\left[B(\alpha^1, \alpha^2) - B(\beta^1, \beta^2)\right] \wedge \omega(\zeta) = \overline{\partial}\gamma = d\gamma.$$

(d) Prove that if $\alpha^1 = \alpha^2 = (w_1, w_2)$ then $B(\alpha^1, \alpha^2)$ simplifies as

$$B(\alpha^1, \alpha^2) \wedge \omega(\zeta) = (2-1)!\eta(w) \wedge \omega(\zeta).$$

(e) Let $\mathcal{S}$ be a small sphere of radius $\epsilon > 0$ centered at 0 such that $\mathcal{S} \subset \Omega$. Use part **(c)** to see that

$$\int_{\partial\Omega} f(\zeta)\eta(w) \wedge \omega(\zeta) = \int_{\mathcal{S}} f(\zeta)\eta(w) \wedge \omega(\zeta).$$

(f) Now use **(c)** and **(d)** to see that

$$\int_{\mathcal{S}} f(\zeta)\eta(w) \wedge \omega(\zeta) = \int_{\mathcal{S}} f(\zeta)\eta(v) \wedge \omega(\zeta),$$

where

$$v(z, \zeta) = \frac{\overline{\zeta}_j - z_j}{|\zeta - z|^2}.$$

[**Warning:** Be careful if you decide to apply Stokes's theorem.] From the theory of the Bochner—Martinelli kernel, we know that the last line is $2 \cdot W(2) \cdot f(0)$.

Now let

$$\Omega = \left\{z \in \mathbb{C}^2 \mid \rho(z) < 0\right\}.$$

We assume that Ω is strongly convex, i.e., that the matrix of (real) second partial derivatives of ρ induce a positive definite quadratic form.

Define the complex tangent line to $\partial\Omega$ at a point $P \in \partial\Omega$ to be the zero set of the function

$$\mathcal{L}_P(z) = \rho(P) + \frac{\partial\rho}{\partial z_1}(P)(z_1 - P_1) + \frac{\partial\rho}{\partial z_2}(P)(z_2 - P_2).$$

Now the Taylor expansion of ρ about a point $P \in \partial\Omega$, expressed in complex notation, is

$$\rho(z) = \rho(P) + \frac{\partial \rho}{\partial z_1}(P)(z_1 - P_1) + \frac{\partial \rho}{\partial z_2}(P)(z_2 - P_2) + \frac{\partial \rho}{\partial \overline{z}_1}(P)(\overline{z}_1 - \overline{P}_1) + \frac{\partial \rho}{\partial \overline{z}_2}(P)(\overline{z}_2 - \overline{P}_2) + (\text{quadratic terms}) + \mathcal{E}(z).$$

Here $\mathcal{E}(z)$ is an error term of order 3 and higher.

This last may be rewritten (because $P \in \partial\Omega$ so $\rho(P) = 0$) as

$$\rho(z) = 2\Re\mathcal{L}_P(z) + (\text{quadratic terms}) + \mathcal{E}(z).$$

And, because Ω is hypothesized to be strongly convex, we find that

$$\rho(z) = 2\Re\mathcal{L}_P(z) + (\text{positive term}).$$

We conclude that the variety $\{z \mid \mathcal{L}_P(z) = 0\}$ only intersects $\overline{\Omega}$ at the point P. For this reason, we may use the notation of the Cauchy—Fantappié formula to define

$$w_1(z, P) = \frac{(\partial \rho/\partial z_1)(P)}{\mathcal{L}_P(z)}$$

and

$$w_2(z, P) = \frac{(\partial \rho/\partial z_2)(P)}{\mathcal{L}_P(z)}.$$

It follows then that

$$\sum_{j=1}^{2} w_j \cdot (z_j - P_j) = 1.$$

So Cauchy—Fantappié theory applies, using this (w_1, w_2), and we get a reproducing kernel.

In the case that Ω is the unit ball B, the reader may check that the resulting kernel is the Szegő kernel.

10.10 The Ideas of Kerzman and Stein

We close our discussion of reproducing kernels with a consideration of how the Szegő kernel is related to the more explicit kernel that arises from the Cauchy—Fantappié formalism. This set of ideas is due to Kerzman and Stein [119].

C. Fefferman [69] made an important contribution in 1974 when he produced an asymptotic expansion for the Bergman kernel of a strongly pseudoconvex domain. Basically he was able to write

$$K(z, \zeta) = P(z, \zeta) + \mathcal{E}(z, \zeta),$$

where P (the principal term) is, in suitable local coordinates, the Bergman kernel of the ball and $\mathcal{E}$ (the error term) is a term of strictly lower order (in some measurable sense). This powerful formula gives one a means for calculating mapping properties of the Bergman integral. Fefferman himself used the formula to calculate the boundary asymptotics of Bergman metric geodesics (for the purpose of proving the smooth boundary extension of biholomorphic mappings). Fefferman states in his paper—although the details have never been worked out—that there is a similar asymptotic expansion for the Szegő kernel of a strongly pseudoconvex domain.

At about the same time, Boutet de Monvel and Sjöstrand [35] used the technique of Fourier integral operators [110] to directly derive an asymptotic expansion for the Szegő kernel of a strongly pseudoconvex domain. This expansion is quite similar to Fefferman's: there is a principal term, which in suitable local coordinates is the Szegő kernel of the ball, and there is an error term which is of lower order. It is not known whether the techniques of [35] can be used to derive an asymptotic expansion for the Bergman kernel.

The main purpose of the present section is to consider another method, due to Kerzman and Stein, for deriving asymptotic expansions for the canonical kernels that is more elementary and uses less machinery. Fefferman's rather complicated argument uses Kohn's solution of the $\overline{\partial}$-Neumann problem as well as the theory of nonisotropic singular integrals. Boutet de Monvel and Sjöstrand's argument uses the theory of Fourier integral operators. The method of Kerzman and Stein [119] that we treat here uses only basic complex function theory and a little functional analysis.

At this time there are virtually no results about asymptotic expansions for the canonical kernels on weakly pseudoconvex domains. Some interesting partial results appear in [92].

10.11 The Relation between Constructible Kernels and Canonical Kernels on Strongly Pseudoconvex Domains

The ideas that we present now have thus far only been developed on strongly pseudoconvex domains. It is an important open problem to determine how to carry out a similar program on finite type domains or more general domains.

In previous sections, we have defined the Szegő projection $S : L^2(\partial\Omega) \to H^2(\Omega)$. We also have a mapping $H : L^2(\partial\Omega) \to H^2(\Omega)$ that is determined by the constructive Henkin kernel discussed above. We note that H defines a bounded operator from $L^2(\partial\Omega)$ to $H^2(\Omega)$ (the Hardy space—see [141, Ch. 8]) for the following reason.

Let Ω be strongly pseudoconvex. This means that the Levi form is positive definite at each boundary point (see [141, Ch. 3] for details). It is known that $\partial\Omega$, when equipped with balls coming from the complex structure and the usual boundary area measure (see [199], [141, Ch. 8]), is a space of homogeneous type in the sense of Coifman and Weiss [53]. Further, it is straightforward to verify that the Henkin operator H satisfies the hypotheses of the David-Journé $T1$ theorem for spaces of homogeneous type (see [50] for a nice exposition of these ideas). Thus we may conclude that the Henkin operator maps $L^2(\partial\Omega)$ to $L^2(\partial\Omega)$. Since the Henkin kernel also obviously maps $L^2(\partial\Omega)$ to holomorphic functions, we may conclude that the Henkin integral maps $L^2(\partial\Omega)$ to $H^2(\partial\Omega)$.

Now of course S, being a projection, is self-adjoint. So $S = S^*$. It is not at all true that $H = H^*$, but one may calculate (see below for the details) that $A \equiv H^* - H$ is small in a measurable sense.

We also have

$$HS = S\,, \qquad SH^* = S\,,$$
$$SH = H\,, \qquad H^*S = H^*\,.$$

Let us discuss these four identities for a moment.

For the first, notice that S is the projection onto H^2, and H preserves holomorphic functions. So certainly $HS = S$. For the second, we calculate that

$$\langle SH^*x, y\rangle = \langle H^*x, Sy\rangle = \langle x, HSy\rangle = \langle x, Sy\rangle$$

(because H preserves holomorphic functions) and thus $= \langle Sx, y\rangle$. Hence $SH^* = S$. For the third, notice that H maps to the holomorphic functions, and S preserves holomorphic functions. And, for the fourth, we calculate that

$$\langle H^*Sx, y\rangle = \langle Sx, Hy\rangle = \langle x, SHy\rangle = \langle x, Hy\rangle = \langle H^*x, y\rangle\,.$$

In conclusion, $H^*S = H^*$.

Now we see that

$$SA = S(H^* - H) = SH^* - SH = S - H\,.$$

As a result,

$$S = H + SA$$

so

$$S(I - A) = H\,.$$

In conclusion,

$$S = H(I - A)^{-1}\,.$$

If indeed we can show that A is norm small in a suitable sense, then $(I - A)^{-1}$ is well defined by a Neumann series. Thus we may write

$$S = H + HA + HA^2 + \cdots + HA^j + HA^{j+1} + \cdots\,.$$

Hence we have expressed the Szegő projection S as an asymptotice expansion in terms of the Henkin projection H. By applying this asymptotic expansion to the Dirac delta mass, this last formula can be translated into saying that the Szegő *kernel S* can be written as an asymptotic expansion in terms of the Henkin *1 kernel.*

Some new ideas connected with the discussion here appear in [144].

It should be noted that Ewa Ligocka [153] has shown that these same ideas may be applied to expand the Bergman kernel in an asymptotic expansion in terms of the Henkin kernel. We shall not treat the details of her argument here.

10.12 Concluding Remarks

Reproducing kernels are part of the bedrock of modern analysis. A detailed treatment of reproducing kernels in the real variable setting appears in [23, 24]. Of course, there are many books about integral kernels in the complex analysis setting. Certainly [178] is one of the most comprehensive. The book [141] also has considerable treatment of the Bergman, Szegő, and Poisson-Szegő kernels.

We expect that the ideas presented here will be central to our subject for a good many years to come.

11

Conformal Mappings

In complex analysis, the classification of domains is always a fundamental problem. When $n = 1$, the problem is completely understood when the domain is simply connected due to the famous Riemann mapping theorem.

Note first that the boundary of a smooth bounded domain Ω in $\mathbb{C}$ is a finite union of smooth closed curves. We do not talk about finite type in that context because it makes no sense in one complex variable. And the behavior of the $\overline{\partial}$-Neumann problem is the same on all domains in that context.

A curve in general does not carry any information induced from the ambient complex structure—just because the tangent space to a curve at a point has no complex structure. Thus it is of less use to study the boundary geometry in treating the classification problem when $n = 1$. In several variables the situation is much more complicated as we see now.

11.1 Classification of Domains in $\mathbb{C}^n$

Let M be a smooth hypersurface in $\mathbb{C}^n$, $n \geq 2$. Then M is a smooth submanifold of real dimension $2n-1$ greater than or equal to three. Let $\mathbb{C}T(M)$ be the complexified tangent bundle on M. In $\mathbb{C}T(M)$, there is a natural subbundle $T^{1,0}(M)$ induced from the complex structure of the ambient space, that is, $T^{1,0}(M) = \mathbb{C}T(M) \cap T^{1,0}(\mathbb{C}^n)$. Hence, $\dim_{\mathbb{C}} T^{1,0}(M) = n - 1$ and

$$\mathbb{C}T(M) = T^{1,0}(M) \oplus T^{0,1}(M) \oplus E$$

where $T^{0,1}(M) = \overline{T^{1,0}(M)}$ and E is a line bundle.

Definition 11.1.1. The subbundle $T^{1,0}(M)$ is called the Cauchy-Riemann structure on M, and we call $\big(M, T^{1,0}(M)\big)$ a CR manifold. Any smooth section of $T^{1,0}(M)$ $\big(T^{0,1}(M)\big)$ is called a type $(1,0)((0,1))$ vector field on M.

See also [32] and [48]. With this CR structure on M one can study the geometry and calculate the invariants resulting from it. For instance, the number of positive eigenvalues of the Levi form associated with the boundaries of pseudoconvex domains is a CR invariant. The geometry on a CR manifold is one of the main research subjects in several complex variables which has been intensively studied by many authors. The paper by S.S. Chern and J. Moser [49]

DOI: 10.1201/9781003619581-11

is an excellent reference for dealing with the CR invariants, a subject that is beyond the scope of this article.

We make the following definitions.

Definition 11.1.2. Let M and N be two smooth hypersurfaces in $\mathbb{C}^n, n \geq 2$, and let $\varphi : M \to N$ be a smooth mapping. φ is said to be a CR mapping if $\varphi_* T^{1,0}(M) \subset T^{1,0}(N)$. If φ is a diffeomorphic CR mapping, we say that M is CR diffeomorphic to N.

Definition 11.1.3. A C^1 function f on M is called a CR function if $\bar{L}f = 0$ for any type $(0, 1)$ vector field $\bar{L}$ on M.

Thus, if $f \in C^1(\bar{\Omega}) \cap \mathcal{O}(\Omega)$, where Ω is a domain in $\mathbb{C}^n, n \geq 2$, with smooth boundary, the restriction of f to the boundary is automatically a CR function on $\partial\Omega$. We also have the following immediate consequence.

Corollary 11.1.4. *Let $\varphi : M \to N$ be a smooth mapping between two smooth hypersurfaces in $\mathbb{C}^n, n \geq 2$. Then, φ is a CR mapping if and only if φ_j, $1 \leq j \leq n$, is a CR function on M, where $\varphi = (\varphi_1, \ldots, \varphi_n)$.*

So far we have done a very brief introduction to the CR manifolds. From here a natural question arises. Since the boundaries carry rich information of the CR geometry, why not use the various CR invariants naturally attached to the boundaries to classify domains? If this is the case, then the investigation of the CR invariants attached to the boundaries will give us a method to tackle the classification problem. It sounds reasonable to do so, provided that one can make clear first what the connection between CR diffeomorphisms of the boundaries and biholomorphisms of the domains is. The remainder of this section is thus devoted to explain this scheme.

From now on, we shall always confine ourselves to smooth bounded pseudoconvex domains Ω in $\mathbb{C}^n$ with $n \geq 2$ unless the contrary is explicitly stated. The extension from a CR diffeomorphism between two boundaries $\partial\Omega_i, i = 1, 2$, to a biholomorphism between D_1 and D_2 is considerably much easier than the other direction. This can be done via the aid of the generalized version of the Hartogs extension theorem.

Theorem 11.1.5. *Let Ω be a smooth bounded domain in $\mathbb{C}^n$ with connected boundary, and let f be a smooth CR function defined on $\partial\Omega$. Then, f extends holomorphically to a function $F \in C^\infty(\bar{\Omega}) \cap \mathcal{O}(\Omega)$ such that $F|_{\partial\Omega} = f$.*

For a proof of Theorem 11.1.5, the reader is referred to [97] or [178]. Using Theorem 11.1.5, one can easily prove the following extension theorem from the boundaries to the domains.

Theorem 11.1.6. *Let $f : \partial\Omega_1 \to \partial\Omega_2$ be a CR diffeomorphism between two connected boundaries. Then f extends smoothly to a CR diffeomorphism between $\bar{\Omega}_1$ and $\bar{\Omega}_2$, that is, f and f^{-1} extend smoothly to $\bar{\Omega}_1$ $(\bar{\Omega}_2)$ such that $f : D_1 \to D_2$ is a biholomorphism.*

Proof. According to Theorem 11.1.5, $f\left(f^{-1}\right)$ extends smoothly to $\bar{\Omega}_1\left(\bar{\Omega}_2\right)$ such that $f\left(f^{-1}\right)$ is holomorphic in $D_1\left(D_2\right)$. Using the fact that f is a CR diffeomorphism between $\partial\Omega_1$ and $\partial\Omega_2$, we deduce that $\det\left(f'\right)(z) \neq 0$ for $z \in D_1$, where (f') is the complex Jacobian of f. Thus, f is locally a biholomorphism. Similar arguments also hold for f^{-1}. It follows that $f\left(f^{-1}\right)$ must map $D_1\left(D_2\right)$ into $D_2\left(D_1\right)$. Since f is one-to-one on the boundary $\partial\Omega_1$, f must be a biholomorphism (see also [47]). This proves the theorem.

Thus, it remains to see how to extend smoothly a biholomorphism from the domains to their closures. Before we proceed any further, let us digress a little to introduce the Bergman projection and Bergman kernel function.

For any domain Ω in $\mathbb{C}^n$, let $\mathcal{H}(\Omega)$ be the space of square-integrable holomorphic functions on Ω. Obviously, $\mathcal{H}(\Omega)$ is a closed subspace of $L^2(\Omega)$, and hence is itself a separable Hilbert space. If $D = \mathbb{C}^n$, then $\mathcal{H}\left(\mathbb{C}^n\right) = \{0\}$. Thus, we are interested in the case when $\mathcal{H}(\Omega)$ is nontrivial, in particular, when Ω is bounded. Since $\mathcal{H}(\Omega)$ is a closed subspace of $L^2(\Omega)$, we may consider the orthogonal projection from $L^2(\Omega)$ onto $\mathcal{H}(\Omega)$.

Definition 11.1.7. The orthogonal projection P from $L^2(\Omega)$ onto $\mathcal{H}(\Omega)$ is called the Bergman projection on Ω.

On the other hand, one may construct a reproducing kernel for $\mathcal{H}(\Omega)$ as follows. For any $w \in D$, let Λ_w be the point evaluation map

$$\begin{aligned} \Lambda_w : \mathcal{H}(\Omega) &\to \mathbb{C} \\ f &\mapsto f(w). \end{aligned}$$

It is easily verified by Cauchy's estimate that Λ_w satisfies

$$|\Lambda_w(f)| = |f(w)| \leq c d(w)^{-n} \|f\|_{L^2(\Omega)}$$

where $d(w)$ is the distance from w to the complement of Ω, and the constant c depends only on the space dimension n. Hence, by the Riesz representation theorem, there is a unique element, denoted by $K_D(\cdot, w)$, in $\mathcal{H}(\Omega)$ such that

$$f(w) = \Lambda_w(f) = (f, K_D(\cdot, w)) = \int_D f(z)\overline{K_D(z,w)}dV_z$$

for all $f \in \mathcal{H}(\Omega)$. It is easy to check that $K(z,w) = \overline{K(w,z)}$ and

$$\|K_D(\cdot, w)\|_{L^2(\Omega)} \leq c d(w)^{-n} \quad \text{for any } w \in D$$

The function $K_D(z,w)$ thus defined is called the Bergman kernel function for D.

For any $f \in L^2(\Omega)$, write $f = f_1 + f_2$, where $f_1 \in \mathcal{H}(\Omega)$ and $f_2 \in \mathcal{H}(\Omega)^{\perp}$. We have

$$Pf = f_1 = \int_D K_D(z,w) f_1(w) dV_w = \int_D K(z,w) f(w) dV_w$$

Thus, we have the following representation of the Bergman projection.

Theorem 11.1.8. *The Bergman projection* $P_D : L^2(\Omega) \to \mathcal{H}(\Omega)$ *is represented by*

$$P_D f(z) = \int_D K(z,w) f(w) dV_w$$

for all $f \in L^2(\Omega)$ *and* $z \in D$.

Now, we return to the problem of how to extend a biholomorphism of domains smoothly up to the boundaries. This extension problem is proved case by case. The first significant achievement in this aspect is due to C. Fefferman [70], who proved the following fundamental result.

Theorem 11.1.9 (Fefferman). *Let* f *be a biholomorphism between two smooth bounded strongly pseudoconvex domains* D_1 *and* D_2. *Then* f *extends smoothly to a* CR *diffeomorphism between* $\bar{\Omega}_1$ *and* $\bar{\Omega}_2$.

It follows from Theorem 11.1.9 that if two strongly pseudoconvex domains have different CR invariants attached to the boundaries, then there is no biholomorphism between them. The proof of Theorem 11.1.9 relies heavily on the detailed analysis of the Bergman kernel function near the strongly pseudoconvex boundary. In this case, the boundary behavior of the Bergman kernel function has the following expression:

$$K_D(z,z) = \frac{\phi(z)}{(-r(z))^{n+1}} + \tilde{\phi}(z)\log(-r(z))$$

where r is a strictly plurisubharmonic defining function for $D, \phi, \tilde{\phi} \in C^\infty(\bar{\Omega})$ and $\phi \neq 0$ near $\partial\Omega$. By a smooth (strictly) plurisubharmonic function we mean a smooth real-valued function such that the complex Hessian is positive (definite) semi-definite at every point of the domain.

However, it seems not easy to extend Fefferman's approach to weakly pseudoconvex domains. Around 1980, S. Bell and E. Ligocka [18] proposed a new approach to this extension problem. We describe it here.

First, we introduce a condition concerning the regularity of the Bergman projection operator which is useful in proving the regularity of a biholomorphic mapping near the boundary.

Definition 11.1.10. A smooth bounded domain Ω in $\mathbb{C}^n$ is said to satisfy condition R if the Bergman projection P associated with Ω maps $C^\infty(\bar{\Omega})$ continuously into $C^\infty(\bar{\Omega}) \cap \mathcal{O}(\Omega)$.

Various equivalent statements of condition R are given in the next theorem.

Theorem 11.1.11. *Let* Ω *be a smooth bounded domain in* $\mathbb{C}^n$ *with Bergman projection* P *and Bergman kernel function* $K(z,w)$. *The following conditions are equivalent:*

(1) Ω *satisfies condition* R.

(2) For each positive integer s, there is a nonnegative integer $m = m_s$ *such that* P *is bounded from* $W_0^{s+m}(\Omega)$ *to* $\mathcal{H}^s(\Omega)$.

(3) For each multi-index α, *there are constants* $c = c_\alpha$ *and* $m = m_\alpha$ *such that*

$$\sup_{z\in D}\left|\frac{\partial^\alpha}{\partial z^\alpha}K(z,w)\right| \leq cd(w)^{-m}$$

where $d(w)$ *is the distance from the point* w *to the boundary* $\partial\Omega$.

For a proof of Theorem 11.1.11, see [21] and [48]. Here $W_0^s(\Omega)$ is the completion of $C_0^\infty(\Omega)$ in $W^s(\Omega)$, the Sobolev space of order s on Ω, and $\mathcal{H}^s(\Omega) = W^s(\Omega) \cap \mathcal{O}(\Omega)$

Condition R in general does not hold on smooth bounded domain. A smooth bounded non-pseudoconvex domain in $\mathbb{C}^2$ on which condition R fails was constructed by D. Barrett [12]. However, the powerfulness of condition R still can be seen from the following important result proved by S. Bell and E. Ligocka [17].

Theorem 11.1.12 (Bell-Ligocka). *Let* D_1 *and* D_2 *be two smooth bounded domains (not necessarily pseudoconvex) in* $\mathbb{C}^n, n \geq 2$, *and let* f *be a biholomorphic mapping from* D_1 *onto* D_2. *Suppose that condition* R *holds on both* D_1 *and* D_2, *then* f *extends smoothly to the boundary.*

Note that if both D_1 and D_2 are assumed to be pseudoconvex, then the assertion of Theorem 11.1.12 is still valid if condition R holds only on one of the domains D_1 and D_2; see [19]. Roughly speaking, to prove Theorem 11.1.12 we first construct, under condition R, local holomorphic coordinates smooth up to the boundary near a boundary point. Then we show that the biholomorphic map becomes linear in this coordinate system. Hence, it extends smoothly up to the boundary.

Theorem 11.1.12 indicates that condition R is sufficient for the extension problem of biholomorphic map. Thus, it becomes extremely important to verify the validity of condition R on a given domain, in particular, on a smooth bounded pseudoconvex domain in view of the non-pseudoconvex counterexample discovered by D. Barrett. Obviously, from Theorem 11.1.11, the regularity of the Bergman projection is closely related to that of the Bergman kernel function. Unfortunately, the Bergman kernel function in general cannot be obtained explicitly. Therefore, it is not easy to get the regularity of the Bergman projection directly from the Bergman kernel function unless on certain very special domains, say, balls or Reinhardt domains (domains that are invariant under rotations in each variable).

Usually, condition R on pseudoconvex domains is verified through the so-called $\bar{\partial}$-Neumann problem which had been intensively studied long before condition R was proposed. The $\bar{\partial}$-Neumann problem can be formulated as follows. For any smooth bounded pseudoconvex domain $\Omega \subset \mathbb{C}^n, n \geq 2$, denote by $L^2_{(p,q)}(\Omega)$ the space of all (p,q)-forms with square integrable coefficients. Hence, any $f \in L^2_{(p,q)}(\Omega)$ can be written as

$$f = \sum_{|I|=p,|J|=q}{}' f_{I,J} dz^I \wedge d\bar{z}^J$$

where $I = (i_1, \ldots, i_p), J = (j_1, \ldots, j_q)$ are increasing multi-indices with $i_s, j_t \in \mathbb{N}, dz^I = dz_{i_1} \wedge \ldots \wedge dz_{i_p}, d\bar{z}^J = d\bar{z}_{j_1} \wedge \ldots \wedge d\bar{z}_{j_q}$, and the prime means that the summation ranges only over those increasing multi-indices. The coefficients $f_{I,J}$ are defined to be anti-symmetric in I and J. The inner product on $L^2_{(p,q)}(\Omega)$ is defined by

$$(f, g) = \sum{}' \int_D f_{I,J} \overline{g_{I,J}} dV$$

for $f, g \in L^2_{(p,q)}(\Omega)$. The $\bar{\partial}$ operator acts on $L^2_{(p,q)}(\Omega)$ in the distribution sense. Hence, $f \in L^2_{(p,q)}(\Omega)$ is in the domain of $\bar{\partial}$ if $\bar{\partial} f \in L^2_{(p,q+1)}(\Omega)$, where

$$\bar{\partial} f = \sum{}' \left(\bar{\partial} f_{I,J}\right) \wedge dz^I \wedge d\bar{z}^J$$

It is easily seen that $\bar{\partial}$ is a closed, linear, densely defined operator, and $\bar{\partial}$ forms a complex, i.e., $\bar{\partial}^2 = 0$. The Hilbert space adjoint

$$\bar{\partial}^* : L^2_{(p,q)}(\Omega) \to L^2_{(p,q-1)}(\Omega)$$

of $\bar{\partial}$ is also a closed, linear, densely defined operator. Hence, $g \in \mathrm{Dom}\left(\bar{\partial}^*\right)$ if there is a $h \in L^2_{(p,q-1)}(\Omega)$ such that for any $\phi \in \mathrm{Dom}(\bar{\partial}) \cap L^2_{(p;q-1)}(\Omega)$, we have

$$(g, \bar{\partial}\phi) = (h, \phi)$$

We then define $\bar{\partial}^* g = h$. Clearly, $\bar{\partial}^*$ also forms a complex, and we have

$$L^2_{(p,q-1)}(\Omega) \underset{\bar{\partial}^*}{\overset{\bar{\partial}}{\rightleftarrows}} L^2_{(p,q)}(\Omega) \underset{\bar{\partial}^*}{\overset{\bar{\partial}}{\rightleftarrows}} L^2_{(p,q+1)}(\Omega).$$

Then we form the complex Laplacian

$$\Box = \overline{\partial\partial}^* + \bar{\partial}^* \bar{\partial} : L^2_{(p,q)}(\Omega) \to L^2_{(p,q)}(\Omega) \tag{11.1}$$

on

$$\mathrm{Dom}(\Box) = \left\{ f \in \mathrm{Dom}(\bar{\partial}) \cap \mathrm{Dom}(\bar{\partial}^*) \;\middle|\; \bar{\partial} f \in \mathrm{Dom}(\bar{\partial}^*) \text{ and } \bar{\partial}^* f \in \mathrm{Dom}(\bar{\partial}) \right\} .$$

$\Box$ is a closed, linear, self-adjoint, densely defined operator.

The $\bar{\partial}$-Neumann problem is, given $f \in L^2_{(p,q)}(\Omega)$, to prove the existence and regularity of the solution u to the equation (11.1), i.e.,

$$\Box u = f$$

Using Hilbert space techniques, the existence of the solution u in the L^2 sense can be obtained. In fact, one can prove

Theorem 11.1.13 (Hörmander [111]). *Let Ω be a bounded pseudoconvex domain in $\mathbb{C}^n, n \geq 2$. For each $0 \leq p \leq n, 1 \leq q \leq n$, there exists a bounded operator $N_{(p,q)} : L^2_{(p,q)}(\Omega) \to L^2_{(p,q)}(\Omega)$ such that:*

(1) $\mathcal{R}\left(N_{(p,q)}\right) \subset \operatorname{Dom}\left(\Box_{(p,q)}\right), N_{(p,q)}\Box_{(p,q)} = \Box_{(p,q)}N_{(p,q)} = I$ *on* $\operatorname{Dom}\left(\Box_{(p,q)}\right)$.

(2) For any $f \in L^2_{(p,q)}(\Omega), f = \bar{\partial}\bar{\partial}^* N_{(p,q)} f \oplus \bar{\partial}^*\bar{\partial} N_{(p,q)} f$.

(3) $\bar{\partial} N_{(p,q)} = N_{(p,q+1)}\bar{\partial}$ *on* $\operatorname{Dom}(\bar{\partial}), 1 \leq q \leq n-1$.

(4) $\bar{\partial}^* N_{(p,q)} = N_{(p,q-1)}\bar{\partial}^*$ *on* $\operatorname{Dom}\left(\bar{\partial}^*\right), 2 \leq q \leq n$.

(5) Let δ be the diameter of Ω. The following estimates hold for any $f \in L^2_{(p,q)}(\Omega)$:

$$\left\|N_{(p,q)} f\right\| \leq \frac{e\delta^2}{q}\|f\|$$

$$\left\|\bar{\partial} N_{(p,q)} f\right\| \leq \sqrt{\frac{e\delta^2}{q}}\|f\|$$

$$\left\|\bar{\partial}^* N_{(p,q)} f\right\| \leq \sqrt{\frac{e\delta^2}{q}}\|f\|$$

The operator $N = N_{(p,q)}$ is called the $\bar{\partial}$-Neumann operator. Now, it is clear how to relate the $\bar{\partial}$-Neumann operator with the Bergman projection. If $f \in \operatorname{Dom}(\bar{\partial})$ and $g \in \mathcal{H}(\Omega)$, then

$$\left(g, \bar{\partial}^* N_{(0,1)}\bar{\partial} f\right) = \left(\bar{\partial} g, N_{(0,1)}\bar{\partial} f\right) = 0.$$

This shows $\bar{\partial}^* N_{(0,1)}\bar{\partial} f \perp \mathcal{H}(\Omega)$. On the other hand, using (1) and (3) of Theorem 11.1.13, we have

$$\begin{aligned}
\bar{\partial}\left(f - \bar{\partial}^* N_{(0,1)}\bar{\partial} f\right) &= \bar{\partial} f - \overline{\partial\partial}^* N_{(0,1)}\bar{\partial} f \\
&= \bar{\partial} f - \left(\overline{\partial\partial}^* + \bar{\partial}^*\bar{\partial}\right) N_{(0,1)}\bar{\partial} f \\
&= \bar{\partial} f - \bar{\partial} f \\
&= 0.
\end{aligned}$$

Hence,

$$Pf = f - \bar{\partial}^* N_{(0,1)} \bar{\partial} f \tag{11.2}$$

for $f \in \text{Dom}(\bar{\partial})$. However, 11.2 also holds for any $f \in L^2(\Omega)$ if $\bar{\partial}^* N_{(0,1)} \bar{\bar{\partial}}$ is first viewed as a bounded operator on the dense subspace $\text{Dom}(\bar{\partial})$ and then extended by continuity to the whole space $L^2(\Omega)$. Thus, using (11.2), the regularity of the Bergman projection P will follow immediately from that of the $\bar{\partial}$-Neumann operator N.

Vast efforts had been devoted to proving the regularity of the $\bar{\partial}$-Neumann problem in the last few decades. The program was first initiated by J.J. Kohn [122] when Ω is a strongly pseudoconvex domain. In this case, a subelliptic 1/2-estimate for (p, q)-forms was established. To be more precise, we make the following definition.

Definition 11.1.14. Let Ω be a smooth bounded pseudoconvex domain in $\mathbb{C}^n, n \geq 2$, and let $z_0 \in \partial\Omega$. The $\bar{\partial}$-Neumann problem on Ω is said to satisfy a subelliptic ϵ-estimate for (p, q)-forms at $z_0, 0 < \epsilon < 1$, if there exists an open neighborhood U of z_0 such that

$$\|f\|_\epsilon^2 \leq C \left(\|\bar{\partial} f\|^2 + \left\|\bar{\partial}^* f\right\|^2 + \|f\|^2 \right) \tag{11.3}$$

for $f \in \mathcal{D}_{(p,q)}(\Omega) = C^\infty_{(p,q)}(\bar{\Omega}) \cap \text{Dom}\left(\bar{\partial}^*\right)$ with $\text{supp}(f)$ contained in U, where $\|\cdot\|_\epsilon$ denotes the Sobolev ϵ-norm and $C > 0$ is independent of f.

Then, we have

Theorem 11.1.15 (Kohn). *Let Ω be a smooth bounded strongly pseudoconvex domain in $\mathbb{C}^n, n \geq 2$. Then subelliptic 1/2-estimate for (p, q)-forms holds on $D, 0 \leq p \leq n$ and $1 \leq q \leq n - 1$.*

A complete characterization of subelliptic 1/2-estimate for (p, q)-forms at z_0 was obtained by Hörmander [109]. That is, 11.3 holds with $\epsilon = 1/2$ for (p, q)-forms at z_0 if and only if the Levi form at z_0 has either at least $n - q$ positive eigenvalues or at least $q + 1$ negative eigenvalues. This is usually called condition $Z(q)$. See also [48] and [72].

In [130], it was shown through a general scheme that if (11.3) holds, then the $\bar{\partial}$-Neumann operator N will gain 2ϵ derivatives in Sobolev scale, that is,

$$\|\zeta N f\|_{s+2\epsilon} \leq C_s \left(\|\zeta' f\|_s + \|f\| \right)$$

where $\zeta, \zeta' \in C_0^\infty(U)$ and $\zeta' = 1$ on the support of ζ.

Thus, we have the following theorem.

Theorem 11.1.16. *The Bergman projection preserves Sobolev spaces $W^s(\Omega)$, $s \geq 0$, on any smooth bounded strongly pseudoconvex domain in $\mathbb{C}^n, n \geq 2$. In particular, condition R holds on any smooth bounded strongly pseudoconvex domain in $\mathbb{C}^n$.*

See also [129] and [48]. Using Theorems 11.1.12 and 11.1.16, we reestablish Fefferman's mapping theorem (Theorem 11.1.9).

Apart from strong pseudoconvexity, we begin to encounter the problem caused by the vanishing of the Levi form. In [124], Kohn launched another program to study the regularity of the $\bar{\partial}$-Neumann problem on weakly pseudoconvex domains. When Ω is a smooth bounded pseudoconvex domain in $\mathbb{C}^2$, the Levi form λ is just a nonnegative function on the boundary. Hence, Ω is strongly pseudoconvex at a boundary point z_0 if and only if $\lambda(z_0) > 0$. Let r be a smooth defining function for Ω, and let

$$L = \frac{\partial r}{\partial z_1}\frac{\partial}{\partial z_2} - \frac{\partial r}{\partial z_2}\frac{\partial}{\partial z_1}$$

be the generator of the tangential type $(1,0)$ vector fields on $\partial\Omega$. Define the subspaces $\mathcal{L}_k$'s inductively by

$$\mathcal{L}_1 = \langle L, \bar{L}\rangle$$

and

$$\mathcal{L}_k = \langle \mathcal{L}_{k-1}, [X,Y]\rangle \quad \text{for} \quad X, Y \in \mathcal{L}_{k-1}, k \geq 2$$

Then we make the following definition of finite type.

Definition 11.1.17. Let Ω be a smooth bounded pseudoconvex domain in $\mathbb{C}^2$. Ω is said to be of finite type at $z_0 \in \partial\Omega$ if $\lambda(z_0) \neq 0$ or $X\lambda(z_0) \neq 0$ for some $X \in \mathcal{L}_k$.

The order of finite type at $z_0 \in \partial\Omega$ is defined to be 2 if $\lambda(z_0) \neq 0$, and be $2+k$ if $\lambda(z_0) = 0$ and k is the smallest positive integer such that $X\lambda(z_0) \neq 0$ for some $X \in \mathcal{L}_k$. Otherwise, Ω is said of infinite type at z_0.

Obviously, finite type is an open condition on the boundary. According to Definition 11.1.17, strictly pseudoconvex boundary points are of type 2. Also, due to pseudoconvexity of the domain, the type is always an even integer. With this setup, Kohn was able to establish the subelliptic estimate at the finite type points.

Theorem 11.1.18. *Let Ω be a smooth bounded pseudoconvex domain in $\mathbb{C}^2$. Suppose that Ω is of finite type m at $z_0 \in \partial\Omega$. Then a subelliptic $1/m$-estimate for the $\bar{\partial}$-Neumann problem holds at z_0.*

The necessity of a subelliptic $1/m$-estimate at z_0 is due to P. Greiner [90]. The type condition in $\mathbb{C}^2$ is equivalent to the largest order of contact that a one-dimensional complex manifold can have with the boundary at z_0. See also [26]. If we move now to higher-dimensional spaces, the situation is becoming more complicated.

In [128], Kohn introduced the concept of finite ideal type for the subellipticity of the $\bar{\partial}$-Neumann problem. Define the ideals I_k^q of germs of C^∞-functions at a boundary point $z_0 \in \partial\Omega$ inductively as follows:

$$I_1^q(z_0) = \sqrt[\mathbb{R}]{\left(r, \operatorname{coeff}\left\{\partial r \wedge \bar{\partial} r \wedge (\partial\bar{\partial} r)^{n-q}\right\}\right)}$$

where r is a smooth defining function for Ω and coeff $\{\cdot\}$ stands for the coefficients of the forms with respect to some holomorphic coordinate system, and, for $k \geq 2$,

$$I_k^q(z_0) = \sqrt[\mathbb{R}]{\left(I_{k-1}^q(z_0), \text{coeff}\left\{\partial f_1 \wedge \ldots \wedge \partial f_j \wedge \partial r \wedge \bar{\partial} r \wedge (\partial\bar{\partial} r)^{n-q-j}\right\}\right)}$$

where $f_1, \ldots, f_j \in I_{k-1}^q$. Here $\sqrt[\mathbb{R}]{I}$ means the radical ideal of I, that is,

$$\sqrt[\mathbb{R}]{I} = \{f \mid \text{ there exists } g \in I \text{ and } m \text{ such that } |f|^m \leq |g|\}$$

Then the following theorem was proved in [128].

Theorem 11.1.19. *Let Ω be a smooth bounded pseudoconvex domain in $\mathbb{C}^n$, $n \geq 2$, and let z_0 be a boundary point. If $1 \in I_k^q(z_0)$ for some k, then a subelliptic ϵ-estimate for the $\bar{\partial}$-Neumann problem for (p, q)-forms holds at z_0, where $0 < \epsilon \leq 1/2$.*

Conversely, if a subelliptic ϵ-estimate for (p, q)-forms holds at z_0, is it necessary that $1 \in I_k^q(z_0)$ for some k? The necessity of the finite ideal type so far is still open. When the defining function r is real analytic near z_0, with the aid of a theorem due to K. Diederich and J.E. Fornaess [60], we have a complete characterization of the subellipticity of the $\bar{\partial}$-Neumann problem at z_0.

Theorem 11.1.20. *Let Ω be a smooth bounded pseudoconvex domain in $\mathbb{C}^n$, $n \geq 2$, and let z_0 be a boundary point. If there exists a defining function r which is real analytic near z_0, then a subelliptic ϵ-estimate for the $\bar{\partial}$-Neumann problem for (p, q)-forms holds at z_0 if and only if there does not exist germs of complex variety of dimension q at z_0 in the boundary.*

In particular, we have

Theorem 11.1.21. *If Ω is a smooth bounded pseudoconvex domain in $\mathbb{C}^n$, $n \geq 2$, with real analytic boundary, then a subelliptic ϵ-estimate for the $\bar{\partial}$ Neumann problem for (p, q)-forms, $0 \leq p \leq n, 1 \leq q \leq n-1$, holds on Ω.*

According to the theorem of Diederich and Fornaess [60], a smooth bounded pseudoconvex domain with real analytic boundary does not contain any nontrivial complex variety in the boundary. Hence, one may apply Theorem 11.1.20 to obtain Theorem 11.1.21.

On the other hand, using the order of contact of complex varieties with the boundary at z_0, D'Angelo [57] also introduced a notion of finite type which we now present. Denote by $\nu(f)$ the order of vanishing at the origin in $\mathbb{C}$ of a smooth vector-valued function f defined in an open neighborhood of the origin in $\mathbb{C}$.

Definition 11.1.22. Let M be a smooth hypersurface of $\mathbb{C}^n$ containing z_0. Let r be a local defining function for M near z_0. Then z_0 is called a point of finite type if

$$\tau^*(r) = \sup_g \frac{\nu(r \circ g)}{\nu(g)} \leq \tau < \infty \tag{11.4}$$

where the supremum is taken over all germs of nonconstant holomorphic maps $g : \mathbb{C} \to \mathbb{C}^n$ with $g(0) = z_0$. The least such number τ is defined to be the type of the point z_0, denoted by $\triangle(M, z_0)$. A smooth bounded domain Ω is of finite type if every boundary point of Ω is of finite type.

See also the book by J. D'Angelo [58]. Thus, the type τ of a point $z_0 \in \partial\Omega$ is equal to the largest order of contact of a one-dimensional complex variety can have with the boundary $\partial\Omega$ at z_0. Then, the following theorem is proved in [26].

Theorem 11.1.23. *The type given by (11.4) is an open condition on the boundary.*

D. Catlin also proposed a notion of type condition closely related to the one in the sense of Definition 11.1.22. Then, he proved the following result.

Theorem 11.1.24. *Let Ω be a smooth bounded pseudoconvex domain in $\mathbb{C}^n$, $\frac{n}{2} \geq 2$, and let z_0 be a boundary point. Then a subelliptic ϵ-estimate for the $\bar{\partial}$-Neumann problem holds at z_0 if and only if Ω is of finite type at z_0.*

Detailed proofs of Theorem 11.1.24 and related results can be found in [40, 41, 42]. So far, the quantitative estimate of ϵ is not very precise. However, the story of seeking a subelliptic ϵ-estimate for the $\bar{\partial}$-Neumann problem almost ended here.

Next, we need to treat the regularity of the $\bar{\partial}$-Neumann problem on pseudoconvex domains with infinite type boundary points. For this situation, in general, there is no subelliptic ϵ-estimate. Thus, we are seeking for global regularity of the $\bar{\partial}$-Neumann problem.

In [43], a condition named property (P) was introduced by D. Catlin for global regularity of the $\bar{\partial}$-Neumann problem. Domains with property (P) may be viewed as a generalization of finite type domains. Here is the definition.

Definition 11.1.25. Let Ω be a smooth bounded domain in $\mathbb{C}^n, n \geq 2$. The boundary of Ω is said to satisfy **property (P)** if for every positive number $M > 0$, there is a plurisubharmonic function $\lambda \in C^\infty(\bar{\Omega})$ with $0 \leq \lambda \leq 1$ such that

$$\sum_{j,k=1}^n \frac{\partial^2 \lambda}{\partial z_j \partial \bar{z}_k}(z) t_j \bar{t}_k \geq M|t|^2 \quad \text{for all} \quad z \in \partial\Omega$$

where $t = (t_1, \ldots, t_n) \in \mathbb{C}^n$.

See also the paper concerning B-regular set by N. Sibony [194]. Obviously, property (P) implies the absence of any nontrivial complex variety on the boundary $\partial\Omega$.

Using property (P), Catlin was able to prove the following compactness estimate for the $\bar{\partial}$-Neumann problem on Ω.

Theorem 11.1.26. *Let Ω be a smooth bounded pseudoconvex domain in $\mathbb{C}^n$, $n \geq 2$. Suppose that $\partial\Omega$ satisfies property (6.30). Then, for any $\epsilon > 0$, there is a $C(\epsilon) > 0$ such that*

$$\|f\|^2 \leq \epsilon \left(\|\bar{\partial} f\|^2 + \left\|\bar{\partial}^* f\right\|^2\right) + C(\epsilon)\|f\|_{-1}^2$$

for $f \in \mathrm{Dom}(\bar{\partial}) \cap \mathrm{Dom}\left(\bar{\partial}^*\right)$, where $\|\cdot\|_{-1}$ denotes the Sobolev norm of order -1 on Ω.

Now, combining Theorem 11.1.26 with a theorem of Kohn and Nirenberg [130] gives the following consequence of property (P).

Theorem 11.1.27. *Let Ω be a smooth bounded pseudoconvex domain in $\mathbb{C}^n$, $n \geq 2$. Suppose that $\partial\Omega$ satisfies property (6.30). Then the $\bar{\partial}$-Neumann problem is globally regular on Ω.*

Typical examples of pseudoconvex domains with property (P) are those whose boundaries can be stratified by submanifolds M of holomorphic dimension zero. M is said of holomorphic dimension zero if the restriction of the Levi form to type $(1,0)$ vector fields tangent to M is positive. Thus, property (P) provides us with a tool for proving the global regularity of the $\bar{\partial}$-Neumann problem when the domain does not contain any complex variety on the boundary. However, it was pointed out by Sibony [194] that there exists a smooth bounded pseudoconvex domain Ω in $\mathbb{C}^2$ with no analytic disc contained in the boundary, yet property (P) fails on this domain. Whether the absence of analytic disc on the boundary implies the global regularity of the $\bar{\partial}$-Neumann problem remains to be unknown.

On the other hand, a vector field technique is developed to prove the global regularity of the $\bar{\partial}$-Neumann problem for general weakly pseudoconvex domains. Let $\Omega \subseteq \mathbb{C}^n, n \geq 2$, be a smooth bounded pseudoconvex domain, and let r be a smooth defining function for Ω. Set

$$L_n = \frac{4}{|\nabla r|^2} \sum_{j=1}^{n} \frac{\partial r}{\partial \bar{z}_j} \frac{\partial}{\partial z_j} \tag{11.5}$$

if $|\nabla r| \neq 0$, and

$$L_{jk} = \frac{\partial r}{\partial z_j} \frac{\partial}{\partial z_k} - \frac{\partial r}{\partial z_k} \frac{\partial}{\partial z_j} \quad \text{for } 1 \leq j < k \leq n \tag{11.6}$$

We have $L_n r = 1$ in a neighborhood of the boundary and the L_{jk}'s are tangent to the level sets of r. Also, the L_{jk}'s span the space of tangential type (1,0) vector fields at every boundary point of Ω.

The main idea of this method is to construct a real tangential vector field T on some open neighborhood of the boundary such that the commutators of T with type $(1,0)$ and $(0,1)$ vector fields have small modulus in L_n direction on the boundary. We formulate the required properties of T in the following condition:

Condition (T). For any given $\epsilon > 0$, there exists a smooth real vector field $T = T_\epsilon$, depending on ϵ, defined in some open neighborhood of $\bar{\Omega}$ and tangent to the boundary with the following properties:

(1) On the boundary, T can be expressed as

$$T = a_\epsilon(z)\left(L_n - \bar{L}_n\right), \quad \text{mod}\left(T^{1,0}(\partial\Omega) \oplus T^{0,1}(\partial\Omega)\right)$$

for some smooth function $a_\epsilon(z)$ with $|a_\epsilon(z)| \geq \delta > 0$ for all $z \in \partial\Omega$, where δ is a positive constant independent of ϵ.

(2) If $\mathcal{V}$ is any one of the vector fields $L_n, \bar{L}_n, L_{jk}$ and $\bar{L}_{jk}, 1 \leq j < k \leq n$, then

$$[T, \mathcal{V}]|_{\partial\Omega} = A_{\mathcal{V}}(z)L_n, \quad \text{mod}\left(T^{1,0}(\partial\Omega) \oplus T^{0,1}(\partial\Omega), \bar{L}_n\right)$$

for some smooth function $A_{\mathcal{V}}(z)$ with $\sup_{\partial\Omega}|A_{\mathcal{V}}(z)| < \epsilon$.

Using Condition (T), we can prove the following theorem.

Theorem 11.1.28. *Let Ω be a smooth bounded pseudoconvex domain in $\mathbb{C}^n$, $n \geq 2$, with a smooth defining function r. Suppose that condition (T) holds on Ω. Then the $\bar{\partial}$-Neumann operator N maps $W^s_{(p,q)}(\Omega), 0 \leq p \leq n, 1 \leq q \leq n$, boundedly into itself for each nonnegative real s.*

A related version of Theorem 11.1.28 was obtained by H.P. Boas and E. Straube [29]. They first proved the regularity for the Bergman projections under a condition similar to condition (T). Then, using the equivalence of regularity for the $\bar{\partial}$-Neumann operators and the Bergman projections proved earlier [28], they deduced the regularity for the $\bar{\partial}$-Neumann operators. See also [45]. Theorem 11.1.28 gives a direct treatment for the regularity of the $\bar{\partial}$-Neumann operators. A detailed proof of Theorem 11.1.28 can be found in [48].

Roughly speaking, to prove Theorem 11.1.28, using elliptic regularization method (see the paper by J.J. Kohn and L. Nirenberg [130]), it suffices to prove an a priori estimate for the solution $u = Nf$ to the $\bar{\partial}$-Neumann problem. Hence, we assume u is smooth up to the boundary and start to estimate

$$\left\|\bar{\partial}T^k u\right\|^2 + \left\|\bar{\partial}^* T^k u\right\|^2 \tag{11.7}$$

for $k \in \{0\} \cup \mathbb{N}$. Commuting T^k with $\bar{\partial}$ and $\bar{\partial}^*$, we see that it suffices to estimate

$$\left\|[\bar{\partial}, T]T^{k-1}u\right\|^2 + \left\|\left[\bar{\partial}^*, T\right]T^{k-1}u\right\|^2 \tag{11.8}$$

modulo some lower-order terms that can be controlled by the induction hypotheses. The most difficult terms to estimate in (11.8) occur when $[\bar{\partial}, T]$ or

$[\bar{\partial}^*, T]$ gives nontrivial component in L_n-direction. The assumption, condition (T), is used here to guarantee that such terms can be absorbed by the left-hand side, namely, (11.7). This proves Theorem 11.1.28.

So far, the vector field method can be well applied to two large classes of pseudoconvex domains to obtain the regularity of the $\bar{\partial}$-Neumann operators. Unlike pseudoconvex domains of finite type, both cases allow the boundaries of the domains to contain a large piece of complex variety. One is the case when the domain has a plurisubharmonic defining function. The other case is when the domain enjoys a circular transverse symmetry. We now describe them below.

Definition 11.1.29. Let Ω be a smooth bounded domain in $\mathbb{C}^n, n \geq 2. r$ is called a smooth plurisubharmonic defining function for Ω if r is a defining function for Ω and plurisubharmonic on the boundary $\partial\Omega$.

Thus, if Ω has a plurisubharmonic defining function, then Ω is automatically pseudoconvex. It is crucial to observe that if r is a plurisubharmonic defining function, then, for each k, the derivatives of $\partial r/\partial z_k$ of type $(0,1)$ in the directions that lie in the null space of the Levi form must vanish. This observation enables us to construct the required real tangential vector fields T such that the commutators of T and type (1,0) (or $(0,1)$) vector fields have small modulus in the L_n-direction. Using Theorem 11.1.28, we have the following theorem.

Theorem 11.1.30. *Let $\Omega \subseteq \mathbb{C}^n, n \geq 2$, be a smooth bounded pseudoconvex domain admitting a plurisubharmonic defining function $r(z)$. Then the $\bar{\partial}$ Neumann operator N is exactly regular on $W^s_{(p,q)}(\Omega)$ for $0 \leq p \leq n, 1 \leq q \leq n$ and all real $s \geq 0$.*

By "exactly regular" we mean that the $\bar{\partial}$-Neumann operator N preserves $W^s_{(p,q)}(\Omega)$ for all $s \geq 0$. For a proof of 11.1.30, the reader is refered to [29] and [48].

Theorem 11.1.30 can be applied to the class of convex domains. Let Ω be a smooth bounded convex domain in $\mathbb{R}^N, N \geq 2$, containing the origin. For any $x \in \mathbb{R}^N$, the Minkowski functional $\mu(x)$ is defined by

$$\mu(x) = \inf \{\lambda > 0 \mid x \in \lambda\Omega\}$$

where $\lambda\Omega = \{\lambda t \mid t \in \Omega\}$. It is not hard to verify that the Minkowski functional μ is a plurisubharmonic defining function for Ω. Thus, we have

Theorem 11.1.31. *Let Ω be a smooth bounded convex domain in $\mathbb{C}^n, n \geq 2$. Then the $\bar{\partial}$-Neumann operator N is exactly regular on $W^s_{(p,q)}(\Omega)$ for $0 \leq p \leq n$, $1 \leq q \leq n$ and all real $s \geq 0$.*

For a proof of Theorem 11.1.31, the reader is referred to [29, 46] and [29].

The other important class of pseudoconvex domains that satisfy the hypotheses of condition (T) are circular domains with transverse symmetries.

Definition 11.1.32. A domain Ω in $\mathbb{C}^n$ is called circular if $e^{i\theta} \cdot z = \left(e^{i\theta} z_1, \ldots e^{i\theta} z_n\right) \in \Omega$ for any $z \in \Omega$ and $\theta \in \mathbb{R}$. Ω is called Reinhardt if $\left(e^{i\theta_1} z_1, \ldots, e^{i\theta_n} z_n\right) \in \Omega$ for any $z \in \Omega$ and $\theta_1, \ldots, \theta_n \in \mathbb{R}$, and Ω is called complete Reinhardt if $z = (z_1, \ldots, z_n) \in \Omega$ implies $(w_1, \ldots, w_n) \in \Omega$ for all $|w_j| \leq |z_j|, 1 \leq j \leq n$.

Thus, a Reinhardt domain is automatically circular. Let Ω be a smooth bounded circular domain in $\mathbb{C}^n, n \geq 2$. Define $r(z)$ by

$$r(z) = \begin{cases} d(z, \partial\Omega) & \text{for } z \notin D, \\ -d(z, \partial\Omega) & \text{for } z \in D, \end{cases}$$

where $d(z, \partial\Omega)$ denotes the distance from z to the boundary $\partial\Omega$. It is easy to see that r is a smooth defining function for Ω such that $r(z) = r\left(e^{i\theta} \cdot z\right)$ and that $|\nabla r| = 1$ on the boundary. Denote by Λ the map of the S^1-action on Ω from $S^1 \times D$ to Ω defined by

$$\begin{aligned} \Lambda : S^1 \times \Omega &\to \Omega \\ \left(e^{i\theta}, z\right) &\mapsto e^{i\theta} \cdot z = \left(e^{i\theta} z_1, \ldots, e^{i\theta} z_n\right). \end{aligned}$$

For each fixed θ, Λ is an automorphism of Ω and Λ can be extended smoothly to a map from $S^1 \times \bar{\Omega}$ to $\bar{\Omega}$. Hence, for each fixed $z \in \bar{\Omega}$, we consider the orbit of z, namely, the map

$$\begin{aligned} \pi_z : S^1 &\to \bar{\Omega} \\ e^{i\theta} &\mapsto e^{i\theta} \cdot z. \end{aligned}$$

Then, π_z induces a vector field T on $\bar{\Omega}$, in fact on $\mathbb{C}^n$, by

$$T_z = \pi_{z,*}\left(\left.\frac{\partial}{\partial\theta}\right|_{\theta=0}\right) = i\sum_{j=1}^{n} z_j \frac{\partial}{\partial z_j} - i\sum_{j=1}^{n} \bar{z}_j \frac{\partial}{\partial \bar{z}_j} \tag{11.9}$$

where $\pi_{z,*}$ is the differential map induced by π_z. Note that T is tangent to the level sets of r. In particular, T is tangent to the boundary of Ω.

Definition 11.1.33. Let K be a compact subset of the boundary of a smooth bounded circular domain Ω. Ω is said to have transverse circular symmetry on K if for each point $z \in K$ the vector field T defined in (11.9) is not contained in $T_z^{1,0}(\partial\Omega) \oplus T_z^{0,1}(\partial\Omega)$.

Transverse symmetry was first introduced by D. Barrett in [12]. It is obvious from (11.9) and Definition 11.1.33 that Ω has transverse circular symmetry on the whole boundary if and only if $\sum_{j=1}^{n} z_j \left(\partial r / \partial z_j\right)(z) \neq 0$ on $\partial\Omega$. Then, we prove

Theorem 11.1.34. *Let $\Omega \subseteq \mathbb{C}^n, n \geq 2$, be a smooth bounded circular pseudoconvex domain with a smooth defining function r defined by (11.1). Suppose that $\sum_{j=1}^{n} z_j \left(\partial r / \partial z_j\right)(z) \neq 0$ on the boundary. Then the $\bar{\partial}$-Neumann problem is exactly regular on Ω.*

Proof. Let T be the vector field defined in (11.9). By assumption, T is transversal to $T^{1,0}(\partial\Omega)\oplus T^{0,1}(\partial\Omega)$ everywhere on the boundary. The key of the proof is to observe that $\mathrm{Tr} \equiv 0, [T, \partial/\partial\bar{z}_j] = i\partial/\partial\bar{z}_j$ and $[T, \partial/\partial z_j] = -i\partial/\partial z_j$. Hence, we have

$$[T, L_{jk}] = -2iL_{jk}$$

for all $1 \leq j < k \leq n$, and

$$[T, L_n]|_{\partial\Omega} = \left[T, \bar{L}_n\right]\big|_{\partial\Omega} = 0$$

where L_n and L_{jk} are defined in (11.5) and (11.6). It follows that condition (T) holds on Ω. By Theorem 11.1.28, this proves Theorem 11.1.34. □

Next a geometric argument shows that a complete Reinhardt domain always enjoys transverse circular symmetry.

Theorem 11.1.35. *Let* $\Omega \subseteq \mathbb{C}^n, n \geq 2$, *be a smooth bounded complete Reinhardt pseudoconvex domain with a smooth defining function* $r(z) = r\left(e^{i\theta_1}z_1, \ldots, e^{i\theta_n}z_n\right)$ *for all* $\theta_1, \ldots, \theta_n \in \mathbb{R}$. *Then, we have* $\sum_{j=1}^n z_j\left(\partial r/\partial z_j\right) \neq 0$ *on* $\partial\Omega$. *In particular, the* $\bar{\partial}$*-Neumann operator* N *is exactly regular on* $W^s_{(p,q)}(\Omega)$ *for* $0 \leq p \leq n, 1 \leq q \leq n$ *and all real* $s \geq 0$.

Theorems 11.1.34 and 11.1.35 can be found in [45] or [48].

Now, it is natural to ask whether the vector field method can be applied to any weakly pseudoconvex domain or not. Unfortunately, the answer is negative. An important counterexample known as "worm domain" was constructed by K. Diederich and J.E. Fornaess in [61]. Let us present it here (see Kiselman [120]).

Let $\beta > \pi/2$. Fix a smooth function $\eta : \mathbb{R} \to \mathbb{R}$ with the following properties:

(1) $\eta(x) \geq 0, \eta$ is even and convex;

(2) $\eta^{-1}(0) = I_{\beta-\pi/2}$, where $I_{\beta-\pi/2} = [-\beta + \pi/2, \beta - \pi/2]$;

(3) there exists an $a > 0$ such that $\eta(x) > 1$ if $x < -a$ or $x > a$;

(4) $\eta'(x) \neq 0$ if $\eta(x) = 1$.

Note that (4) follows from (1) and (2). The existence of such a function is obvious.

Define

$$D_\beta = \left\{(z_1, z_2) \in \mathbb{C}^2 \mid r(z) < 0\right\}$$

where $r(z) = \left|z_1 + e^{i\log|z_2|^2}\right|^2 + \eta\left(\log|z_2|^2\right) - 1$ is the defining function for D_β.

It is not hard to see that D_β has smooth boundary and is defined locally by a plurisubharmonic function. Thus, we have:

Proposition 11.1.36. *For each fixed $\beta > \pi/2, D_\beta$ is a smooth bounded pseudoconvex domain in $\mathbb{C}^2$.*

D_β is strictly pseudoconvex everywhere except on the closure of the annulus A, where

$$A = \left\{(0, z_2) \in \mathbb{C}^2 \;\middle|\; |\log |z_2|^2 | < \beta - \pi/2\right\}$$

The following result shows that for each fixed $\beta > \pi/2$, there is no C^2 global defining function which is plurisubharmonic on the boundary of D_β.

Theorem 11.1.37. *For any $\beta > \pi/2$, there is no C^2 defining function $\rho(z)$ for D_β such that $\rho(z)$ is plurisubharmonic on the boundary of D_β.*

Proof. Let $\rho(z)$ be such a C^2 defining function for D_β that is plurisubharmonic on the boundary $\partial\Omega_\beta$. Then there is a C^1 positive function h defined in some neighborhood of $\partial\Omega_\beta$ such that $\rho(z) = hr$. A direct calculation shows that the complex Hessian of $\rho(z)$ acting on any $(\alpha, \beta) \in \mathbb{C}^2$ for any point $p \in A \subset \partial\Omega_\beta$ is given by

$$\begin{aligned}\mathcal{L}_{\rho(z)}(p; (\alpha, \beta)) =& 2\,\mathrm{Re}\left[\bar{\alpha}\beta\left(\frac{ih}{z_2} + \frac{\partial h}{\partial z_2}\right) e^{i\log|z_2|^2}\right] \\ &+ \left[h + 2\,\mathrm{Re}\left(\frac{\partial h}{\partial z_1} e^{i\log|z_2|^2}\right)\right] |\alpha|^2\end{aligned} \tag{11.10}$$

Since, by assumption, (11.10) is always nonnegative, we must have

$$\left(\frac{ih}{z_2} + \frac{\partial h}{\partial z_2}\right) e^{i\log|z_2|^2} \equiv 0$$

on A, or, equivalently,

$$\frac{\partial}{\partial \bar{z}_2}\left(he^{-i\log|z_2|^2}\right) \equiv 0$$

on A. Consequently,

$$g\,(z_2) = h\,(0, z_2)\, e^{-i\log|z_2|^2}$$

is a holomorphic function on the annulus A. It follows that

$$g\,(z_2)\, e^{i\log z_2^2} = h\,(0, z_2)\, e^{-2\arg z_2} = c$$

is also locally a holomorphic function on A, and hence it must be a constant c, since the right-hand side is real. This implies that

$$h\,(0, z_2) = ce^{2\arg z_2}$$

is a well-defined, C^1 positive function on A, which is impossible. This proves Theorem 11.1.37. □

In particular, Theorem 11.1.30 is not applicable to worm domains. Actually, D. Barrett [13] was able to show the following:

Theorem 11.1.38. *For any $\beta > \pi/2$, the Bergman projection on D_β does not map $W^k(D_\beta)$ into $W^k(D_\beta)$ when $k \geq \pi/(2\beta - \pi)$.*

Basically, the idea of the proof is that we first dilate the z_1-component of D_β to an unbounded worm domain D'_β for z_2 belonging to $\{z_2 \in \mathbb{C} \mid |\log|z_2|^2| < \beta - \pi/2\}$. The Bergman kernel function on D'_β can be related to the weighted Bergman kernel function on a horizontal strip I_β of the complex plane. Now, using the explicit asymptotic expansion of the weighted Bergman kernel function on I_β, we deduce the assertion of the theorem.

With Barrett's result at hand, it was finally proved by M. Christ [51] that the Bergman projection is not globally regular on worm domains. Namely, we have

Theorem 11.1.39. *Condition R fails on worm domain D_β for any $\beta > \pi/2$.*

To prove Theorem 11.1.39, he showed that for each $\beta > \pi/2$, there is a sequence of positive numbers $\{s_j = s(\beta)_j\}_{j=1}^\infty$ tending to infinity such that if $f \in C^\infty_{(p,1)}(\bar{\Omega}_\beta)$ and $Nf = N_{(p,1)}f \in C^\infty_{(p,1)}(\bar{\Omega}_\beta)$ (this would be the case if condition R holds on D_β; see [27]), then, for each j, the following estimate holds:

$$\|Nf\|_{W^{s_j}(D_\beta)} \leq C_j \|f\|_{W^{s_j}(D_\beta)} \tag{11.11}$$

for some $C_j > 0$. Since $C^\infty(\bar{\Omega}_\beta)$ is dense in every $W^k(D_\beta), k \geq 0$, (11.11) will violate Barrett's result. This proves Theorem 11.1.39.

It might be interesting if one can reprove Theorem 11.1.39 via a direct treatment of the Bergman kernel function on D_β. This will shed more insights into the behavior of the Bergman projection near the weakly pseudoconvex boundary points on worm domains.

Although condition R fails on weakly pseudoconvex domains in general, yet the classification problem still remains. A complete understanding of the classification problem of domains in $\mathbb{C}^n, n \geq 2$, is still far from reach. It is possible that any biholomorphic map between two smooth bounded domains extends automatically smoothly up to the boundary without any additional assumption. New ideas and techniques are needed for further investigations. Hopefully, one can get a clearer picture of it in the near future. See also the survey papers by S. Bell [16] and H.P. Boas and E. Straube [30]. Finally, we should also note that if the metric is changed, then it is always possible to obtain Sobolev estimates for the weighted $\bar{\partial}$-Neumann problem. See [125, 127].

I think it is about time to stop at this stage. If, after reading this article, the reader wishes to learn more about several complex variables, I highly recommend the books by S.-C. Chen and M.-C. Shaw [48], G.B. Folland and J.J. Kohn [72], L. Hörmander [111], S. Krantz [139], R. Narasimhan [162] and R.M. Range [178].

12

Fundamental Solutions in Complex Analysis: The Cauchy Riemann Operator

In this chapter we explore complex analysis from the partial differential equations point of view. We create and study fundamental solutions for certain important operators. We shall find that there are many important consequences and applications of these ideas.

This chapter is based on the articles [99, 100].

12.1 The Fundamental Equation and the Homotopy Formula

We make our definitions first on $\mathbb{R}^n$. Let $C_c^\infty(\mathbb{R}^n)$ be the vector space of compactly supported smooth functions on $\mathbb{R}^n$. If $x = (x_1, \ldots, x_n)$ are coordinates on $\mathbb{R}^n$, we let $D_i = \partial/\partial x_i$ and $D^\alpha = D_1^{\alpha_1} \ldots D_1^{\alpha_n}$ for $\alpha = (\alpha_1, \ldots, \alpha_n) \in (\mathbb{Z}^+)^n$. The C^p-topology is defined on $C_c^\infty(\mathbb{R}^n)$ by saying that a sequence $\varphi_n \to 0$ in case there is a compact set K with all $\operatorname{supp} \varphi_n \subset K$ and with

$$D^\alpha \varphi_n(x) \longrightarrow 0$$

uniformly for $x \in K$ and all α satisfying $[\alpha] = \alpha_1 + \cdots + \alpha_n \leqslant p$. The C^∞ topology is defined by saying that $\varphi_n \to 0$ in case all $\operatorname{supp} \varphi_n \subset K$ and $\varphi_n \to 0$ in the C^p topology for each p.

Definition 12.1.1. A distribution on $\mathbb{R}^n$ is a linear map $T : C_c^\infty(\mathbb{R}^n) \to \mathbb{C}$ that is continuous in the C^∞ topology. The vector space of distributions on $\mathbb{R}^n$ is denoted $\mathcal{D}(\mathbb{R}^n)$.

Now any linear map from a topological vector space V to $\mathbb{C}$ is continuous $\Leftrightarrow$ the inverse of the unit ball in $\mathbb{C}$ is open in the topology of V. Since the space $C_c^\infty(K)$ of functions supported in a compact set K the C^∞ topology is the union of the C^p topologies, we see that any distribution is locally of finite order.

DOI: 10.1201/9781003619581-12

Definition 12.1.2. A k-form on an open subset U of $\mathbb{R}^p$ is a function $\omega : U \to \bigwedge^k(\mathbb{R}^p)$. The set of all C^∞ k-forms on U is denoted by $\mathcal{E}^k(U)$. The set of all C^∞ k-forms with compact supports in U (here $\operatorname{spt}(\sum \omega_\alpha dx^\alpha) = \bigcup \operatorname{spt} \omega_\alpha$) is denoted by $\mathcal{D}^k(U)$. The dual space of $\mathcal{D}^k(U)$, $\mathcal{D}_k(U)$, is called the *space of k-currents.*

The space $\mathcal{D}^k(U)$ is endowed with the usual topology so that a sequence $\{\omega_j\} \subset \mathcal{D}^k(U)$ converges to $\omega \in \mathcal{D}^k(U)$ as $j \to \infty$ if and only if $U \cap$ closure $\left(\bigcup_{j=1}^\infty \operatorname{spt}(\omega_j)\right)$ is compact and $\left|D^\beta \omega_j - D^\beta \omega\right| \to 0$ uniformly on U for all β as $j \to \infty$.

Definition 12.1.3. A *k-dimensional current* (briefly called an *k-current*) in U is a continuous linear map $T : \mathcal{D}^k(U) \to \mathbb{C}$ that is continuous in the C^∞ topology. The set of such k-currents will be denoted $\mathcal{D}_k(U)$.

Note that in case $k = 0$ the k-currents in U are just the Schwartz distributions on U. More importantly though, the k-currents, $k \geq 1$, can be interpreted as a generalization of the k-dimensional oriented submanifolds M having locally finite H^k-measure in U. Indeed, given such an $M \subset U$ with orientation ξ (thus $\xi(x)$ is continuous on M with $\xi(x) = \pm\tau_1 \wedge \ldots \wedge \tau_k$ for all $x \in M$, where $\tau_1, \ldots, \tau_k$ is an orthonormal basis for $T_x M$)[1], then there is a corresponding k-current $\|M\| \in \mathcal{D}_k(U)$ defined by

$$\|M\|(\omega) = \int_M \langle \omega(x), \xi(x) \rangle dH^k(x), \qquad \omega \in D^k(u),$$

where $\langle\ ,\ \rangle$ denotes the dual pairing for $\Lambda^k(\mathbb{R}^p), \Lambda_n(\mathbb{R}^p)$. (That is, the k-current $\|M\|$ is obtained by integration of k-forms over M in the usual sense of differential geometry: $\|M\|(\omega) = \int_M \omega$ in the usual notation of differential geometry.)

Suppose that X is an oriented real C^∞ manifold of dimension n. Let $\mathcal{D}^p(X)$ denote the space of C^∞ forms of degree p on X which have compact support, and let $\mathcal{E}^p(X)$ denote the space of C^∞ forms of degree p on X. The dual space of $\mathcal{D}^p(X)$ is called the space of p-currents. Let $\mathcal{D}'^r(X)$ denote the space of currents of degree r on X. Then $\mathcal{D}'^r(X)$ can be considered either as the space of forms of degree r with distribution coefficients, or as the dual of $\mathcal{D}^{n-r}(X)$. These two points of view are unified by the duality pairing

$$\langle u, \varphi \rangle_X = \int_X u \wedge \varphi.$$

Similarly $\mathcal{E}'^r(X)$, the space of currents of degree r with compact support on X, is the dual of $\mathcal{E}^{n-r}(X)$.

[1]Thus $\xi(x) \in \Lambda_k(T_x M)$; notice this differs from the usual convention of differential geometry where we would take $\xi(x) \in \Lambda^k T_x M$.

Let Y be an oriented real C^∞ manifold of dimension m. Then $Y \times X$ has an orientation induced by those on Y and X. For $\varphi \in \mathcal{D}^p(Y)$ and $\psi \in \mathcal{D}^q(X)$, we define $\varphi \otimes \psi \in \mathcal{D}^{p+q}(Y \times X)$ by

$$\varphi \otimes \psi = \pi_1^* \varphi \wedge \pi_2^* \psi \,,$$

where π_1 and π_2 are the projections of $Y \times X$ onto its factors. A current $K \in \mathcal{D}'^{m+r}(Y \times X)$ will be referred to as a *kernel of degree* $m + r$ on $Y \times X$. Such a kernel defines a continuous linear operator (which we will again denote by K) from $\mathcal{D}^p(Y)$ into $\mathcal{D}'^{p+r}(X)$ for all p, by

$$\langle K\varphi, \psi \rangle_X = \langle K, \varphi \otimes \psi \rangle_{Y \times X} \tag{12.1}$$

for $\varphi \in \mathcal{D}^p(Y)$ and $\psi \in \mathcal{D}^{n-p-r}(X)$. If K is continuous on $Y \times X$ then

$$K\varphi(x) = \int_Y K(y, x) \wedge \varphi(y) = \langle K, \varphi \rangle_Y \tag{12.2}$$

where the integrand in (12.2) is required to be of degree m in the variables of Y, and it is required that all of the differentials in the variables of X be on the right prior to integration. Thus for example

$$\int_{t \in \mathbb{R}} K(s,t) ds \wedge \varphi(t) dt = -\left[\int K(s,t) \varphi(t) dt \right] ds$$

It will frequently be convenient to use (12.2) to indicate the relationship between the kernel K and the operator K even when the kernel is a distribution.

An operator $K : \mathcal{D}^*(Y) \to \mathcal{D}'^*(X)$ will be said to have *type* r if $K : \mathcal{D}^p(Y) \to \mathcal{D}'^{p+r}(X)$ for all p. Thus a kernel of degree $m + r$ on $Y \times X$ induces an operator of type r by (12.1). By an extension of the Schwartz kernel theorem it can be seen that every operator of type r comes from a unique kernel of degree $m + r$ in this manner. Since the emphasis in this book is on the construction of kernels, we will not use this fact in any essential way.

For an operator $K : \mathcal{D}^*(Y) \to \mathcal{D}'^*(X)$, we define the *adjoint operator* $K' : \mathcal{D}^*(X) \to \mathcal{D}'^*(Y)$ by

$$\langle K\varphi, \psi \rangle_X = \langle \varphi, K'\psi \rangle_Y$$

If K is of type r, then K' is of type $r + m - n$. Clearly the identity operator is self-adjoint and $(AB)' = B'A'$; however, it is not true in general that $(K')' = K$.

Let $s : X \times Y \to Y \times X$ be the natural "switch" map $s(x, y) = (y, x)$. The pullback s^* is formally the identity map on forms; however, if dy and dx are orientation forms on Y and X, then $dy \wedge dx$ and $dx \wedge dy$ are orientation forms on $Y \times X$ and $X \times Y$ respectively, and

$$s^*(dy \wedge dx) = dy \wedge dx = (-1)^{mn} dx \wedge dy \,.$$

Similarly, if $\varphi \in \mathcal{D}^p(Y)$ and $\psi \in \mathcal{D}^q(X)$, then

$$s^*(\varphi \otimes \psi) = (-1)^{p \cdot q} \psi \otimes \varphi .$$

Now if K is an operator (kernel) of type r, then for $\varphi \in \mathcal{D}^p(Y)$ and $\psi \in \mathcal{D}^{n-p-r}(X)$.

$$\begin{aligned} \langle K\varphi, \psi \rangle_X &= \langle K, \varphi \otimes \psi \rangle_{Y \times X} \\ &= (-1)^{mn} \langle s^* K, s^*(\varphi \otimes \psi) \rangle_{X \times Y} \\ &= (-1)^{mn+p(n+m+r)} \langle \varphi, (s^* K)\, \psi \rangle_Y \end{aligned}$$

Thus if we define the operator $c : \mathcal{D}'^*(Y) \to \mathcal{D}'^*(Y)$ by

$$c\psi = (-1)^{m-q} \psi, \quad \psi \in \mathcal{D}'^q(Y) .$$

Then, since $(s^* K)\, \psi \in \mathcal{D}'^{m-p}(Y)$,

$$K' \psi = c^{n+m+r} (-1)^{mn} \, (s^* K)\, \psi .$$

Definition 12.1.4. For a kernel $K \in \mathcal{D}'^*(Y \times X)$ we define the transpose ${}^tK \in \mathcal{D}'^*(X \times Y)$ by

$${}^tK = (-1)^{mn} s^* K .$$

Then it follows that, if K is of type r,

$$K' = c^{n+m+r\, t} K .$$

Notice that ${}^tI = I$, and ${}^t\,({}^tK) = K$; however, it is not generally true that ${}^t(AB) = {}^tB^tA$. For the purposes of this book the transpose is more important than the adjoint. As a first indication of this notice that

$$d' = dc = -cd$$

while

$${}^td = -d .$$

Definition 12.1.5. Suppose $K \in \mathcal{D}'^{m-r}(Y \times X)$.

(a) If $K : \mathcal{D}^p(Y) \to \mathcal{E}^{p-r}(X)$ for all p, then K is called a *regular kernel.*

(b) If $K : \mathcal{D}^p(Y) \to \mathcal{D}'^{p-r}(X)$ extends to a continuous linear operator $K : \mathcal{E}'^p(Y) \to \mathcal{D}'^{p-r}(X)$, then K is called an *extendable kernel.*

(c) If K is both regular and extendable, then K is called a *biregular kernel.*

Alternative characterizations of regular and extendable kernels are available as standard versions of the Schwartz kernel theorem, but will not be needed in the sequel. We will, however, need the next result.

Proposition 12.1.6. *A kernel $K \in \mathcal{D}'^{m+r}(Y \times X)$ is regular if and only if ${}^tK$ (or K') is extendable.*

Proposition 12.1.6 is a standard result from functional analysis and a proof will not be given. Notice that a kernel K is biregular if and only if both K and ${}^tK$ are regular.

As a first example we will identify the kernel of the identity operator $I : \mathcal{D}^*(X) \to \mathcal{D}^*(X)$. Let $\Delta = \{(x,x) \in X \times X \mid x \in X\}$ denote the diagonal submanifold. Then $[\Delta] \in \mathcal{D}'^n(X \times X)$ will denote the current defined by

$$\langle [\Delta], \varphi \rangle_{X \times X} = \int_\Delta \varphi$$

Lemma 12.1.7. *The kernel $[\Delta]$ induces the identity operator on $\mathcal{D}^*(X)$.*

Proof. Let $\delta(x) = \langle x, x \rangle$ denote the parameterization of Δ. Let K denote the operator induced by $[\Delta]$. Then by (12.1)

$$\begin{aligned}\langle K\varphi, \psi \rangle = \langle [\Delta], \varphi \otimes \psi \rangle &= \int_\Delta \varphi \otimes \psi \\ &= \int_X \delta^*(\varphi \otimes \psi) = \int_X \varphi \wedge \psi \\ &= \langle \varphi, \psi \rangle .\end{aligned}$$

□

Remark 12.1.8. Suppose $f : X \to Y$ is a C^∞ map, and let $G_f = \{\langle f(x), x \rangle \mid x \in X\}$ denote the graph of f. Then $[G_f] \in \mathcal{D}'^m(Y \times X)$, and the associated operator is $\varphi \to f^*\varphi$. The special case when $X = Y$ and f is the identity map is Lemma 12.1.7.

If X and Y are complex manifolds of (complex) dimension n and m respectively and if $K \in \mathcal{D}'^{m+r,m+s}(Y \times X)$, then K induces an operator $K : \mathcal{D}^{p,q}(Y) \to \mathcal{D}'^{p+r,q+s}(X)$, for all p, q. Such an operator is said to be of type (r, s). The above discussion extends to such operators with obvious changes. For example

$${}^tK = s^*K$$

and

$$K' = c^{r+s\,t}K$$

The next result, although trivial, is basic to our approach.

Theorem 12.1.9 (The General Homotopy Formula).

(a) *Suppose $K \in \mathcal{D}'^{m+r}(Y \times X)$. Then $dK \in \mathcal{D}'^{m+r+1}(Y \times X)$, and the operators K and dK are related by the formula*

$$dK)\varphi = (-1)^m \left[dK\varphi + (-1)^{r+1} K d\varphi\right]$$

(b) *Suppose X and Y are complex manifolds and $K \in \mathcal{D}'^{m+r,m+s}(Y \times X)$. Then*

$$(\bar{\partial}K)\varphi = \bar{\partial}K\varphi + (-1)^{r+s+1}K\bar{\partial}\varphi\,.$$

Proof. We will only prove (a) since the proof of (b) is similar. By definition if $\varphi \in \mathcal{D}^p(Y)$ and $\psi \in \mathcal{D}^{n-p-r}(X)$, then

$$\begin{aligned}\langle (dK)\varphi, \psi\rangle &= \langle dK, \varphi \otimes \psi\rangle = (-1)^{m+r+1}\langle K, d(\varphi \otimes \psi)\rangle \\ &= (-1)^{m+r+1}\left[\langle K, d\varphi \otimes \psi\rangle + (-1)^p\langle K, \varphi \otimes d\psi\rangle\right] \\ &= (-1)^{m+r+1}\left[\langle K d\varphi, \psi\rangle + (-1)^p\langle K\varphi, d\psi\rangle\right] \\ &= (-1)^m\left[\langle dK\varphi, \psi\rangle + (-1)^{r+1}\langle K d\varphi, \psi\rangle\right]\,.\end{aligned}$$

The reader may find that that this proof is more intuitive using (12.2). Then

$$\begin{aligned}(dK)\varphi &= \int_Y (d_x K + d_y K) \wedge \varphi \\ &= (-1)^m d\int_Y K \wedge \varphi + (-1)^{m+r+1}\int_Y K \wedge d\varphi\end{aligned}$$

The $(-1)^m$ occurs in the first term because the differential involved in d_x must be moved past m dy's before the integration is performed. □

The most important case of the general homotopy formula is obtained when $Y = X$ and $dK = [\Delta](\bar{\partial}K = [\Delta])$.

Definition 12.1.10.

(a) A solution E to the *fundamental equation*

$$dE = (-1)^n[\Delta]$$

on a manifold X of dimension n is called a *fundamental solution for exterior differentiation on X.*

(b) A solution E to the fundamental equation

$$\bar{\partial}E = [\Delta]$$

on a complex manifold X is called a *fundamental solution to the Cauchy Riemann equations on X.*

Definition 12.1.11. Suppose X is a complex manifold. An operator E of type $(0, -1)$ is said to satisfy the *chain homotopy equation* on X if

$$\bar{\partial}E\varphi + E\bar{\partial}\varphi = \varphi \quad \text{for all} \quad \varphi \in \mathcal{D}^*(X)\,.$$

The chain homotopy equation is simply the operator version of the fundamental equation.

Theorem 12.1.12. *Let X be a complex manifold of dimension n. A kernel $E \in \mathcal{D}'^{n,n-1}(X \times X)$ is a fundamental solution for the Cauchy Riemann equations if and only if the associated operator E satisfies the chain homotopy equation.*

Proof. Apply 12.1.11 to the kernel E. □

Any extendable fundamental solution has the reproducing property for holomorphic functions. Suppose $D \subset\subset X$ has a continuously differentiable boundary ∂D. Let $[\partial D] \in \mathcal{D}'^1(X)$ be the current defined by

$$\langle [\partial D], \varphi \rangle = \int_{\partial D} \varphi$$

Theorem 12.1.13. *Suppose X is a complex manifold and let E be an extendable fundamental solution for the Cauchy Riemann equations on X. Let $D \subset\subset X$ has a continuously differentiable boundary. Then for every function G which is holomorphic in D and continuous on $\bar{D}$ we have*

$$E(-G[\partial D])(z) = \begin{cases} G(z) & z \in D \\ 0 & z \in X \setminus \overline{D} \end{cases}$$

Proof. Let χ_D denote the characteristic function of D. By Stokes' Theorem $d\chi_D = -[\partial D]$. Therefore $\overline{\partial}\chi_D = -[\partial D]^{0,1}$, the 0,1 piece of $[\partial D]$. Since G is continuous on $\overline{D}$, $\chi_D G \in \mathcal{E}'(X)$ and $\overline{\partial}\,(\chi_D G) = -G[\partial D]^{0,1}$. Since E is of type $(0,-1)$, $E\,(\chi_D G) = 0$, and $E\left(G[\partial D]^{1,0}\right) = 0$. Thus, by the chain homotopy formula,

$$\begin{aligned} \chi_D G &= E\overline{\partial}\,(\chi_D G) = E\left(-G[\partial D]^{0,1}\right) \\ &= E(-G[\partial D])\,. \end{aligned}$$

□

Remark 12.1.14. Notice that if E is continuous on $\partial D \times D$ then we have

$$G(z) = \int_{\partial D} G(\zeta)E(\zeta, z) \quad , \quad z \in D\,.$$

Finally we notice that the transpose of a fundamental solution is also a fundamental solution.

Proposition 12.1.15.

(a) *If $E \in \mathcal{D}'^{n,n-1}(X)$ is a fundamental solution for $\overline{\partial}$ then so is ${}^tE$.*

(b) *If $E \in \mathcal{D}'^{n-1}(X)$ is a fundamental solution for d, then so is ${}^tE$.*

Proof. We will only prove (a). Since the switch map s is a biholomorphism, it follows from 12.1.4 that $\overline{\partial}^t K = {}^t(\overline{\partial}K)$ for any kernel K. Since ${}^t[\Delta] = [\Delta]$ on even-dimensional manifolds, the result follows. □

12.2 Fundamental Solutions of Convolution Type

An important class of kernels on $\mathbb{R}^m$ is obtained by extending the notion of convolution to forms (and/or currents). Let $\tau(y,x) \equiv x - y$ and $\tau : \mathbb{R}^m \times \mathbb{R}^m \to \mathbb{R}^m$. Then $\tau^* : \mathcal{E}^p(\mathbb{R}^m) \to \mathcal{E}^p(\mathbb{R}^m \times \mathbb{R}^m)$. We adopt the notation $f(x-y)$ for $(\tau^* f)(y,x)$ (which should not be confused with evaluation of the coefficients of f at the point $x-y$). The linear map τ may be considered as a projection of $\mathbb{R}^{2m}$ onto a linear subspace. Consequently the pullback $\tau^* f$ is (well) defined for any current $f \in \mathcal{D}'^p(\mathbb{R}^m)$, by tensoring with 1 in the normal plane. In this general case we still denote $(\tau^* f)(y,x)$ by $f(x-y)$.

Definition 12.2.1. Given $f \in \mathcal{D}'^p(\mathbb{R}^m)$ and $\varphi \in \mathcal{D}^q(\mathbb{R}^m)$,

$$(f\#\varphi)(x) \equiv \int_y f(y) \wedge \varphi(x-y) = (-1)^m \int_y f(x-y) \wedge \varphi(y) . \tag{12.3}$$

By making the change of variables $y = r, x = s + r$ in either of the above two expressions for $f\#\varphi$, we see that

$$\langle f\#\varphi, \psi\rangle = \langle f(r) \wedge \varphi(s), \psi(r+s)\rangle .$$

Hence the two formulae in (12.3) for $f\#\varphi$ are equal and

$$\langle f\#\varphi, \psi\rangle = \langle f(y) \wedge \varphi(x), \psi(x+y)\rangle . \tag{12.4}$$

The formula (12.4) has several advantages over (12.3).

Definition 12.2.2. Given $f, g \in \mathcal{D}'^*(\mathbb{R}^n)$, with one having compact support,

$$\langle f\#g, \psi\rangle \equiv \langle f(y) \wedge g(x), \psi(x+y)\rangle$$

for all $\psi \in \mathcal{D}^*(\mathbb{R}^m)$.

Notice that

$$\deg(f\#g) = \deg(f) + \deg(g) - n$$

or, said differently, that convolution by a current f of dimension r lowers degree by r.

Example 12.2.3. If $k = [0]$ then

$$[0]\#g = g \quad \text{for all} \quad g \in \mathcal{D}'^*(\mathbb{R}^m) .$$

More generally if $k(y) = u(y)dy$, with $u \in \mathcal{D}'(\mathbb{R}^m)$, is of degree m, then

$$k\#\left(v dy^I\right) = (u * v) dx^I$$

where $u * v$ is ordinary convolution.

Theorem 12.2.4. *The convolution product #, when defined, is commutative and associative:*

$$f\#g = (-1)^{pq} g\#f$$

where $p = \deg f$ *and* $q = \deg(g)$*; and*

$$f\#(g\#h) = (f\#g)\#h\,. \tag{12.5}$$

Proof. Theorem 12.2.4 follows immediately from (12.4):

$$\begin{aligned}\langle f\#(g\#h), \psi\rangle =& \langle f(y) \wedge (g\#h)(x), \psi(x+y)\rangle \\ =& \langle f(y) \wedge g(w) \wedge h(x-w), \psi(x+y)\rangle \quad (\text{ in } y, w, x \text{ space }) \\ =& \langle f(r) \wedge g(s) \wedge h(t), \psi(r+s+t)\rangle \quad (\text{ in } r, s, t \text{ space })\end{aligned}$$

by the change of variables $y = r, w = s, x = t + s$. Similarly, one can show that $(f\#g)\#h$ is given by the same expression. □

Definition 12.2.5. Given $k \in \mathbb{D}'^{m-r}(\mathbb{R}^m)$,

$$K(y,x) = (-1)^m k(x-y) \in \mathcal{D}'^{m-r}(\mathbb{R}^m \times \mathbb{R}^m)$$

is called the *convolution kernel associated with* k. The corresponding operator K is said to be of *convolution type.*

It is clear that operators of convolution type are biregular.

If K is the convolution kernel associated with k, then ${}^tK$ is the convolution kernel associated with $(-1)^m \check{k}$, where $\check{k}(x) = k(-x)$. Thus, if $k \in \mathcal{D}'^{m-1}(\mathbb{R}^m)$, then

$$K'\varphi = c^{rt}K\varphi = (-1)^m c^r \check{k}\#\varphi\,.$$

Consequently the adjoint of a convolution kernel is not a convolution kernel unless r is even.

If K is a kernel of convolution type on $\mathbb{R}^m$, say $K\varphi = k\#\varphi$ where $k \in \mathcal{D}'^{m-r}(\mathbb{R}^m)$, then the current k can be recovered from K by the formula

$$k = (-1)^{mr} K([0]) \tag{12.6}$$

since $K([0]) = k\#[0] = (-1)^{mr}[0]\#k = (-1)^{mr}k$. If $K_j = k_j\#\varphi, j = 1, 2$, then the composition $K_1 \circ K_2$ is also of convolution type with $(K_1 \circ K_2)(\varphi) = (k_1\#k_2)\#\varphi$ by (12.5).

Example 12.2.6. The contraction operator $Kf = \partial/\partial y_j \lrcorner f$ is of convolution type, and corresponds to the current

$$k = (-1)^{m+j-1}\delta_0(y)dy_1 \wedge \cdots \wedge \widehat{dy_j} \wedge \cdots \wedge dy_m\,.$$

Similarly the coefficient-wise differentiation $\partial f/\partial y_j$ is of convolution type, and corresponds to the current $(-1)^m \partial/\partial y_j \delta_0 dy_1 \wedge \cdots \wedge dy_n$. Consequently d^*, the adjoint of d with respect to the Euclidean metric, is also of convolution type, since

$$d^*g = -\sum_{j=1}^{m} \frac{\partial}{\partial y_j}\left(\frac{\partial}{\partial y_j} \lrcorner\, g\right).$$

Thus by (12.6),

$$d^*g = (-1)^m \left(d^*[0]\right) \# g \tag{12.7}$$

Notice, however, that the operators $g \to dy_j \wedge g$ and $g \to dg$ are not of convolution type, since a convolution operator cannot be of positive type.

We state without proof the following characterization of convolution operators and kernels. For $w \in \mathbb{R}^m$, let $\tau_w(y) = y - w$.

Theorem 12.2.7.

(a) An operator K of type r on $\mathbb{R}^m$ is of convolution type if and only if

(1) $\tau_w^* K\varphi = K\tau_w^*\varphi$ for all $w \in \mathbb{R}^m$ and $\varphi \in \mathcal{D}^*(\mathbb{R}^m)$ and

(2) $X \lrcorner K\varphi = (-1)^r K(X \lrcorner \varphi)$ for all vector fields X with constant coefficients.

(b) A kernel $K \in \mathcal{D}'^{m-r}(\mathbb{R}^m \times \mathbb{R}^m)$ is a convolution kernel if and only if

(1) $\tau_{(w,-w)}^* K = K$ for all $w \in \mathbb{R}^m$ and

(2) $(\partial/\partial y_j + \partial/\partial x_j) \lrcorner K = 0$ for $1 \leqslant j \leqslant m$.

Theorem 12.2.7 can be proved easily using the characterization of convolution operators acting on functions. Condition (a1) of Theorem 12.2.7 by itself characterizes those operators K such that each coefficient of $K\varphi$ is a linear combination of ordinary convolutions of distributions with coefficients of φ.

Now we wish to consider the case $m = 2n$ with $\mathbb{R}^{2n} = \mathbb{C}^n$. Suppose $K\varphi = k\#\varphi$ is of convolution type. Since $K = \tau^*k, [\Delta] = \tau^*[0]$, and $\overline{\partial}$ commutes with τ^*, $\overline{\partial}K = [\Delta]$ on $\mathbb{C}^n \times \mathbb{C}^n$ if and only if

$$\overline{\partial}k = [0] \tag{12.8}$$

on $\mathbb{C}^n$. That is, a kernel K of convolution type associated with k is a fundamental solution for $\overline{\partial}$ if and only if (12.8) is satisfied.

Next we turn to describing specific fundamental solutions of convolution type.

The Cauchy Kernel in One Variable. Since

$$\frac{1}{w} \in L^1_{\text{loc}}(\mathbb{C})$$

and

$$\frac{1}{2\pi i}\overline{\partial}\frac{dw}{w} = [0] \text{ on } \mathbb{C}, \tag{12.9}$$

the convolution kernel

$$C(\zeta, z) \equiv \frac{1}{2\pi i} \frac{d(z-\zeta)}{z-\zeta}$$

associated with $(1/2\pi i)dw/w$, is a fundamental solution. This, the most basic of all fundamental solutions, is called the Cauchy kernel and denoted by C.

Since every convolution kernel is biregular, the operator C extends to all compactly supported currents $\varphi \in \mathcal{E}'^{*}(\mathbb{C})$.

The homotopy formula is divided into two cases. First, suppose $\varphi \in \mathcal{E}'^{p,0}(\mathbb{C})$ is of bidegree $p, 0$ (with $p = 0$ or 1). Then $C\varphi$ must vanish since $C(\zeta, z) \wedge \varphi(\zeta)$ cannot be of bidegree $(1,1)$ in ζ. Therefore C is a left inverse for $\overline{\partial}$:

$$C\overline{\partial}\varphi = \frac{dw}{2\pi i w} \# \overline{\partial}\varphi = \varphi \tag{12.10}$$

for all $\varphi \in \mathcal{E}'^{p,0}(\mathbb{C})$. Second, suppose $\varphi \in \mathcal{E}'^{p,1}(\mathbb{C})$ (with $p = 0$ or 1). Then $\overline{\partial}\varphi$ must vanish. Therefore, C is also a right inverse for $\overline{\partial}$:

$$\overline{\partial} C\varphi = \overline{\partial}\left\{\frac{dw}{2\pi i w} \# \varphi\right\} = \varphi$$

for all $\varphi \in \mathcal{E}'^{p,1}(\mathbb{C})$.

Let χ_Ω denote the characteristic function of a bounded open set Ω with smooth boundary $\partial\Omega$. If F is holomorphic in $\overline{\Omega}$, then application of (12.10) to $\varphi = F\chi_\Omega$ yields

$$F(z)\chi_\Omega(z) = C\overline{\partial}(F\chi_\Omega) = C\left(F\overline{\partial}\chi_\Omega\right).$$

However, since $\overline{\partial}\chi_\Omega = -[\partial\Omega]^{0,1}$, this becomes

$$F(z)\chi_\Omega(z) = \frac{1}{2\pi i}\int_{\partial\Omega} \frac{F(\zeta)d\zeta}{\zeta - z}$$

which is the Cauchy integral formula.

On the other hand if $G \in \mathcal{E}^{0,1}(\overline{\Omega})$, then we may extend G to have compact support in $\mathbb{C}$.

If we set

$$F = CG = \frac{1}{2\pi i}\int \frac{G(\zeta)d\zeta}{\zeta - z}$$

then using (12.10) we have $\overline{\partial}F = G$ and $F \in \mathcal{E}^{0,0}(\overline{G})$. This is just the usual proof of local existence of solutions to the equation $\overline{\partial}F = G$ which at the same time proves (globally) smoothness up to the boundary.

Of course, these are just a few of the many standard uses of the Cauchy kernel. They are included here because the formalism and the simplicity of the above proofs carry over to several variables once we have made the appropriate generalization of the Cauchy kernel in Section 12.45.

The Slice Cauchy Kernel. Given a complex line $\sigma \in \mathbb{CP}^{n-1}$ in $\mathbb{C}^n$ through the origin, let w denote a coordinate on $\mathbb{C}$ with the property that $\tau(w) \equiv w\widetilde{\sigma}$

parameterizes the line, where $\widetilde{\sigma}$ is a (unit) vector on the line σ. Let $[\sigma]$ denote the current given by integration of forms over $\sigma \subset \mathbb{C}^n$. Then

$$c_\sigma \equiv \frac{1}{2\pi i}\frac{dw}{w} \wedge [\sigma] = \frac{1}{2\pi i}\tau_*\left(\frac{dw}{w}\right)$$

is a well-defined current on $\mathbb{C}^n$ which satisfies

$$\frac{1}{2\pi i}\overline{\partial}\left(\frac{dw}{w} \wedge [\sigma]\right) = [0] \quad \text{on} \quad \mathbb{C}^n$$

because of (12.9) and the fact that the push forward τ_* commutes with $\overline{\partial}$. Therefore the associated convolution kernel, denoted $\mathbb{C}_\sigma$, is a fundamental solution on $\mathbb{C}^n$. This fundamental solution is called the *slice Cauchy-kernel* on the line σ. In standard coordinates, if we take σ to be the z_1-axis (so $w \equiv z_1$) and let $z = (z_1, z'')$ then

$$C_\sigma = \frac{1}{2\pi i}\frac{d(\zeta_1 - z_1)}{\zeta_1 - z_1} \wedge [z'' = \zeta'']$$

on $\mathbb{C}^n \times \mathbb{C}^n$.

It is natural to try and construct another fundamental solution by averaging the slice Cauchy kernels C_σ over $\sigma \in \mathbb{CP}^{n-1}$ with respect to the Fubini-Study volume, denoted by $d\lambda(\sigma)$, on $\mathbb{CP}^{n-1}$. This gives rise to a natural fundamental solution, since

$$\begin{aligned}\overline{\partial}\int C_\sigma d\lambda(\sigma) &= \int \overline{\partial}C_\sigma d\lambda(\sigma)\\ &= \int [\Delta] d\lambda(\sigma) = [\Delta]\end{aligned}$$

on $\mathbb{C}^n \times \mathbb{C}^n$. This kernel $B \equiv \int C_\sigma d\lambda(\sigma)$ is called the *Bochner-Martinelli kernel.*

Now consider the Newtonian potential (as a $2n$-form) on $\mathbb{C}^n$

$$e \equiv \gamma_n |z|^{2-2n}(dz \cdot d\overline{z})^n ,$$

where $dz \cdot d\overline{z} = \sum dz_j \wedge dz_j$ and $\gamma_n = 2^n/n(n-1)(i\pi)^n$. Let E denote the operator of the convolution type corresponding to e. It is a standard fact that

$$\overline{\partial\partial} * e = [0] .$$

Define

$$b \equiv \overline{\partial} * e .$$

Then

$$\overline{\partial} b = [0]$$

and if B denotes the corresponding kernel, then

$$\overline{\partial} B = [\Delta]$$

by Theorem 12.2.7 and (12.8).

Thus we have another fundamental solution of convolution type. In fact, the above two fundamental solutions are equal by the next theorem.

Let

$$\alpha_n \equiv (n-1)!(-1)^{n(n-1)/2}(2\pi i)^{-n},$$
$$\overline{w} \cdot dw \equiv \sum \overline{w}_j dw_j ,$$

and

$$d\overline{w} \cdot dw \equiv \sum d\overline{w}_j \wedge dw_j .$$

Also let $\widehat{d\overline{w}_j}$ denote $d\overline{w}_1 \wedge \cdots \wedge d\overline{w}_n$ with $d\overline{w}_j$ deleted.

Theorem 12.2.8. *The following four currents are equal and hence will be denoted jointly by $b \in \mathcal{D}'^{n,n-1}(\mathbb{C}^n)$.*

$$b = \int_{\sigma \in \mathbb{CP}^{n-1}} c_\sigma d\lambda(\sigma) , \tag{c1}$$

$$b = \overline{\partial}^* e , \tag{c2}$$

$$b = \alpha_n \frac{\sum (-1)^{j-1} \overline{w}_j \widehat{d\overline{w}_j} dw}{|w|^{2n}} , \tag{c3}$$

$$b = \frac{1}{(2\pi i)^n} \frac{\overline{w} \cdot dw}{|w|^2} \wedge \left[\frac{d\overline{w} \cdot dw}{|w|^2} \right]^{n-1} . \tag{c4}$$

This current b satisfies

$$\overline{\partial} b = [0] \quad on \quad \mathbb{C}^n$$

so that the associated convolution kernel (denoted B and called the Bochner-Martinelli kernel) is a biregular fundamental solution.

Moreover, B can be expressed in any of the following ways.

$$B = \int_{\sigma \in \mathrm{CP}^{n-1}} C_\sigma d\lambda(\sigma) , \tag{w1}$$

$$B(\psi) \equiv \left(\overline{\partial}^* e \right) \# \psi = \overline{\partial}^* (e \# \psi) = e \# \left(\overline{\partial}^* \psi \right) , \tag{w2}$$

$$B(\zeta, z) = \alpha_n \frac{\sum (-1)^{j-1} \left(\overline{z}_j - \overline{\zeta}_j \right) \widehat{d(\overline{z}_j - \overline{\zeta}_j)} \wedge d(z-\zeta)}{|z-\zeta|^{2n}} , \tag{w3}$$

$$B(\zeta, z) = \frac{1}{(2\pi i)^n} \frac{(\overline{z} - \overline{\zeta}) \cdot d(z-\zeta)}{|z-\zeta|^2} \wedge \left[\frac{d(\overline{z} - \overline{\zeta}) \cdot d(z-\zeta)}{|z-\zeta|^2} \right]^{n-1} . \tag{w4}$$

Because of the classical nature of this theorem, the proof is only sketched. The equality $\overline{\partial}^* (e \# \psi) = \left(\overline{\partial}^* e \right) \# \psi$ appears to be new. It will be used in Section 12.6.

Proof. As we have seen above, using either expression (c1) or (c2) for b we have that $\overline{\partial} b = [0]$ on $\mathbb{C}^n$. The standard calculation that (c2) and (c3) are equal is omitted. The proof that (c1) and (c3) are equal is straightforward using the $\mathbb{CP}^{n-1}$ version of polar coordinates to express an integral over $\mathbb{C}^n$ as an iterated integral over $\mathbb{CP}^{n-1}$ and $\mathbb{C} - \{0\}$. Namely,

$$\int_{C^n} \varphi(w) d\lambda(w) = \frac{\pi^{n-1}}{(n-1)!} \int_{\mathbb{CP}^{n-1}} \int_{\mathbb{C}} \varphi(t\widetilde{\sigma}) t^{2n-2} d\lambda(t) d\lambda(\sigma)$$

for $\varphi \in C_0^\infty(\mathbb{C}^n)$.

The proof that (c3) and (c4) are equal is omitted here, since it is given in a more general context in the next section. The equivalence of (w1–w4) can be deduced from the fact that if b is given by (c1–c4) respectively, then the associated convolution kernel B is given by (w1–w4) respectively.

It remains to prove the last two equalities in (w2). First, $\left(\overline{\partial}^* e\right) \# \psi = \overline{\partial}^* (e \# \psi)$. Exactly as in (12.7) above one can show that $\overline{\partial}^*$ is an operator of convolution type. Since $E(\psi) = e \# \psi$ is of convolution type the operator composition $\overline{\partial}^* E$ must also be of convolution type. Now $\overline{\partial}^* E([0]) = \overline{\partial}^* (e \# [0]) = \overline{\partial}^* e$ so that, by (12.6), $\left(\overline{\partial}^* \circ E\right)(\psi) = \left(\overline{\partial}^* e\right) \# \psi$.

Second, since $\overline{\partial} e = 0$, we have that the kernel E satisfies $\overline{\partial} E = 0$ on $\mathbb{C}^n \times \mathbb{C}^n$. The corresponding operator equation is $\overline{\partial} \circ E = E \circ \overline{\partial}$. Finally, because $E = E^*$ is self adjoint, $E \circ \overline{\partial}^* = \overline{\partial}^* \circ E$ or $e \# \left(\overline{\partial}^* \psi\right) = \overline{\partial}^* (e \# \psi)$.

The procedure used to obtain (c1) and (w1) above can be generalized as follows. Let $d_\mu(\sigma)$ denote any positive measure on $\mathbb{CP}^{n-1}$ with total mass equal to one. We define the Cauchy-Bochner-Martinelli kernel with weight μ to be

$$B_\mu = \int_{\mathbb{CP}^{n-1}} C_\sigma d_\mu(\sigma)$$

It follows that each such B_μ is a biregular fundamental solution on $\mathbb{C}^n$ of convolution type, and that if $\mu \in L^1\left(\mathbb{CP}^{n-1}\right)$ then B_μ has $L^1_{\text{loc}}(\mathbb{C}^n)$ coefficients. Although these kernels B_μ (other than the special cases C_σ and B) will not be used here we mention one example intermediate to C_σ and B. Choose a complex linear subspace V of $\mathbb{C}^n$ and then express $\mathbb{C}^n$ as the product $\mathbb{C}^p \times \mathbb{C}^{n-p}$ with coordinates $z = (z', z'')$, so that $V = \mathbb{C}^p \times [0] = \{z \mid z'' = 0\}$. Let $b_p(z')$ denote the current occurring in (c1–c4) of Theorem 12.2.8 with $\mathbb{C}^n$ replaced by $\mathbb{C}^p$. Then

$$k \equiv b_p(z') \wedge [z'' = 0]$$

corresponds to taking the weight μ to be the volume measure on the $\mathbb{CP}^{p-1} \subset \mathbb{CP}^{n-1}$ determined by $\mathbb{C}^p \times [0] \subset \mathbb{C}^n$. Hence, $\overline{\partial} k = [0]$ on $\mathbb{C}^n$. This fact can also be deduced immediately from the equation $\overline{\partial} b_p = [0]$ on $\mathbb{C}^p$, by pushing it forward into $\mathbb{C}^n$. □

We conclude this section by briefly mentioning some of the applications of the above fundamental solutions. The slice Cauchy kernel was used by Hans

Lewy [151, 150], and Hörmander [111]. Moreover, Hörmander uses this kernel to prove that on $\mathbb{C}^n$ the equation $\overline{\partial}F = G$ where G satisfies $\overline{\partial}G = 0$ and G has compact support has a solution F with compact support in the special case where G has degree $p, 1$. These two related applications were further developed in Harvey-Lawson [97] and [98]. For a look at Hans Lewy's two proofs (mentioned above) from the point of view of fundamental solutions and the homotopy formula see Harvey-Polking.

If we try to use the Bochner-Martinelli kernel (or the slice Cauchy kernel) to carry out in $\mathbb{C}^n$ the program we used successfully in C we meet with only partial success. In exactly the same way we can show that

$$\chi_\Omega(z)F(z) = \int_{\partial\Omega} B(\zeta, z)F(\zeta) \qquad , \quad z \in \mathbb{C}^n \partial\Omega$$

for every $F \in \mathcal{O}(\overline{\Omega})$ when Ω is an open subset of $\mathbb{C}^n$ with a smooth boundary. (In fact the same proof works with B replaced by any extendable fundamental solution E on a neighborhood of $\overline{\Omega}$.) However, if we try to use B to solve the equation $\overline{\partial}F = G$ locally when $\partial\overline{G} = 0$ in Ω, we come across a snag. If we extend G to have compact support in $\mathbb{C}^n$, then by the homotopy formula,

$$G = \overline{\partial}BG + B\overline{\partial}G\,.$$

Since it is not clear how to extend G so that it has compact support and satisfies $\overline{\partial}G = 0$ globally, this formula is of little use.

Notice in addition that the kernel $B(\zeta, z)$ is not holomorphic in z if $n \geqslant 2$.

All of the examples of fundamental solutions considered in this section are convolution kernels. Some of the major deficiencies of these kernels are eliminated in the new fundamental solutions introduced in this book (which are not of convolution type).

12.3 A Formal Identity

We now begin the development of the new fundamental solution for the Cauchy Riemann equations. In this section we will be primarily concerned with a certain algebraic identity in the complement of the singular set of the kernels involved. The development will be concluded in the next section by analyzing the behavior across the singular set.

The results of this section have their origin in the fundamental work of Leray [149]. Later Koppelman [133, 132], and Harvey [96] gave different proofs of Leray's formula. We shall reformulate and extend these results, giving in particular a fourth proof of Leray's formula. It has come to our attention that some of this work has been obtained independently by Roos [185].

Let V be an open subset of $\mathbb{C}^n \times \mathbb{C}^n$ with coordinates (ζ, z). For $1 \leqslant j \leqslant N$ let $u^j(\zeta, z) = \left(u_1^j(\zeta, z), \ldots, u_n^j(\zeta, z)\right)$ be a C^∞ map from V to C^n. We will use the notation

$$
\begin{aligned}
u^j \cdot (\zeta - z) &= \sum_{i=1}^{n} u_i^j(\zeta, z)\,(\zeta_i - z_i) \\
u^j \cdot d(\zeta - z) &= \sum_{i=1}^{n} u_i^j(\zeta, z) d\,(\zeta_i - z_i) \\
\overline{\partial} u^j \cdot d(\zeta - z) &= \overline{\partial}\left[u^j \cdot d(\zeta - z)\right] \\
&= \sum_{i,k=1}^{n} \left(\frac{\partial u_i^j}{\partial \overline{\zeta}_k} d\overline{\zeta}_k + \frac{\partial u_i^j}{d\overline{z}_k} d\overline{z}_k \right) \wedge d\,(\zeta_i - z_i)
\end{aligned}
$$

For $1 \leqslant j \leqslant N$ we define the 1 form

$$
\omega^j = \frac{u^j \cdot d(\zeta - z)}{u^j \cdot (\zeta - z)} \tag{12.13}
$$

The form ω^j is smooth on $V - A_j$, where

$$
A_j = \left\{(\zeta, z) \in V \;\middle|\; u^j \cdot (\zeta - z) = 0\right\}
$$

is the j th singular set, and

$$
\overline{\partial}\omega^j = \frac{\overline{\partial} u^j \cdot d(\zeta - z)}{u^j \cdot (\zeta - z)} - \frac{\overline{\partial} u^j \cdot (\zeta - z) \wedge u^j \cdot d(\zeta - z)}{\left[u^j \cdot (\zeta - z)\right]^2}\,. \tag{12.14}
$$

Proposition 12.3.1. *For $k \geqslant 0$*

$$
\begin{aligned}
\omega^j \wedge \left(\overline{\partial}\omega^j\right)^k &= \left[u^j \cdot (\zeta - z)\right]^{-(k+1)} u^j \cdot d(\zeta - z) \wedge \left[\overline{\partial} u^j \cdot d(\zeta - z)\right]^k \\
&= \frac{u^j \cdot d(\zeta - z)}{u^j \cdot (\zeta - z)} \wedge \left[\frac{\overline{\partial} u^j \cdot d(\zeta - z)}{u^j \cdot (\zeta - z)}\right]^k .
\end{aligned}
$$

Proof. This identity follows immediately from (12.13) and (12.14), since ω^j wedge the second term in (12.14) vanishes. □

We will denote by **N** the non negative integers. Then

$$
\mathbf{N}^N = \{\alpha = (\alpha_1 \cdots \alpha_N) \mid \alpha_j \in \mathbf{N}\}
$$

is the set of multi-indices. For $\alpha \in \mathbf{N}^N$ we set $|\alpha| = \alpha_1 + \cdots + \alpha_N$. We define

$$
(\overline{\partial}\omega)^\alpha = \left(\overline{\partial}\omega^1\right)^{\alpha_1} \wedge \cdots \wedge \left(\overline{\partial}\omega^N\right)^{\alpha_N}
$$

Let $I = \{i_1, \ldots, i_p\}$ be a subset of $\{1, 2, \ldots, n\}$. Then $\mathbf{N}^I$ is the set of maps from I to N. We will make the identification

$$
\mathbf{N}^I = \left\{\alpha \in \mathbf{N}^N \;\middle|\; \alpha_j = 0 \quad \text{if} \quad j \notin I\right\}
$$

Finally, we define the forms

$$\omega^I = \omega^{i_1} \wedge \cdots \wedge \omega^{i_p}\,, \tag{12.15}$$

$$E_I = \omega^I \wedge \sum_{\substack{\alpha \in N^I \\ |\alpha| = n-p}} (\overline{\partial}\omega)^\alpha\,. \tag{12.16}$$

Notice that the forms ω^I and E_I are skew-symmetric in the index set I. In addition it is clear that $E_I = 0$ if $p > n$. We will set $E_I = 0$ if $I = \emptyset$. Notice the special cases

$$\begin{aligned} E_j &= \omega^j \wedge \left(\overline{\partial}\omega^j\right)^{n-1} && p = 1\,, \\ E_{ij} &= \omega^i \wedge \omega^j \wedge \sum_{k=1}^{n-1} \left(\overline{\partial}\omega^i\right)^{k-1} \left(\overline{\partial}\omega^j\right)^{n-k-1} && p = 2\,, \\ E_I &= \omega^I && p = n\,. \end{aligned}$$

For $1 \leqslant j \leqslant p$ we set

$$I_j = \left\{i_1, \ldots, \widehat{i_j}, \ldots, i_p\right\}.$$

The primary result of this section is the following identity.

Theorem 12.3.2. *Let* $|I| = p \geqslant 1$. *Then*

$$\overline{\partial} E_I = \sum_{j=1}^{p} (-1)^j E_{I_j}$$

in $V \setminus \cup_{k \in I} A_k$.

The special case where $N = 2$ *and* $p = 2$ *is particularly important. It says that:*

$$\overline{\partial} E_{12} = E_1 - E_2,$$

or

$$\overline{\partial}\left[\omega^1 \wedge \omega^2 \wedge \sum_{k=1}^{n-1} \left(\overline{\partial}\omega^1\right)^{k-1} \wedge \left(\overline{\partial}\omega^2\right)^{n-k-1}\right] = \omega^1 \wedge \left(\overline{\partial}\omega^1\right)^{n-1} - \omega^2 \wedge \left(\overline{\partial}\omega^2\right)^{n-1}$$

on $V \setminus (A_1 \cup A_2)$.

Proof. By (12.16) and (12.15),

$$\begin{aligned} \overline{\partial} E_I &= \overline{\partial}\omega^I \wedge \sum_{\substack{\alpha \in N^I \\ |\alpha| = n-p}} (\overline{\partial}\omega)^\alpha \\ &= \sum_{j=1}^{p} (-1)^{j-1} \omega^{I_j} \wedge \sum_{\substack{\alpha \in \mathbb{N}^I \\ |\alpha| = n-p+1 \\ \alpha_{ij} > 1}} (\overline{\partial}\omega)^\alpha\,. \end{aligned}$$

Thus

$$\overline{\partial} E_I = \sum_{j=1}^{p} (-1)^j E_{I_j} + \sum_{j=1}^{p} (-1)^{j-1} \omega^{I_j} \wedge \sum_{\substack{\alpha \in \mathbf{N}^I \\ |\alpha| = n-p+1}} (\overline{\partial}\omega)^\alpha .$$

It remains to show that the second term on the right vanishes.

In fact, for each fixed α with $|\alpha| = n - p + 1$ we have

$$\sum_{j=1}^{p} (-1)^{j-1} \wedge \omega^{I_j} \wedge (\overline{\partial}\omega)^\alpha = 0 \tag{12.17}$$

which is also a useful result. If $N = p = 2$, (12.17) says

$$\omega^1 \wedge (\overline{\partial}\omega)^\alpha = \omega^2 \wedge (\overline{\partial}\omega)^\alpha$$

for all α with $|\alpha| = n - 1$. To prove (12.17) we use the Euler vector field

$$\theta = \sum_{j=1}^{n} (\zeta_i - z_i) \frac{\partial}{\partial \zeta_i} .$$

Inspection of the formulas (12.13) for ω^j and (12.14) for $\overline{\partial}\omega^j$ shows that

$$\begin{aligned} \theta \lrcorner \omega^j &= 1 \qquad 1 \leqslant j \leqslant N , \\ \theta \lrcorner \overline{\partial}\omega^j &= 0 \qquad 1 \leqslant j \leqslant N . \end{aligned} \tag{12.18}$$

Notice that

$$\omega^I \wedge (\overline{\partial}\omega)^\alpha = 0$$

since this form is of degree $n + 1$ in $d(\zeta - z)$. Contraction with θ is an anti-derivation, so using (12.19) and (12.18) we have

$$\begin{aligned} 0 &= \theta \lrcorner \left(\omega^I \wedge (\overline{\partial}\omega)^\alpha \right) \\ &= \left(\theta \lrcorner \omega^I \right) \wedge (\overline{\partial}\omega)^\alpha \\ &= \sum_{j=1}^{p} (-1)^{j-1} \omega^{I_j} \wedge (\overline{\partial}\omega)^\alpha \end{aligned}$$

and the result is proved. □

In the typical application of Proposition 12.3.2 in this book we will have $N = 3$, $u^1 = u$ a given map, $u^2(\zeta, z) = {}^t u(\zeta, z) = u(z, \zeta)$, and $u^3(\zeta, z) = \overline{\zeta} - \overline{z}$. In this case

$$\begin{aligned} (2\pi i)^{-n} E_3(\zeta, z) &= (2\pi i)^{-n} \frac{(\overline{\zeta} - \overline{z}) \cdot d(\zeta - z)}{|\zeta - z|^2} \wedge \left[\frac{d(\overline{\zeta} - \overline{z}) \cdot d(\zeta - z)}{|\zeta - z|^2} \right]^{n-1} \\ &= B(\zeta, z) \end{aligned} \tag{12.19}$$

where $B(\zeta, \mathrm{z})$ is the Bochner-Martinelli kernel. We will set

$$L(\zeta, z) = (2\pi i)^{-n} E_1(\zeta, z) = (2\pi i)^{-n} \frac{u \cdot d(\zeta - z)}{u \cdot (\zeta - z)} \wedge \left[\frac{\overline{\partial} u \cdot d(\zeta - z)}{u \cdot (\zeta - z)}\right]^{n-1}$$

and refer to this kernel as the *Leray kernel.* We also set

$$\begin{aligned} H(\zeta, z) &= (2\pi i)^{-n} E_{13}(\zeta, z) \\ &= (2\pi i)^{-n} \frac{u \cdot d(\zeta - z)}{u \cdot (\zeta - z)} \wedge \frac{(\overline{\zeta} - \overline{z}) \cdot d(\zeta - z)}{|\zeta - z|^2} \\ &\quad \wedge \sum_{p=1}^{n-1} \left[\frac{\overline{\partial} u \cdot d(\zeta - z)}{u \cdot (\zeta - z)}\right]^{p-1} \wedge \left[\frac{d(\overline{\zeta} - \overline{z}) \cdot d(\zeta - z)}{|\zeta - z|^2}\right]^{n-p-1} \end{aligned}$$

and refer to this kernel as the Henkin kernel. Also let

$$\begin{aligned} R(\zeta, z) &= (2\pi i)^{-n} E_{12}(\zeta, z) \\ &= (2\pi i)^{-n} \frac{u \cdot d(\zeta - z)}{u \cdot (\zeta - z)} \wedge \frac{{}^t u \cdot d(\zeta - z)}{{}^t u \cdot (\zeta - z)} \\ &\quad \wedge \sum_{p=1}^{n-1} \left[\frac{\overline{\partial} u \cdot d(\zeta - z)}{u \cdot (\zeta - z)}\right]^{p-1} \wedge \left[\frac{\overline{\partial}^t u \cdot d(\zeta - z)}{{}^t u \cdot (\zeta - z)}\right]^{n-p-1} \end{aligned} \tag{12.20}$$

and

$$\begin{aligned} K(\zeta, z) &= (2\pi i)^{-n} E_{123} \\ &= (2\pi i)^{-n} \frac{u \cdot d(\zeta - z)}{u \cdot (\zeta - z)} \wedge \frac{{}^t u \cdot d(\zeta - z)}{{}^t u \cdot (\zeta - z)} \wedge \frac{(\overline{\zeta} - \overline{z}) \cdot d(\zeta - z)}{|\zeta - z|^2} \\ &\quad \wedge \sum_{i+j+k=n-3} \left[\frac{\overline{\partial} u \cdot d(\zeta - z)}{u \cdot (\zeta - z)}\right]^i \wedge \left[\frac{\overline{\partial}^t u \cdot d(\zeta - z)}{{}^t u \cdot (\zeta - z)}\right]^j \\ &\quad \wedge \left[\frac{(\overline{\zeta} - \overline{z}) \cdot d(\zeta - z)}{|\zeta - z|^2}\right]^k . \end{aligned}$$

Then

$$\begin{aligned} {}^t L &= (2\pi i)^{-n} E_2 , \\ {}^t H &= (2\pi i)^{-n} E_{23} . \end{aligned}$$

The following formulas are important special cases of Theorem 12.3.2 which will be used in later sections.

$$\begin{aligned} \overline{\partial} H &= L - B && \text{on } V \setminus A_1 , && (12.21) \\ \overline{\partial}^t H &= {}^t L - B && \text{on } V \setminus A_2 , \\ \overline{\partial} R &= L - {}^t L && \text{on } V \setminus (A_1 \cup A_2) , \\ \overline{\partial} K &= H - {}^t H - R && \text{on } V \setminus (A_1 \cup A_2) . && (12.22) \end{aligned}$$

The following four examples are typical. A special class of the maps u are the complex gradients of convex functions. If ρ is a convex function defined on an open subset $U \subset \mathbb{C}^n$, then

$$u(\zeta, z) = \frac{\partial \rho}{\partial \zeta}(\zeta) = \left(\frac{\partial \rho}{\partial \zeta_1}(\zeta), \ldots, \frac{\partial \rho}{\partial \zeta_n}(\zeta)\right)$$

is defined on $V = U \times \mathbb{C}^n$.

Example 12.3.3. Let $\rho(\zeta) = |\zeta|^2 - R^2$, the defining function for the sphere of radius R. Then

$$u = \overline{\zeta}, \quad {}^t u = \overline{z}$$

on $\mathbb{C}^n \times \mathbb{C}^n$, and

$$\begin{aligned} L =& (2\pi i)^{-n} \frac{\overline{\zeta} \cdot d(\zeta - z)}{\overline{\zeta} \cdot (\zeta - z)} \wedge \left[\frac{d\overline{\zeta} \cdot d(\zeta - z)}{\overline{\zeta} \cdot (\zeta - z)}\right]^{n-1}, \\ H =& (2\pi i)^{-n} \frac{\overline{\zeta} \cdot d(\zeta - z)}{\overline{\zeta} \cdot (\zeta - z)} \wedge \frac{(\overline{\zeta} - \overline{z}) \cdot d(\zeta - z)}{|\zeta - z|^2} \\ & \wedge \sum_{p=1}^{n-1} \left[\frac{d\overline{\zeta} \cdot d(\zeta - z)}{\overline{\zeta} \cdot (\zeta - z)}\right]^{p-1} \wedge \left[\frac{d(\overline{\zeta} - \overline{z}) \cdot d(\zeta - z)}{|\zeta - z|^2}\right]^{n-p-1}, \\ R =& (2\pi i)^{-n} \frac{\overline{\zeta} \cdot d(\zeta - z)}{\overline{\zeta} \cdot (\zeta - z)} \wedge \frac{\overline{z} \cdot d(\zeta - z)}{\overline{z} \cdot (\zeta - z)} \\ & \wedge \sum_{p=1}^{n-1} \left[\frac{d\overline{\zeta} \cdot d(\zeta - z)}{\overline{\zeta} \cdot (\zeta - z)}\right]^{p-1} \wedge \left[\frac{d\overline{z} \cdot d(\zeta - z)}{\overline{z} \cdot (\zeta - z)}\right]^{n-p-1}. \end{aligned}$$

In this case $\omega^1 \wedge \omega^2 \wedge \omega^3 \equiv 0$ and hence $K \equiv 0$ so by (12.22)

$$R = H - {}^t H .$$

Example 12.3.4. Let $\rho(\zeta) = |\zeta'|^2 - \operatorname{Im} \zeta_n$, where $\zeta' = (\zeta_1, \ldots, \zeta_{n-1})$ and $\zeta = (\zeta', \zeta_n)$. Then

$$u(\zeta, z) = \left(\overline{\zeta}', i/2\right) \quad \text{and} \quad {}^t u(\zeta, z) = (\overline{z}', i/2) .$$

Note that for this example $K \neq 0$ and $R \neq H - {}^t H$. Let $M = \{\zeta \in \mathbb{C}^n \mid \rho(\zeta) = 0\}$ denote the hyperquadric, or boundary of the unbounded ball. Suppose $\zeta, z \in M$, i.e., $\zeta = \left(\zeta', \operatorname{Re} \zeta_n + i |\zeta'|^2\right)$ and $z = \left(z', \operatorname{Re} z_n + i |z'|^2\right)$. Then under the multiplication

$$z \cdot \zeta \equiv \left(z' + \zeta', \operatorname{Re}(z_n + \zeta_n) + 2 \operatorname{Im} z' \cdot \overline{\zeta}' + i |z' + \zeta'|^2\right),$$

M is isomorphic to the Heisenberg group. Let w_n denote the nth complex coordinate function on $\mathbb{C}^n$ restricted to M. By direct calculation one can show that for $\zeta, z \in M$

$$2iu(\zeta, z) \cdot (\zeta - z) = w_n\left(\zeta^{-1}z\right)$$

and

$$2i^t u(\zeta, z) \cdot (\zeta - z) = \overline{w_n\left(\zeta^{-1}z\right)}.$$

This example will be discussed further at the end of Section 12.8.

Example 12.3.5. Let $\rho(\zeta) = 2\operatorname{Re}\zeta_n$. The equation $\rho(\zeta) = 0$ defines a hyperplane in $\mathbb{C}^n$. In this case

$$u(\zeta, z) = (0, \ldots, 0, 1) = {}^t u.$$

We also have

$$L = {}^t L = R = 0$$

and

$$H = {}^t H = (2\pi i)^{-n} \frac{d(\zeta_n - z_n)}{\zeta_n - z_n} \wedge \frac{\left(\overline{\zeta}' - \overline{z}'\right) \cdot d(\zeta' - z')}{|\zeta - z|^2} \wedge \left[\frac{d\left(\overline{\zeta}' - \overline{z}'\right) \cdot d(\zeta' - z')}{|\zeta - z|^2}\right]^{n-2},$$

where $z' = (z_1, \ldots, z_{n-1})$. From (12.21) we have

$$\overline{\partial} H = -B \quad \text{in} \quad \mathbb{C}^{2n} \setminus \{(\zeta, z) \mid \zeta_n = z_n\}. \tag{12.23}$$

The kernel H is locally integrable across the singular set, so if χ_ϵ is the characteristic function of the set $\{(\zeta, z) \mid |\zeta_n - z_n| > \epsilon\}$, then

$$H = \lim_{\epsilon \to 0+} \chi_\epsilon H.$$

Using (12.23) we see that

$$\overline{\partial} H = \lim_{\epsilon \to 0+} \left(\overline{\partial}\chi_\epsilon \wedge H - \chi_\epsilon B\right).$$

Since the Bochner-Martinelli kernel is locally integrable, it follows that the limit

$$S = \lim_{\epsilon \to 0+} \overline{\partial}\chi_\epsilon \wedge H$$

exists, and

$$\overline{\partial} H = S - B \quad \text{in} \quad \mathbb{C}^{2n}. \tag{12.24}$$

An easy calculation shows that

$$\begin{aligned} S &= (2\pi i)^{1-n}\,[\zeta_n = z_n] \wedge \frac{\left(\overline{\zeta}' - \overline{z}'\right)\cdot d\,(\zeta' - z')}{|\zeta' - z'|^2} \wedge \frac{d\left(\overline{\zeta}' - \overline{z}'\right)\cdot d\,(\zeta' - z')}{|\zeta' - z'|^2} \\ &= [\zeta_n = z_n] \wedge B\,(\zeta', z') \end{aligned}$$

which is the Bochner-Martinelli kernel on the hyperplane $\zeta_n = z_n$. Thus (12.24) provides the $\overline{\partial}$ homology between two fundamental solutions for the Cauchy Riemann equations: the Bochner-Martinelli kernel on $\mathbb{C}^n$ and the Bochner-Martinelli kernel on a hyperplane.

$$\overline{\partial} H = [\zeta_n = z_n] \wedge B\,(\zeta', z') - B(\zeta, z)\,.$$

The process used here of passing from (12.23) to (12.24) is a prototype of the general procedure we will discuss in the next section.

Example 12.3.6. This example generalizes the previous one. Let $1 \leqslant k < n$ and set

$$u(\zeta, z) = \left(\overline{\zeta}_1 - \overline{z}_1, \ldots, \overline{\zeta}_k - \overline{z}_k, 0, \ldots, 0\right)\,.$$

Then

$$\begin{aligned} {}^t u &= -u\,, \\ L = {}^t L &= R = 0\,. \end{aligned}$$

If we set $z' = (z_1 \cdots z_k)\,, z'' = (z_{k+1}, \ldots, z_n)$, then

$$\begin{aligned} H = {}^t H =& (2\pi i)^{-n} \frac{\left(\overline{\zeta}' - \overline{z}'\right)\cdot d\,(\zeta' - z')}{|\zeta' - z'|^2} \wedge \frac{(\overline{\zeta} - \overline{z})\cdot d(\zeta - z)}{|\zeta - z|^2} \\ &\wedge \sum_{p=1}^{k} \left[\frac{d\left(\overline{\zeta}' - \overline{z}'\right)\cdot d\,(\zeta' - z')}{|\zeta' - z'|^2}\right]^{p-1} \wedge \left[\frac{d(\overline{\zeta} - \overline{z})\cdot d(\zeta - z)}{|\zeta - z|^2}\right]^{n-p-1}. \end{aligned}$$

From (12.21) we have

$$\overline{\partial} H = -B \quad \text{in} \quad \mathbb{C}^{2n} \setminus \{(\zeta, z) \mid \zeta' = z'\}$$

Again H is locally integrable on $\mathbb{C}^n$, so if χ_ϵ is the characteristic function of the set $\{(\zeta, z) \mid |\zeta' - z'| > \epsilon\}$ then

$$H = \lim_{\epsilon \to 0^+} \chi_\epsilon H \quad \text{exists}$$

and

$$\overline{\partial} H = S - B \quad \text{on} \quad \mathbb{C}^{2n} \tag{12.25}$$

where

$$S = \lim_{\epsilon \to 0+} \overline{\partial}\chi_\epsilon \wedge H$$
$$= (2\pi i)^{k-n} [\zeta' = z'] \wedge \frac{\left(\overline{\zeta}'' - \overline{z}''\right) \cdot d(\zeta'' - z'')}{|\zeta'' - z''|^2}$$
$$\wedge \left[\frac{d\left(\overline{\zeta''} - \overline{z}''\right) \cdot d(\zeta'' - z'')}{|\zeta'' - z''|^2}\right]^{n-h-1}$$
$$= [\zeta' = z'] \wedge B(\zeta'', z'')$$

is the Bochner-Martinelli kernel on the linear subspace $\zeta' = z'$. In particular, when $k = n - 1$,

$$S = [\zeta' = z'] \wedge \frac{d(\zeta_n - z_n)}{2\pi i(\zeta_n - z_n)}$$

is the Cauchy kernel on a line. Thus (12.25) provides the $\overline{\partial}$ homology between the Cauchy kernel on a line and the Bochner-Martinelli kernel.

12.4 A New Fundamental Solution for the Cauchy-Riemann Operator

Let $u(\zeta, z)$ be a smooth $\mathbb{C}^n$-valued function which is defined on an open subset $V \subset \mathbb{C}^{2n}$. Under suitable restrictions, such a function generates kernels on any product set $U_1 \times U_2 \subset V$, and in particular, gives rise to a fundamental solution for the Cauchy Riemann equations.

Definition 12.4.1. Suppose $V \subset \mathbb{C}^n \times \mathbb{C}^n$ is open. An infinitely differentiable map $u : V \to \mathbb{C}^n$ is called a *regular (kernel) generating function* on V if, whenever $(\zeta_0, z_0) \in V$ and $u(\zeta_0, z_0) \circ (\zeta_0 - z_0) = 0$, then $f(\zeta) = u(\zeta, z_0) \cdot (\zeta - z_0)$ has maximal rank at $\zeta = \zeta_0$.

An infinitely differentiable map $u : V \to \mathbb{C}^n$ is called an *extendable generating function* on V if, whenever $(\zeta_0, z_0) \in V$ and $u(\zeta_0, z_0) \cdot (\zeta_0 - z_0) = 0$, then $g(z) = u(\zeta_0, z) \cdot (\zeta_0, z)$ has maximal rank at $z = z_0$.

If u is both regular and extendable, then u is called a *biregular generating function* on V.

Notice that f and g in Definition 12.4.1 are complex-valued functions. Thus maximal rank means the rank is two. Equivalently we can use the real and imaginary parts of f (or of g) as independent coordinates near ζ_0 (or z_0).

Proposition 12.4.2. *Suppose $V \subset \mathbb{C}^n \times \mathbb{C}^n$ is open and $u : V \to \mathbb{C}^n$ is infinitely differentiable. If $u(z, z) \neq 0$ for all $(z, z) \in V$, then u is a biregular generating function on a neighborhood of $\Delta \cap V = \{(\zeta, z) \in V \mid \zeta = z\}$.*

Proof. Suppose $\zeta_0 = z_0$. Then

$$\overline{\partial} f(\zeta_0) = 0, \quad \partial f(\zeta_0) = u(\zeta_0, \zeta_0) \cdot d\zeta \neq 0$$

so f has maximal rank at each point of $\Delta \cap V$. The same is obviously true for g. Since maximal rank is an open condition the result follows. □

Suppose u is a generating function (either regular or extendable) on V. Let

$$A = \{(\zeta, z) \in V \mid u(\zeta, z) \cdot (\zeta - z) = 0\} .$$

Using the notation of Section 12.3, we set

$$u^1(\zeta, z) = u(\zeta, z), \quad u^2(\zeta, z) = \overline{\zeta} - \overline{z}$$

and we define in the domain $V - A$

$$\begin{aligned} L(\zeta, z) &= (2\pi i)^{-n} E_1(\zeta, z) \\ &= (2\pi i)^{-n} \frac{u \cdot d(\zeta - z)}{u \cdot (\zeta - z)} \wedge \left[\frac{\overline{\partial} u \cdot d(\zeta - z)}{u \cdot (\zeta - z)} \right]^{n-1} \end{aligned} \tag{12.26}$$

and

$$\begin{aligned} H(\zeta, z) &= (2\pi i)^{-n} E_{12}(\zeta, z) \\ &= (2\pi i)^{-n} \frac{u \cdot d(\zeta - z)}{u \cdot (\zeta - z)} \wedge \frac{(\overline{\zeta} - \overline{z}) \cdot d(\zeta - z)}{|\zeta - z|^2} \\ &\quad \wedge \sum_{k=1}^{n-1} \left[\frac{\overline{\partial} u \cdot d(\zeta - z)}{u \cdot (\zeta - z)} \right]^{k-1} \wedge \left[\frac{d(\overline{\zeta} - \overline{z}) \cdot d(\zeta - z)}{|\zeta - z|^2} \right]^{n-k-1} . \end{aligned}$$

Notice that

$$\begin{aligned} B(\zeta - z) &= (2\pi i)^{-n} E_2(\zeta, z) \\ &= (2\pi i)^{-n} \frac{(\overline{\zeta} - \overline{z}) \cdot d(\zeta - z)}{|\zeta - z|^2} \wedge \left[\frac{d(\overline{\zeta} - \overline{z}) \cdot d(\zeta - z)}{|\zeta - z|^2} \right]^{n-1} \end{aligned}$$

is the Bochner-Martinelli kernel, which is defined and biregular on all of $\mathbb{C}^n \times \mathbb{C}^n$. Finally, as a result of Theorem 12.3.2 (see (12.21)), we have

$$\overline{\partial} H = L - B \quad \text{on} \quad V - A . \tag{12.27}$$

Let $\chi_\epsilon(\zeta, z)$ denote the characteristic function of the set

$$\{(\zeta, z) \in V \mid |u(\zeta, z) \cdot (\zeta - z)| \geq \epsilon\} .$$

The next theorem is our primary result concerning the existence of fundamental solutions. Together with Theorem 12.4.7, it forms the nucleus of this book.

Theorem 12.4.3. *Suppose that u is a generating function defined on $V \subset \mathbb{C}^n \times \mathbb{C}^n$.*

(a) *The forms L and H, which are smooth in $V - A$, define Cauchy principal value currents on V by the formulas*

$$L = \lim_{\epsilon \to 0+} \chi_\epsilon L \tag{12.28}$$

$$H = \lim_{\epsilon \to 0+} \chi_\epsilon H \,. \tag{12.29}$$

Furthermore the limit

$$S = \lim_{\epsilon \to 0+} \overline{\partial}\chi_\epsilon \wedge H \tag{12.30}$$

exists and defines a current $S \in \mathcal{D}'^{n,n-1}(V)$ with

$$\operatorname{supp} S \subset A \,. \tag{12.31}$$

(b) *We have*

$$\overline{\partial} H = L + S - B \quad \text{on} \quad V \tag{12.32}$$

$$\overline{\partial}(L + S) = [\Delta] \text{ on } V \,. \tag{12.33}$$

(c) *Suppose that U_1 and U_2 are open subsets of $\mathbb{C}^n$ and $U_1 \times U_2 \subset V$. The kernels L, H, and S when restricted to $U_1 \times U_2$ are regular, extendable, or biregular provided that the generating function u is regular, extendable, or biregular.*

Proof. We will only discuss the case when u is regular since the other case is similar. Suppose we have shown that the principal value currents L and H exist on V and are regular kernels on $U_1 \times U_2$ whenever $U_1 \times U_2 \subset V$. Then by (12.29), (12.27), and (12.28)

$$\begin{aligned} \overline{\partial} H &= \lim_{\epsilon \to 0+} \overline{\partial}\chi_\epsilon \wedge H + \lim_{\epsilon \to 0+} \chi_\epsilon \overline{\partial} H \\ &= \lim_{\epsilon \to 0+} \overline{\partial}\chi_\epsilon \wedge H + L - B \,. \end{aligned}$$

Thus the limit S in (12.30) exists and we have (12.31) and (12.32). Formula (12.33) follows from (12.32), since $\overline{\partial} B = [\Delta]$. Finally if $U_1 \times U_2 \subset V$, and $\varphi \in 0^p p, q\,(U_1)$, then by (12.32) and the general homotopy formula (Theorem 12.2.4)

$$\begin{aligned} S\varphi &= B\varphi - L\varphi + (\overline{\partial} H)\varphi \\ &= B\varphi - L\varphi + \overline{\partial} H\varphi - H\overline{\partial}\varphi \,. \end{aligned}$$

Since B, L, and H are regular, it follows that $S\varphi \in \mathcal{D}^{p,q-1}$ so S is regular on $U_1 \times U_2$. □

Thus everything follows easily provided we show that the limits in (12.28) and (12.29) exist, and represent regular kernels. A slightly more precise result is contained in the following theorem.

Theorem 12.4.4. *Suppose that u is a regular generating function in V, and $U_1 \times U_2 \subset V$. Then for all $\varphi \in \mathcal{D}^{p,q}(U_1)$ the principal value integrals*

$$L\varphi(z) = \lim_{\epsilon \to 0+} \int_{|u\cdot(\zeta - z)| > \epsilon} L(\zeta, z) \wedge \varphi(\zeta)$$

$$H\varphi(z) = \lim_{\epsilon \to 0+} \int_{|u\cdot(\zeta - z)| > \epsilon} H(\zeta, z) \wedge \varphi(\zeta)$$

exist. Furthermore $L\varphi \in \mathcal{D}^{p,q-1}(U_2)$ and $H\varphi \in \mathcal{D}^{p,q-2}(U_2)$.

By analyzing the coefficients of L and H, Theorem 12.4.4 follows easily from the next result.

Proposition 12.4.5. *Suppose u is a regular generating function in V, and suppose that $k(\zeta, z) \in C^\infty(V)$ satisfies*

$$|k(\zeta, z)| \leqslant C|\zeta - z|^r .$$

Suppose $U_1 \times U_2 \subset V$. For $\varphi \in C_0^\infty(U_1)$ and $\epsilon > 0$ define

$$K_\epsilon\varphi(z) = \int_{|u\cdot(\zeta - z)| > \epsilon} k(\zeta, z)|\zeta - z|^{-2q}[u(\zeta, z)\cdot(\zeta - z)]^{-p}\varphi(\zeta)d\lambda(\zeta), \quad (12.34)$$

where $p, q,$ and r are non-negative integers. If either $q = 0$, or $q > 0$ and $2(n - q) + r - p > 1$, then

$$K\varphi(z) = \lim_{\epsilon \to 0+} K_\epsilon\varphi(z)$$

exists and $K\varphi \in C^\infty(U_2)$.

Proof. It suffices to show that every point $(\zeta_0, z_0) \in V$ has a product neighborhood $U_1 \times U_2$ where Proposition 12.4.5 is true. The result then followed by a standard covering and partition of the unity argument.

There are three cases.

(1) $(\zeta_0, z_0) \in V \setminus A$,

(2) $(\zeta_0, z_0) \in A \setminus \Delta$ if $q > 0$ or $(\zeta_0, z_0) \in A$ if $q = 0$,

(3) $(\zeta_0, z_0) \in \Delta \cap V$ and $q > 0$.

The first case is trivial; simply choose U_1 and U_2 such that $U_1 \cap U_2 = \varnothing$ and $U_1 \times U_2 \cap A = \varnothing$. Then the integrand in (12.34) is infinitely differentiable in $U_1 \times U_2$.

In case (2), if $q > 0$ we can choose U_1 and U_2 disjoint. Consequently we need not worry about the $|\zeta - z|^{-2q}$ term in (12.34) in this case. Since u is regular, $f(\zeta) = u(\zeta, z_0) \cdot (\zeta - z_0)$ has rank 2 at $\zeta = \zeta_0$. Thus if U_1 and U_2 are sufficiently small we can find an infinitely differentiable map $\Phi : U_1 \times U_2 \to \mathbb{C}^n$ such that

$$w = \Phi_z(\zeta) = \Phi(\zeta, z)$$

is a diffeomorphism on U_1 for each $z \in U_2$, and

$$w_n = u(\zeta, z) \cdot (\zeta - z)$$

In the w coordinates the integrand in (12.34) becomes

$$k(\zeta, z)|\zeta - z|^{-2q}[u \cdot (\zeta - z)]^{-p}\varphi(\zeta)d\lambda(\zeta) = F(w, z)w_n^{-p}d\lambda(w)$$

where F is infinitely differentiable in both variables and is compactly supported in w. Thus

$$K_\epsilon\varphi(z) = \int_{|w_n|>\epsilon} F(w, z)w_n^{-p}d\lambda(w) \tag{12.35}$$

If $p = 0$ or 1, then $K_\epsilon\varphi(z) \to K\varphi(z)$ where

$$K\varphi(z) = \int F(w, z)w_n^{-p}d\lambda(w)$$

It is clear that $K\varphi \in C^\infty(U_2)$, since we may differentiate under the integral sign arbitrarily often. □

If $p \geqslant 2$ we need the following Lemma.

Lemma 12.4.6. *For each positive integer* p*, there is a constant* M_p *such that*

$$\left| \int_{|w|\geq\varepsilon} \left[w^{-p}\varphi(w) - \frac{1}{(p-1)!} w^{-1} \frac{\partial^{p-1}}{\partial w^{p-1}}\varphi(w) \right] d\lambda(w) \right| \leq M_p \in \| \varphi \|_{p-1}$$

for all $\varphi \in C_0^\infty(\mathbb{C})$*, where*

$$\|\varphi\|_{p-1} = \sup_{\substack{|\alpha|\leqslant p-1 \\ w\in C}} |D^\alpha\varphi(w)| .$$

We will proceed with the proof of Proposition 12.4.5 and prove the lemma later. It follows from Lemma 12.4.6 and (12.35), that $K_\epsilon\varphi(z) \to K\varphi(z)$, where

$$K\varphi(z) = \frac{1}{(p-1)!} \int w_n^{-1} \frac{\partial^{p-1}}{\partial w_n^{p-1}} F(w, z)d\lambda(w) .$$

Again it is clear that $K\varphi \in C^\infty(U_2)$.

In case (3) we once more choose a map

$$w = \Phi_z(\zeta) = \Phi(\zeta, z)$$

which is a diffeomorphism on U_1 for each $z \in U_2$, with

$$w_n = u(\zeta, z) \cdot (\zeta - z),$$

but now, since $\zeta_0 = z_0$, we can insist that

$$\Phi(z, z) \equiv 0. \tag{12.36}$$

Then

$$|\zeta - z|^2 = h(w, z),$$

where $h \in C^\infty(U_1 \times U_2)$, and, because of (12.36), $h(w, z)$ vanishes to second order at $w = 0$. Thus, by Taylor's Theorem,

$$h(w, z) = \operatorname{Re} \sum_{i,j=1}^{n} a_{ij}(w, z) w_i w_j + \sum_{i,j=1}^{n} b_{ij}(w, z) w_i \overline{w}_j.$$

It follows that

$$\left|D_z^\alpha D_w^\beta h(z, w)\right| \leqslant \begin{cases} C_{\alpha\beta}|w|^{2-|\beta|} & |\beta| \leqslant 2 \\ C_{\alpha\beta} & |\beta| \geqslant 2. \end{cases} \tag{12.37}$$

In the w coordinates, the integrand in (12.34) becomes

$$k(\zeta, z)|\zeta - z|^{-2q}[u \cdot (\zeta - z)]^{-p} \varphi(\zeta) d\lambda(\zeta) = w_n^{-p} h(w, z)^{-q} F(w, z) d\lambda(w),$$

where F is infinitely differentiable in both variables, is compactly supported in w, and satisfies

$$|F(w, z)| \leqslant C|w|^r.$$

Just as in (12.37), it follows that

$$\left|D_z^\alpha D_w^\beta F(w, z)\right| \leqslant \begin{cases} C_{\alpha\beta}|w|^{r-|\beta|} & |\beta| \leqslant r \\ C_{\alpha\beta} & |\beta| \geqslant r. \end{cases} \tag{12.38}$$

Now

$$K\varphi(z) = \int_{|w_n| \geqslant \epsilon} w_n^{-p} h(w, z)^{-q} F(w, z) d\lambda(w).$$

If $p = 0, 1$, and $2(n - q) + r - p > 0$, the integrand is integrable across $w_n = 0$. Thus $K_\epsilon \varphi(z) \to K\varphi(z)$ where

$$K\varphi(z) = \int w_n^{-p} h(w, z)^{-q} F(w, z) d\lambda(w).$$

From (12.37) and (12.38) it follows that we may differentiate under the integral arbitrarily often, so $K\varphi \in C^{\infty}(U_2)$.

If $p \geqslant 2$, we again use Lemma 12.4.6. It follows that

$$\left| K_\epsilon \varphi(z) - \frac{1}{(p-1)!} \int_{|w_n|>\epsilon} w_n^{-1} \left(\frac{\partial}{\partial w_n}\right)^{p-1} \left[h(w,z)^{-q} F(w,z)\right] d\lambda(w) \right|$$
$$\leqslant M\epsilon \int_{w' \in C^{n-1}} \sup_{|\beta| \leqslant p-1} \left| D^\beta \left[h(w,z)^{-q} F(w,z)\right]\right| d\lambda(w'), \quad (12.39)$$

where $w = (w', w_n)$. If $F(w,z) = 0$ for $|w| \geqslant R$, then by (12.37) and (12.38), the right-hand side of (12.39) is bounded by

$$M\epsilon \int_{|w'|<R} |w'|^{-2q+r-p+1} \, d\lambda(w'),$$

which converges to zero as $\epsilon \to 0$ provided $2(n-q)+r-p > 1$. On the other hand

$$w_n^{-1} |w|^{-2q+r-p+1} \in L^1_{\text{loc}}(\mathbb{C}^n)$$

provided $2(n-q)+r-p > 0$, so the integral on the left-hand side of (12.39) converges as $\epsilon \to 0$. Thus $K_\epsilon \varphi(z) \to K\varphi(z)$, whereby (12.39)

$$K\varphi(z) = \frac{1}{(p-1)!} \int w_n^{-1} \left(\frac{\partial}{\partial w_n}\right)^{p-1} \left[h(w,z)^{-q} F(w,z)\right] d\lambda(w).$$

Again by (12.37) and (12.38), we may differentiate arbitrarily often under the integral sign, so $K_\varphi \in C^{\infty}(U_2)$.

Proof of Lemma 12.4.6. By induction, it suffices to prove that

$$\left| \int_{|w|>\epsilon} \left[(p-1) w^{-p} \varphi(w) - w^{1-p} \frac{\partial \varphi}{\partial w}(w)\right] dw \wedge d\overline{w} \right| \leqslant M\epsilon \|\varphi\|_{p-1}.$$

However,

$$\left[w^{1-p} \frac{\partial \varphi}{\partial w} - (p-1) w^{-p} \varphi(w)\right] dw \wedge d\overline{w} = d\left[w^{1-p} \varphi \, d\overline{w}\right]$$

so by Stokes's Theorem

$$\int_{|w| \geqslant \epsilon} \left[(p-1) w^{-p} \varphi - w^{1-p} \frac{\partial \varphi}{\partial w}\right] dw \wedge d\overline{w} = \int_{|w|=\epsilon} w^{1-p} \varphi(w) \, d\overline{w}.$$

By Taylor's Theorem

$$\varphi(w) = P(w) + R(w)$$

where

$$P(w) = \sum_{\alpha+\beta \leqslant p-2} \frac{1}{\alpha! \beta!} \frac{\partial^{\alpha+\beta}}{\partial w^\alpha \partial \overline{w}^\beta} \varphi(0) w^\alpha \overline{w}^\beta$$

and

$$|R(w)| \leqslant M_p \|\varphi\|_{p-1} |w|^{p-1}.$$

Since

$$\int_{|w|=\epsilon} w^\alpha \overline{w}^\beta d\overline{w} = \begin{cases} -2\pi i \epsilon^{\alpha+\beta+1} & \text{if} \quad \alpha = \beta + 1 \\ 0 & \text{if} \quad \alpha \neq \beta + 1 \end{cases}$$

it follows that

$$\int_{|w|=\epsilon} P(w) d\overline{w} = 0.$$

Thus

$$\left| \int_{|w|=\epsilon} w^{1-p} \varphi(w) d\overline{w} \right| = \left| \int_{|w|=\epsilon} w^{1-p} R(w) d\overline{w} \right| \leqslant 2\pi M_p \epsilon \|\varphi\|_{p-1}$$

and the lemma follows. □

As will become clear in the next section, a generating function satisfying certain crucial additional properties will be available only in an open subset $V \subset \Omega \times \Omega$. In order to get a fundamental solution on Ω we will use the next theorem.

Theorem 12.4.7. *Suppose $\Omega \subset \mathbb{C}^n$ is open and suppose $u(\zeta, z)$ is a generating function defined on an open subset $V \subset \Omega \times \Omega$. Then if $\varphi \in C^\infty(\Omega \times \Omega)$ and* $\operatorname{supp} \varphi \subset V$*, the kernel*

$$\begin{aligned} E &= B + \overline{\partial}(\varphi H) \\ &= \varphi(L + S) + (1 - \varphi) B + \overline{\partial}\varphi \wedge H \end{aligned} \tag{12.40}$$

is a fundamental solution in Ω. The fundamental solution E will be regular, extendable, or biregular provided that the generating function u is regular, extendable, or biregular.

Proof. Since $\operatorname{supp} \varphi \subset V, \varphi H$ is defined in $\Omega \times \Omega$. That E is a fundamental solution on Ω is clear since $\overline{\partial} E = \overline{\partial} B = [\Delta]$. The expression (12.40) for E follows from (12.32). Since L, H, and S are regular, extendable, or biregular provided that u is, the same is true for E. □

Next, we prove a result which indicates that nontrivial generating functions do exist.

Proposition 12.4.8. *Suppose $U \subset \mathbb{C}^n$ is open and that ρ is an infinitely differentiable convex function on U such that $d\rho$ never vanishes on U. Then the map $u(\zeta, z)$ defined by $u_j(\zeta, z) = (\partial\rho/\partial\zeta_j)(\zeta), 1 \leqslant j \leqslant n$, is a biregular generating function on $U \times \mathbb{C}^n$.*

Proof. We will set

$$u = \frac{\partial \rho}{\partial \zeta} = \left(\frac{\partial \rho}{\partial \zeta_1}, \ldots, \frac{\partial \rho}{\partial \zeta_n} \right)$$

Then u never vanishes on $U \times \mathbb{C}^n$, so by Proposition 12.4.2, u is a biregular generating function near the diagonal. For $\zeta \in U$ fixed, $g(z) = (\partial\rho/\partial z)(\zeta)\cdot(\zeta - z)$ is a holomorphic function of z, and $\partial g/\partial z = -\partial\rho/\partial\zeta \neq 0$. Thus g has maximal rank and consequently, u is an extendable generating function on $U \times \mathbb{C}^n$.

With $z \in \mathbb{C}^n$ fixed, set $f(\zeta) = (\partial\rho/\partial\zeta)(\zeta)\cdot(\zeta - z)$. We must show that f has rank 2 at each point $\zeta \in A_z^* = \{\zeta \in U | (\partial\rho/\partial\zeta)(\zeta)\cdot(\zeta - z) = 0\}$. Since convexity is a real property it is convenient to replace the complex coordinates ζ and $z \in \mathbb{C}^n$ by real coordinates y and $x \in \mathbb{R}^{2n}$. Let J be the linear map on $\mathbb{R}^{2n}$ corresponding to multiplication by i on $\mathbb{C}^n$. Then

$$2f(\zeta) = \varphi(y) + i\psi(y)$$

where

$$\varphi(y) = \frac{\partial\rho}{\partial y}\cdot(y - x)$$
$$\psi(y) = J\left(\frac{\partial\rho}{\partial y}\right)\cdot(y - x)$$

and the inner product is the standard inner product on $\mathbb{R}^{2n}$. Let

$$H(y) = \left(\frac{\partial^2\rho}{\partial y_i \partial y_j}(y)\right)$$

be the (real) Hessian of ρ evaluated at y. Since ρ is a convex function, $H(y)$ is a positive semi-definite matrix at each point $y \in U$. Let $w = y - x$. Then

$$\frac{\partial\varphi}{\partial y} = \frac{\partial\rho}{\partial y} + Hw$$
$$\frac{\partial\psi}{\partial y} = J\frac{\partial\rho}{\partial y} - HJw$$

To show that f has maximal rank, it suffices to find a vector $v \in \mathbb{R}^{2n}$ such that the determinant

$$D(v) = \begin{vmatrix} \frac{\partial\varphi}{\partial y}\cdot v & \frac{\partial\varphi}{\partial y}\cdot Jv \\ \frac{\partial\psi}{\partial y}\cdot v & \frac{\partial\psi}{\partial y}\cdot Jv \end{vmatrix} \tag{12.41}$$

is nonzero. Furthermore, we only have to do this for points $y \in U$ such that $\varphi(y) = \psi(y) = 0$, and such that $w = y - x \neq 0$.

Suppose first that H is positive definite on the plane spanned by w and Jw. In this case take $v = w$ in 12.41. Then

$$\begin{aligned} D(w) &= \begin{vmatrix} Hw\cdot w & Hw\cdot Jw \\ -HJw\cdot w & -HJw\cdot Jw \end{vmatrix} \\ &= -\left[(Hw\cdot w)(HJw\cdot JW) - (HJw\cdot w)^2\right] \end{aligned}$$

which is negative by the Cauchy-Schwartz inequality.

If H is not positive definite on the plane spanned by w and Jw, then there is a vector

$$w_1 = \alpha w + \beta Jw, \quad \alpha^2 + \beta^2 = 1$$

such that

$$Hw_1 = 0$$

Set

$$w_2 = Jw_1 = -\beta w + \alpha Jw,$$

and

$$v = \frac{\partial \rho}{\partial y} + t\,(\alpha w_1 + \beta w_2)$$

where t is a real number to be chosen. A calculation shows that

$$D(v) = \left|\frac{\partial \rho}{\partial y}\right|^4 - \left|\frac{\partial \rho}{\partial y}\right|^2 \left(Hw_2 \cdot \left[\beta \frac{\partial \rho}{\partial y} + \alpha J \frac{\partial \dot{\rho}}{\partial y}\right] + tHw_2 \cdot w_2\right)$$

If $Hw_2 = 0$, then $D(y) = |\partial\rho/\partial y|^4 > 0$. If on the other hand $Hw_2 \neq 0$, then $Hw_2 \cdot w_2 > 0$ so we can choose t so that $D(v) \neq 0$. $\square$

Finally, we will describe the propagation of singularities by the operators L, H, and S. Roughly speaking, these operators propagate singularities along the singular set A. Let

$$A_\zeta = \{z \mid (\zeta, z) \in A\}\ ,$$
$$A_z^* = \{\zeta \mid (\zeta, z) \in A\}\ .$$

Theorem 12.4.9. *Suppose u is a biregular generating function defined in $V \subset \mathbb{C}^{2n}$, and $U_1 \times U_2 \subset V$. Let $v \in \mathcal{E}'^{p,q}(U_1)$, and let $w \in \mathcal{E}'^{p,q}(U_2)$. Then*

(a) $\operatorname{supp} Sv \subset \{z \mid A_z^* \cap \operatorname{supp} v \neq \varnothing\}$
$\operatorname{supp} {}^tSw \subset \{\zeta_0 \mid A_\zeta \cap \operatorname{supp} w \neq \varnothing\}$

(b) If T is one of the operators L, H, or S then

$$\text{sing supp}\, Tv \subset \{z \mid A_z^* \cap \text{sing supp}\ \ c \neq \varnothing\}$$
$$\text{sing supp}\, {}^tTw \subset \{\zeta \mid A_\zeta \cap \text{sing supp}\ \ w \neq \varnothing\}\ .$$

Proof. We will prove the result only for L, H, and S. The result for the transposes follows easily.

(a) It suffices to prove the result when $v \in \|^{p,q}(U_1)$.
Let $\varphi \in \mathcal{D}^{n-p,n-q+1}(U_2)$ and suppose

$$\operatorname{supp} \varphi \cap \{z \mid A_z^* \cap \operatorname{supp} v \neq \varnothing\} = \varnothing$$

This implies that

$$A \cap (\operatorname{supp} v) \times (\operatorname{supp} \varphi) = \varnothing$$

Since $\operatorname{supp} S \subset A$,

$$(Sv, \varphi) = (S, v \otimes \varphi) = 0$$

and the result follows.

(b) We will first prove that

$$\operatorname{sing\,supp} Tv \subset \{z \mid A_z^* \cap \operatorname{supp} v \neq \varnothing\} \tag{12.42}$$

Suppose $A_{z_0}^* \cap \operatorname{supp} v = \varnothing$. Then $u(\zeta, z_0) \cdot (\zeta - z_0) \neq 0$ for $\zeta \in \operatorname{supp} v$. Since $\operatorname{supp} v$ is compact, it follows that there is an $\epsilon > 0$ and a neighborhood ω of $\operatorname{supp} v$ such that $u(\zeta, z) \cdot (\zeta - z) \neq 0$ for $|z - z_0| < \epsilon$ and $\zeta \in \omega$. But then the kernel $T(\zeta, z) \in C^\infty(\omega \times B(z_0, \epsilon))$, so $Tv \in C^\infty(B(z_0, \epsilon))$. Thus (12.42) is proved.

Now suppose $A_{z_0}^* \cap \operatorname{sing\,supp} v = \varnothing$. As before there is an $\epsilon > 0$ and a neighborhood ω of $\operatorname{sing\,supp} v$ such that $u(\zeta, z) \cdot (\zeta - z) \neq 0$ for $|z - z_0| < \epsilon$ and $\zeta \in \omega$. Choose $\varphi \in C_0^\infty(\omega)$ such that $\varphi \equiv 1$ on a neighborhood of $\operatorname{sing\,supp} v$. Then $(1 - \varphi)v \in \mathcal{D}^{p,q}(U_1)$ and $A_{z_0} \cap \operatorname{supp}(\varphi v) = \varnothing$. Since

$$Tv = T[\varphi v] + T[(1 - \varphi)v]$$

$\operatorname{sing\,supp} Tv \subset \operatorname{sing\,supp} T[\varphi v]$. It follows from (12.42) that Tv is infinitely differentiable near z_0.

□

For real analytic singularities, we get a weaker result which is followed by the proof of (12.42).

Theorem 12.4.10. *Suppose u is a real analytic biregular generating function defined in $V \subset \mathbb{C}^{2n}$ and $U_1 \times U_2 \subset V$. Let $v \in \mathcal{E}'^{p,q}(U_1), w \in \mathcal{E}'^{p,q}(U_2)$. If T is one of the operators L, H, or S then*

(a) Tv is real analytic in $U_2 \setminus \{z \mid A_z^* \cap \operatorname{supp} v \neq \varnothing\}$

(b) ${}^tTw$ is real analytic in $U_1 \setminus \{\zeta \mid A_\zeta \cap \operatorname{supp} w \neq \varnothing\}$.

12.5 Selected Applications

We present here three applications of the techniques developed previously. Subsections A, B, and C may be read independently.

12.5.1 A. Solving Smoothly up to the Boundary

Throughout this subsection, we assume that D is a bounded open subset of $\mathbb{C}^n$ with an infinitely differentiable boundary. The problem we address is the following. Given $G \in \mathcal{E}^{p,q}(\overline{D})$ $(q \geqslant 1)$, satisfying the compatibility condition $\overline{\partial} G \equiv 0$, find $F \in \mathcal{E}^{p,q-1}(\overline{D})$ which solves the equation $\overline{\partial} F = G$ in D. Using the techniques of the $\overline{\partial}$-Neumann problem J.J. Kohn has shown that this is always possible if D is only assumed to be pseudoconvex, [126].

In order to solve this problem we will need to make an assumption about the pseudoconvexity of ∂D. We will do this by assuming the existence of a generating function of a special type. Let $\Omega \subset \mathbb{C}^n$ be open and suppose $M = \{z \in \Omega \mid \rho(z) = 0\}$ where $\rho \in C^\infty(\Omega)$ and $d\rho(z) \neq 0$ on M. Let $\Omega^\pm = \{z \in \Omega \mid \pm\rho(z) > 0\}$ (in the application in this subsection $M = \partial D$ and $\Omega^- = D$).

Definition 12.5.1. A generating function u defined in a neighborhood V of $M \times M$ which satisfies

(a) $u(\zeta, z)$ is holomorphic for $z \in \Omega^-$ for each fixed $\zeta \in \Omega^+$, and

(b) $u(\zeta, z) \cdot (\zeta - z) \neq 0$ for $(\zeta, z) \in V \cap \left(\overline{\Omega}^+ \times \Omega^-\right)$ is called a *support function for M (with respect to Ω^-).*

If the generating function u satisfies (a) and

(b') $u(\zeta, z) \cdot (\zeta - z) \neq 0$ for $(\zeta, z) \in V \cap \left(\overline{\Omega}^+ \times \overline{\Omega}^-\right) - \Delta$: then u is called a *strong support function for M (with respect to Ω^-).*

If this generating function u is defined only in a neighborhood V of $\Delta \cap (M \times M)$, then u is called a local support function for M (with respect to Ω^-).

In the case when $D = \Omega^- \subset\subset \Omega$ and $V \supset \partial D \times \overline{D}$, we will call u a *support function for ∂D.*

Remark 12.5.2. If u is a local support function for M with respect to Ω^-, then for each $\zeta \in M$ the function $q(z) = u(\zeta, z) \cdot (\zeta - z)$ is holomorphic in $U \cap \Omega^-$ for some neighborhood U of ζ by (a), and q never vanishes in $U \cap \Omega^-$ by (b). It follows easily that M is pseudoconvex as a piece of the boundary of Ω^-.

Remark 12.5.3. If u is a strong local support function for ∂D defined on a neighborhood V of $\Delta \cap (\partial D \times \partial D)$, then necessarily $u(z, z) \neq 0$ for $z \in \partial D$. By Proposition 12.4.2 u is a biregular generating function on perhaps a smaller open set. Thus we may consider any local support function to be biregular.

There are some easily verified sufficient conditions for the existence of support functions.

Proposition 12.5.4. *Suppose ρ is a convex function. Then*

$$u(\zeta) = \frac{\partial \rho}{\partial \zeta}(\zeta)$$

is a biregular support function for M with respect to Ω^-.

Remark 12.5.5. If $D \subset\subset \mathbb{C}^n$ is convex and has an infinitely differentiable boundary, then

$$\rho(z) = \begin{cases} \operatorname{dist}(z, \partial D) & z \notin D \\ -\operatorname{dist}(z, \partial D) & z \in \overline{D} \end{cases}$$

is an infinitely differentiable convex function in some neighborhood of ∂D and Proposition 12.5.4 applies.

Proof. By Proposition 12.4.8, $u(\zeta)$ is a biregular generating function. Since u is independent of z property (a) of a support function is automatic. Let $\Omega_c = \{\zeta \mid \rho(\zeta) < c\}$. Then $A_\zeta = \{z \mid u(\zeta) \cdot (\zeta - z) = 0\}$ is the holomorphic tangent space to $\partial\Omega_c$ with $c = \rho(\zeta)$. Since ρ is convex, $A_\zeta \cap \Omega_{\rho(\zeta)} = \phi$ for all ζ. In particular, $A_\zeta \cap \Omega^- = \varnothing$ for all $\zeta \in \overline{\Omega}^+$, and this is property (b) of a support function. Suppose the defining function ρ has the property that

$$\rho(z) = \rho(\zeta) + \operatorname{Re} Q(\zeta, z) + r(\zeta, z) \tag{12.43}$$

where $Q(\zeta, z)$ is holomorphic in Ω for each fixed $\zeta \in \Omega$, and

$$Q(z, z) \equiv 0 \tag{12.44}$$

$$|r(\zeta, z)| \leqslant C|\zeta - z|^2 \tag{12.45}$$

Notice that Taylor's Theorem together with (12.43—12.45) imply that

$$Q(\zeta, z) = 2\sum_{j=1}^{n} \frac{\partial \rho}{\partial \zeta_j}(\zeta)\,(z_j - \zeta_j) + O\left(|\zeta - z|^2\right)$$

Thus (12.43) may be considered to be a form of Taylor's expansion with remainder for ρ. □

Definition 12.5.6. The hypersurface M is said to be geometrically pseudoconvex with respect to Ω^- if M has a defining function ρ, such that $\rho(z) < 0$ for $z \in \Omega^-$, and ρ has a Taylor expansion (12.43) with the additional property that

$$r(\zeta, z) \geqslant 0$$

whenever

$$Q(\zeta, z) = 0, \quad |\rho(\zeta)| < \epsilon, \quad |\rho(z)| < \epsilon, \quad 0 < |\zeta - z| < \epsilon$$

for some $\epsilon > 0$.

We will say that M is strictly geometrically pseudoconvex with respect to Ω^- if

$$r(\zeta, z) > 0$$

whenever

$$Q(\zeta, z) = 0, \quad |\rho(\zeta)| < \epsilon, \quad |\rho(z)| < \epsilon, \quad 0 < |\zeta, -z| < \epsilon$$

for some $\epsilon > 0$.

Clearly, if M is strictly pseudoconvex as the boundary of Ω^-, then M is strictly geometrically pseudoconvex with respect to Ω^-.

Proposition 12.5.7. *Suppose $D \subset\subset \mathbb{C}^n$ has an infinitely differentiable boundary and that ∂D is strictly geometrically pseudoconvex with respect to D. Then there is a strong biregular support function u for ∂D defined in a neighborhood V of $\partial D \times \overline{D}$ with the additional property that there is a neighborhood V' of $\Delta \cap V$ and a function $F(\zeta, z) \in C^\infty(V')$ which is holomorphic in z for each fixed ζ and*

(a) $u(\zeta, z) \cdot (\zeta - z) = F(\zeta, z) Q(\zeta, z)$ *on* V'

(b) $u(\zeta, z) \cdot (\zeta - z)$ *is never zero on* $V \setminus V'$.

Proof. This follows by a construction due to Henkin [103] (also see Ovrelid [165]).

Suppose now that u is a regular support function for ∂D. We may assume that u is defined on a set $V = (\Omega - K) \times \Omega$ where K is compact, Ω is open, and $K \subset D \subset\subset \Omega$. Choose $\varphi \in C^\infty(\Omega)$ such that φ vanishes identically on a neighborhood of K. Then we can apply Theorem 12.4.7 to see that

$$E(\zeta, z) = B(\zeta, z) + \overline{\partial}[\varphi(\zeta) H(\zeta, z)] \tag{12.46}$$

is a regular fundamental solution in Ω. We will make the additional assumption that

$$1 - \varphi \in C_0^\infty(D) \tag{12.47}$$

□

Theorem 12.5.8. *Suppose u is a regular support function for ∂D, and the kernel E is defined by (12.46) where φ satisfies (12.47). If $q \geqslant 1$ and $G \in \mathcal{D}^{p,q}(\Omega)$ satisfies $\overline{\partial} G \equiv 0$ on D, then*

$$G = \overline{\partial} E G \text{ on } \quad D,$$

and $EG \in \mathcal{E}^{p,q+1}(\Omega)$.

Remark 12.5.9. The solution $F_1 = EG$ to $\overline{\partial} F = G$ on D can be written in the following two ways. First, by the general homotopy formula (Theorem 12.2.4) for the operator φH,

$$F_1 = B(G) - H(\overline{\partial} G) + \overline{\partial} H(\varphi G) \tag{12.48}$$

since $\varphi\overline{\partial}G = \overline{\partial}G$ by (12.47). Second, substituting $\overline{\partial}H = L+S-B$ into (12.46),

$$F_1 = B((1-\varphi)G) + (L+S)(\varphi G) + H(\overline{\partial}\varphi \wedge G)\,. \tag{12.49}$$

In particular, (12.48) implies that

$$F_2 = B(G) - H(\overline{\partial}G)$$

also satisfies $\overline{\partial}F_2 = G$ on D, and again $F_2 \in \mathcal{E}^{p,q+1}(\Omega)$. This solution does not depend on the cutoff function φ. Notice that $F_1 = F_2$ if $q = 1$.

The implications of properties (a) and (b) of a support function are isolated in the following lemma.

Lemma 12.5.10. *Suppose u is a support function for ∂D and $G \in \mathcal{D}^{p,q}(\Omega)$ vanishes in D*

(a) If $q \neq 1, L(\varphi G) \equiv 0$ in D

(b) $S(\varphi G) \equiv 0$ in D.

Proof.

(a) Since u is holomorphic for $z \in D$ and $\zeta \in \Omega \backslash D$, $\overline{\partial}u = \overline{\partial}_\zeta u$ in $(\Omega\backslash D)\times D$. Thus the kernel $L(\zeta, z)$ (see (12.26)) is of degree $n-1$ in $d\overline{\zeta}$ in $(\Omega\backslash D)\times D$. Since $\operatorname{supp} G \subset \Omega \setminus D$ part (a) follows.

(b) Because of (12.40)

$$\begin{aligned}\operatorname{supp} S(\varphi G) &\simeq \{z \mid A_z^* \cap \operatorname{supp} \varphi G \neq \varnothing\}\\ &\subset \{z \mid A_z^* \cap (\Omega \setminus D) \neq \varnothing\}\\ &\subset \Omega \setminus D\end{aligned}$$

by property (b) of a support function.

□

Proof of Theorem 12.5.8. By Theorem 12.4.7 E is a regular fundamental solution, so $F = EG \in \varepsilon^{p,q+1}(\Omega)$. In addition

$$G = \overline{\partial}EG + E\overline{\partial}G$$

on Ω. Thus it suffices to show that $E\overline{\partial}G$ vanishes in D. By (12.40)

$$E\overline{\partial}G = B[(1-\varphi)\overline{\partial}G] + (L+S)(\varphi\overline{\partial}G) + H(\overline{\partial}\varphi \wedge \overline{\partial}G)$$

Since $\overline{\partial}G \equiv 0$ in D, it follows from (12.47) that $(1-\varphi)\overline{\partial}G = \overline{\partial}\varphi \wedge \overline{\partial}\mathrm{G} \equiv 0$. Finally,

$$E\overline{\partial}G = L(\varphi\overline{\partial}G) + S(\varphi\overline{\partial}G)$$

vanishes in D by Lemma 12.5.10. That completes the proof of the theorem. □

Remark 12.5.11. Suppose that ∂D has a support function which is extendable instead of regular. Then the conclusions of Theorem 12.5.8, as extended in Remark 12.5.9, remain valid for any $G \in \mathcal{E}'^{p,q}(\Omega)$, except that we cannot conclude that F_1 and F_2 are smooth even if G is smooth. If the support function u is biregular we can make a more precise statement.

Corollary 12.5.12. *Suppose u is a biregular support function for ∂D. Suppose $G \in \mathcal{E}'^{p,q}(\Omega)$ and $F_1 = E(G), F_2 = B(G) - H(\overline{\partial} G)$. Then*

$$\begin{aligned}
&\operatorname{sing\,supp} F_1 \subset \operatorname{sing\,supp} G \cup \{z \in \Omega \mid A_z^* \cap \operatorname{sing\,supp} \varphi G \neq \varnothing\} \\
&\operatorname{sing\,supp} F_2 \subset \operatorname{sing\,supp} G \cup \left\{z \mid A_z^* \cap \operatorname{sing\,supp} \overline{\partial} G \neq \varnothing\right\} .
\end{aligned}$$

Notice in particular that if $\overline{\partial} G \equiv 0$ in D then

$$D \cap \operatorname{sing\,supp} F_2 = D \cap \operatorname{sing\,supp} G .$$

We cannot make the same statement about F_1. However, if $G \in \mathcal{G}^{p,q}(D)$, then both F_1 and F_2 belong to $\mathcal{E}^{p,q-1}(D)$.

Now suppose u is an extendable support function for ∂D. If $G \in \mathcal{F}^{p,q}(\overline{D})$, let $\chi_D G \in \mathcal{E}'^{p,q}(\Omega)$ denote any extension of G multiplied by the characteristic function of D. Then, by Stokes's Theorem, $\overline{\partial} \chi_D = -[\partial \dot{D}]^{0,1}$, where $[\partial D]^{0,1}$ is the (0,1) piece of $[\partial D]$, the current which consists of integrating $2n-1$ forms over ∂D. Consequently

$$\overline{\partial}\left(\chi_D G\right) = \chi_D \overline{\partial} G - [\partial D]^{0,1} \wedge G .$$

Suppose that $\overline{\partial} G \equiv 0$ in D. Then by Remark 12.5.9

$$F_1 = E\left(\chi_D G\right) \qquad \text{and} \qquad F_2 = B\left(\chi_D G\right) + H\left([\partial D]^{0,1} \wedge G\right)$$

both solve the equation $\overline{\partial} F = G$ on D. The solution F_2 is due to Henkin [104] for $q = 1$ and Lieb [152] for $q > 1$. If the support function u is biregular, then by Corollary 12.5.12, both solutions are smooth in D. We will prove in Theorem 12.7.2 that F_2 is continuously differentiable on $\overline{D}$. For the solution F_1 we can say more.

Theorem 12.5.13. *Suppose u is a biregular support function for ∂D. If $q \geqslant 1$ and $G \in \mathcal{E}^{p,q}(\overline{D})$, then $F_1 = E\left(\chi_D G\right) \in \mathcal{E}^{p,q-1}(\overline{D})$.*

Proof. We need only show that F_1 extends smoothly across ∂D. We have from (12.49)

$$F_1 = B[(1-\varphi)G] + (L+S)\left(\varphi \chi_D G\right) + H(\overline{\partial} \varphi \wedge G) \tag{12.50}$$

Let $\widetilde{G} \in \mathcal{D}^{p,q}(\Omega)$ be an extension of G. By Lemma 12.5.10 (a) $L\left[\varphi\left(1-\chi_D\right) \widetilde{G}\right] \equiv 0$ in D if $q \neq 1$. Thus on D (with $q \neq 1$)

$$L\left(\varphi \chi_D G\right) \equiv L(\varphi \widetilde{G}) \quad \text{which is smooth in } \overline{D}$$

We will prove in Proposition 12.7.4 that $L\left[\varphi\left(1-\chi_D\right)\widetilde{G}\right]$ is smooth in $\overline{D}$ even if $q=1$. Thus, assuming this result,

$$L\left(\varphi\chi_D G\right) = L(\varphi\widetilde{G}) - L\left[\varphi\left(1-\chi_D\right)G\right]$$

is smooth in $\overline{D}$ since L is regular. It follows similarly from Lemma 12.5.10 (b) that

$$S\left(\varphi\chi_D G\right) \equiv S(\varphi\widetilde{G}) \quad \text{in} \quad D$$

Consequently, $S\left(\varphi\chi_D G\right) \in \overleftarrow{C}^{p,q-1}(\overline{D})$ since S is regular. Because B and H are also regular kernels, each term in (12.50) belongs to $\overleftarrow{cm}^{p,q-1}(\overline{D})$, and the result follows. □

12.5.2 B. Solving with no Growth Restrictions and with Compact Supports

We wish to consider the following two problems on a given open subset Ω of $\mathbb{C}^n$.

No Growth Restrictions. Suppose $G \in \mathcal{E}^{p,q}(\Omega)(q \geqslant 1)$ and $\overline{\partial}G = 0$ on Ω. Find $F \in \mathcal{E}^{p,q-1}(\Omega)$ satisfying $\overline{\partial}F = G$ on Ω.

Compact Supports. Suppose $G \in \mathcal{E}'^{p,q}(\Omega)(q \geqslant 1)$ and $\overline{\partial}G = 0$ on Ω. (If $q=n$ also assume G annihilates all holomorphic p-forms on Ω.) Find $F \in \mathcal{E}'^{p,q-1}(\Omega)$ satisfying $\overline{\partial}F = G$ on Ω.

Moreover, we wish to solve these problems with a kernel. However, the case $q=1$ with no growth restrictions cannot be solved with a kernel. More precisely, a result of Grothendieck says that it is impossible to find a continuous linear map $E : \mathcal{Z}^{p,1}(\Omega) \to \mathcal{E}(\Omega)$ with $\overline{\partial} = \text{id}$ (the identity). Here $\mathcal{Z}^{p,1}(\Omega) \equiv \left\{G \in \mathcal{E}^{p,1}(\Omega) \mid \partial G = 0\right\}$. As we shall see below, in this case ($q=1$) we can solve on any slightly smaller set with a kernel (on an appropriate Ω).

The problem of solving the inhomogeneous Cauchy-Riemann equations without growth restrictions has been discussed by Hortmann [112] using kernels of a different type.

In order to solve the above two problems using the fundamental solutions constructed in Section 12.4, we will need a generating function with some additional properties.

Definition 12.5.14. Let $\Omega \subset \mathbb{C}^n$ be open. A generating function defined on $(\Omega - K) \times \Omega$ where $K \subset \Omega$ is compact is called a *support function* for Ω if for every compact set $K_1 \subset \Omega$ there is a compact set $K_2, K \subset K_2 \subset \Omega$ such that

(a) $u(\zeta, z)$ is holomorphic on K_1 for each fixed $\zeta \in \Omega \setminus K_2$

(b) $A_\zeta \cap K_1 = \varnothing$ for all $\zeta \in \Omega \setminus K_2$.

Remark 12.5.15. Condition (b) is equivalent to (b') $A_z^* \subset K_2$ for all $z \in K_1$.

Lemma 12.5.16. *Suppose Ω is a convex open subset of $\mathbb{C}^n$, and $\rho(\zeta)$ is a strictly convex exhaustion function. Then $u(\zeta) \equiv (\partial\rho/\partial\zeta)(\zeta)$ is a biregular support function for Ω.*

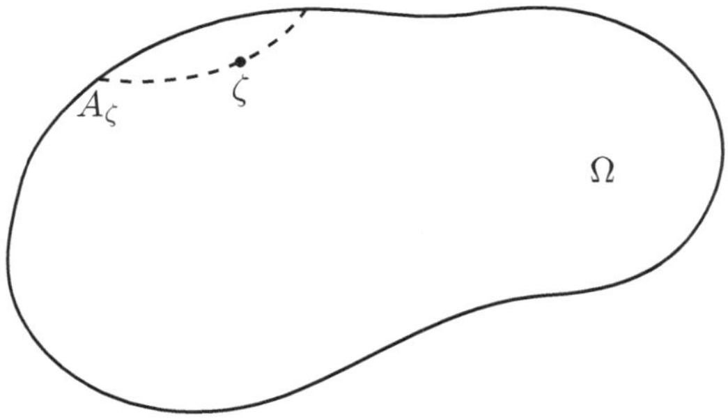

(a) Condition (b')

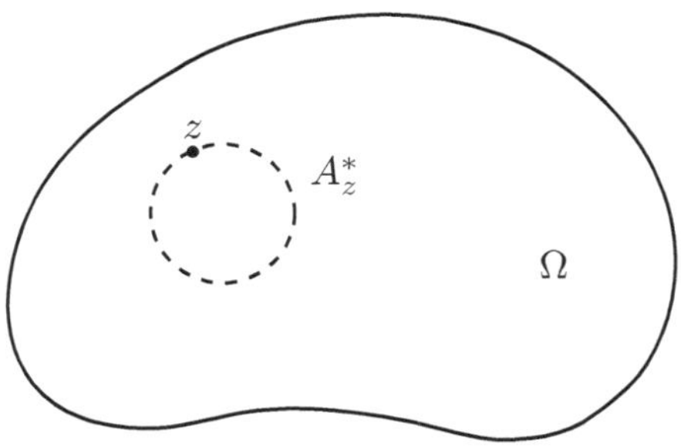

(b) Condition (b)

FIGURE 12.1

Proof. Since ρ is a strictly convex exhaustion, there exists a unique minimum, say ζ_0, for ρ in Ω. Restricting ρ to the ray through ζ_0 and $\zeta \in \Omega - \{\zeta_0\}$ we see that ρ has a derivative which is zero at ζ_0 and strictly increasing along the ray towards ζ. In particular, $d\rho(\zeta) \neq 0$. Hence $d\rho$ is non-zero on $\Omega - \{\zeta_0\}$. Now taking the compact set K in Definition 12.5.14 to be $\{\zeta_0\}$ we have that:

$$u(\zeta) \equiv \frac{\partial\rho}{\partial\zeta}(\zeta) : (\Omega - \{\zeta_0\}) \times \mathbb{C}^n \to \mathbb{C}^n$$

is a biregular support function for Ω. This is because of Proposition 12.4.8 and the convexity of the prelevel sets $\{\rho(\zeta) < c\}$. □

Remark 12.5.17. The above Lemma remains valid if we have $\rho(\zeta) < c$ for all $\zeta \in \Omega$ and $\lim_{\zeta \to \partial\Omega} \rho(\zeta) = c$ where $c \in (-\infty, +\infty)$. (The case $c = +\infty$ is just the lemma as it is.)

If u is a support function for Ω and $\varphi \in C^\infty(\Omega)$ vanishes in a neighborhood of K, then by Theorem 12.4.7

$$\begin{aligned} E(\zeta, z) &= B(\zeta, z) + \overline{\partial}[\varphi(\zeta)H(\zeta, z)] \\ &= (1 - \varphi(\zeta))B(\zeta, z) + \varphi(\zeta)(L(\zeta, z) + S(\zeta, z)) \\ &\quad + \overline{\partial}\varphi(\zeta) \wedge H(\zeta, z) \end{aligned} \tag{12.51}$$

is a fundamental solution on Ω. We will assume in addition that

$$1 - \varphi \in C_0^\infty(\Omega)\,. \tag{12.52}$$

Theorem 12.5.18. *Suppose u is a regular support function for Ω and the kernel E is defined by (12.51), where φ satisfies (12.52). Then the operator E extends to a map*

$$E : (\mathcal{E})^{p,q}(\Omega) \to (\mathcal{E})^{p,q-1}(\Omega) \quad q \geqslant 2.$$

For all $G \in \mathcal{E}^{p,q}(\Omega)(q \geq 2)$ we have the homotopy formula

$$G = \overline{\partial}EG + E\overline{\partial}G$$

In particular if $\overline{\partial}G \equiv 0$ in Ω, then $G = \overline{\partial}EG$.

If $G \in \mathcal{E}^{p,1}(\Omega)$ satisfies $\overline{\partial}G \equiv 0$ on Ω and $D \subset\subset \Omega$, then

$$G = \overline{\partial}E(\chi G) \quad \textit{in} \quad D$$

provided $\chi \in C_0^\infty(\Omega)$ is identically equal to one on a sufficiently large compact subset of Ω.

Theorem 12.5.18 enables us to solve the Cauchy-Riemann equations $\overline{\partial}F = G$ without any restriction on the growth of G near $\partial\Omega$ provided $G \in \|p, q(\Omega)$, $q \geqslant 2$. By dualizing we can solve with compact support.

Theorem 12.5.19. *Suppose u is a regular support function for Ω and the kernel E is defined by (12.51) where φ satisfies (12.52). Then if $q \leqslant n-1$ and $G \in \mathcal{E}'^{p,q}(\Omega)$, ${}'EG \in \mathcal{E}'^{p,q-1}(\Omega)$. Furthermore, we have the homotopy formula*

$$G = \overline{\partial}^t EG + {}^tE\overline{\partial}G$$

where ${}^tEG$ and ${}^tE\overline{\partial}G$ have compact support in Ω. In particular if $\overline{\partial}G \equiv 0$ then $G = \overline{\partial}' EG$ where ${}'EG$ has compact support in Ω.

Again we will state a lemma which elucidates the effect of properties (a) and (b) of a support function.

Lemma 12.5.20. *Suppose u is a support function for Ω. For every compact set $K_1 \subset \Omega$ there is a compact set $K_2 \subset \Omega$ such that if $G \in \mathcal{D}^{p,q}(\Omega)$ vanishes on K_2, then* *(a)* $q \neq 1, L(\varphi G) \equiv 0$ *on* $\overset{\circ}{K}_1$

(b) $S(\varphi G) \equiv 0$ *on* $\overset{\circ}{K}_1$.

The proof is similar to that of Lemma 12.5.10 and will be omitted.

Proof of Theorem 12.5.18 Let $D \subset\subset \Omega$ and for $K_1 = \overline{D}$ choose K_2 by Lemma 12.5.20. Then for $G \in \mathcal{E}^{p,q}(\Omega), q \geqslant 2$ define

$$EG(z) = E(\psi G)(z) \quad z \in D$$

where $\psi \in C_0^\infty(\Omega)$ is identically equal to one on $K_2 \cup K_1$. By 12.5.20, this definition is independent of the choice of ψ. Furthermore, for $z \in D$ we have by the homotopy formula

$$\begin{aligned} G(z) &= \psi(z)G(z) = \overline{\partial}E(\psi G) + E\overline{\partial}(\psi G) \\ &= \overline{\partial}EG + E\overline{\partial}G \end{aligned}$$

since $\overline{\partial}(\psi G) = \overline{\partial}\psi \wedge G + \psi\overline{\partial}G \equiv \psi\overline{\partial}G$ in K_2 and we can again apply 12.5.20. Now if $q = 1$ and $\overline{\partial}G \equiv 0$ then $\overline{\partial}(\psi G) \equiv 0$ in K_2 so

$$G = \overline{\partial}E(\psi G) \quad \text{in } D$$

That completes the proof.

Proof of Theorem 12.5.19 That ${}^tEG$ has compact support for $G \in \mathcal{E}'^{p,q}$ $q \leqslant n-1$ follows by dualizing Proposition 12.5.4. It only remains to show that ${}^tE\overline{\partial}G$ has compact support for $G \in \mathcal{E}'^{p,n-1}(\Omega)$. This reduces easily to showing that $\varphi(z){}^tL\overline{\partial}G(z)$ has compact support. Suppose $\operatorname{supp} G \subset K_1$. By Definition 12.5.14 there is a compact set K_2 such that ${}^tL(\zeta, z)$ is a holomorphic function of ζ for each $z \in \Omega - K_2$. Integration by parts now shows that

$$\varphi(z){}^tL\overline{\partial}G(z) = \varphi(z)\int {}^tL(\zeta, z) \wedge \overline{\partial}G(\zeta) \equiv 0$$

for $z \in \Omega - K_2$. That completes the proof.

12.5.3 The E.E. Levi Problem

The E.E. Levi Problem is to show that a strictly pseudoconvex domain is a domain of holomorphy. Here we will present an elementary solution to this problem using the technique developed in this book. Actually, instead of assuming that the domain D is strictly pseudoconvex, we will only assume that ∂D has a strong local support function (see Definition 12.5.1).

Theorem 12.5.21. *Suppose* $D \subset\subset \mathbb{C}^n$ *has an infinitely differentiable boundary, and that there is a strong local support function for* ∂D. *Then* D *is a domain of holomorphy.*

Suppose ρ is a defining function for D; i.e., suppose ρ is a C^∞ function defined in a neighborhood of $\overline{D}$, $D = \{z \mid \rho(z) < 0\}$, and $d\rho(z) \neq 0$ for $z \in \partial D$. Define $u(\zeta, z) = (u_1(\zeta, z), \ldots, u_n(\zeta, z))$ where

$$u_j(\zeta, z) = 2\frac{\partial \rho}{\partial \zeta_j}(\zeta) + \sum_{k=1}^{n} \frac{\partial^2 \rho}{\partial \zeta_j \partial \zeta_k}(\zeta)\,(z_k - \zeta_k)\,. \tag{12.53}$$

The following result is essentially due to E.E. Levi.

Lemma 12.5.22. *If D is strictly pseudoconvex, then $u(\zeta, z)$ defined by (12.53) is a strong local biregular support function for ∂D, when restricted to a suitably small neighborhood V of $\Delta \cap (\partial D \times \partial D)$.*

Proof. By Taylor's Theorem

$$\rho(z) = \rho(\zeta) - \mathrm{Re}(u(\zeta, z) \cdot (\zeta - z)) + r(\zeta, z) \tag{12.54}$$

where

$$r(\zeta, z) = \sum_{j,k=1}^{n} \frac{\partial^2 \rho}{\partial \zeta_j \partial \overline{\zeta}_k}(\zeta)\,(z_j - \zeta_j)\,(\overline{z}_k - \overline{\zeta}_k) + O\left(|\zeta - z|^3\right)$$

Since D is strictly pseudoconvex, the Levi form

$$\sum_{j,k=1}^{n} \frac{\partial^2 \rho}{\partial \zeta_j \partial \overline{\zeta}_k}(\zeta) w_j \overline{w}_k$$

is positive definite when restricted to $H_\zeta(\partial D)$, the holomorphic tangent space to ∂D. However, $H_\zeta(\partial D) = \{w \mid \Sigma\,(\partial\rho/\partial\zeta_j)\,(\zeta) w_j = 0\}$ is equal to the tangent space at ζ of the manifold $A_\zeta = \{z \mid u(\zeta, z) \cdot (\zeta - z) = 0\}$. Consequently, there is an $\epsilon > 0$ such that $r(\zeta, z) \geqslant C|\zeta - z|^2$ if $z \in A_\zeta$ and $|\zeta - z| < \epsilon$. Thus from (12.54) we see that $\rho(z) - \rho(\zeta) > 0$ if $(z, \zeta) \in A$ and $0 < |z - \zeta| < \epsilon$. This immediately implies condition (b') of Definition 12.5.1. Condition (a) is evident. Finally, u is biregular because of the second remark after Definition 12.5.1. $\square$

Proof of Theorem 12.5.21. Let $\zeta \in \partial D$. Then $f(z) = [u(\zeta, z) \cdot (\zeta - z)]^{-1}$ is a holomorphic function in $U \cap D$, where U is a neighborhood of ζ and $|f(z)| \to \infty$ as $z \to \zeta$. A standard proof (see J.J. Kohn [123]) will then show that D is a domain of holomorphy, provided we know that $\overline{\partial} : \mathcal{E}^{0,0}(\overline{D}) \to \mathcal{Z}^{0,0}(\overline{D})$ had closed range with finite codimension $\left(Z^{p,q}(\overline{D}) = \left\{G \in \mathcal{E}^{p,q}(\overline{D}) \mid \overline{\partial} G \equiv 0\right\}\right)$.

It will be our purpose to prove the above fact about the operator $\overline{\partial} : \mathcal{E}^{0,0}(\overline{D}) \to Z^{0,1}(\overline{D})$ by using the technique of this book to construct a parametrix for $\overline{\partial}$. That is, we will find a current $P \in$ on $n, n-1(\Omega \times \Omega)$, where Ω is a neighborhood of $\overline{D}$, such that

$$\overline{\partial} P = [\Delta] - K$$

where

$$K \in \mathcal{E}^{n,n}(\Omega \times \Omega)$$

From this point the statements about the range $\overline{\partial}$ follow easily.

Suppose u is a strong local support function defined in a neighborhood V of $\Delta \cap (\partial D \times \partial D)$. For some $\epsilon > 0, V$ contains the compact $K = \{(\zeta, z) \in \partial D \times \overline{D} \mid |\zeta - z| = \epsilon\}$, and $K \cap A = \varnothing$. Since A is closed, K has an open neighborhood $U \subset\subset V$, such $\overline{U} \cap A = \varnothing$. If ρ is a defining function for D we can take this neighborhood to be of the form

$$U = \{(\zeta, z) \mid |\rho(\zeta)| < \delta, \rho(z) < \delta, \alpha < |\zeta - z| < \beta\}$$

where α, β, and δ are positive. We may also assume that

$$V \supset \{(\zeta, z) \mid |\rho(\zeta)| < \delta, \rho(z) < \delta, |\zeta - z| < \beta\} \tag{12.55}$$

Set

$$\Omega = \{z \mid \rho(z) < \delta\}$$

and choose

$$\begin{gathered} \psi \in C_0^\infty(B(0,\beta)), \quad \psi \equiv 1 \quad \text{on} \quad \overline{B(0,\alpha)}, \\ \varphi \in C^\infty(\Omega), \quad 1 - \varphi \in C_0^\infty(D), \quad \varphi(\zeta) \equiv 0 \quad \text{on} \quad \{\zeta \mid \rho(\zeta) \leqslant -\delta\} \end{gathered} \tag{12.56}$$

Then

$$P(\zeta, z) = \psi(\zeta - z)[B(\zeta, z) + \overline{\partial}(\varphi(\zeta) H(\zeta, z))] \tag{12.57}$$

is a well defined current in $\mathcal{D}'^{n,n-1}(\Omega \times \Omega)$, since $\operatorname{supp}[\psi(\zeta - z)\varphi(\zeta)] \subset V$ by (12.55). Furthermore,

$$\overline{\partial} P = [\Delta] - K \tag{12.58}$$

where

$$K = -\overline{\partial}\psi \wedge [B + \overline{\partial}(\varphi H)] \tag{12.59}$$

Since

$$\begin{aligned} \operatorname{sing\,supp} K &\subset \operatorname{supp}(\varphi(\zeta)\overline{\partial}\psi(\zeta - z)) \cap A \\ &\subset \overline{U} \cap A \\ &= \varnothing \end{aligned}$$

$K \in \mathcal{S}^{n,n}(\Omega \times \Omega)$. Differentiating (12.58) we find that

$$\overline{\partial} K \equiv 0 \tag{12.60}$$

For $G \in \mathcal{E}^{p,q}(\overline{D})$ we define

$$\begin{aligned} P_1 G &= P\,(\chi_D G) \\ K_1 G &= K\,(\chi_D G) \end{aligned}$$

□

Proposition 12.5.23. *We have that*

(a) $P_1 : \mathcal{E}^{p,q}(\overline{D}) \to \mathcal{E}^{p,q-1}(\overline{D})$

(b) $K_1 : E^{p,q}(\overline{D}) \to \mathcal{E}^{p,q}(\overline{D})$ *is compact*

(c) $K_1 : Z^{p,q}(\overline{D}) \to Z^{p,q}(\overline{D})$

(d) *For* $G \in \mathcal{E}^{p,q}(\overline{D}), q \geqslant 1$, *we have*

$$\overline{\partial} P_1 G + P_1 \overline{\partial} G = G - K_1 G .$$

In particular, if $\overline{\partial} G \equiv 0$ *in* $\overline{D}$ *then*

$$\overline{\partial} P_1 G = G - K_1 G .$$

Corollary 12.5.24. *For* $q \geqslant 1, \overline{\partial} : \mathcal{E}^{p,q-1}(\overline{D}) \to Z^{p,q}(\overline{D})$ *has a closed range of finite codimension.*

Proof. The range of $\overline{\partial}$ contains the range of $\overline{\partial} P_1 : Z^{p,q}(\overline{D}) \to Z^{p,q}(\overline{D})$ or, equivalently by part (d) of Proposition 12.5.23, the range of $I - K_1 : Z^{p,q}(\overline{D}) \to Z^{p,q}(\overline{D})$. By part (b) K_1 is compact. Consequently, the Fredholm Theorem for Frechet spaces $I - K_1$ has closed range of finite codimension. It follows that the same is true for $\overline{\partial}$. □

Corollary 12.5.24 completes the proof of Theorem 12.5.21.

Proof of Proposition 12.5.23. By (12.57)

$$\begin{aligned} P(\zeta, z) =& \psi(\zeta - z)[(1 - \varphi(\zeta))B(\zeta, z) + \varphi(\zeta)(L(\zeta, z) \\ &+ S(\zeta, z)) + \overline{\partial}\varphi(\zeta) \wedge H(\zeta, z)] \end{aligned}$$

By (12.56)

$$P(\chi_D G) = (\psi B)((1 - \varphi)G) + (\psi H)(\overline{\partial}\varphi \wedge G) + (\psi L)(\varphi \chi_D G) + (\psi S)(\varphi \chi_D G)$$

By property (b') of Definition 12.5.1 and Theorem 12.4.9

$$(\psi S)(\varphi(1 - \chi_D) G) \equiv 0$$

in D. Thus $(\psi S)(\varphi \chi_D G) \equiv (\psi S)(\varphi G)$ in D so $(\psi S)(\varphi \chi_D G) \in \mathcal{E}^{p,q-1}(\overline{D})$. By property (a) of Definition 12.5.1 L is of degree $n - 1$ in $d\overline{\zeta}$ for $\zeta \notin D$ and $z \in D$. Thus if $q \geqslant 2$ $(\psi L)(\varphi \chi_D G) \equiv (\psi L)(\varphi G)$ in D and again $(\psi L)(\varphi \chi_D G) \in \mathcal{E}^{p,q-1}(\overline{D})$. The other terms in (12.81) are obviously smooth in $\overline{D}$. For $q = 1$ we refer to Proposition 12.7.4 from which it follows that $(\psi L)(\varphi(1 - \chi_D) G)$ is smooth in $\overline{D}$. Thus $(\psi L)(\varphi \chi_D G) = (\psi L)(\varphi G) - (\psi L)(\varphi(1 - \chi_D) G)$ is smooth in $\overline{D}$. This proves part (a).

Since $K \in \mathcal{E}^{n,n}(\Omega \times \Omega)$, part (b) is clear.

By the general homotopy formula and (12.60), we see that

$$\overline{\partial}KF = K\overline{\partial}F$$

for all $F \in \mathcal{E}'^{*}(\Omega)$. Thus if $\overline{\partial}G \equiv 0$ in D and $G \in \mathcal{E}^{p,q}(\overline{D})$,

$$\overline{\partial}K\left(\chi_D G\right) = -K\left([\partial D]^{0,1} \wedge G\right)$$

By (12.59)

$$-K\left([\partial D]^{0,1} \wedge G\right) = \left(\overline{\partial}\psi \wedge (L+S)\right)\left([\partial D]^{0,1} \wedge G\right)$$

Using the properties of a support function, it follows that $K\left([\partial D]^{0,1} \wedge G\right) \equiv 0$ in D. Thus $\overline{\partial}K_1 G \equiv 0$ if $\overline{\partial}G \equiv 0$, and part (c) follows.

If $G \in \mathcal{E}^{p,q}(\overline{D})$, then by the homotopy formula and (12.58) it follows that

$$\chi_D G - K\left(\chi_D G\right) = \overline{\partial}P\left(\chi_D G\right) + P\overline{\partial}\left(\chi_D G\right) \tag{12.61}$$

Now $\overline{\partial}\left(\chi_D G\right) = \chi_D \overline{\partial}G - [\partial D]^{0,1} \wedge G$, and

$$P\left([\partial D]^{0,1} \wedge G\right) = [\psi(L+S)]\left([\partial D]^{0,1} \wedge G\right)$$

It follows from the properties of a support function that $P\left([\partial D]^{0,1} \wedge G\right) \equiv 0$ in D, provided $q \geqslant 1$. Thus restricting (12.61) to D we have

$$G - K_1 G = \overline{\partial}P_1 G + P_1\overline{\partial}G$$

which is (d). □

Remark 12.5.25. The boundary behavior that was quoted from Section 12.7 in the above proof can be circumvented by introducing a continuous linear operator which extends $G \in \mathcal{E}^{p,q}(\overline{D})$ to $\widetilde{G} \in \mathcal{D}^{p,q}(\Omega)$ and then alternatively defining $P_1 G = P\widetilde{G}$ and $K_1 G = K\widetilde{G}$.

12.6 Jump Behavior of the Bochner-Martinelli Transform

In this section we will discuss the boundary behavior of the Bochner-Martinelli transform of a form on a hypersurface. The results generalize the Plemelj jump relations for the Cauchy transform in $\mathbb{C}$ and are analogous to the jump relations for double and single-layer potentials in $\mathbb{R}^n$.

First, we will briefly discuss the induced Cauchy-Riemann complex. It was originally introduced in the fundamental work of Kohn-Rossi [131]. Let M denote a closed hypersurface defined implicitly by $M \equiv \{z \in \Omega \mid \rho(z) = 0\}$

where Ω is an open subset of $\mathbb{C}^n$ and $\rho \in C^\infty(\Omega)$ with $|d\rho| \equiv 1$ on M. Let $\Omega^\pm \equiv \{z \in \Omega \mid \pm\rho(z) > 0\}$, and give M the orientation induced as the boundary of Ω^-, so that the pair (Ω^-, M) is an oriented submanifold with boundary.

The *bundle of* p, q *forms on* M denoted by $\Lambda^{p,q}|_M$ is, by definition, the restriction of the bundle $\Lambda^{p,q}$ of ambient p, q forms to the submanifold M. Thus a section f of $\Lambda^{p,q}|_M$ is obtained from an ambient p, q form by restricting the coefficients of the p, q form to M. Let $N^* \equiv \partial\rho$ and $\bar{N}^* = \bar{\partial}\rho$. Suppose a 1,0 vector field N is given along M with the property that $N \rfloor N^* = 1$. For example, N can be chosen dual to N^* using the standard inner product on $\mathbb{C}^n$. Then each section f of $\Lambda^{p,q}|_M$ has the (unique) decomposition

$$f = \bar{N} \lrcorner (\bar{N}^* \wedge f) + \bar{N}^* \wedge (\bar{N} \lrcorner f)$$

or

$$f = f_t + \bar{N}^* \wedge f_n$$

where $f_t \equiv \bar{N} \lrcorner \left(\bar{N}^* \wedge f\right)$ is called the *complex tangential part of* f and $f_n \equiv \bar{N} \lrcorner f$ is called the complex normal part of f. If $f = f_t$ (i.e., $\bar{N} \lrcorner f = 0$) then f is called a *complex tangential* p, q *form on* M, or a *boundary* p, q *form on* M. The sub bundle of $\Lambda^{p,q}|_M$ whose sections are complex tangential will be denoted by $\Lambda^{p,q}(M)$. The space of smooth sections of this sub bundle (i.e., the space of boundary p, q forms) will be denoted $\mathcal{E}^{p,q}(M)$.

Definition 12.6.1. The induced Cauchy-Riemann operator

$$\bar{\partial}_b : \mathcal{E}^{p,q}(M) \to \mathcal{E}^{p,q+1}(M)$$

is defined as follows. Given $f \in \mathcal{E}^{p,q}(M)$, choose $\tilde{f} \in \mathcal{E}^{p,q}(\Omega)$ such that $\tilde{f}_t = f$. Then

$$\bar{\partial}_b f = (\bar{\partial}\tilde{f})_t$$

the complex tangential part of $\bar{\partial}\tilde{f}$.

Note that $\bar{\partial}_b f$ does not depend on the extension $\tilde{f}$; since if $\tilde{f}_1$ and $\tilde{f}_2$ denote two different extensions of f then $\tilde{f}_1 - \tilde{f}_2 = \rho g_1 + \bar{\partial}\rho \wedge g_2$ where $g_1 \in \mathcal{E}^{p,q}(\Omega)$ and $g_2 \in \mathcal{E}^{p,q-1}(\Omega)$. Hence $\bar{\partial}\tilde{f}_1 - \bar{\partial}\tilde{f}_2 = \bar{\partial}\rho \wedge \left(g_1 - \bar{\partial}g_2\right)$ on M so that $\bar{\partial}\tilde{f}_1$ and $\bar{\partial}\tilde{f}_2$ have the same complex tangential part on M. Of course, $\bar{\partial}_b^2 f = 0$ since $\bar{\partial}^2 \tilde{f} = 0$. Thus

$$\mathcal{E}^{p,0}(M) \xrightarrow{\bar{\partial}_b} \mathcal{E}^{p,1}(M) \to \cdots \to \mathcal{E}^{p,n-2}(M) \xrightarrow{\bar{\partial}_b} \mathcal{E}^{p,n-1}(M) \to 0$$

is a differential complex, called the *induced Cauchy-Riemann complex.*

The important mapping $\mathcal{E}^{p,q}(\Omega) \to \mathcal{E}^{p,q}(M)$ defined by sending f to f_t (the complex tangential part of f on M) will also be denoted by τ, so that $\tau f = f_t$. (The notation τf is more convenient than f_t if f has a complicated expression.) As an immediate consequence of Definition 12.6.17.2 we have that for $f \in \mathcal{E}^{p,q}(\Omega)$

$$(\bar{\partial}f)_t = \bar{\partial}_b f_t \quad \text{or} \quad \tau\bar{\partial}f = \bar{\partial}_b \tau f.$$

It is useful to reinterpret the equation $\bar{\partial}_b f = g$ on M as a current equation on Ω. This is accomplished in Lemma 12.6.2 below. Let $\chi^{\pm}$ denote the characteristic function of $\Omega^{\pm}$. Since M is the oriented boundary of Ω^- (by hypothesis),

$$d\chi^+ = -d\chi^- = [M] \quad \text{on } \Omega$$

The splitting $d = \partial + \bar{\partial}$ induces the decomposition

$$[M] = d\chi^+ = \partial\chi^+ + \bar{\partial}\chi^+ = [M]^{1,0} + [M]^{0,1}$$

Let σ denote the surface measure on M. Then

$$[M] = d\chi^+ = \sigma d\rho$$

if we assume that $|d\rho| \equiv 1$ on M. Therefore

$$\begin{aligned} [M]^{1,0} &= \partial\chi^+ = \sigma\partial\rho \\ [M]^{0,1} &= \bar{\partial}\chi^+ = \sigma\bar{\partial}\rho \end{aligned}$$

Now given $f \in \mathcal{E}^{p,q}(M)$ we will let $[M]^{0,1} \wedge f$ denote the current $\sigma\bar{\partial}\rho \wedge f$. Let $\tilde{f} \in \mathcal{E}^{p,q}(\Omega)$ satisfy $\tilde{f}_t = f$. Then the current $[M]^{0,1} \wedge \tilde{f}$ equals $[M]^{0,1} \wedge f$ and obviously does not depend on the extension of f. Thus we have a well-defined map $\mathcal{E}^{p,q}(M) \to \mathcal{D}_M'^{p,q+1}(\Omega)$ where $\mathcal{D}_M'^{p,q}(\Omega)$ denotes the space of currents of bidegree p, q on Ω which are supported in M. This map is an inclusion. That is, given $f \in \mathcal{E}^{p,q}(M)$, if $[M]^{0,1} \wedge f = 0$ then $f = 0$. This is because $[M]^{0,1} \wedge f = \sigma\bar{\partial}\rho \bigwedge f$ and $f \in \mathcal{E}^{p,q}(M)$ vanishes if and only if $\bar{\partial}\rho \wedge f$ vanishes on M (note $f = \bar{N} \lrcorner (\bar{\partial}\rho \wedge f)$).

Frequently, it is very convenient to replace the equation $\bar{\partial}_b f = g$ on M by an ambient current equation as follows.

Lemma 12.6.2. *Suppose $f \in \mathcal{E}^{p,q}(M)$ and $g \in \mathcal{E}^{p,q+1}(M)$. The equation $\bar{\partial}_b f = g$ is valid on M if and only if the equation $\bar{\partial}\left(f \wedge [M]^{0,1}\right) = g \wedge [M]^{0,1}$ is valid on Ω.*

Proof. Let $\tilde{f}$ denote an extension of f

$$\begin{aligned} \bar{\partial}\left(f \wedge [M]^{0,1}\right) = \bar{\partial}\left(\tilde{f} \wedge [M]^{0,1}\right) &= \bar{\partial}\left(\tilde{f} \wedge \bar{\partial}\chi^+\right) = \bar{\partial}\tilde{f} \wedge \bar{\partial}\chi^+ \\ &= \bar{\partial}\tilde{f} \wedge \sigma\bar{\partial}\rho = (\bar{\partial}\tilde{f})_t \wedge \sigma\bar{\partial}\rho \\ &= \bar{\partial}_b f \wedge [M]^{0,1} \end{aligned}$$

Since $\bar{\partial}_b f = g$ if and only if $\bar{\partial}_b f \wedge [M]^{0,1} = g \wedge [M]^{0,1}$ the proof is complete. □

Remark 12.6.3. The above discussion of the induced Cauchy-Riemann complex can be easily extended from the case of $f \in \mathcal{E}^{p,q}(M)$ smooth to more general f as follows. Let $\mathcal{D}'^{p,q}(M)$ denote the distribution sections of the vector bundle $\Lambda^{p,q}(M)$ over the manifold M.

First approach. Of course, the differential operator

$$\bar{\partial}_b : \mathcal{E}^{p,q}(M) \to \mathcal{E}^{p,q+1}(M)$$

also acts on distribution sections and we have

$$\bar{\partial}_b : \mathcal{D}'^{p,q}(M) \to \mathcal{D}'^{p,q+1}(M).$$

Moreover, given $f \in \mathcal{D}'^{p,q}(M)$, we may also define $[M]^{0,1} \wedge f \equiv [M]^{0,1} \wedge \tilde{f}$, where $\tilde{f}$ is an ambient extension of f independent of the normal variable. In fact, if σ denotes surface measure on M and f is a distribution on M we need only define $\sigma \cdot f$. First localize. Second, by using an ambient diffeomorphism we can assume M is a hyperplane, say given by $t_m = 0$ in $\mathbb{R}^m$. Let $t = (t', t_m)$ with $t' \in \mathbb{R}^{m-1}$. Then $\sigma = \delta_0(t_m)$ and $f(t')$ is a distribution on $\mathbb{R}^{m-1}$. It is customary to denote the distribution $\tilde{f}(t) \equiv f(t') \otimes 1$ by $f(t')$. Now $\delta_0(t_n)\tilde{f}(t) \equiv f(t') \otimes \delta_0(t_n)$ is well-defined and thus $\sigma \cdot f$ is well defined (one can easily check that this definition of $\sigma \cdot f$ is independent of the choices made). Obviously $[M]^{0,1} \wedge f = 0$ if and only if $f = 0$; and essentially the same proof as in the smooth case shows that the important Lemma 12.6.2 remains valid with $f \in \mathcal{D}'^{p,q}(M)$.

Second approach. The map $\tau : \mathcal{E}^{n-p,n-1-q}(\Omega) \to \mathcal{E}^{n-p,n-q-1}(M)$ has adjoint $\tau' : \mathcal{D}'^{p,q}(M) \to \mathcal{D}'^{p,q+1}_M(\Omega)$ where $\mathcal{D}'^{*}_M(\Omega)$ denotes the space of currents on Ω which are supported in M. If $f \in \mathcal{E}^{p,q}(M)$ then

$$\tau'(f) = [M]^{0,1} \wedge f$$

because $(\tau'(f), \tilde{g})_M = \int_M (f \wedge \tilde{g}_t)^{n,n-1} = ([M]^{0,1}, \tilde{f} \wedge \tilde{g}) = ([M]^{0,1} \wedge f, \tilde{g})$ for $\tilde{g} \in \mathcal{D}^{n-p,n-1-q}(\Omega)$. Consequently, we may define $[M]^{0,1} \wedge f$ to be $\tau'(f)$ for all $f \in \mathcal{D}'^{p,q}(M)$. The dual of the equation $\bar{\partial}_b \tau = \tau \bar{\partial}$ is $-\tau' \hat{\partial}_b = \bar{\partial}\tau'$, which applies to $f \in \mathcal{D}'^{p,q}(M)$ to yield

$$\bar{\partial}\left(f \wedge [M]^{0,1}\right) = \bar{\partial}_b f \wedge [M]^{0,1}$$

Remark 12.6.4. The induced boundary complex can also be described intrinsically (see Polking-Wells [173] for more details) as follows. The complexified cotangent bundle to M denoted $\mathcal{C}T^*(M)$ decomposes orthogonally into $T^{*1,0}(M) \oplus T^{*0,1}(M) \oplus F^*$. Let $\Lambda^{p,q}(M)$ denote the bundle

$$\Lambda^p\left(T^{*1,0} \oplus F^*\right) \otimes \Lambda^q\left(T^{*0,1}M\right)$$

and define $\bar{\partial}_b$ mapping sections φ of $\Lambda^{p,q}(M)$ into sections of $\Lambda^{p,q+1}(M)$ by letting $\bar{\partial}_b\varphi$ denote the orthogonal projection of $d\varphi$ onto $\Lambda^{p,q+1}(M)$. This intrinsic $\bar{\partial}_b$ complex is isomorphic but not equal to the $\bar{\partial}_b$-complex defined previously.

The Cauchy transform of a function f defined on a curve in the complex plane $\mathcal{C}$ can be generalized as follows.

Definition 12.6.5. Given $f \in \mathcal{D}^{p,q}(M)$, the Bochner-Martinelli transform of f is

$$B\left([M]^{0,1} \wedge f\right) \in \mathcal{D}'^{p,q}(\Omega)$$

Since B is a convolution operator with a kernel which is smooth away from the origin, it follows that $B\left([M]^{0,1} \wedge f\right) \in \mathcal{E}^{p,q}(\Omega \setminus M)$. Furthermore for $z \in \Omega - M$ we can write

$$B\left([M]^{0,1} \wedge f\right)(z) = -\int_M [B(\zeta, z) \wedge f(\zeta)]^{n,n-1}$$

where $[B(\zeta, \mathrm{z}) \wedge f(\zeta)]^{n,n-1}$ is the piece of $B \wedge f$ which is of bidegree $n, n-1$ in ζ. Let

$$B\left([M]^{0,1} \wedge f\right)\big|_{\Omega^\pm}$$

denote the Bochner-Martinelli transform of f restricted to $\Omega^\pm$ respectively. Then

$$B\left([M]^{0,1} \wedge f\right)\big|_{\Omega^\pm} \in \mathcal{E}^{p,q}\left(\Omega^\pm\right)$$

In fact, the Bochner-Martinelli transform is smooth up to M from each side, and the complex tangential boundary values can be computed in terms of a Cauchy-principal value integral on M, defined for $z \in M$ by

$$B_b f(z) = \tau \lim_{\epsilon \to 0^+} -\int_{\substack{|\zeta - z| > \epsilon \\ \zeta \in M}} [B(\zeta, z) \wedge f(\zeta)]^{n,n-1}$$

whenever the limits exists.

Theorem 12.6.6. *Suppose $f \in \mathcal{D}^{p,q}(M)$.*

(a) $B\left([M]^{0,1} \wedge f\right)\big|_{\Omega^\pm}$ *has an infinitely differentiable extension to* $\bar{\Omega}^\pm$.

(b) For each $z \in M$ the Cauchy principal value integral $B_b f(z)$ exists and $B_b f \in \mathcal{E}^{p,q}(M)$.

(c) If we define

$$B^\pm f = \tau\left(B\left([M]^{0,1} \wedge f\right)\right)\big|_{\Omega^\pm}$$

then $B^\pm f \in \mathcal{E}^{p,q}(M)$, and $B^\pm f$ and $B_b f$ are related by the Plemelj jump relations which can be expressed either as

$$B^+ f - B^- f = f \tag{12.62}$$

$$B^+ f + B^- f = 2B_b f$$

or as

$$B^\pm f = B_b f \pm \frac{1}{2} f \tag{12.63}$$

The proof will proceed in two stages. First, we will prove that $B\left([M]^{0,1} \wedge f\right)|_{\Omega^\pm}$ has a C^∞ extension to $\bar{\Omega}^\pm$ and the jump relation 12.62, by using a C^∞ version of the Cauchy-Kowalewski theorem for $\bar{\Box} \equiv \bar{\partial}\bar{\partial}^* + \bar{\partial}^*\bar{\partial}$ in conjunction with a current version of Green's identity for $\Box$. The Cauchy data can be expressed most naturally as follows.

Lemma 12.6.7. *Let* $Cf \equiv \left(f_t, f_n, \left(\bar{\partial}^* f\right)_t, (\bar{\partial} f)_n\right)$ *define*

$$C : \mathcal{E}^{p,q}(\Omega) \to \mathcal{E}^{p,q}(M) \oplus \mathcal{E}^{p,q-1}(M) \oplus \mathcal{E}^{p,q-1}(M) \oplus \mathcal{E}^{p,q}(M)$$

The map C *is onto and* $Cf = 0$ *if and only if* f *vanishes to second order on* M. *Thus the fourtuple* $\left(f_t, f_n, \left(\bar{\partial}^* f\right)_t, (\bar{\partial} f)_n\right)$ *forms a representation of the Cauchy data of* f *on* M *for* $\bar{\square}$.

Proof. Of course, $f \in \mathcal{E}^{p,q}(\Omega)$ vanishes to second order on M if and only if $f = \rho^2 g$ for some $g \in \mathcal{E}^{p,q}(\Omega)$.

Suppose f vanishes to second order on M. Let $\star$ denote the Hodge star operator. Then $f = \rho^2 g$ and $\star f = \rho^2 \star g$. Since

$$\bar{\partial}^* f = - \star \partial \star f = - \star \partial \left(\rho^2 \star g\right) = - \star \left(2\rho\partial\rho \wedge \star g + \rho^2 \partial \star g\right)$$

and

$$\bar{\partial} f = 2\rho\bar{\partial}\rho \wedge f + \rho^2 \bar{\partial} f$$

both vanish when (the coefficients are) restricted to M it follows that $Cf = 0$.

Suppose $Cf = 0$. Since $f = f_t + \bar{\partial}\rho \wedge f_n$ on M, f must vanish on M. Therefore $f = \rho h$ for some $h \in \mathcal{E}^{p,q}(\Omega)$. Since the operation sending h to $\star[\partial\rho \wedge \star h]$ is the adjoint of the operation of wedging with $\bar{\partial}\rho$ it must equal the operation of contracting with N. Therefore on M the expression

$$\bar{\partial} * f = - \star \partial \star (\rho h) = - \star [\partial\rho \wedge \star h] + \rho\bar{\partial} * h$$

reduces to $\bar{\partial}^* f = -N \rfloor h$. Consequently, on M,

$$\left(\bar{\partial}^* f\right)_t = -h_n \quad \text{if} \quad f = \rho h \tag{12.64}$$

(Note $\bar{\partial}^* f$ is purely tangential on M). Also on M

$$\bar{\partial} f = \bar{\partial}(\rho h) = \bar{\partial}\rho \wedge h + \rho\bar{\partial} h = \bar{\partial}\rho \wedge h_t$$

so that

$$(\bar{\partial} f)_n = h_t \quad \text{if} \quad f = \rho h \tag{12.65}$$

Finally, to prove that C is onto, suppose $(\psi_1, \psi_2, \psi_3, \psi_4)$ is given and let $(\tilde{\psi}_1, \tilde{\psi}_2, \tilde{\psi}_3, \tilde{\psi}_4)$ denote smooth extensions to Ω. First let $f = \tilde{\psi}_1 + \bar{\partial}\rho \wedge \tilde{\psi}_2$. Then $f_t = \psi_1$ and $f_n = \psi_2$. Therefore we may assume that $\psi_1 = \psi_2 = 0$. In this case let $f \equiv \rho\left(\tilde{\psi}_4 + \bar{\partial}\rho \wedge \psi_3\right)$. Then $f_t = 0, f_n = 0, (\bar{\partial} * f)_t = \psi_3$, and $(\bar{\partial} f)_n = \psi_4$ by (12.64) and (12.65). □

Remark 12.6.8. The tangential part of the Cauchy data, $\left(f_t, \left(\bar{\partial}^* f\right)_t\right)$ is called the $\bar{\partial}$-Dirichlet data and the normal part of the Cauchy data $\left(f_n, (\bar{\partial} f)_n\right)$ is called the $\bar{\partial}$-Neumann data.

The next result will be referred to as the C^∞ version of the CauchyKowalewski theorem for $\bar{\square} = \bar{\partial}\bar{\partial}^* + \bar{\partial} * \bar{\partial}$.

Theorem 12.6.9. *Given $\psi_1 \in \mathcal{E}^{p,q}(M), \psi_2 \in \mathcal{E}^{p,q-1}(M), \psi_3 \in \mathcal{E}^{p,q-1}(M)$, $\psi_4 \in \mathcal{E}^{p,q}(M)$ and $g \in \mathcal{E}^{p,q}(\Omega)$ there is an $f \in \mathcal{E}^{p,q}(\Omega)$ such that*

$$Cf = (\psi_1, \psi_2, \psi_3, \psi_4) \tag{12.66}$$

and

$$\overline{\Box} f - g \text{ vanishes to infinite order on } M. \tag{12.67}$$

Furthermore, if f_1 and f_2 both satisfy (12.66) and (12.67) then $f_1 - f_2$ vanishes to infinite order on M.

Remark 12.6.10. One can obtain, as an easy corollary, the C^∞ Cartan-Kähler theorem for $\bar\partial$ with the initial surface a hypersurface. See Lemma 12.6.15 below.

Proof. Conditions (12.66) and (12.67) determine uniquely all of the derivatives of f on M. Thus the second part of the theorem follows. On the other hand, once we know all of the derivatives of f on M, the Whitney extension theorem assures us of the existence of $f \in \mathcal{E}^{p,q}(M)$ with these derivatives, so the first part follows as well. □

Green's identity for $\bar\Box$ on Ω^+ can be stated in the following current form, with the four layers involving the four pieces of the Cauchy data. Let σ denote surface measure on M.

Lemma 12.6.11. *Suppose $F \in \mathcal{E}^{p,q}(\Omega)$. Then on Ω:*

$$\begin{aligned}\bar\Box\left(\chi^+ F\right) - \chi^+ \bar\Box F = \bar\partial * \left([M]^{0,1} \wedge F_t\right) - \bar\partial\left(F_n \cdot \sigma\right) \\ + [M]^{0,1} \wedge \left(\bar\partial^* F\right)_t - (\bar\partial F)_n \cdot \sigma\end{aligned} \tag{12.68}$$

Proof.

$$\bar\partial\left(\chi^+ F\right) = \chi^+ \bar\partial F + \bar\partial\chi^+ \wedge F = \chi^+ \bar\partial F + [M]^{0,1} \wedge F \tag{12.69}$$

while, because $\partial\chi^+ = \partial\rho \cdot \sigma$,

$$\begin{aligned}\bar\partial^*\left(\chi^+ F\right) &= - \star \partial \star \left(\chi^+ F\right) \\ &= \chi^+ \bar\partial^* F - \star\partial\rho \star F \cdot \sigma \\ &= \chi^+ \bar\partial^* F - (X \lrcorner F)\sigma \\ &= \chi^+ \bar\partial^* F - F_n \cdot \sigma.\end{aligned} \tag{12.70}$$

Repeated use of (12.69) and (12.70) implies (12.68) above. □

Now we can complete stage one of the proof of Theorem 12.6.6 Assume $f \in \mathcal{D}^{p,q}(M)$. By the C^∞ version of the Cauchy-Kowalewski theorem for $\overline{\Box}$ there exists $F \in \mathcal{E}^{p,q}(\Omega)$ such that:

$$F_t = f, \quad F_n = 0, \quad \left(\bar\partial^* F\right)_t = 0, \quad \text{and} \quad (\bar\partial F)_n = 0$$

and

$$\bar{\square} F \text{ vanishes to infinite order on } M. \tag{12.71}$$

Using Green's identity we have:

$$\bar{\square}\left(\chi^{+} F\right)=\chi^{+} \bar{\square} F+\bar{\partial} *\left([M]^{0,1} \wedge f\right) \tag{12.72}$$

since $F_t = f$ is the only non-zero piece of the Cauchy data for F. Because of (12.71), the term

$$\chi^{+} \bar{\square} F \in \mathcal{E}^{p,q}(\Omega)$$

Next we compute $\bar{\square}$ applied to $B\left([M]^{0,1} \wedge f\right)$ and then compare the answer with $\square$ applied to $\chi^{+}F$. Because of Theorem 12.2.8 (2),

$$B\left([M]^{0,1} \wedge f\right)=b \#\left([M]^{0,1} \wedge f\right)=e \#\left[\bar{\partial}^{*}\left([M]^{0,1} \wedge f\right)\right]$$

where e is the standard (Newtonian) convolution fundamental solution for $\bar{\square} = \frac{1}{4}\Delta$ in $\mathbb{C}^n = \mathbb{R}^{2n}$. Thus,

$$\bar{\square}\left[B\left([M]^{0,1} \wedge f\right)\right]=\bar{\partial} *\left([M]^{0,1} \wedge f\right) \tag{12.73}$$

Comparison of (12.73) with (12.72) shows that $\bar{\square}$ applied to $B([M]^{0,1} \wedge f) - \chi^{+}F$ equals $\chi^{+}\bar{\square}F$ which is C^{∞}. Therefore

$$G \equiv B\left([M]^{0,1} \wedge f\right)-\chi^{+} F \in \mathcal{E}^{p,q}(\Omega) \tag{12.74}$$

by interior regularity for $\bar{\square}$. Consequently, $B\left([M]^{0,1} \wedge f\right)\big|_{\Omega_{+}}$ has a C^{∞} extension to $\bar{\Omega}^{+}$. Furthermore, by (12.74)

$$0=G_t-G_t=B^{+}(f)-F_t-B^{-}(f) \quad \text{on} \quad M$$

so that the "tangential jump" of the Bochner-Martinelli transform of f is equal to f. This completes the first stage of the proof of Theorem 12.6.6.

Remark 12.6.12. The remaining pieces of the Cauchy data for $B\left([M]^{0,1} \wedge f\right)\big|_{\Omega^{\pm}}$ have no jumps. That is $\left[B\left([M]^{0,1} \wedge f\right)\right]_n$, $\left[\bar{\partial}^{*} B\left([M]^{0,1} \wedge f\right)\right]_t$, and $\left[\bar{\partial} B\left([M]^{0,1} \wedge f\right)\right]_n$ all have continuous extensions across M. This follows immediately from (12.74).

The second stage of the proof of Theorem 12.6.6 consists of proving the Cauchy principal value integral exists, and the one-sided formulae (12.63). Both of these facts are proven in Harvey-Lawson [97] Appendix B in the special case where $f \in \mathcal{D}^{0,0}(M)$ is a function. We will reduce the general case, $f \in \mathcal{D}^{p,q}(M)$, to this special case.

Now consider the special case where $f(z_0) = 0$ (i.e., the coefficients of f vanish at $z_0 \in M$). Then

$$|B(\zeta, z+\epsilon d\rho(z))| \leqslant C|\zeta-z|^{1-2n}$$

for $z, \zeta \in M$ and all small ϵ. It follows that the Cauchy principal value integral exists since it reduces to an ordinary integral; and

$$B_b f(z_0) = B^{\pm} f(z_0) = -\tau \int_M [B(\zeta, z_0) \wedge f(\zeta)]^{n,n-1}$$

so that (12.63) is valid at z_0. By this observation we have reduced our task to proving the theorem for forms of the sort

$$f(z) = \left[\psi(z) d\bar{z}^J \wedge dz^I\right]_t$$

where $\psi \in \mathbb{C}_0^\infty(\Omega)$, and $\psi(z) \equiv 1$ in a neighborhood of z_0.

The next result enables us to reduce the general case to the case where f is a function.

Lemma 12.6.13.

$$\begin{aligned} B(\zeta, z) \wedge \bar{\partial}\rho(\zeta) \wedge d\bar{\zeta}^J \wedge d\zeta^I =& B(\zeta, z) \wedge \bar{\partial}\rho(\zeta) \wedge d\bar{z}^J \wedge dz^I \\ &+ A(\zeta, z) + \bar{\partial}\rho(z) \wedge C(\zeta, z) \end{aligned}$$

where $|A(\zeta, z + \epsilon d\rho(z))| \leqslant C|\zeta - z|^{2-2n}$ *for* $\zeta, z \in M$ *and* ϵ *small*

Proof. Let $N = N(z)$ denote the 0,1 vector field on Ω dual to $\bar{\partial}\rho(z)$, as discussed previously. Let

$$D(\zeta, z) \equiv B(\zeta, z) \wedge \bar{\partial}\rho(\zeta) \wedge d\bar{\zeta}^J \wedge d\zeta^I - B(\zeta, z) \wedge \bar{\partial}\rho(\zeta) \wedge d\bar{z}^J \wedge dz^I .$$

Then

$$D(\zeta, z) = A(\zeta, z) + \bar{\partial}\rho(z) \wedge C(\zeta, z)$$

where $A(\zeta, z) = N \lrcorner (\bar{\partial}\rho(z) \wedge D(\zeta, z))$, and $C(\zeta, z) = N \lrcorner D(\zeta, z)$. Now

$$B(\zeta, z) = \alpha(\zeta, z) d(\zeta - z)$$

where

$$d(\zeta - z) = d(\zeta_1 - z_1) \wedge \cdots \wedge d(\zeta_n - z_n)$$

Since $d(\zeta - z) \wedge (d\zeta_j - dz_j) = 0$ it follows that $B(\zeta, z) \wedge d\zeta^I = B(\zeta, z) \wedge dz^I$. Thus

$$D(\zeta, z) = B(\zeta, z) \wedge \bar{\partial}\rho(\zeta) \wedge d\left(\bar{\zeta}^J - d\bar{z}^J\right) \wedge dz^I$$

and

$$\bar{\partial}\rho(z) \wedge D(\zeta, z) = \bar{\partial}\rho(z) \wedge B(\zeta, z) \wedge [\bar{\partial}\rho(\zeta) - \bar{\partial}\rho(z)] \wedge d\left(\bar{\zeta}^J - d\bar{z}^J\right) \wedge dz^I$$

and

$$\begin{aligned} \bar{\partial}\rho(\zeta) - \bar{\partial}\rho(z) =& \sum_{j=1}^n \left(\frac{\partial\rho}{\partial\bar{\zeta}_j}(\zeta) - \frac{\partial\rho}{\partial\bar{z}_j}(z)\right) d\bar{z}_j \\ &+ \sum_{j=1}^n \frac{\partial\rho}{\partial\bar{\zeta}_j}(\zeta) d\left(\bar{\zeta}_j - \bar{z}_j\right) \end{aligned}$$

Notice that $B(\zeta, z)$ wedge the first term satisfies the desired estimate for $A(\zeta, z)$, while $B(\zeta, z)$ wedge the second term is of the form $\alpha(\zeta, z) \wedge d(\bar{\zeta} - \bar{z})$. Since $d(\bar{\zeta} - \bar{z}) \wedge \left(d\bar{\zeta}^J - d\bar{z}^J\right) = 0$, this proves the Lemma. □

Since $[M]^{0,1} = \bar{\partial}\rho\sigma$ where σ is surface measure on M,

$$\begin{aligned} &-\int_{\substack{M \\ |\zeta - z| > \epsilon}} \left[B(\zeta, z) \wedge \psi(\zeta) d\bar{\zeta}^J \wedge d\zeta^I\right]^{n,n-1} \\ &= \int_{|\zeta - z| > \epsilon} \psi(\zeta) B(\zeta, z) \wedge \bar{\partial}\rho(\zeta) \wedge d\bar{\zeta}^J \wedge d\zeta^I d\sigma(\zeta) \end{aligned}$$

By Lemma 12.6.13, if $f = \left[\psi d\bar{z}^J \wedge dz^I\right]_t$ then

$$\begin{aligned} B_b f(z) = &\left[-\lim_{\epsilon \to 0} \int_{\substack{\zeta \in M \\ |z - \zeta| > \epsilon}} \psi(\zeta) B(\zeta, z)\right] \left(d\bar{z}^J \wedge dz^I\right)_t \\ &+ \tau \int \psi(\zeta) A(\zeta, z) d\sigma(\zeta) \end{aligned}$$

Because Theorem 12.6.6 is valid in the case of a function, the above limit exists and we have

$$B_b f(z) = [B_b \psi(z)] \left(d\bar{z}^J \wedge dz^I\right)_t + \tau \int \psi(\zeta) A(\zeta, z) d\sigma(\zeta) \tag{12.75}$$

This completes the proof that the Cauchy principal value in Theorem 12.6.6 always exists. Similarly, we can show that

$$B^+ f(z) = \left[B^+ \psi(z)\right] \left(d\bar{z}^J \wedge dz^I\right)_t + \tau \int \psi(\zeta) A(\zeta, z) d\sigma(\zeta) \tag{12.76}$$

Comparison of (12.75) and (12.76) shows that the boundary formula (12.63) is true for $f = \left[\psi d\bar{z}^J \wedge dz^I\right]_t$. This completes the proof of Theorem 12.6.6.

In addition to analyzing the boundary or jump behavior of $B\left([M]^{0,1} \wedge f\right)$ as was done in Theorem 12.6.6, we will need the boundary behavior of $B\left(\chi_{\Omega^-} G\right)$.

Lemma 12.6.14. *Given $G \in \mathcal{G}^{p,q}\left(\bar{\Omega}^-\right)$, both $F^\pm \equiv B\left(\chi_{\Omega^-} G\right)|_{\Omega^\pm}$ have C^∞ extensions to M. Moreover, the jump:*

$$\begin{aligned} F^+(z) - F^-(z) &= 0 \text{ on } \quad M \\ \left(\bar{\partial}^* F^+\right)_t - \left(\bar{\partial}^* F^-\right)_t &= 0 \text{ on } \quad M \\ \left(\bar{\partial} F^+\right)_n - \left(\bar{\partial} F^-\right)_n &= G_n \text{ on } \quad M. \end{aligned}$$

Proof. Let $\chi \equiv \chi_{\Omega^-}$. Then because of Theorem 12.2.8 (2),

$$\bar{\Box} B(\chi \cdot G) = \bar{\Box}\left(\left(\bar{\partial}^* e\right) \# (\chi G)\right) = \bar{\Box}\left(e \# \bar{\partial}^* (\chi \cdot G)\right) = \bar{\partial}^* (\chi \cdot G).$$

Hence by (12.70),

$$\bar{\Box}B(\chi \cdot G) = \chi\bar{\partial} * G + G_n \cdot \sigma \tag{12.77}$$

Because of the C^∞ version of the Cauchy-Kowalewski Theorem for $\bar{\Box}$ we may choose $F_1 \in \mathcal{E}^{p,q-1}(\Omega)$ with $CF_1 = (0, 0, 0, G_n)$, and $\bar{\Box}F_1 = \bar{\partial}^*G$ to infinite order on M. The current form of Green's identity (Lemma 12.6.11) implies

$$\bar{\Box}\,(\chi \cdot F_1) = \chi \cdot \bar{\Box}F_1 + G_n \cdot \sigma \tag{12.78}$$

since $\left(\bar{\partial}F_1\right)_n = G_n$. Comparing (12.77) and (12.78) yields

$$\bar{\Box}\,(B(\chi \cdot G) - \chi \cdot F_1) = \chi\left(\bar{\partial}^*G - \bar{\Box}F_1\right)$$

is smooth on Ω across M. Therefore $B(\chi \cdot G) - \chi F_1$ is smooth across M. Since $\chi F_1|_{\Omega^\pm}$ is smooth up to $M, B(\chi \cdot G)_{\Omega^\pm}$ is also smooth up to M. Moreover, the jump behavior of $B(\chi \cdot G)$ and $\chi \cdot F_1$ must agree. □

We conclude this section with two results that will not be used elsewhere in this paper. First, we deduce the C^∞ version of the Cartan-Kähler theorem for $\bar{\partial}$ as a corollary of the C^∞ Cauchy-Kowalewski theorem for $\bar{\Box}$.

Lemma 12.6.15. *Suppose $G \in \mathcal{E}^{p,q}(\Omega)$ (the inhomogeneous term) and $f \in \mathcal{E}^{p,q-1}(M)$ (the initial data) are given and satisfy: $\bar{\partial}_b f = G_t$ on M (the initial differential equation) and $\bar{\partial}G$ vanishes to infinite order on M (the compatibility condition). Then there exists an $F \in \mathcal{E}^{p,q-1}(\Omega)$ such that*

$$F_t = f \quad \textit{on} \quad M$$

and

$$\bar{\partial}F - G \textit{ vanishes to infinite order on } M$$

See Andreotti-Hill [4] and Polking-Wells [173].

Proof. By Theorem 12.6.9, there exists $H \in \mathcal{F}^{p,q}(\Omega)$ satisfying the initial conditions

$$H_t = 0, \quad H_n = 0, \quad \left(\bar{\partial}^*H\right)_t = f, \quad (\bar{\partial}H)_n = 0$$

and $\bar{\Box}H - G$ vanishes to infinite order on M. Let $F \equiv \bar{\partial}^*H$. Then $F_t = f$ and $\bar{\partial}F - G = \bar{\Box}H - G - \bar{\partial} * \bar{\partial}H$. Hence it remains to show that $\bar{\partial} * \bar{\partial}H$ vanishes to infinite order on M. In fact $\bar{\partial}H$ vanishes to infinite order on M. Note that $(\bar{\partial}H)_t = \bar{\partial}_b H_t = 0$, $(\bar{\partial}H)_n = 0$, $(\bar{\partial}(\bar{\partial}H))_n = 0$, and $(\bar{\partial} * (\bar{\partial}H))_t = (\bar{\Box}H - \bar{\partial}\bar{\partial} * H)_t = G_t - \bar{\partial}_b f = 0$. That is $\bar{\partial}H$ has zero Cauchy data. Moreover, $\bar{\Box}\bar{\partial}H = \bar{\partial}\Box H = \bar{\partial}(\bar{\Box}H - G) + \bar{\partial}G$ vanishes to infinite order on M. Therefore $\bar{\partial}H$ must vanish to infinite order on M by the last part of Theorem 12.6.9.

Finally, we will study the relationship between $\bar{\partial}*$ and $\bar{\partial}_b^*$ and $\bar{\Box} = \bar{\partial}\bar{\partial} * + \bar{\partial} * \bar{\partial}$ and $\bar{\Box}_b = \bar{\partial}_b\bar{\partial}_b^* + \bar{\partial}_b^*\bar{\partial}_b$ on a Hermitian complex manifold X. We will assume that M has the metric induced by that on X. The vector field N used to define the $\bar{\partial}_b$ complex on M is chosen with the additional property that N is orthogonal to the holomorphic tangent space of M. As before $M = \{z \in X \mid \rho(z) = 0\}$, but we will now assume that $|\bar\partial\rho(z)| = 1$ for all $z \in M$, where the length of $\bar\partial\rho(z)$ is calculated with respect to the metric on X. Since $\bar N \lrcorner \bar\partial\rho \equiv 1$ we have

$$\begin{aligned} f &= \bar N \lrcorner(\bar\partial\rho \wedge f) + \bar\partial\rho \wedge (\bar N \lrcorner f) \\ &= f_t + \bar\partial\rho \wedge f_n \end{aligned}$$

as before. We will consider f_t and f_n to be defined on X (near M) by this equation.

It will be more convenient for our purpose here to use the vector field

$$\bar Z = |\bar\partial\rho|^2 \bar N$$

For example, if F is a $p, q+1$ form and G is a p, q form on X then, since $F = |\bar\partial\rho|^{-2}[\bar\partial\rho \wedge (\bar Z \lrcorner F) + \bar Z \lrcorner (\bar\partial\rho \wedge F)]$, we have

$$\begin{aligned} \langle F, \bar\partial\rho \wedge G\rangle &= |\bar\partial\rho|^{-2}\langle \bar\partial\rho \wedge (\bar Z \lrcorner F), \bar\partial\rho \wedge G\rangle \\ &= \langle \bar Z \lrcorner F, G\rangle \end{aligned} \tag{12.79}$$

where the pairing $\langle\, ,\rangle$ is the pointwise metric. Clearly

$$\bar\partial(\bar\partial\rho \wedge G) = -\bar\partial\rho \wedge \bar\partial G$$

By duality it follows using (12.79) that

$$\bar\partial^*(\bar Z \lrcorner G) = -\bar Z \lrcorner \bar\partial^* G \tag{12.80}$$

Other commutation relations are also important. We define the first-order differential operator L acting on forms on X by

$$\bar Z \lrcorner \bar\partial G = -\bar\partial(\bar Z \lrcorner G) + L(G) \tag{12.81}$$

Then by duality,

$$\bar\partial^*(\bar\partial\rho \wedge G) = -\bar\partial\rho \wedge \bar\partial^* G + L^*(G) \tag{12.82}$$

where L^* is the dual of L on X with respect to the L^2 inner product. □

Example 12.6.16. When $X = \mathbb{C}^n$ with the standard Euclidean metric and M is the unit sphere, we have

$$\begin{aligned} \rho(z) &= |z|^2 - 1\,, \\ \bar Z &= \sum \bar z_j \partial/\partial \bar z_j\,, \\ Z &= \sum z_j \partial/\partial z_j\,, \\ L(G) &= \bar Z(G) + qG\,, \\ L^*(G) &= [Z(G) + (n-q)G] \end{aligned}$$

when $G \in \mathcal{E}^{p,q}(\mathbb{C}^n)$. Here $Z\left(\sum G_{IJ} dz^I \wedge d\bar z^J\right) = \sum Z\left(G_{IJ}\right) dz^I \wedge d\bar z^J$.

Proposition 12.6.17.

(a) *Suppose $g \in \mathcal{E}^{p,q}(M)$ and $G \in \mathcal{E}^{p,q}(X)$ satisfy $G_t = g$ on M and G_n vanishes to second order on M. Then $\bar{\partial}_b^* g = \bar{\partial}^* G$ on M.*

(b) *Suppose $g \in \mathcal{E}^{p,q}(M)$ and $G \in \mathcal{F}^{p,q}(X)$ satisfy $G_t = g$ on M and G_n vanishes to third order on M. Then*

$$\bar{\Box}_b g = \bar{\Box} G - L^* L(G) - \bar{Z}\left(|\bar{\partial}\rho|^2\right) L(G) + \bar{\partial}\rho \wedge \left[\bar{\partial}^*, L\right] G\,.$$

Remark 12.6.18. It can be seen by Lemma 12.6.7 that it is not enough to assume that G_n vanishes to first order on M in part (a).

In the case of the sphere in $\mathbb{C}^n$ the above formula for $\bar{\partial}_b^*$ is used by Dadok-Harvey [56]. In this same case part (b) is due to Geller [78].

Proof.

(a) First suppose $F \in \mathcal{E}^{p,q}(X)$ and $f \in \mathcal{E}^{p,q-2}(M)$. Then

$$\begin{aligned}
\left\langle \bar{\partial}\left(f \wedge [M]^{0,1}\right), F\right\rangle_X &= \left\langle f \wedge [M]^{0,1}, \bar{\partial}^* F\right\rangle_X \\
&= (-1)^{p+q} \left\langle [M]^{0,1} \wedge f, \bar{\partial}^* F\right\rangle_X \\
&= (-1)^{p+q} \left\langle f, \left(\bar{\partial}^* F\right)_n\right\rangle_M
\end{aligned}$$

On the other hand

$$\begin{aligned}
\left\langle \bar{\partial}\left(f \wedge [M]^{0,1}\right), F\right\rangle_X &= \left\langle \bar{\partial}_b f \wedge [M]^{0,1}, F\right\rangle_X \\
&= (-1)^{p+q-1} \left\langle \bar{\partial}_b f, F_n\right\rangle_M \\
&= (-1)^{p+q-1} \left\langle f, \bar{\partial}_b^* F_n\right\rangle_M
\end{aligned}$$

Thus

$$\bar{\partial}_b^* F_n = -\left(\bar{\partial}^* F\right)_n \tag{12.83}$$

Letting $F = \bar{\partial}\rho \wedge G$ we see from (12.83) that

$$\begin{aligned}
\bar{\partial}_b^* g &= -(\bar{\partial} * (\bar{\partial}\rho \wedge G))_n \\
&= -|\bar{\partial}\rho|^{-2} \bar{Z} \lrcorner \bar{\partial} * (\bar{\partial}\rho \wedge G) \\
&= |\bar{\partial}\rho|^{-2} \bar{\partial} * [\bar{Z} \lrcorner (\bar{\partial}\rho \wedge G)] \\
&= |\bar{\partial}\rho|^{-2} \bar{\partial} * \left[|\bar{\partial}\rho|^2 G - \bar{\partial}\rho \wedge (\bar{Z} \lrcorner G)\right]
\end{aligned}$$

Since $\bar{Z} \lrcorner G = |\bar{\partial}\rho|^2 G_n$ vanishes to second order on M

$$\bar{\partial}_b^* g = |\bar{\partial}\rho|^{-2} \bar{\partial} * \left(|\bar{\partial}\rho|^2 G\right) \quad \text{on} \quad M$$

Now $|\bar{\partial}\rho|^2 = 1$ on M so there is a function h such that $|\bar{\partial}\rho|^2 = 1 + \rho h$. Thus

$$\begin{aligned}\left\langle \bar{\partial} * \left(|\bar{\partial}\rho|^2 G\right), \varphi\right\rangle_X &= \langle G, (1+\rho h)\bar{\partial}\varphi\rangle_X \\ &= \left\langle |\bar{\partial}\rho|^2 \bar{\partial} * G, \varphi\right\rangle_X - \langle G, h\bar{\partial}\rho \wedge \varphi + \rho\bar{\partial}h \wedge \varphi\rangle_X \\ &= \left\langle |\bar{\partial}\rho|^2 \bar{\partial} * G - hZ \lrcorner G - \rho A \lrcorner G, \varphi\right\rangle_X\end{aligned}$$

where the vector field A is defined by $\langle A \lrcorner G, \varphi\rangle = \langle G, \bar{\partial}h \wedge \varphi\rangle$. Thus

$$\bar{\partial} * \left(|\bar{\partial}\rho|^2 G\right) = |\bar{\partial}\rho|^2 \bar{\partial} * G - h\bar{Z} \lrcorner G - \rho A \lrcorner G \tag{12.84}$$

and since $\bar{Z} \lrcorner G = 0$, on M we have

$$\bar{\partial}_b g = \bar{\partial}^* G \quad \text{on} \quad M$$

(b) Since $G_t = g$ we have by (12.81)

$$\begin{aligned}\bar{\partial}_b g = (\bar{\partial}G)_t &= \bar{\partial}G - |\bar{\partial}\rho|^{-2}\bar{\partial}\rho \wedge (\bar{Z} \lrcorner \bar{\partial}G) \\ &= \bar{\partial}G + |\bar{\partial}\rho|^{-2}\bar{\partial}\rho \wedge [\bar{\partial}(\bar{Z} \lrcorner G) - L(G)]\end{aligned}$$

Since the right-hand side is purely tangential, it follows from part (a) that on M

$$\bar{\partial}_b^* \bar{\partial}_b g = \bar{\partial} * \bar{\partial}G - \bar{\partial} * \left[|\bar{\partial}\rho|^{-2}\bar{\partial}\rho \wedge (\bar{Z} \lrcorner \bar{\partial}G)\right]$$

Since $|\bar{\partial}\rho| \equiv 1$, there is a function h such that $|\bar{\partial}\rho|^{-2} = 1 + \rho h$. Applying $\bar{Z}$ to this equation we see that

$$\begin{aligned}\bar{Z}\left(|\bar{\partial}\rho|^{-2}\right) &= \rho\bar{Z}(h) + h\bar{Z}(\rho) \\ &= h \quad \text{on} \quad M.\end{aligned}$$

On the other hand

$$\begin{aligned}\bar{Z}\left(|\bar{\partial}\rho|^{-2}\right) &= -|\bar{\partial}\rho|^{-2}\bar{Z}\left(|\bar{\partial}\rho|^2\right) \\ &= -\bar{Z}\left(|\bar{\partial}\rho|^2\right) \quad \text{on} \quad M.\end{aligned}$$

Consequently using (12.84) we see that on M

$$\begin{aligned}\bar{\partial}_b^* \bar{\partial}_b g =& \bar{\partial} * \bar{\partial}G - \bar{\partial} * [\bar{\partial}\rho \wedge (\bar{Z} \lrcorner \bar{\partial}G)] \\ & - \bar{Z}\left(|\bar{\partial}\rho|^2\right) \bar{Z} \lrcorner [\bar{\partial}\rho \wedge (\bar{Z} \lrcorner \bar{\partial}G)].\end{aligned}$$

Finally using (12.81) and (12.82) and the fact that $\bar{Z} \lrcorner G$ vanishes to third order on M we have

$$\bar{\partial}_b^* \bar{\partial}_b g = \bar{\partial}^* \bar{\partial}G + \bar{\partial}\rho \wedge \bar{\partial}^* L(G) - L^* L(G) - \bar{Z}\left(|\bar{\partial}\rho|^2\right) L(G) \tag{12.85}$$

By part (a) $\bar{\partial}_b^* g = \bar{\partial}^* G$ on M, so

$$\bar{\partial}_b \bar{\partial}_b^* g = (\bar{\partial}\bar{\partial} * G)_t = \bar{\partial}\bar{\partial} * G - |\bar{\partial}\rho|^{-2} \rho \wedge (\bar{Z} \lrcorner \bar{\partial}\bar{\partial}^* * G)$$

Using (12.80), (12.81) and the fact that $\bar{Z} \lrcorner G$ vanishes to third order on M we see that $\bar{Z} \rfloor \bar{\partial}\bar{\partial} * G = L(\bar{\partial} * G)$. Thus

$$\bar{\partial}_b \bar{\partial}_b^* g = \bar{\partial}\bar{\partial}^* G - \bar{\partial}\rho \wedge L(\bar{\partial} * G)$$

Putting this together with (12.85) we get

$$\bar{\Box}_b g = \bar{\Box} G - L^* L(G) - \bar{Z}\left(|\bar{\partial}\rho|^2\right) L(G) + \bar{\partial}\rho \wedge \left[\bar{\partial}^*, L\right] G.$$

□

12.7 Boundary Behavior of the Kernels H and L

Our next project will be to imitate the results of the last section for the kernels H and L are defined in Section 12.4, in so far as this is possible. Let $\Omega \subset \mathbb{C}^n$ be open, and let $M = \{z \in \Omega \mid \rho(z) = 0\}$, where $\rho \in C^\infty(\Omega)$ and $d\rho(z) \neq 0$ for $z \in M$. Let $\Omega^\pm = \{z \in \Omega \mid \pm\rho(z) > 0\}$. It will be necessary here to make an assumption about the pseudo-convexity of M. As we did in Section 12.5, we will do this by assuming the existence of a support function Definition 12.5.1. The main results of this section are contained in the following two theorems.

Theorem 12.7.1. *Suppose $u(\zeta, z)$ is a biregular support function for M relative to Ω^- defined in $\Omega \times \Omega$. Let L and H be the kernels generated by u.*

(a) If $\varphi \in \mathcal{D}^{p,q}(M)$, then $H\left([M]^{0,1} \wedge \varphi\right) \in \mathcal{E}^{p,q-1}\left(\Omega^-\right)$ and all tangential derivatives of $H\left([M]^{0,1} \wedge \varphi\right)$ and of $\bar{\partial} H\left([M]^{0,1} \wedge \varphi\right)$ are continuously differentiable up to M from Ω^-.

(b) If $\varphi \in \mathcal{D}^{p,1}(M)$ and $\bar{\partial}_b \varphi = 0$, then $H\left([M]^{0,1} \wedge \varphi\right) \in \mathcal{E}^{p,0}\left(\bar{\Omega}^-\right)$.

(c) If $\varphi \in \mathcal{D}^{p,0}(M)$, then $L\left([M]^{0,1} \wedge \varphi\right) \in \mathcal{E}^{p,0}\left(\bar{\Omega}^-\right)$.

(d) If $\varphi \in \mathcal{D}^{p,q}(M)$, then ${}^tH\left([M]^{0,1} \wedge \varphi\right) \in \mathcal{E}^{p,q-1}\left(\Omega^+\right)$ and all tangential derivatives of ${}^tH\left([M]^{0,1} \wedge \varphi\right)$ and of $\bar{\partial}^t H\left([M]^{0,1} \wedge \varphi\right)$ are continuously differentiable up to M from Ω^+.

(e) If $\varphi \in \mathcal{D}^{p,1}(M)$ and $\bar{\partial}_b \varphi = 0$, then ${}^tH\left([M]^{0,1} \wedge \varphi\right) \in \mathcal{E}^{p,0}\left(\bar{\Omega}^+\right)$.

(f) If $\varphi \in \mathcal{D}^{p,n-1}(M)$, then ${}^tL\left([M]^{0,1} \wedge \varphi\right) \in \mathcal{E}^{p,n-1}\left(\bar{\Omega}^+\right)$.

Let $\chi^\pm$ denote the characteristic function of $\Omega^\pm$.

Theorem 12.7.2. *Suppose $u(\zeta, z)$ is a biregular support function for M relative to Ω^- defined in $\Omega \times \Omega$. Let H and L be the kernels generated by u.*

(a) If $\varphi \in \mathcal{D}^{p,q}(\Omega)$, then $H\left(\chi^+\varphi\right) \in \mathcal{E}^{p,q-2}\left(\Omega^-\right)$ and all tangential derivatives of $H\left(\chi^+\varphi\right)$ are twice continuously differentiable in $\bar{\Omega}^-$.

(b) If $\varphi \in \mathcal{D}^{p,1}(\Omega)$, then $L\left(\chi^+\varphi\right) \in \mathcal{E}^{p,0}\left(\bar{\Omega}^-\right)$.

(c) If $\varphi \in \mathcal{D}^{p,q}(\Omega)$, then ${}^tH\left(\chi^-\varphi\right) \in \mathcal{E}^{p,q-2}\left(\Omega^+\right)$ and all tangential derivatives of ${}^tH\left(\chi^-\varphi\right)$ are twice continuously differentiable in $\bar{\Omega}^+$.

(d) If $\varphi \in \mathcal{D}^{p,n}(\Omega)$, then ${}^tL\left(\chi^+\varphi\right) \in \mathcal{E}^{p,n-1}\left(\bar{\Omega}^+\right)$.

Since all of the techniques of the proof of Theorem 12.7.2 are used in the proof of Theorem 12.7.1 we will only prove Theorem 12.7.1.

Before proving Theorem 12.7.1 it is appropriate to point out a consequence. This will provide a contrast to Theorem 12.5.13. Suppose $D \subset\subset \mathbb{C}^n$ and u is a biregular support function for ∂D. If $G \in \mathcal{E}^{p,q}(\bar{D})$ and $\bar{\partial} G \equiv 0$ in D then by Remark 12.5.11 $G = \bar{\partial}\left[B\left(\chi_D G\right) + H\left(\left[\partial D\right]^{0,1} \wedge G\right)\right]$ in D. This solution,

$$B\left(\chi_D G\right) + H\left(\left[\partial D\right]^{0,1} \wedge G\right)$$

is called the *Henkin-Lieb solution* to the Cauchy-Riemann equations in D.

Theorem 12.7.3. *Suppose $D \subset\subset \mathbb{C}^n$ and u is a biregular support function for ∂D. Suppose $G \in \mathcal{E}^{p,q}(\bar{D})$ satisfies $\bar{\partial} G = 0$ in D.*

(a) If $q = 1$, the Henkin-Lieb solution

$$B\left(\chi_D G\right) + H\left(\left[\partial \Omega\right]^{0,1} \wedge G\right) \in \mathcal{E}^{p,0}(\bar{D}).$$

(b) If $q \geqslant 2$, the Henkin-Lieb solution is continuously differentiable in $\bar{D}$.

Proof. By Theorem 12.7.1 $H\left(\left[\partial D\right]^{0,1} \wedge G\right)$ has the required regularity, while by Lemma 12.6.14 $B\left(\chi_D G\right) \in \mathcal{E}^{p,q-1}(\bar{D})$. □

For the case when D is strictly pseudoconvex Henkin [105] has proven precise estimates for the Henkin-Lieb solution when $G \in L^r(D)$, and when G has measure coefficients. Part (a) of Theorem 12.7.3 was proved by Siu [195] when D is strictly pseudoconvex, and in this case Krantz [136] has proved precise estimates on the derivatives when $q = 1$. Some weakly pseudo-convex cases have been studied by Range [176]. The major portion of the proof of Theorem 12.7.1 is contained in the following proposition. Let σ denote the surface measure on M.

Proposition 12.7.4. *Suppose that u is a biregular support function for M with respect to Ω^- defined in $\Omega \times \Omega$. Suppose $k \in C^\infty(\Omega \times \Omega)$ satisfies*

$$|k(\zeta, z)| \leqslant C|\zeta - z|^r$$

For $\varphi \in C_0^\infty(M)$ *set*

$$K_1\varphi(z) = \int_{\zeta\in M} k(\zeta,z)|\zeta - z|^{-2q}[u(\zeta,z)\cdot(\zeta - z)]^{-p}\varphi(\zeta)d\sigma(\zeta), z \in \Omega^-$$

$$K_2\varphi(\zeta) = \int_{z\in M} k(\zeta,z)|\zeta - z|^{-2q}[u(\zeta,z)\cdot(\zeta - z)]^{-p}\varphi(z)d\sigma(z), \zeta \in \Omega^+ \tag{12.86}$$

(a) *If* $q = 0$, *then* $K_1\varphi \in C^\infty\left(\bar{\Omega}^-\right)$ *and* $K_2\varphi \in C^\infty\left(\bar{\Omega}^+\right)$.

(b) *If* $q > 0$, *then all tangential derivatives of* $K_1\varphi$ *belong to* $C^s\left(\bar{\Omega}^-\right)$ *and all tangential derivatives of* $K_2\varphi$ *belong to* $C^s\left(\bar{\Omega}^+\right)$, *where* $s = 2(n - q - 1) + r - p$.

(c) *If* $q = 0$ *or if* $q > 0$ *and* $s = 2(n - q - 1) + r - p \geqslant 0$ *then*

$$K_1\varphi(z) = \lim_{\epsilon\to 0^+} \int_{\rho(\zeta)=\epsilon} k(\zeta,z)|\zeta - z|^{-2q}[u(\zeta,z)\cdot(\zeta - z)]^{-p}\varphi(\zeta)d\sigma(\zeta)$$

$$K_2\varphi(\zeta) = \lim_{\epsilon\to 0^-} \int_{\rho(z)=\epsilon} k(\zeta,z)|\zeta - z|^{-2q}[u(\zeta,z)\cdot(\zeta - z)]^{-p}\varphi(z)d\sigma(z)$$

uniformly on compact subsets of M.

Remark 12.7.5. Let X be a vector field on Ω. We say that X is tangential if $X\rho \equiv 0$ in Ω. The conclusion in part (b) means that if $X_1, \ldots, X_N$ are tangential vector fields, then $X_1X_2\cdots X_NK_1\varphi$ belongs to $C^s\left(\bar{\Omega}^-\right)$.

Proof. Since u is a support function, the integrand in (12.86) is an infinitely differentiable function of $z \in \Omega^-$ for $\zeta \in M$. Therefore $K_1\varphi \in C^\infty\left(\Omega^-\right)$. Similarly $K_2\varphi \in C^\infty\left(\Omega^+\right)$. We will prove the theorem for $K_1\varphi$. The proof for $K_2\varphi$ is similar.

Since $K_1\varphi \in C^\infty\left(\Omega^-\right)$, it suffices to show that every point $(\zeta_0, z_0) \in M\times M$ has a product neighborhood $U \times V$ in $\Omega\times\Omega$ such that $K_1\varphi$ has the required smoothness in $\bar{V}^- = V \cap \bar{\Omega}^-$ for each function $\varphi \in C_0^\infty(U\cap M)$. We may also assume that φ is defined in U and $\varphi \in C_0^\infty(U)$.

First suppose $(\zeta_0, z_0) \in M \times M$ and $u(\zeta_0, z_0)\cdot(\zeta_0 - z_0) \neq 0$. Then in particular $\zeta_0 \neq z_0$ so we can choose U and V so that $U \cap V = \varnothing$, and $u\cdot(\zeta, z) \neq 0$ in $U\times V$. Then the integrand in (12.86) is infinitely differentiable in $U \times V$. The result is trivial in this case. □

When $(\zeta_0, z_0) \in M \times M$ and $u(\zeta_0, z_0)\cdot(\zeta_0 - z_0) = 0$ we need the change of variables described in the following lemma.

Lemma 12.7.6. *Suppose* u *is a regular support function for* M *with respect to* Ω^-. *Given* $(\zeta_0, z_0) \in M \times M$ *such that* $u(\zeta_0, z_0)\cdot(\zeta_0 - z_0) = 0$, *there is a neighborhood* $U \times V$ *of* (ζ_0, z_0) *in* $\Omega\times\Omega$ *and a* C^∞ *map* $\Phi : U \times V \to \mathbb{C}^n$ *such that for each* $z \in V, w = \Phi_z(\zeta) = \Phi(\zeta, z)$ *is a diffeomorphism on* U *with the following properties:*

(a) $\Phi(\zeta_0, z_0) = 0$

(b) $w_n(\zeta, z) = \lambda(\zeta, z)u(\zeta, z) \cdot (\zeta - z)$ *where* $\lambda \in C^\infty(U \times V)$ *and does not vanish in* $U \times V$

(c) $\operatorname{Re} w_n(\zeta, z) > 0$ *if* $\rho(\zeta) > 0$ *and* $\rho(z) < 0$

(d) The function $R(w, z) = \rho\left(\Phi_z^{-1}(w)\right)$ *can be written*

$$R(w, z) = \operatorname{Re} w_n + G(w', z)$$

where $w' = (w_1, \ldots, w_{n-1})$.

If in addition $\zeta_0 = z_0$, *then we can choose* Φ *with the properties (b) and (c) and (a')* $\Phi(z, z) \equiv 0$

(d') the function $R(w, z)$ *can be written*

$$R(w, z) = \rho(z) + \operatorname{Re} w_n + H(w', z)$$

where

$$|H(w', z)| \leqslant C\,|w'|^2 .$$

Proof. Since u is a regular generating function we can choose Ψ defined near (ζ_0, z_0) such that

$$\tilde{w} = \Psi(\zeta, z) = \Psi_z(\zeta)$$

is a diffeomorphism for z near z_0 and satisfies

$$\Psi(\zeta_0, z_0) = 0$$
$$\tilde{w}_n = u(\zeta, z) \cdot (\zeta - z)$$

Let

$$\tilde{\rho}(\tilde{w}, z) = \rho\left(\Psi_z^{-1}(\tilde{w})\right)$$

The fact that u is a support function for M with respect to Ω^- implies that

$$\frac{\partial}{\partial \tilde{w}_j}\tilde{\rho}(0, z_0) = 0 \quad j = 1, 2, \ldots, n-1 \tag{12.87}$$

To see this, suppose for example that $(\partial\tilde{\rho}/\partial x)(0, z_0) \neq 0$ where $x = \operatorname{Re} w_1$. Then by the implicit function theorem, we can solve the equation $\tilde{\rho}(w, z) = t$ for x near $(w, z, t) = (0, z_0, 0)$; i.e.,

$$\operatorname{Re} w_1 = x = h(\operatorname{Im} w_1, w_2, \ldots, w_n, z, t)$$

But then with $t > 0$ small and z close to z_0 with $\rho(z) < 0$, the point $\tilde{w} = (h(0, z, t), 0, \ldots, 0)$ satisfies $\tilde{\rho}(\tilde{w}, z) = t$; or if $\zeta = \Psi_z^{-1}(\tilde{w}), \rho(\zeta) > 0$. Thus $\zeta \in \Omega^+, z \in \Omega^-$ and $\tilde{w}_n = u(\zeta, z) \cdot (\zeta - z) = 0$ which is a contradiction. Therefore (12.87) is true.

As a result of (12.87) the Taylor expansion of first order of $\tilde{\rho}$ at $(0, z_0)$ has the form

$$\tilde{\rho}(\tilde{w}, z) = \operatorname{Re}(\beta\tilde{w}_n) + g(\tilde{w}, z) \tag{12.88}$$

where β is a non-zero constant and g contains the terms linear in $z - z_0$ together with the Taylor remainder. Thus

$$|g(\tilde{w}, z_0)| \leqslant C|\tilde{w}|^2 \tag{12.89}$$

We introduce new variables by

$$\begin{aligned} \hat{w}_j &= \tilde{w}_j \quad 1 \leqslant j \leqslant n-1 \\ \hat{w}_n &= \beta\tilde{w}_n \end{aligned}$$

Then (12.88) becomes

$$\tilde{\rho}(\hat{w}, z) = \operatorname{Re}\hat{w}_n + g(\hat{w}, z) \tag{12.90}$$

We now consider the Taylor expansion of $g(\hat{w}, z)$ of order zero in the $\hat{w}_n$ variable only

$$g(\hat{w}, z) = g(\hat{w}', 0, z) + \operatorname{Re}(\alpha(\hat{w}, z)\hat{w}_n) \tag{12.91}$$

Clearly, by (12.89), $\alpha(0, z_0) = 0$. Thus if we introduce new variables

$$\begin{aligned} w_j &= \hat{w}_j \quad 1 \leqslant j \leqslant n-1 \\ w_n &= (1 + \alpha(\hat{w}, z))\hat{w}_n \\ &= \lambda(\zeta, z)u(\zeta, z) \cdot (\zeta - z) \end{aligned} \tag{12.92}$$

we have $w = \Phi(\zeta, z)$ and if U and V are chosen suitably small, (a) and (b) are satisfied. Furthermore, if we substitute (12.92) and (12.91) into (12.90) we have

$$R(w, z) = \rho\left(\Phi_z^{-1}(w)\right) = \operatorname{Re} w_n + g(w', 0, z)$$

thus (d) follows with $G(w', z) = g(w', 0, z)$.

The set $\{(\zeta, z) \in U \times V \mid \rho(\zeta) > 0, \rho(z) < 0\}$ is connected and therefore its image under Φ, the set $W = \{(w, z) \mid \rho(z) < 0, R(w, z) > 0\}$, is also connected. Notice that $R(0, w_n, z_0) = \operatorname{Re} w_n$. Thus $\operatorname{Re} w_n > 0$ on at least a subset of W. Notice also that the definition of W does not depend on $\operatorname{Im} w_n$, and that w_n never vanishes on W. Thus we must have $\operatorname{Re} w_n > 0$ everywhere on W, and (c) follows.

If $z_0 = \zeta_0$ we modify the argument slightly. First, we insist at the beginning that $\Psi(z, z) \equiv 0$. Then we look at the Taylor expansion of $\tilde{\rho}(\tilde{w}, z)$ in $\tilde{w}$ only about the point $(0, z)$. We get

$$\tilde{\rho}(\tilde{w}, z) = \rho(z) + \operatorname{Re}\beta(z)\tilde{w}_n + \operatorname{Re} l_z(\tilde{w}') + Q(\tilde{w}, z) \tag{12.93}$$

where l_z is a complex linear function of $\tilde{w}'$, $\beta(z)$ is a complex number and

$$|Q(\tilde{w}, z)| \leqslant C|\tilde{w}|^2 \tag{12.94}$$

By (12.87), which is still valid, $l_{z_0} = 0$ and $\beta(z_0) \neq 0$. Furthermore, as in (12.91) we have

$$Q(\tilde{w}, z) = Q(\tilde{w}', 0, z) + \operatorname{Re}(\alpha(\tilde{w}, z)\tilde{w}_n)$$

where $\alpha(0, z_0) = 0$. Now we introduce new variables $w = \Phi(\zeta, z)$ by

$$\begin{aligned} w_j &= \tilde{w}_j \quad 1 \leqslant j \leqslant n-1 \\ w_n &= l_z(\tilde{w}') + (\beta(z) + \alpha(\tilde{w}, z))\tilde{w}_n \end{aligned}$$

If U and V are suitably restricted, (a') and (b) will be true. If we substitute into (12.93) we get

$$R(w, z) = \rho(z) + \operatorname{Re} w_n + H(w', z)$$

where $H(w', z) = Q(w', 0, z)$. Thus by (12.94) (d') is valid. Part (c) follows as before.

We now return to the proof of Theorem 12.7.3.

Suppose that $(\zeta_0, z_0) \in M \times M$ and $u(\zeta_0, z_0) \cdot (\zeta_0 - z_0) = 0$. Suppose in addition that either $\zeta_0 \neq z_0$, or $q = 0$. We will apply Lemma 12.7.6 and if $\zeta_0 \neq z_0$, we make the additional restriction that $U \cap V = \varnothing$. Then if $\varphi \in C_0^\infty(U)$, we can make the change of variables given by Lemma 12.7.6 and we see that

$$\begin{aligned} &k(\zeta, z)|\zeta - z|^{-2q}[u(\zeta, z) \cdot (\zeta - z)]^{-p}\varphi(\zeta)d\sigma(\zeta) \\ &= F(w, z)w_n^{-p}d\sigma(w) \end{aligned}$$

where F is infinitely differentiable. Thus if $M_z = \{w \mid R(w, z) = 0\}$ we have

$$K_1\varphi(z) = \int_{w \in M_z} F(w, z)w_n^{-p}d\sigma(w) \tag{12.95}$$

Let $w_n = x_n + iy_n$. Then

$$w_n^{-p} = \frac{(-i)^p}{(p-1)!}\frac{\partial^p}{\partial y_n^p}\log w_n \tag{12.96}$$

where we are using the principal branch of $\log w_n$. By part (d) of Lemma 12.7.6, $\partial/\partial y_n$ is tangential to M_z. Thus we can integrate by parts in (12.95) using (12.96)

$$K_1\varphi(z) = \frac{i^p}{(p-1)!}\int_{M_z} \log w_n \frac{\partial^p}{\partial y_n^p}F(w, z)d\sigma(w) \tag{12.97}$$

Set

$$\begin{aligned} U_z &= \{w \mid R(w, z) > 0\} \\ \nabla(w, z) &= 2\,|d_w R(w, z)| \end{aligned}$$

Then using Gauss' theorem and part (d) of Lemma 12.7.6, we can change (12.97) to an integral over U_z.

$$\begin{aligned} K_1\varphi(x) &= -\frac{i^p}{(p-1)!}\int_{U_z} \log w_n \frac{\partial}{\partial \bar{w}_n}\left[\nabla(w,z)\frac{\partial^p}{\partial y_n^p}F(w,z)\right] d\lambda(w) \\ &= \int_{U_z} \log w_n P_p(D)F(w,z)d\lambda(w) \end{aligned} \tag{12.98}$$

where

$$P_p(D) = -\frac{i^p}{(p-1)!}\frac{\partial}{\partial \bar{w}_n}\nabla(w,z)\frac{\partial^p}{\partial y_n^p}$$

Let $\chi(w,z)$ denote the characteristic function of the set $\{(w,z) \mid R(w,z) > 0\}$. Then from (12.98) we see that

$$K_1\varphi(z) = \int \chi(w,z)\log w_n P_p(D)F(w,z)d\lambda(w) \tag{12.99}$$

Now let X be a vector field in the z variables. Then

$$\left[X - 2(XR)(w,z)\frac{\partial}{\partial \bar{w}_n}\right] R \equiv 0$$

so

$$\left[X - 2(XR)(w,z)\frac{\partial}{\partial \bar{w}_n}\right] \chi \equiv 0$$

Differentiating (12.99) under the integral sign, we see that

$$XK_1\varphi(z) = \int \chi(w,z)\log w_n \tilde{X}P_p(D)F(w,z)d\lambda(w) \tag{12.100}$$

where $\tilde{X}$ is the first order differential operator

$$\tilde{X}f(w,z) = Xf(w,z) - 2\frac{\partial}{\partial \bar{w}_n}[(XR)(w,z)f(w,z)]$$

By iterating (12.100) it follows that

$$X_1\cdots X_N K_1\varphi(z) = \int \chi(w,z)\log w_n \tilde{X}_1\cdots\tilde{X}_N P_p(D)F(w,z)d\lambda(w) \tag{12.101}$$

Since the integrand in (12.101) is bounded by $C\chi_1(w)\,|\log w_n|$ uniformly for $z \in \bar{V}^- = V \cap \bar{\Omega}^-$ where $\chi_1(w)$ is the characteristic function of $\{w \mid \operatorname{Re} w_n > 0\}$ it is apparent that $K_1\varphi \in C^\infty\left(\bar{V}^-\right)$ in this case.

Finally we must consider the case when $\zeta_0 = z_0 \in M$ and $q > 0$. We proceed as before with modifications that are necessary because $q > 0$ and because we will use the stronger aspects of Lemma 12.7.6.

In the (w,z) coordinates we have

$$|\zeta - z|^2 = f(w,z)$$

and by part (a′) of 12.7.6

$$f(w,z) \leqslant C|w|^2$$

Thus all first-order derivatives of f in the w variables vanish when $w=0$, and the Taylor expansion of order 1 for f in the w variables only has the form

$$f(w,z) = \mathrm{Re} \sum_{i,j=1}^{n} g_{ij}(w,z)w_i w_j + \sum_{i,j=1}^{n} h_{ij}(w,z)w_i\bar{w}_j$$

Consequently

$$\left|D_z^\alpha D_w^\beta f(w,z)\right| \leqslant \begin{cases} C_{\alpha\beta}|w|^{2-|\beta|} & |\beta| \leqslant 2 \\ C_{\alpha\beta} & |\beta| \geqslant 2 \end{cases} \tag{12.102}$$

Making the change of variables $w = \Phi(\zeta, z)$, the integrand in (12.86) becomes

$$k(\zeta,z)|\zeta - z|^{-2q}[u(\zeta,z)\cdot(\zeta - z)]^{-p}\varphi(\zeta)d\sigma(\zeta) = w_n^{-p} f(w,z)^{-q}F(w,z)d\sigma(w)$$

where $F(w,z)$ is infinitely differentiable, and, because

$$|k(\zeta,z)| \leqslant C|\zeta - z|^r$$
$$|F(w,z)| \leqslant C|w|^r$$

Just as in (12.102) it follows that

$$\left|D_z^\alpha D_w^\beta F(w,z)\right| \leqslant \begin{cases} C_{\alpha\beta}|w|^{r-|\beta|} & |\beta| \leqslant r \\ C_{\alpha\beta} & |\beta| \geqslant r \end{cases} \tag{12.103}$$

Corresponding to (12.95) we have

$$K_1\varphi(z) = \int_{w\in M_z} w_n^{-p} f(w,z)^{-q}F(w,z)d\sigma(w)$$

The argument leading up to (12.99) yields

$$K_1\varphi(z) = \int \chi(w,z)\log w_n P_p(D)\left[f(w,z)^{-q}F(w,z)\right]d\lambda(w) \tag{12.104}$$

Examination of the integrand in (12.104), together with (12.102) and (12.103), shows that the worst behavior comes about when all of the $p+1$ derivatives in $P_p(D)$ are applied to f^{-q}. Consequently, the integrand is bounded by

$$C\chi_1(w)\left|\log w_n\right| |w|^{r-2q-p-1} \tag{12.105}$$

uniformly for $z \in \bar{V}^-$. The function in (12.105) is integrable if $2(n-q-1)+r-p \geqslant 0$. In this case $K_1\varphi \in C\left(\bar{V}^-\right)$.

Corresponding to (12.101) we have

$$X_1X_2\cdots X_nK_1\varphi(z) = \int \chi(w,z)\log w_n\tilde{X}_1\cdots\tilde{X}_N P_p(D)\left[f^{-q}F\right]d\lambda(w)$$

Notice that by (12.102) and (12.103), the derivatives with respect to z do not worsen the integrand in (12.105). However, since

$$\tilde{X} = X - 2\frac{\partial}{\partial \bar{w}_n}(XR)(w, z)$$

each vector field adds a derivative with respect to w. In general there is no help for this problem, but if X is tangential, then $X\rho \equiv 0$, so by part (d′) of Lemma 12.7.6

$$|XR| = |XH| \leqslant C|w|^2$$

Thus tangential derivatives do not worsen the integrand. Therefore if at most s of the vector fields $X_1, \ldots, X_N$ are non-tangential the integrand in (12.105) is bounded by

$$\chi_1(w)\,|\log w_n|\,|w|^{r-2q-p-s-1}$$

uniformly for $z \in \bar{V}^-$. Thus all tangential derivatives of $K_1\varphi$ belong to $C^s\left(\bar{V}^-\right)$ where $s = 2(n-q-1)+r-p$.

To prove part (c) we define

$$K_1^\epsilon\varphi(z) = \int_{\rho(\zeta)=\epsilon} k(\zeta, z)|\zeta - z|^{-2q}[u(\zeta, z)\cdot(\zeta - z)]^{-p}\varphi(\zeta)d\sigma(\zeta)$$

for $\epsilon > 0$ and $z \in \bar{\Omega}^-$. Then with $\chi^\epsilon(w, z)$ denoting the characteristic function of the set $\{(w, z) \mid R(w, z) > \epsilon\}$ we see that locally $K_1^\epsilon\varphi(z)$ is given by (12.104) or (12.101) with χ replaced by χ^ϵ. The integrands are still dominated by $C\chi_1(w)\,|\log w_n|$ or by $C\chi_1(w)\,|\log w_n|\,|w|^{r-2q-p-1}$ uniformly in $\epsilon \geqslant 0$. Thus under the hypothesis of part (c) $K_1^\epsilon\varphi \to K_1\varphi$ uniformly on compact subsets of M as $\epsilon \to 0^+$. □

Proof of Theorem 12.7.1. Since

$$L(\zeta, z) = (2\pi i)^{-n}\frac{u(\zeta, z)\cdot d(\zeta - z)}{u\cdot(\zeta - z)} \wedge \left[\frac{\bar{\partial}u\cdot d(\zeta - z)}{u\cdot(\zeta - z)}\right]^{n-1}$$

each coefficient in L is of the form of the kernel in Proposition 12.7.4 or equation (12.86) with $q = 0$. Thus parts (c) and (f) of Theorem 12.7.1 follow immediately from Theorem 12.7.3

We will only prove parts (a) and (b), since parts (d) and (e) are similar. The kernel H has the form

$$H = \sum_{p=1}^{n-1} H_p$$

where

$$\begin{aligned} H_p(\zeta, z) =& (2\pi i)^{-n}\frac{u\cdot d(\zeta - z)}{u\cdot(\zeta - z)} \wedge \frac{(\bar{\zeta} - \bar{z})\cdot d(\zeta - z)}{|\zeta - z|^2} \wedge \left[\frac{\bar{\partial}u\cdot d(\zeta - z)}{u\cdot(\zeta - z)}\right]^{p-1} \\ &\wedge \left[\frac{d(\bar{\zeta} - \bar{z})\cdot d(\zeta - z)}{|\zeta - z|^2}\right]^{n-p-1} \end{aligned}$$

Examination shows that each coefficient of H_p is of the form of the kernel in (12.86) with $q = n - p$ and $r = 1$. Thus all tangential derivatives of $H_p\left([M]^{0,1} \wedge \varphi\right)$ belong to $C^{p-1}\left(\bar{\Omega}^-\right)$. We will abbreviate this statement by saying, for example, that

$$H_p\left([M]^{0,1} \wedge \varphi\right) \in \tilde{C}^{p-1}\left(\bar{\Omega}^-\right)$$

Thus

$$H\left([M]^{0,1} \wedge \varphi\right) \in \tilde{C}^0\left(\bar{\Omega}^-\right)$$

We get the extra derivative from the homotopy formula for H. It follows from the homotopy formula Theorem 12.2.4, that

$$[\bar{\partial} H]\left([M]^{0,1} \wedge \varphi\right) = \bar{\partial} H\left([M]^{0,1} \wedge \varphi\right) + H\left([M]^{0,1} \wedge \bar{\partial}_b \varphi\right)$$

On the other hand since $\bar{\partial} H = (L + S) - B$, we have

$$[\bar{\partial} H]\left([M]^{0,1} \wedge \varphi\right) = (L + S)\left([M]^{0,1} \wedge \varphi\right) - B\left([M]^{0,1} \wedge \varphi\right)$$

Since u is a support function for M with respect to Ω^-, $S\left([M]^{0,1} \wedge \varphi\right)$ vanishes in Ω^-. Thus in Ω^-

$$\begin{aligned} \bar{\partial} H\left([M]^{0,1} \wedge \varphi\right) = L\left([M]^{0,1} \wedge \varphi\right) - B\left([M]^{0,1} \wedge \varphi\right) \\ - H\left([M]^{0,1} \wedge \bar{\partial}_b \varphi\right) \end{aligned} \tag{12.106}$$

It follows from part (c) of Theorem 12.7.1, equation (12.64), and what has already been proved about H, that $\bar{\partial} H\left([M]^{0,1} \wedge \varphi\right) \in \tilde{C}^0\left(\bar{\Omega}^-\right)$. If $\bar{\partial}_b \varphi = 0$, it follows that $\bar{\partial} H\left([M]^{0,1} \wedge \varphi\right) \in \mathcal{E}^{p,q}\left(\bar{\Omega}^-\right)$.

We can now prove part (b). In this case $H\left([M]^{0,1} \wedge \varphi\right) \in \mathcal{E}^{p,0}\left(\Omega^-\right)$ and if f is the coefficient of dz^I in $H\left([M]^{0,1} \wedge \varphi\right)$, we know that $f \in \tilde{C}^0\left(\bar{\Omega}^-\right)$, and that $\bar{\partial} f \in C^\infty\left(\bar{\Omega}^-\right)$. Since every derivative of f can be written as the sum of a tangential derivative of f and derivatives of $\partial f / \partial \bar{z}_j, 1 \leqslant j \leqslant n$, it follows that $f \in C^\infty\left(\bar{\Omega}^-\right)$.

To prove (a) we must examine $\bar{\partial} H\left([M]^{0,1} \wedge \varphi\right)$ more closely. Let

$$Z = \sum_{j=1}^{n} u_j(z, z) \frac{\partial}{\partial \bar{z}_j}$$

We first show that

$$Z(\rho) = Z \rfloor \bar{\partial} \rho \neq 0$$

To see this let $z^0 \in M$ and suppose $u_n\left(z^0, z^0\right) \neq 0$. Then choose coordinates $w = \Phi(z)$ by

$$\begin{aligned} w_j &= z_j - z_j^0, \quad 1 \leqslant j \leqslant n \leqslant 1 \\ w_n &= u\left(z, z^0\right) \cdot \left(z - z^0\right) \end{aligned}$$

By the same proof as was used to prove (12.87) we see that

$$\left.\frac{\partial}{\partial \bar{w}_j}\rho\left(\Phi^{-1}(w)\right)\right|_{w=0} = 0 \quad 1 \leqslant j \leqslant n-1$$
$$\left.\frac{\partial}{\partial \bar{w}_n}\rho\left(\Phi^{-1}(w)\right)\right|_{w=0} \neq 0.$$

But then by the chain rule

$$Z(\rho)\left(z^0\right) = \left|u\left(z^0, z^0\right)\right|^2 \left.\frac{\partial}{\partial \bar{w}_n}\rho\left(\Phi^{-1}(w)\right)\right|_{w=0} \neq 0$$

Now let

$$X = [Z(\rho)]^{-1} Z$$

If Φ is any form, then, since $X \lrcorner \bar{\partial}\rho \equiv 1$, we have

$$\Phi = X \lrcorner (\bar{\partial}\rho \wedge \Phi) + \bar{\partial}\rho \wedge X \lrcorner \Phi = \Phi_1 + \bar{\partial}\rho \wedge \Phi_2$$

In particular if $\Phi = H\left([M]^{0,1} \wedge \varphi\right)_2$ then $\Phi_2 = \sum_{p=1}^{m-1} X \lrcorner H_p\left([M]^{0,1} \wedge \varphi\right)$. We have shown that $H_p\left([M]^{0,1} \wedge \varphi\right) \in \tilde{C}^{p-1}\left(\bar{\Omega}^-\right)$. Now consider the kernel

$$Z \lrcorner H_1(\zeta, z) = -(n-1)(2\pi i)^{-n} \frac{u(\zeta, z) \cdot d(\zeta - z)}{u \cdot (\zeta - z)} \wedge \frac{(\bar{\zeta} - \bar{z}) \cdot d(\zeta - z)}{|\zeta - z|^2}$$
$$\wedge \frac{u(z, z) \cdot d(\zeta - z)}{|\zeta - z|^2} \wedge \left[\frac{d(\bar{\zeta} - \bar{z}) \cdot d(\zeta - z)}{|\zeta - z|^2}\right]^{n-1}$$

Since $u(\zeta, z) = u(z, z) + 0(|\zeta - z|)$ we see that every coefficient in $X \lrcorner H_1$ is of the form of the kernel in (12.86) with $q = n-1, p = 1$, and $r = 2$. Thus by Proposition 12.7.4 $X \lrcorner H_1\left([M]^{0,1} \wedge \varphi\right) \in \tilde{C}^1\left(\bar{\Omega}^-\right)$ and so $\Phi_2 = X \lrcorner H\left([M]^{0,1} \wedge \varphi\right) \in \tilde{C}^1\left(\bar{\Omega}^-\right)$. Thus

$$\Phi = \Phi_1 + \bar{\partial}\rho \wedge \Phi_2 \tag{12.107}$$

where $\Phi_1 \in \tilde{C}^0\left(\bar{\Omega}^-\right)$, $\Phi_2 \in \tilde{C}^1\left(\bar{\Omega}^-\right)$, and $\bar{\partial}\Phi \in \tilde{C}^0\left(\bar{\Omega}^-\right)$. However,

$$\bar{\partial}\Phi = \bar{\partial}\Phi_1 - \bar{\partial}\rho \wedge \bar{\partial}\Phi_2 \tag{12.108}$$

so $\bar{\partial}\Phi_1 \in \tilde{C}^0\left(\bar{\Omega}^-\right)$. Since $\Phi_1 = X \lrcorner (\bar{\partial}\rho \wedge \Phi)$ contains no $\bar{\partial}\rho$ term, $\bar{\partial}\Phi_1$ involves a normal derivative of each coefficient of Φ. Therefore $\Phi_1 \in \tilde{C}^1\left(\bar{\Omega}^-\right)$ and consequently $\Phi \in \tilde{C}^1\left(\bar{\Omega}^-\right)$.

Thus for every $\varphi \in \mathcal{D}^*(M)$ all tangential derivatives of $H\left([M]^{0,1} \wedge \varphi\right)$ are continuously differentiable in $\bar{\Omega}^-$. It follows immediately from (12.106) that all tangential derivatives of $\bar{\partial}H\left([M]^{0,1} \wedge \varphi\right)$ are continuously differentiable in $\bar{\Omega}^-$. □

Remark 12.7.7. At first glance it seems possible to iterate the argument and thereby show that $H\left([M]^{0,1}\wedge\varphi\right)\in\mathcal{E}^{p,q-1}\left(\bar{\Omega}^-\right)$. However, in the decomposition (12.107) and (12.108) information about the smoothness of $\bar{\partial}\Phi$ yields no information about normal derivatives of Φ_2. However, in two special cases we can say more. If $\rho(z)=|z|^2-R^2$, so that $\Omega^-=B(0,R)$, and $u(\zeta,z)=\bar{\zeta}$, or if $\rho(z)=\operatorname{Re}z_n$, so that $\Omega^-=\{z\mid\operatorname{Re}z_n<0\}$ with $u(\zeta,z)\equiv(0,\ldots,0,1)$, then an easy calculation shows that $X\lrcorner H(\zeta,z)\equiv 0$. Thus in the decomposition (12.107) for $\Phi=H\left([M]^{0,1}\wedge\varphi\right),\Phi_2=X\lrcorner\Phi=0$ and we can conclude in these two cases that $H\left([M]^{0,1}\wedge\varphi\right)\in\mathcal{F}^{p,q-1}\left(\bar{\Omega}^-\right)$ for all $\varphi\in\mathcal{D}^{p,q}(M)$.

According to Theorem 12.7.1 we can make the following definitions.

Definition 12.7.8. Suppose $u(\zeta,z)$ is a biregular support function for M relative to Ω^-. Let L and H be the kernels generated by u. For $\varphi\in Q_D^{p,q}(M)$ we define

$$\begin{aligned} H^-\varphi&=\tau\chi^-H\left([M]^{0,1}\wedge\varphi\right)\\ {}^tH^+\varphi&=\tau\chi^{+t}H\left([M]^{0,1}\wedge\varphi\right)\\ L^-\varphi&=\tau\chi^-L\left([M]^{0,1}\wedge\varphi\right)\\ {}^tL^+\varphi&=\tau\chi^{+t}L\left([M]^{0,1}\wedge\varphi\right)\end{aligned}$$

It follows from 12.7.1 that these operators (kernels) are regular operators (kernels) on M. It follows from the next result that these operators are biregular on M.

Proposition 12.7.9. *Suppose $u(\zeta,z)$ is a biregular support function for M relative to Ω^-. Then*

$$^t\left(H^-\right)=-^tH^+;\quad ^t\left(L^-\right)=-^tL^+$$

Proof. We will provide the proof of the first identity only. By part (c) of Theorem 12.7.3

$$H^-\varphi(z)=\lim_{\epsilon\to0+}\tau\int_{p(\zeta)=\epsilon}H(\zeta,z)\wedge\varphi(\zeta)=\lim_{\epsilon\to0+}\tau H\left(\left[M_\epsilon\right]^{0,1}\wedge\varphi\right)$$

Thus if $\varphi\in\mathcal{D}^{p,q}(M)$ and $\psi\in\mathcal{D}^{n-p,n-q+1}(M)$

$$\begin{aligned}\left(H^-\varphi,\psi\right)_M&=\int_{z\in M}H^-\varphi(z)\wedge\psi(z)\\ &=\left([M]^{0,1}\wedge H^-\varphi,\psi\right)_\Omega\\ &=\lim_{\epsilon\to0+}\left([M]_{0,1}\wedge H\left(\left[M_\epsilon\right]^{0,1}\wedge\varphi\right),\psi\right)_\Omega\\ &=\lim_{\epsilon\to0+}(-1)^{p+q+1}\left(H\left(\left[M_\epsilon\right]^{0,1}\wedge\varphi\right),[M]^{0,1}\wedge\psi\right)_\Omega\\ &=\lim_{\epsilon\to0+}(-1)^{p+q+1}\left(\left[M_\epsilon\right]^{0,1}\wedge\varphi,{}^tH\left([M]^{0,1}\wedge\psi\right)\right)_\Omega\\ &=\lim_{\epsilon\to0+}(-1)^{p+q+1}\int_{M_\epsilon}\varphi(\zeta)\wedge{}^tH\left([M]^{0,1}\wedge\psi\right)\\ &=(-1)^{p+q+1}\int_M\varphi\wedge^tH^+\psi\\ &=\left(\varphi,-c^tH^+\psi\right)_M\end{aligned}$$

So $(H^-)' = -c^t H^+$ where $cf = (-1)^{\dim f} f$ (see (12.4)). On the other hand $(H^-)' = c^t\,(H^-)$ so the result follows.

Next we will carry through the result in Lemma 12.6.11 to the kernels L, H, etc. The situation in general is quite involved so we will limit our considerations to the strictly pseudo-convex case. In addition we will require that the support function $u(\zeta, z)$ and the defining function ρ of M are related. Specifically we will assume that ρ has the expansion

$$\rho(z) = \rho(\zeta) + \mathrm{Re}[\lambda(\zeta, z)u(\zeta, z) \cdot (\zeta - z)] + r(\zeta, z) \tag{12.109}$$

for z and ζ near M where:

$$\lambda(\zeta, z) \text{ never vanishes and is holomorphic in } z \text{ for each fixed } \zeta \tag{12.110}$$

$$|r(\zeta, z)| \leqslant C\,|\zeta - z|^2 \tag{12.111}$$

$$r(\zeta, z) \geqslant C|\zeta - z|^2 \quad \text{if} \quad u(\zeta, z) \cdot (\zeta - z) = 0 \tag{12.112}$$

It will be proved below (12.115) and (12.116) that the holomorphic tangent space $H_z(M)$ is given by

$$H_z(M) = \{w \mid u(z, z) \cdot w = 0\}$$

Thus (12.112) implies that M is strictly pseudo-convex as the boundary of Ω^-. By (12.111) $r(\zeta, z)$ vanishes to second order on the diagonal. Let $Q(z, \zeta - z)$ denote the quadratic part of $r(\zeta, z)$ expanded in ζ about $\zeta = z$ with z held fixed. Then

$$r(\zeta, z) = Q(z, \zeta - z) + O\left(|\zeta - z|^3\right)$$

Theorem 12.7.10. *Suppose $M = \{z \in \Omega \mid \rho(z) = 0\}$ and suppose $u(\zeta, z)$ is a support function for M relative to $\Omega^- = \{z \in \Omega \mid \rho(z) < 0\}$ which satisfies (12.109—12.112).*

(a) *With either z or ζ fixed each coefficient of $H(\zeta, z)$ has a locally integrable restriction to M. Thus for $\varphi \in \mathcal{D}^{p,q}(M)$ and for $z \in M$ we have*

$$H^-\varphi(z) = \tau \int_M [H(\zeta, z) \wedge \varphi(\zeta)]^{n,n-1}$$

$$^tH^+\varphi(z) = \tau \int_M \left[{}^tH(\zeta, z) \wedge \varphi(\zeta)\right]^{n,n-1}$$

(b) *Let $\varphi \in \mathcal{D}^{p,0}(M)$. For each $z \in M$, the principal value integral*

$$L_b\varphi(z) = \lim_{\epsilon \to 0^+} \int_{|u\cdot(\zeta - z)| \geqslant \epsilon} L(\zeta, z) \wedge \varphi(\zeta)$$

exists. Furthermore

$$L^-\varphi(z) = L_b\varphi(z) - c(z)\varphi(z)$$

where

$$c(z) = \frac{1}{2}(2\pi i)^{1-n} \int_{\substack{w\in H_z(M)\\ Q(z,w)\leqslant 1}} \left[\sum \frac{\partial\rho(z)}{\partial z_i \partial \bar{z}_j} d\bar{w}_j \wedge dw_i\right]^{n-1} \tag{12.113}$$

(c) Let $\varphi \in \mathbb{D}^{p,n-1}(M)$*. For each* $z \in M$ *the principal value integral*

$$^tL_b\varphi(z) = \lim_{\epsilon\to 0+} \tau \int_{|u\cdot(\zeta - z)|\geqslant\epsilon} {}^tL(\zeta, z)\wedge\varphi(\zeta)$$

exists. Furthermore

$$^tL^+\varphi(z) = {}^tL_b\varphi(z) + c(z)\varphi(z)$$

The integrand in (12.113) is essentially the Levi form for M. An easy calculation shows that $c(z)$ is also given by

$$c(z) = \frac{1}{2}\frac{\lambda_{2n-2}\{w \in H_z(M) \mid Q(z,w) \leqslant 1\}}{\lambda_{2n-2}\left\{w\in H_z(M) | \sum \frac{\partial^2\rho}{\partial z_i\partial \bar{z}_j} w_i \bar{w}_j \leqslant 1\right\}} \tag{12.114}$$

where λ_{2n-2} is Hausdorff measure in dimension $2n-2$.

The following case of part (b) of Theorem 12.7.10 follows immediately from (12.114).

Corollary 12.7.11. *Suppose* $M = \{z \mid \rho(z) = 0\}$ *is strictly pseudo-convex and* $u(\zeta, z)$ *is a support function for* M *with respect to* $\Omega^- = \{z \mid \rho(z) < 0\}$ *such that*

$$u(\zeta, z)\cdot(\zeta - z) = F(\zeta, z) \sum_{1\leqslant|\alpha|\leqslant 2} \frac{1}{\alpha!}\frac{\partial^\alpha\rho}{\partial\zeta^\alpha}(\zeta)(z-\zeta)^\alpha + O\left(|z-\zeta|^3\right)$$

where $F(\zeta, z)$ *is holomorphic in* z *for fixed* ζ *and never vanishes. Then*

$$L^-\varphi(z) = L_b\varphi(z) - \frac{1}{2}\varphi(z)$$

$$^tL^+\varphi(z) = {}^tL_b\varphi(z) + \frac{1}{2}\varphi(z)$$

The special case of Corollary 12.7.11 where M is the Heisenberg group and L is the Cauchy-Szego kernel due to Koranyi-Vagi [134].

Proof of Theorem 12.7.10. Part (a) is a standard result (see Henkin [104] or Ovrelid [168]) and the proof will not be given. However, the key idea is to find a change of variables similar to that constructed in Lemma 12.7.6. In this case there is a map $w = \Phi(\zeta, z)$ such that (a'), (b), (c), (d') of Lemma 12.7.6 are satisfied and in addition $H(w', z) \geqslant C|w'|^2$. The details are left to the reader. It also follows from this argument that if $\varphi(z) = 0$, then $L(\zeta, z) \wedge \varphi(\zeta)$ is integrable on M. Thus

$$L^{-}\varphi(z) = L_b\varphi(z)$$

Thus we have reduced part (b) to the case where $\varphi \in C_0^\infty(M)$ and $\varphi \equiv 1$ near z. Let χ^+ denote the characteristic function of $\Omega^+ = \{z \mid \rho(z) > 0\}$. Let $B_\epsilon = \{\zeta \mid |u(\zeta, z) \cdot (\zeta - z)| \leqslant \epsilon\}$ and let χ_ϵ^+ denote the characteristic function of $\Omega_\epsilon^+ = \Omega^+ \setminus B_\epsilon$. Then

$$\bar{\partial}\chi^+ = [M]^{0,1}$$
$$\bar{\partial}\chi_\epsilon^+ = [M_\epsilon]^{0,1} + [S_\epsilon]^{0,1}$$

where $M_\epsilon = M \setminus B_\epsilon$ and $S_\epsilon = \partial B_\epsilon \cap \Omega^+$.

It follows from (12.109) and (12.112) that $B_\epsilon \cap \Omega^+ \subset\subset \Omega$ for ϵ small. Suppose that ϵ is so small that $\varphi \equiv 1$ on $B_\epsilon \cap \Omega^+$. Then from the homotopy formula for $L + S$ applied to $\chi^+\varphi$ it follows that for $w \in \Omega^-$.

$$O = L\left(\varphi[M]^{0,1}\right)(w) + L\left(\chi^+\bar{\partial}\varphi\right)(w)$$
$$O = L\left(\varphi\,[M_\epsilon]^{0,1}\right)(w) + L\left([S_\epsilon]^{0,1}\right)(w) + L\left(\chi^+\bar{\partial}\varphi\right)(w)$$

Thus for $w \in \Omega^-$

$$L\left(\varphi[M]^{0,1}\right)(w) = L\left(\varphi\,[M_\epsilon]^{0,1}\right)(w) + L\left([S_\epsilon]^{0,1}\right)(w)$$

The right hand side obviously continues to $\bar{\Omega}^- \cap B_\epsilon$ so

$$L^-\varphi(z) = L\left(\varphi\,[M_\epsilon]^{0,1}\right)(z) + L\left([S_\epsilon]^{0,1}\right)(Z)$$

Since $L_b\varphi(z) = \lim_{\epsilon\to 0^+} L\left(\varphi\,[M_\epsilon]^{0,1}\right)(z)$ it suffices to prove that

$$\lim_{\epsilon\to 0^+} L\left([S_\epsilon]^{0,1}\right)(z) = -c(z)$$

Let

$$c_\epsilon(z) = -L\left([S_\epsilon]^{0,1}(z)\right) = \int_{\zeta\in S_\epsilon} L(\zeta, z)$$

Set

$$v(\zeta, z) = -\lambda(\zeta, z)u(\zeta, z) \tag{12.115}$$

Then, by an easy calculation we see that

$$\begin{aligned} L(\zeta, z) &= (2\pi i)^{-n}(u \cdot (\zeta - z))^{-n} u \cdot d(\zeta - z) \wedge [\bar{\partial} u \cdot d(\zeta - z)]^{n-1} \\ &= (2\pi i)^{-n}(v \cdot (\zeta - z))^{-n} v \cdot d(\zeta - z) \wedge [\bar{\partial} v \cdot d(\zeta - z)]^{n-1} \end{aligned}$$

By (12.109) and (12.111)

$$\rho(z) = \rho(\zeta) + \mathrm{Re}[v(\zeta, z) \cdot (z - \zeta)] + O\left(|\zeta - z|^2\right)$$

By Taylor's Theorem

$$\rho(z) = \rho(\zeta) + 2\,\mathrm{Re}\left[\sum \frac{\partial \rho}{\partial \zeta_j}(\zeta)\,(z_j - \zeta_j)\right] + O\left(|\zeta - z|^2\right)$$

Thus

$$\mathrm{Re}[v(\zeta, z) \cdot (z - \zeta)] = 2\,\mathrm{Re}\left[\sum \frac{\partial \rho}{\partial \zeta}(\zeta)\,(z_j - \zeta_j)\right] + O\left(|\zeta - z|^2\right)$$

Since both functions are holomorphic in z and vanish for $z = \zeta$, we conclude that the first derivatives at $z = \zeta$ must also agree.Thus

$$v_j(\zeta, \zeta) = 2\frac{\partial \rho}{\partial \zeta_j}(\zeta) \tag{12.116}$$

Differentiating (12.116), noticing that $v(\zeta, z)$ is holomorphic in z, we see that

$$\frac{\partial v_j}{\partial \bar{\zeta}_k}(\zeta, \zeta) = 2\frac{\partial^2 \rho}{\partial \zeta_j \partial \bar{\zeta}_k}(\zeta) \tag{12.117}$$

Let

$$\frac{\partial v_j}{\partial \zeta_k}(\zeta, \zeta) = a_{jk}(\zeta)$$

Define the matrices

$$\begin{aligned} h(z) &= \left(\frac{\partial^2 \rho}{\partial z_j \partial \bar{z}_k}(z)\right) \\ A(z) &= (a_{jk}(z)) \end{aligned} \tag{12.118}$$

Then from (12.116), (12.117), and (12.118) it follows that

$$v(\zeta, z) = 2\partial\rho(z) + A(z)(\zeta - z) + 2h(z)(\bar{\zeta} - \bar{z}) + O\left(|\zeta - z|^2\right) \tag{12.119}$$

With $z \in M$ fixed we make a change of variables of the form

$$w = B(\zeta)(\zeta - z) \tag{12.120}$$

where $B(\zeta)$ is an invertible matrix, and we insist that

$$w_n = v(\zeta, z) \cdot (\zeta - z) \tag{12.121}$$

(for example if $v_n(z, z) \neq 0$ we can set $w_j = (\zeta_j - z_j)$ for $1 \leqslant j \leqslant n - 1$). The inverse of (12.120) will have the form

$$\zeta - z = C(w)w \tag{12.122}$$

By (12.121) and (12.122),

$$v \cdot d(\zeta - z) = dw_n - dv \cdot C(w)w \tag{12.123}$$

Using (12.119) we see that

$$\bar{\partial} v = \bar{\partial}_\zeta v = 2h(z)d(\bar{\zeta} - \bar{z}) + O(d(\bar{\zeta} - \bar{z}), |\zeta - z|)$$

The notation $O(d(\bar{\zeta}-\bar{z}), |\zeta-z|)$ means each term in the expression contains a $d\left(\bar{\zeta}_j - \bar{z}_j\right)$ with a coefficient which is $O(|\zeta - z|)$. Using (12.122) this becomes in terms of the w coordinate

$$\bar{\partial} v = 2h(z)\bar{C}(w)d\bar{w} + O(d\bar{w}, |w|)$$

and hence

$$\bar{\partial} v \cdot d(\zeta - z) = 2\left[{}^tC(w)h(z)\bar{C}(w)d\bar{w}\right] \cdot dw + O(dw, d\bar{w}, |w|) \tag{12.124}$$

By (12.109) and (12.109) we see that in the w coordinates

$$S_\epsilon = \{w \mid |w_n| = \epsilon|\lambda(z + Cw, z)|, R(z + Cw, z) < \operatorname{Re} w_n\} \tag{12.125}$$

Now we make the change of variables

$$w_j = \epsilon^{1/2} w_j' \quad 1 \leqslant j \leqslant n-1; \quad w_n = \epsilon w_n'$$

Then by (12.123) and (12.124)

$$\epsilon^{-1} v \cdot d(\zeta - z) = dw_n' + \sum_{j=1}^{n-1} \left[a_j(z)dw_j' + b_j(z)d\bar{w}_j'\right] + O\left(\epsilon^{1/2}\right) \tag{12.126}$$

$$\epsilon^{-1}\bar{\partial} v \cdot d(\zeta - z) = 2h_n(z)d\bar{w}' \cdot dw' + O\left(\epsilon^{1/2}\right) \tag{12.127}$$

The exact values of the coefficients a_j and b_j in (12.126) are unimportant; what is important is that the sum in (12.126) does not contain dw_n' or $d\bar{w}_n'$. The matrix $h_n(z)$ in (12.127) is the matrix ${}^tC(0)h(z)\bar{C}(0)$ with the nth row and column replaced by zeroes.

We will now drop the primes and let $\epsilon \to 0$. Then by (12.126) and (12.127)

$$\begin{aligned} L(\zeta, z) \to &(2\pi i)^{-n} w_n^{-n} \left(dw_n + \sum_{j=1}^{n-1} \left[a_j dw_j + b_j d\bar{w}_j\right]\right) \wedge (2h_n d\bar{w} \cdot dw)^{n-1} \\ =&\frac{1}{2}(\pi i)^{-n} w_n^{-n} dw_n \wedge (h_n(z)d\bar{w} \cdot dw)^{n-1} \end{aligned}$$

By (12.125) and

$$S_\epsilon \to S_0 = \{w \mid |w_n| = |\lambda(z, z)|, Q\left(z, C(0)w'\right) < \operatorname{Re}\left(w_n\right)\}$$

here $w' = (w_1, \ldots, w_{n-1}, 0)$.

Thus $c_\epsilon \to c(z)$ where

$$c(z) = \frac{1}{2}(\pi i)^{-n} \int_{S_0} w_n^{-n} dw_n \wedge (h_n d\bar{w} \cdot dw)^{n-1}$$

Set $\zeta_j = [\operatorname{Re} w_n]^{1/2} w_j, 1 \leqslant j \leqslant n-1, \zeta_n = w_n$, then

$$c(z) = \frac{1}{2}(\pi i)^{-n} \int_{\substack{|\zeta_n|=\lambda \\ \operatorname{Re}\zeta_n>0}} \zeta_n^{-n} (\operatorname{Re}\zeta_n)^{n-1} d\zeta_n \int_{\substack{Q(z,C(0)\zeta)<1 \\ \zeta_n=0}} \left(h_n d\bar{\zeta} \cdot d\zeta\right)^{n-1}$$

$$= \frac{1}{2}(2\pi i)^{1-n} \int_{Q(z,C(0)s)<1} \left[{}^t C(0)h(z)(\bar{C}(0)d\bar{\zeta}) \cdot d\zeta\right]^{n-1}$$

Finally let $w = C(0)\zeta$. When $\zeta_n = 0$ it follows from (12.121) and (12.122) that $v(z,z) \cdot w = 0$. Thus $w \in H_z(M)$ and (12.7.10) follows.

Part (c) follows from part (b) by duality, using Proposition 12.7.9 and the fact that $(L_b)' = -{}^t L_b$. □

12.8 Fundamental Solutions for the Induced Cauchy-Riemann Operator

Here we will present two fundamental solutions for the induced Cauchy Riemann equations on a pseudo-convex hyper-surface which has a biregular support function. In the strictly pseudo-convex case the first fundamental solution was first presented by Romanov [183, 184] and the second independently by Henkin [105] and Skoda [196]. Both can be constructed easily using the results of the previous sections.

Theorem 12.8.1. *Let $u(\zeta, z)$ be a biregular support function for M with respect to Ω^- defined in a neighborhood of $M \times M$. Let H, L, and S be the kernels generated by u. Set*

$$E = {}^t H^+ - H^-$$
$$P = {}^t L^+ - L^-$$

on $M \times M$. Then the differential equations

$$\bar{\partial}_b E = -[\Delta_M] + P \tag{12.128}$$

and

$$\bar{\partial}_b P = 0, \tag{12.129}$$

are satisfied on $M \times M$. The corresponding operator equations are

$$\varphi = \bar{\partial}_b E\varphi + E\bar{\partial}_b\varphi + P\varphi \tag{12.130}$$

$$\bar{\partial}_b P\varphi = P\bar{\partial}_b\varphi = 0 \tag{12.131}$$

for all $\varphi \in \mathcal{D}^*(M)$.

Remark 12.8.2. Since the generating function u is holomorphic in z, L acts non-trivially only on $p, 1$ forms and ${}^tL$ acts non-trivially only on p, n forms. Consequently, P acts non-trivially only on $p, 0$ forms and $p, n-1$ forms. Thus (12.130) has three distinct cases

$$\begin{aligned} q &= 0 & \varphi &= E\bar{\partial}_b\varphi - L^-\varphi \\ 1 \leqslant q &\leqslant n-2 & \varphi &= \bar{\partial}_b E\varphi + E\bar{\partial}_b\varphi \\ q &= n-1 & \varphi &= \bar{\partial}_b E\varphi + {}^tL^+\varphi \end{aligned} \tag{12.132}$$

Remark 12.8.3. The kernel E is called a fundamental solution for $\bar{\partial}_b$ on M in view of (12.132).

Proof. Since u is a support function for M with respect to Ω^-,

$$\begin{aligned} S\left([M]^{0,1} \wedge \varphi\right) &\equiv 0 \quad \text{in} \quad \Omega^- \\ {}^tS\left([M]^{0,1} \wedge \varphi\right) &\equiv 0 \quad \text{in} \quad \Omega^+ \end{aligned}$$

for all $\varphi \in \mathcal{D}^{p,q}(M)$. Thus we have

$$\begin{aligned} [\bar{\partial}H]\left([M]^{0,1} \wedge \varphi\right) &= L\left([M]^{0,1} \wedge \varphi\right) - B\left([M]^{0,1} \wedge \varphi\right) \quad \text{in } \Omega^- \\ \left[\bar{\partial}^t H\right]\left([M]^{0,1} \wedge \varphi\right) &= {}^tL\left([M]^{0,1} \wedge \varphi\right) - B\left([M]^{0,1} \wedge \varphi\right) \quad \text{in } \Omega^+ \end{aligned} \tag{12.133}$$

On the other hand, by the general homotopy formula, Theorem 12.2.4,

$$\begin{aligned} [\bar{\partial}H]\left([M]^{0,1} \wedge \varphi\right) &= \bar{\partial}H\left([M]^{0,1} \wedge \varphi\right) + H\left([M]^{0,1} \wedge \bar{\partial}_b\varphi\right) \\ \left[\bar{\partial}^t H\right]\left([M]^{0,1} \wedge \varphi\right) &= \bar{\partial}^t H\left([M]^{0,1} \wedge \varphi\right) + {}^tH\left([M]^{0,1} \wedge \bar{\partial}_b\varphi\right) \end{aligned} \tag{12.134}$$

Taking boundary values in (12.133) and (12.134) we have

$$\begin{aligned} B^-\varphi &= L^-\varphi + \bar{\partial}_b H^-\varphi + H^-\bar{\partial}_b\varphi \\ B^+\varphi &= {}^tL^+\varphi + \bar{\partial}_b{}^tH^+\varphi + {}^tH^+\bar{\partial}_b\varphi \end{aligned}$$

Finally by (12.62) we see that

$$\begin{aligned} \varphi &= B^+\varphi - B^-\varphi \\ &= \left({}^tL^+ - L^-\right)\varphi + \bar{\partial}_b\left({}^tH^+ - H^-\right)\varphi + \left({}^tH^+ - H^-\right)\bar{\partial}_b\varphi \\ &= P\varphi + \bar{\partial}_b E\varphi + E\bar{\partial}_b\varphi \end{aligned}$$

This proves (12.130), which is equivalent to (12.128) because of the general homotopy formula Theorem 12.2.4.

Since the generating function $u(\zeta, z)$ is holomorphic in z, it follows that L acts non-trivially only on $p, 1$ forms, and ${}^tL$ acts non-trivially only on p, n forms. Thus we have three cases:

$$\begin{aligned} P\varphi &= 0 & \varphi &\in \mathcal{D}^{p,q}(M), \quad q \neq 0, n-1 \\ P\varphi &= -L^-\varphi & \varphi &\in \mathcal{D}^{p,0}(M) \\ P\varphi &= L^t L^+\varphi & \varphi &\in \mathcal{D}^{p,n-1}(M). \end{aligned}$$

Consequently, the operator equations (12.131) (and hence the equivalent differential equation (12.129) say that

$$\bar{\partial}_b L^- \varphi = 0 \quad \text{for} \quad \varphi \in \mathcal{D}^{p,0}(M)$$

and

$${}^t L^+ + \bar{\partial}_b \varphi = 0 \quad \text{for} \quad \varphi \in \mathcal{D}^{p,n-2}(M)$$

Since $u(\zeta, z)$ is holomorphic in z, we see by differentiating under the integral that

$$L\left([M]^{0,1} \wedge \varphi\right)(z) = -\int_M L(\zeta, z) \wedge \varphi(\zeta)$$

is holomorphic in Ω^-. Thus $\bar{\partial}_b L^- \varphi = 0$.

Since ${}^t u(\zeta, z)$ is homomorphic in ζ, we have that

$$\int_{\zeta \in M} {}^t L(\zeta, z) \wedge \bar{\partial}_b \psi(\zeta) = -\int_{\zeta \in M} \bar{\partial}_b \left[{}^t L(\zeta, z) \wedge \psi(\zeta)\right] = 0$$

for $z \in \Omega^+$ and $\psi \in \mathcal{Q}^{n,n-1}(M)$. Therefore ${}^t L^+ \bar{\partial}_b \psi = 0$ which completes the proof of Theorem 12.8.1. □

We will consider two applications of Theorem 12.8.1. First suppose M is compact and is the boundary of Ω^-.

Theorem 12.8.4. *Suppose $M = \partial\Omega^-$ is compact and that u is a support function for M with respect to Ω^- defined in a neighborhood of $M \times M$. Let E and P denote the kernels on M generated by u.*

(a) $q = 0$. $\varphi \in \mathcal{E}^{p,0}(M)$ satisfies $\partial_b \varphi = 0$ if and only if $\varphi = P\varphi$.

(b) $1 \leqslant q \leqslant n - 2$. Given $\varphi \in \mathcal{E}^{p,q}(M)$, there is a $\psi \in \mathcal{E}^{p,q-1}(M)$ such that $\bar{\partial}_b \psi = \varphi$ if and only if $\bar{\partial}_b \varphi = 0$.

(c) $q = n - 1$. Given $\varphi \in \mathcal{E}^{p,n-1}(M)$, there is a $\psi \in \mathcal{E}^{p,n-2}(M)$ such that $\bar{\partial}_b \psi = \varphi$ if and only if $P\varphi = 0$.

Theorem 12.8.4 is an immediate consequence of Theorem 12.8.1.

As a second consequence, we will prove a generalization of a theorem of Sato [191]. His startling result characterizes the range (locally) of the Hans Lewy operator ($\bar{\partial}_b$ on the hyperquadric $y_2 = |z_1|^2$ in $\mathbb{C}^2$) in a situation where a priori estimates would seem ineffective because of the well known denseness of the range. Sato's theorem was extended by Greiner-Kohn-Stein [88] to a characterization of the range (locally) of $\bar{\Box}_b$ on the hyperquadric. The result we present extends Sato's theorem to $\bar{\partial}_b$ on certain pseudoconvex manifolds (including, in particular, cases which are not strongly pseudoconvex.) Obtaining this result was the initial impetus for the present paper. The strongly pseudoconvex case first appears in Henkin [105].

Definition 12.8.5. A localization of a current $v \in \mathcal{D}'^{p,q}(M)$ near a point $z_0 \in M$ is a current $\tilde{v} \in \mathcal{E}'^{p,q}(M)$ such that $\tilde{v} = v$ near z_0.

Theorem 12.8.6. *Let M be a real analytic hypersurface in Ω and suppose $u(\zeta, z)$ is a real analytic support function for M with respect to Ω^-. Let E and P be the operators on M generated by u.*

(a) A current $v \in \mathcal{D}'^{p,n-1}(M)$ is in the range of $\bar{\partial}_b$ near a point $z_0 \in M$ if and only if there is a localization $\tilde{v}$ of v near z_0 such that $P\tilde{v}$ is real analytic near z_0.

(b) If u is a strict support function for M with respect to Ω^- and $P\tilde{v}$ is real analytic near z_0 for one localization $\tilde{v}$ of v near z_0, then this is true for all localizations of v near z_0.

Proof.

(a) If there is a localization $\tilde{v}$ such that $P\tilde{v}$ is real analytic near z_0, then by the Cauchy-Kowalewski Theorem there is a real analytic form φ such that $\bar{\partial}_b\varphi = P\tilde{v}$ near z_0. But then by (12.130) we have near z_0

$$\begin{aligned} v = \tilde{v} &= \bar{\partial}_b E\tilde{v} + P\tilde{v} \\ &= \bar{\partial}_b[E\tilde{v} + \varphi] \end{aligned}$$

Suppose on the other hand that there is a $w \in \mathcal{D}'^{p,n-2}(M)$ such that $\bar{\partial}_b w = v$ near z_0. We may assume $w \in \mathcal{E}'^{p,n-2}(M)$, and then $\tilde{v} = \bar{\partial}_b w$ is a localization of v near z_0. Since $P\tilde{v} = P\bar{\partial}_b w \equiv 0$ it is certainly real analytic near z_0.

(b) Let $\tilde{v}_1$ be another localization of v near z_0. Then

$$w = \tilde{v} - \tilde{v}_1 \in \mathcal{G}'p, n-2(M) \text{ and } w \equiv 0 \text{ near } z_0 \,.$$

By Theorem 12.4.9, ${}^tL\left([M]^{0,1} \wedge w\right)$ is real analytic in the compliment of the set

$$F = \{z \mid \text{ there is a } \zeta \in \operatorname{supp} w \text{ such that } u(\zeta, z) \cdot (\zeta - z) = 0\} \,.$$

Since u is a strict support function for $M, z_0 \notin F$ because $w \equiv 0$ near z_0. Thus ${}^tL\left([M]^{0,1} \wedge w\right)$ is real analytic near z_0, and so $Pw = \tau^t L\left([M]^{0,1} \wedge w\right)$ is also real analytic near z_0. Finally,

$$P\tilde{v}_1 = P\tilde{v} + Pw$$

is real analytic near z_0.

□

We will describe the second fundamental solution for $\bar{\partial}_b$ only for a strongly pseudo-convex hypersurface. As we did for Theorem 12.7.10, we will assume that the support function $u(\zeta, z)$ and the defining function ρ of the hypersurface M are related by (12.109—12.112). As we did in Section 12.3 we let $u^1(\zeta, z) = u(\zeta, z)$, $\quad u^2(\zeta, z) =^1 u(\zeta, z) = u(z, \zeta)$ and $u^3(\zeta, z) = \bar{\zeta} - \bar{z}$. All are defined in a neighborhood of $M \times M$ and we have (see 12.22)

$$\bar{\partial} K = H - {}^t H - R \tag{12.135}$$

where

$$\begin{aligned} R(\zeta, z) =& (2\pi i)^{-n} E_{12} \\ =& (2\pi i)^{-n} \frac{u \cdot d(\zeta - z)}{u \cdot (\zeta - z)} \wedge \frac{{}^t u \cdot d(\zeta - z)}{{}^t u \cdot (\zeta - z)} \\ & \wedge \sum_{p=1}^{n-1} \left[\frac{\bar{\partial} u \cdot d(\zeta - z)}{u \cdot (\zeta - z)} \right]^{p-1} \wedge \left[\frac{\bar{\partial}^t u \cdot d(\zeta - z)}{t^t u \cdot (\zeta - z)} \right]^{n-p-1} \end{aligned}$$

and $K = (2\pi i)^{-n} E_{123}$ (see 12.20). Equation 12.135) holds only in the complement of the singular set where either $u \cdot (\zeta - z)$ or ${}^t u \cdot (\zeta - z)$ vanishes. We will also denote by R the tangential piece of R on $M \times M$, so that $R \in \mathcal{D}^{n,n-2}(M \times M)$. Notice that $R \in \mathcal{E}^{n,n-2}(M \times M - \Delta_M)$.

Theorem 12.8.7. *Suppose* $M = \{z \mid \rho(z) = 0\}$ *and suppose* $u(\zeta, z)$ *is a support function for* M *relative to* $\Omega^- = \{z \in \Omega \mid \rho(z) < 0\}$ *which satisfies (12.109—12.112).*

The kernel $R(\zeta, z)$ *is locally integrable in the variable* ζ *on* M *uniformly for* $z \in M$*; and locally integrable in* z *on* M *uniformly for* $z \in M$*. Moreover, the fundamental equation*

$$\bar{\partial}_b R = -[\Delta_M] + P$$

for $\bar{\partial}_b$ *is satisfied.*

Remark 12.8.8. Thus R has exactly the same properties as the kernel E of Theorem 12.8.1, and consequently is also a fundamental solution for $\bar{\partial}_b$ on M.

Proof. Suppose $z_0 \in M$. Then we can find a change of variables $w = \Phi_z(\zeta) = \Phi(\zeta, z)$ which satisfies properties (a'), (b), (c), and (d') of Lemma 12.7.6 has the additional features that the remainder term, $H(w', z) \geqslant C|w'|^2$. We use (w', t) as coordinates on M where $t = \operatorname{Im} w_n$. Then on M

$$|w_n| \approx |t| + |w'|^2$$

so on M

$$|u \cdot (\zeta - z)| \approx |t| + |w'|^2$$

Next we will prove that

$$|'u \cdot (\zeta - z)| \approx |t| + |w'|^2$$

Note that $\left|\operatorname{Re}^t u \cdot (\zeta - z)\right| \geqslant k\left|w'\right|^2$ because, by hypothesis, (12.109–12.112) are valid. Now

$$\left|\operatorname{Im}^t u \cdot (\zeta - z)\right| \geqslant \left|\operatorname{Im} u \cdot (\zeta - z)\right| - \left|\operatorname{Im}^t u \cdot (\zeta - z) - \operatorname{Im} u \cdot (\zeta - z)\right| .$$

We have $\left|'u \cdot (\zeta - z) - u \cdot (\zeta - z)\right| \approx \left|w'\right|^2$ because

$$^t u(\zeta, z) = u(\zeta, z) + O((\zeta - z)) \tag{12.136}$$

Therefore

$$\left|\operatorname{Im}^t u \cdot (\zeta - z)\right| \geqslant |t| - c\left|w'\right|^2$$

Consequently $\left|'u \cdot (\zeta - z)\right| \geq \max\left\{k\left|w'\right|^2, |t| - c\left|w'\right|^2\right\}$ which is $\geqslant (1/(2 + c/k)).\left(|t| + k\left|w'\right|^2\right)$ by the elementary inequality $\max\{\alpha, \beta - \gamma\} \geqslant 1/(2 + \gamma/\alpha)(\alpha + \beta)$ (see [83]).

It also follows from (12.136) that

$$u \cdot d(\zeta - z) \wedge^t u \cdot d(\zeta - z) = O(|\zeta - z|)$$

Thus each coefficient in R is dominated by

$$\left(|t| + |w'|\right)\left(|t| + \left|w'\right|^2\right)^{-n}$$

which is locally integrable on $\mathbb{C}^{n-1} \times \mathbb{R}$.

Similarly each coefficient of E_{123} is dominated by a term of the form

$$\left(|t| + |w'|\right)^{2-q}\left(|t| + \left|w'\right|^2\right)^{q-n}$$

where $1 \leqslant q \leqslant n-2$. Furthermore each derivative (in z or ζ) of each coefficient of E_{123} is dominated by a term of the form

$$\left(|t| + |w'|\right)^{1-q}\left(|t| + \left|w'\right|^2\right)^{q-n}$$

where $0 \leqslant q \leqslant n-2$. All of these expressions are locally integrable on $\mathcal{C}^{n-1} \times \mathbb{R}$. Thus all first derivatives of E_{123} are locally integrable on $M \times M$, and consequently the restriction of equation (12.135) to $M \times M$ extends across the diagonal Δ_M. Since $E = {}^tH^+ - H^-$ is also an integrable kernel on $M \times M$ Theorem 12.7.10 we have

$$\bar{\partial}_b\left[(2\pi i)^{-n} E_{123}\right] = E - R \quad \text{on} \quad M \times M$$

From this equation and (12.128), it follows that

$$\bar{\partial}_b R = \bar{\partial}_b E = -\left[\Delta_M\right] + P$$

□

Remark 12.8.9. Let K denote the fundamental solution for $\bar{\Box}_b$ on the hypersurface $M = \left\{\zeta = (\zeta', \zeta_n) \;\middle|\; |\zeta'|^2 = \operatorname{Im}\zeta_n\right\}$ obtained by Folland-Stein [73] and Greiner-Kohn-Stein [89]. Then $R = \bar{\partial}_b^* \circ K$. To see this, use the facts that $2iu(\zeta, z) \cdot (\zeta - z) = w_n\left(\zeta^{-1}z\right)$, and $2i^t u(\zeta, z) \cdot (\zeta - z) = \overline{w_n\left(\zeta^{-1}z\right)}$ mentioned in Example 2 of Section 12.3 to obtain

$$R(\zeta, z) = (2i\pi^n)^{-1} \sum_{p=1}^{n-1} w_n\left(\zeta^{-1}z\right)^{-(n-p)} W_n\overline{(\zeta^{-1}z)}^{-p} (\zeta' - z') \cdot d(\zeta' - z')$$
$$\wedge \left[d\bar{\zeta}' \cdot d(\zeta' - z')\right]^{n-p-1} \wedge \left[d\bar{z}' \cdot d(\zeta' - z')\right]^{n-1} \frac{1}{2i} d(\zeta_n - z_n)$$

Next compute that

$$\begin{aligned} K(\zeta, z) = & \frac{-1}{4\pi^n} \sum_{q=1}^{n-2} \frac{1}{(n-q-1)q} \frac{1}{w_n^{n-q-1}\bar{w}_n^p} \left[d\bar{z}' \cdot d(\zeta' - z')\right]^q \\ & \wedge \left[d\bar{\zeta}' \cdot d(\zeta' - z')\right]^{n-q-1} d(\zeta_n - z_n) \\ & - \frac{1}{(n-1)4\pi^n} \frac{\log w_n}{\bar{w}_n^{n-1}} \left[d\bar{z}' \cdot d(\zeta' - z')\right]^{n-1} \wedge d(\zeta_n - z_n) \\ & - \frac{1}{(n-1)4\pi^n} \frac{\log \bar{w}_n}{w_n^{n-1}} \left[d\bar{\zeta}' \cdot d(\zeta' - z')\right]^{n-1} \wedge d(\zeta_n - z_n) \end{aligned}$$

Finally calculate directly that for each $\zeta \in M$ fixed $\bar{\partial}_b^* K = R$.

13

Appendix

In this last chapter we treat some background material that will aid the reader in handling the ideas developed in this book. These include the Ascoli-Arzela theorem, the Baire category theorem, the Riesz transformation, and many other key ideas that come from real analysis.

Readers will want to consult this chapter selectively, each one based on his/her background.

13.1 The Space of Functions in an Open Subset Ω of $\mathbb{C}^n$

For any open subset $\Omega \subset \mathbb{C}^n$, the ring $\mathcal{C}(\Omega)$ of continuous, complex valued functions in Ω has the natural topology of a function space. The topology is defined as follows. For any compact subset $K \subset \Omega$ and any real number $\varepsilon > 0$ let

$$U(\Omega, \varepsilon) = \{f \in \mathcal{C}(\Omega) | |f(z)| < \varepsilon \text{ whenever } z \in K\}$$

the sets $U(K, \varepsilon)$, for all such K and ε, are taken as a basis for the open neighborhoods of the zero element of $\mathcal{C}(\Omega)$, the function $f(z) \equiv 0$. It is a straightforward verification to show that these sets do determine a topology under which $\mathcal{C}(\Omega)$ becomes a topological ring; the details will be left to the reader. It is also quite clear that a sequence of functions $f_j \in \mathcal{C}(\Omega)$ converges to a limit function $f \in \mathcal{C}(\Omega)$ if and only if the functions f_j converge uniformly to f on any compact subset of U; the topology we have introduced is sometimes called "the topology of uniform convergence on compact subsets," for this reason. The ring $\mathcal{C}(\Omega)$ is of course complete in this topology, as is familiar from elementary analysis.

This topology can also be introduced in the following equivalent, but rather more concrete, form.

For any compact subset $K \subset \Omega$ and any $f \in \mathcal{C}(\Omega)$ put

$$\|f\|_K = \sup_{z \in K} |f(z)|$$

The value $\|f\|_K$ is a well-defined, finite real number since f is continuous and K is compact, and the mapping $f \longrightarrow \|f\|_K$ is a pseudo-norm on the ring

DOI: 10.1201/9781003619581-13

$\mathcal{C}(\Omega)$, since clearly

$$\|f+g\|_K \le \|f\|_K + \|g\|_K$$

and

$$\|fg\|_K \le \|f\|_K \cdot \|g\|_K$$

for any elements $f, g \in \mathcal{C}(\Omega)$. Note that for a constant c,

$$\|cf\|_K = |c| \cdot \|f\|_K$$

The mapping is not a norm, since $\|f\|_K = 0$ does not imply that f is the zero element of $\mathcal{C}(\Omega)$; the values of the function f can be rather general on the subset $\Omega \setminus K$. However, $\|f\|_K = 0$ for every K if and only if f is the zero element. The basic open neighborhoods of the zero element, in the topology on $\mathcal{C}(\Omega)$, can be described as

$$U(K,\varepsilon) = \{f \in \mathcal{C}(\Omega) \mid \|f\|_K < \varepsilon\}$$

Now let $K_j \subset \Omega$ be a sequence of compact subsets of Ω such that

$$K_j \subset K_{j+1} \quad \text{and} \quad \bigcup_{j=1}^{\infty} K_j = \Omega$$

and for any elements $f, g \in \mathcal{C}(\Omega)$, define

$$d(f,g) = \sum_{j=1}^{\infty} \frac{1}{2^j} \frac{\|f-g\|_{K_j}}{1+\|f-g\|_{K_j}} \tag{13.1}$$

It is an easy exercise to show that (13.1) defines a metric on the ring $\mathcal{C}(\Omega)$, and that the topology on $\mathcal{C}(\Omega)$ is that given by this metric. It is clear that $\mathcal{C}(\Omega)$ is locally convex and complete in this metric; thus $\mathcal{C}(\Omega)$ is a Frechet space.

Compactness and sequential compactness are the same for metric spaces.

13.2 The Ascoli-Arzelà Theorem

Ascoli and Arzelà proved what was probably the first substantial mathematical theorem about functional analysis, published in 1883–1884 Ascoli [10, p. 545] defined the notion of equicontinuity of a set of continuous functions (of a real variable) at a fixed point; this concept was also used by Arzelà [8] about the same time. This fundamental concept can readily be extended to functions with range in a metric space, or still more general functions (see Bourbaki [34, Chapter 10]). The important Theorem 13.2.1 is known as the Arzelà–Ascoli theorem, although many authors use only one of these names. In the case of

$C[0,1]$, Ascoli [10, pp. 545–549] used a construction which essentially amounts to the sufficiency of the condition for compactness. Ascoli [10] proved the necessity of this condition. Both of these papers are phrased in geometrical terms and it is difficult to resurrect this result from them. However, an extremely clear presentation of this and related theorems was given by Arzelà [9, pp. 55–60].

This theorem essentially generalizes the Bolzano–Weierstrass theorem to the infinite-dimensional function space $C[0,1]$. The latter asserts that any bounded sequence $\{x_n\}$ of real numbers contains a convergent subsequence and more generally that the same is true in n-dimensional space $\mathbb{R}^n$. Actually, Ascoli's theorem continues to hold in more general settings. For instance, it holds with $[0,1]$ replaced by any closed, bounded subset of $\mathbb{R}^n$.

Let X be a metric space and Y complete metric space. Note that every metric space can be metrically embedded in a Banach space, essentially by using the distance function. This is trivial, but useful notationally since instead of writing the distance between two points as $d(x,y)$ one can write $|x-y|$ with the norm notation.

Let X be a subset of a metric space, and let F be a Banach space. Let $\mathcal{F}$ be a subset of the space of continuous maps $C(X,F)$. We shall say that $\mathcal{F}$ is (or its elements are) *equicontinuous* at a point $x_0 \in X$ if, given ε, there exists δ such that whenever $x \in X$ and $d(x, x_0) < \delta$, then

$$|f(x) - f(x_0)| < \varepsilon$$

for all $f \in \mathcal{F}$. We say that $\mathcal{F}$ is equicontinuous on X if it is equicontinuous at every point of X.

We now formulate the Ascoli-Arzelà theorem in a way that suits our purpose.

Theorem 13.2.1 (Ascoli-Arzelà). *Let X be a compact subset of a metric space, and let F be a Banach space. Let $\mathcal{F}$ be a subset of the space of continuous maps $C(X,F)$ with sup norm. Then $\mathcal{F}$ is relatively compact in $C(X,F)$ if and only if the following two conditions are satisfied:*

1. *$\mathcal{F}$ is equicontinuous.*

2. *For each $x \in X$, the set $\mathcal{F}(x)$ consisting of all values $f(x)$ for $f \in \mathcal{F}$ is relatively compact.*

Proof. Assume that $\mathcal{F}$ satisfies the two conditions.

A subset S of a metric space, or a normed vector space, which can be covered by a finite number of open balls of given radius $r > 0$ is said to be totally bounded. S is compact if and only if it is complete and totally bounded. A subset of a topological space is said to be relatively compact if its closure is compact.

We shall prove that $\mathcal{F}$ is relatively compact. For this it is sufficient to show that $\mathcal{F}$ can be covered by a finite number of balls of prescribed radius. Let

$r > 0$. By equicontinuity, for each $x \in X$ we select an open neighborhood $V(x)$ such that if $y \in V(x)$, then $|f(y) - f(x)| < r$ for all $f \in \mathcal{F}$. Then a finite number $V(x_1), \ldots, V(x_n)$ cover X. Each set

$$\mathcal{F}(x_1), \ldots, \mathcal{F}(x_n)$$

is relatively compact, and hence so is their union

$$Y = \mathcal{F}(x_1) \cup \cdots \cup \mathcal{F}(x_n).$$

Let $B(a_1), \ldots, B(a_m)$ be open balls of radius r centered at points $a_1, \ldots, a_m$ which cover Y. Then $f(x_1), \ldots, f(x_n)$ lie in these balls. In fact, for each $i = 1, \ldots, n$ we have

$$f(x_i) \in B\left(a_{\sigma(i)}\right)$$

where $\sigma : \{1, \ldots, n\} \to \{1, \ldots, m\}$ is some mapping.

For each such map σ let $\mathcal{F}_\sigma$ be the set of $f \in \mathcal{F}$ such that for all i, we have

$$\left|f(x_i) - a_{\sigma(i)}\right| < r.$$

Then the finite number of $\mathcal{F}_\sigma$ cover $\mathcal{F}$. It suffices now to prove that each $\mathcal{F}_\sigma$ has diameter $< 4r$. But if $f, g \in \mathcal{F}_\sigma$ and $x \in X$, then x lies in some $V(x_i)$, and then:

$$\begin{aligned} |f(x) - g(x)| &\leqq |f(x) - f(x_i)| + \left|f(x_i) - a_{\sigma(i)}\right| \\ &\quad + \left|a_{\sigma(i)} - g(x_i)\right| + |g(x_i) - g(x)| \\ &< 4r. \end{aligned}$$

This proves our implication and the part of Ascoli's theorem which is used in the applications. The converse is trivial and left to the reader. □

13.3 The Baire Category Theorem

In 1899, Baire's doctoral thesis "Sur les fonctions de variables reelles" appeared in Annali di Matematica Pura ed Applicata 3(3), l-122 (by invitation of Dini)[11]. In order to characterize limits of convergent sequences of continuous functions (and their limits, etc.), Baire defined [p. 65] a subset of $\mathbb{R}$ to be of the first category in $\mathbb{R}$ when it is the union of countably many nowhere dense sets in Iw. As the result basic to functional analysis, he proved "Baire's theorem" that $\mathbb{R}$ is of the "second category" (i.e., not of the first) in itself, a result which he extended to $\mathbb{R}$ in 1904.

For an outstanding exposition, see J.C. Oxtoby's Measure and Category (Berlin: Springer, 1971) [169]. Some of Baire's results for $\mathbb{R}$ had been obtained before by Osgood [167].

The power of such an easy result is somewhat surprising. Baire's theorem is qualified by Körner as a "profound triviality" [115]. Difficult constructions can be replaced by a convenient choice of space X and the dense open sets G_n.

The Baire Category theorem is a very innocent-looking statement which is, furthermore, not too hard to prove. It can, however, be used to obtain some rather interesting results.

One of the main advantages of complete metric spaces, which lies in the foundation of many applications is the "principle of nested closed subsets," which can be formulated as follows: if M is a complete metric space, and $N_1 \supset N_2 \supset \cdots$ is a sequence of nested non-empty closed subsets with diameter tending to zero, then these sets have a unique common point. Many arguments leaning on the principle of nested closed subsets become short and elegant when we use an important corollary of this principle—the so-called Baire theorem.

It requires one piece of terminology that we did not introduce in the main part of the book:

Definition 13.3.1. A subset of a metric space X is said to be *dense* if it intersects every open set $U \subseteq X$. A subset A of a metric space X is called *nowhere dense* (in X) if there is no neighborhood in X on which A is dense.

The following fundamental theorem is proved by Renè-Louis Baire in [11] (see also [186, p. 139]).

Theorem 13.3.2 (Baire Category). *A complete metric space is not the union of a countable collection of nowhere dense sets.*

Proof. The idea of the proof is simple. Suppose that M is the complete metric space and $M = \bigcup_{j=1}^{\infty} A_j$ with each A_j nowhere dense. We will construct a Cauchy sequence $\{x_m\}$ which stays away from each A_j so that its limit point x (which is in M by completeness) is in no A_j, thereby contradicting the statement $M = \bigcup_{n=1}^{\infty} A_j$.

Since A_1 is nowhere dense, we can find $x_1 \notin \overline{A}_1$. Pick an open ball B_1 about x_1 so that $B_1 \cap A_1 = \varnothing$ and so that the radius of B_1 is smaller than one. Since A_2 is nowhere dense, we can find $x_2 \in B_1 \backslash \overline{A}_2$. Let B_2 be an open ball about x_2 so that $\overline{B}_2 \subset B_1, B_2 \cap A_2 = \varnothing$, and with radius smaller than $\frac{1}{2}$. Proceeding inductively, we pick $x_j \in B_{j-1} \backslash \overline{A}_j$ and choose an open ball B_j about x_j satisfying $\overline{B}_j \subset B_{j-1}, B_j \cap A_j = \varnothing$, and having a radius smaller than 2^{1-j}. Now $\{x_j\}_{j=1}^{\infty}$ is a Cauchy sequence since $j, m \geq N$ implies that x_j, $x_m \in B_j$ so

$$\rho(x_j, x_m) \leq 2^{1-N} + 2^{1-N} = 2^{2-N} \to 0$$

as $N \to \infty$. Let $x = \lim_{j \to \infty} x_j$. Since $x_j \in B_j$ for $j \geq N$, we have

$$x \in \overline{B}_j \subset B_{N-1}$$

Thus $x \notin A_{N-1}$ for any N which contradicts $M = \bigcup_{j=1}^{\infty} A_j$. □

It should be remarked that the Baire theorem is a hybrid in that a topological conclusion (the intersection of a countable number of dense open sets is dense) is deduced from a non-topological premise (that the space is complete pseudo-metric).

One may take complements to obtain the following equivalent formulation of the result.

Theorem 13.3.3 (Baire Category again). *Suppose that X is a complete metric space and that $S_1, S_2, S_3, \ldots$ are closed sets, none containing any open ball $B_\varepsilon(x)$. Then $\bigcup_{j=1}^{\infty} S_j$ is not the whole of X.*

Proof. Set $A_j = X \backslash S_j$; then A_j is open and dense, and we may apply the previous formulation. □

The Baire category theorem tells us that if $M = \bigcup_{n=1}^{\infty} A_n$, then some of the sets $\overline{A}_n$ must have nonempty interior. In practice, one rarely uses the Baire category theorem directly but rather one of its consequences.

A terminology has been devised which is very convenient in discussing questions related to the Baire theorem. A subset A of a topological space is nowhere dense in X iff the interior of the closure of A is void; otherwise stated, A is nowhere dense in X iff the open set $X \setminus A$ is dense in X. It is evident that the finite union of nowhere dense sets is nowhere dense. A subset A of X is indexset!meager*meager* in X or of the *first category* in X iff A is the union of a countable family of nowhere dense sets.

The Baire theorem can then be stated: the complement of a meager subset of a complete metric space is dense. (The complement of a meager set is sometimes called *co-meager* or residual in X.) A set A is non-meager or of the *second category* in X iff it is not meager in X (see Kelley [118]).

It is a standard application of the Baire category theorem that most continuous functions are nowhere differentiable. To be precise, the set of all continuous but nowhere differentiable functions on the real line $\mathbb{R}$ is of the second category in $\mathbb{R}$.

The Baire category theorem has many interesting and important applications throughout analysis.

13.3.1 Analytic and $\mathbb{C}^\infty$-Functions

A metric space M is called complete (or a Cauchy space) if every Cauchy sequence of points in M has a limit that is also in M. The completeness of the space $\mathbb{R}$ of real numbers is known from analysis. A closed subset of a complete space is a complete space.

By $\mathcal{C}^\infty([0,1])$ we denote, as usual, the set of all infinitely differentiable real-valued functions on the closed unit interval $[0,1]$. We say that a function f is *analytic* in the interval $(x_0 - d, x_0 + d)$ if f is the limit of a power series in $x - x_0$ with constant coefficients in this interval.

It is convenient to define the metric d_∞ on $\mathcal{C}^\infty([0,1])$ first. We put

$$d_\infty(f_1, f_2) = \sum_{k=0}^{\infty} 2^{-k} \sup_{0 \leqslant x \leqslant 1} \left(\left| f_1^{(k)}(x) - f_2^{(k)}(x) \right| \right) \left[1 + \left| f_1^{(k)}(x) - f_2^{(k)}(x) \right| \right]^{-1}$$

and verify that d_∞ is a complete metric on $\mathcal{C}^\infty([0,1])$. (We understand that $f^{(0)}$ means f.) Plainly $d_\infty(f_1, f_2) \geqslant 0$ and $d_\infty(f_1, f_2) = 0$ if and only if $f_1 = f_2$. Moreover, $d_\infty(f_1, f_2) = d(f_2, f_1)$.

For positive numbers a and b, the inequality

$$a(1+a)^{-1} + b(1+b)^{-1} \geqslant (a+b)(1+a+b)^{-1}$$

holds. This can be verified by clearing fractions. Moreover, the function $u \to u(1+u)^{-1} = 1 - (1+u)^{-1}$ increases with u for u positive. Consequently $d_\infty(f_1, f_2) + d_\infty(f_2, f_3) \geqslant d_\infty(f_1, f_3)$ holds.

It remains to prove that d_∞ is complete on $\mathcal{C}^\infty([0,1])$. Let $\{f_j\}$ be a Cauchy sequence of functions in $\mathcal{C}^\infty([0,1])$; that is, $\lim_{j,m\to\infty} d_\infty(f_j, f_m) = 0$. Then

$$\sup_{0 \leqslant x \leqslant 1} \left(\left| f_j^{(k)}(x) - f_m^{(k)}(x) \right| \right) \left[1 + \left| f_j^{(k)}(x) - f_m^{(k)}(x) \right| \right]^{-1} \leqslant d_\infty(f_j, f_m)$$

for each k, so the sequence of continuous functions $f_j^{(k)}$ converges uniformly on $[0,1]$, say to the continuous function g_k. Then

$$\lim_{j\to\infty} \sup_{0 \leqslant x \leqslant 1} \left| f_j^{(k)}(x) - g_k(x) \right| = 0$$

$$\lim_{j\to\infty} \sup_{0 \leqslant x \leqslant 1} \left| f_j^{(k)}(x) - g_k(x) \right| \left[1 + \left| f_j^{(k)}(x) - g_k(x) \right| \right]^{-1} = 0$$

for each $k \geqslant 0$, and hence

$$\lim_{j\to\infty} \sum_{k=0}^{\infty} 2^{-k} \sup_{0 \leqslant x \leqslant 1} \left| f_j^{(k)}(x) - g_k(x) \right| \left[1 + \left| f_j^{(k)}(x) - g_k(x) \right| \right]^{-1} = 0 \quad (13.2)$$

It suffices to prove that $g_0 \in \mathcal{C}^\infty([0,1])$ and $g_0^{(k)} = g_k$ for all $k = 0, 1, 2, \ldots$, by induction on k. For $k = 0$, there is nothing to prove. We assume that $g_0^{(N-1)} = g_{N-1}$ for fixed $N \geqslant 1$ in all such situations. Clearly the sequence $\{f_j'\}$ is also a Cauchy sequence in $\mathcal{C}^\infty([0,1])$ because $d_\infty\left(f_j', f_m'\right) \leqslant 2d(f_j, f_m)$ by the definition of d_∞. Moreover, $f_j^{(i)}$ converges uniformly to g_{i+1} for $i \geqslant 0$. By the induction hypothesis, $g_0^{(N)} = g_0^{(N-1)} = g_N$. This completes the induction. Thus $g_0 \in \mathcal{C}^\infty([0,1])$ and $\lim_{j\to\infty} d(f_j, g_0) = 0$ follows from (13.2).

We say that a function f is analytic in the interval $(x_0 - d, x_0 + d)$ if f is the limit of a power series in $x - x_0$ with constant coefficients in this interval. We say that f is *nowhere analytic* if it is not analytic in any interval. Baire's theorem illustrates the difference between analytic functions and $\mathcal{C}^\infty$-functions. To be a $\mathcal{C}^\infty$-function, the kth derivative must be continuous for every k. We can show using the Baire Category Theorem that most $\mathcal{C}^\infty$-functions are not analytic at any point. This construction was outlined by Morgenstern in his paper [157] (the proof in [157] contains two gaps, see [64, p. 356]).

Theorem 13.3.4 (Morgenstern). *Typical $\mathcal{C}^\infty([0,1])$-functions are not analytic at any point.*

Proof. In order that $f \in \mathcal{C}^\infty([0,1])$ be analytic at $a \in [0,1]$, we must have

$$\sup\left\{ \sqrt[k]{\left|f^{(k)}(a)\right| / k!} \;\middle|\; k \in \mathbb{Z}^+ \right\} = c < \infty.$$

Letting

$$T(a;c) = \left\{ f \in \mathcal{C}^\infty \;\middle|\; |f^{(k)}(a)| \leqslant k!c^k \text{ for all } k \in \mathbb{Z}^+ \right\}$$

it follows that if f is analytic at a, then f belongs to $T(a;c)$ for some $c \geqslant 0$. Since analyticity at a point implies analyticity in some neighborhood of that point, we find that the functions analytic anywhere in $[0,1]$ are contained in

$$\bigcup \left\{ T(a;c) \mid a \text{ rational }; c \in \mathbb{Z}^+ \right\}.$$

We will now prove this union to be a set of the first category, by showing that each $T(a;c)$ is nowhere dense.

Each $T(a;c)$ is closed: Indeed, $\left\{ f \in \mathcal{C}^\infty([0,1]) \mid |f^{(k)}(a)| \leq k!c^k \right\}$ is clearly closed for each k, and therefore so also is their intersection.

Each $T(a;c)$ has empty interior. Given $f \in T(a;c)$ and any neighborhood $B(f, 2\varepsilon)$ of f, choose j so large that $\sum_j^\infty 2^{-j} < \varepsilon$, and then select a $b > 2$ so that $\varepsilon b^j > (2j)!c^{2j}$. Let

$$s(x) = f(x) + \varepsilon b^{-j} \cos b(x-a).$$

Note that the absolute value of the kth-derivative of the function $\varphi(x) := \cos(b(x-a)x)$ is $b^m|\cos(b(x-a))|$ (if k is even) or $b^k|\sin(b(x-a))|$ (if k is odd). On the other hand, we have for any $x \in [0,1]$ that either $|\sin(b(x-a))| \geq 1/2$ or $|\cos(b(x-a))| \geq 1/2$, because of the basic law $\sin^2 t + \cos^2 t = 1$. Then $s \in \mathcal{C}^\infty([0,1])$ and $d_\infty\left(f^{(k)}, s^{(k)}\right) \leqslant \varepsilon b^{k-j} < \varepsilon \cdot 2^{k-j}$ for each $k < j$, consequently $s \in B(f, 2\varepsilon)$. However, $\left|s^{(2j)}(a) - f^{(2j)}(a)\right| = \varepsilon b^j > (2j)!c^{2j}$, so that $s \notin T(a;c)$. This proof is due to H. Salzmann and K. Zeller [190]. □

13.4 Analytic Arc

Definition 13.4.1. Consider a curve $\gamma : [0,1] \to \mathbb{C}$ which has the equation

$$z = z(t), \quad 0 \leq t \leq 1$$

where the function $z(t)$ satisfies the following condition: For any point t_0, $0 < t_0 < 1$, we can find a positive number δ such that inside the region $t_0 - \delta < t < t_0 + \delta$, $z(t)$ can be expanded into a convergent power series

$$z(t) = \sum_{n=0}^{\infty} c_n (t - t_0)^n, \quad c_1 \neq 0$$

Then γ is called an *analytic arc*.

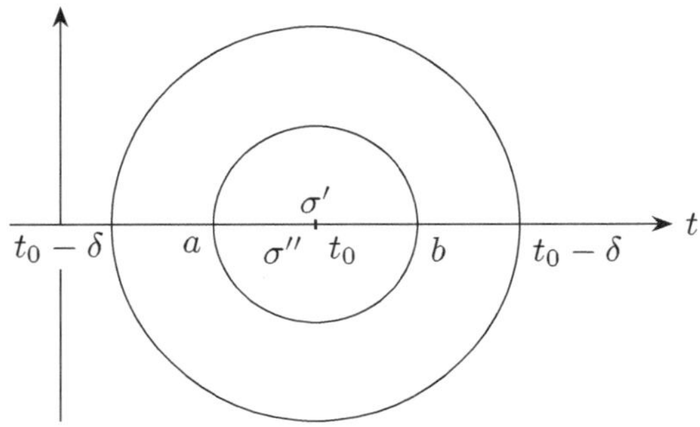

FIGURE 13.1

If in the series above, t is treated as a complex variable, then by the properties of power series, this series converges absolutely on the disc $|t - t_0| < \delta$, thereby determining an analytic function $z(t)$. Since $z'(t_0) = c_1 \neq 0$, we then get a disc $\sigma : |t - t_0| < \delta' \, (\delta' \leq \delta)$, so that the function $z(t)$ is univalent on σ.

Theorem 13.4.2. *Suppose U is a bounded domain whose boundary is a Jordan curve γ which is a union of k analytic arcs $\gamma_j (j = 1, \ldots, k)$, and $w = f(z)$ is a univalent analytic function on U which maps U onto the unit disc $\mathbb{D}$. Then, except at the endpoints, the function $w = f(z)$ is analytic on each $\gamma_j (j = 1, \ldots, k)$, and $f'(z) \neq 0$.*

Proof. Consider any analytic arc γ_j defined by an equation as given in Definition 13.4.1, and suppose $z_0 = z(t_0) \, (0 < t_0 < 1)$ is a point on γ_j. We only need to prove the existence of a neighborhood d containing z_0 and a function $F(z)$ satisfying the following conditions:

1. $F(z)$ is analytic in d, and $F'(z_0) \neq 0$.
2. In a domain d' lying in the intersection of U and d and having z_0 as a boundary point, $F(z) = f(z)$.

The function $z = z(t)$ maps the disc $\sigma : |z - z_0| < \delta' (\leq \delta)$ Figure 13.1 onto a domain d containing the point z_0 (Fig. 13.2), and it maps the upper half-disc σ' and the lower half-disc σ'' onto the two parts d' and d'' of d, respectively. Now, supposing that d' lies inside U, the function

$$\varphi(t) = f\{z(t)\}$$

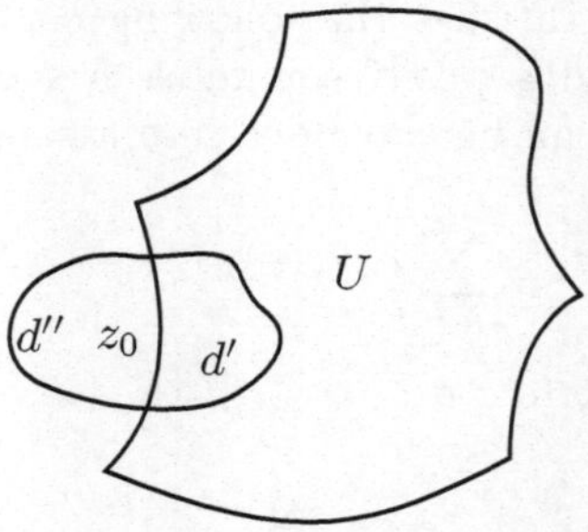

FIGURE 13.2

is then analytic in the upper half-disc σ' and continuous on the diameter $\overline{ab}$. Moreover, this function maps σ' into a domain of $\mathbb{D}$, and maps the diameter $\overline{ab}$ into an arc of the circle $\partial\mathbb{D}$ (Figure 13.3). By the reflection principle, we can find a univalent analytic function $\Phi(t)$ inside σ such that the equation $\Phi(t) = \varphi(t)$ holds inside σ'. Now suppose that $t = t(z)$ is the inverse function of $z = z(t)$. Then the function

$$F(z) = \Phi\{t(z)\}$$

is univalent and analytic inside $d, F'(z_0) \neq 0$, and the equation $F(z) = f(z)$ holds inside the domain d'. □

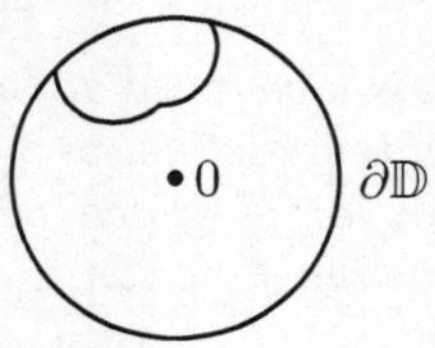

FIGURE 13.3

Now we give a sufficient condition for the univalence of analytic functions.

Theorem 13.4.3. *Suppose U and V are bounded simply connected domains enclosed by piecewise smooth, analytic closed curves γ and β, respectively, and $w = f(z)$ is a function satisfying the following conditions:*

1. *$w = f(z)$ is analytic inside U and continuous on $\bar{U} = U \cup \gamma$,*
2. *$w = f(z)$ maps γ bijectively onto β.*

Then $w = f(z)$ is univalent in U and conformally maps U onto V.

Proof. Suppose the curve γ has equation $z = z(t) \quad (a \leq t \leq b)$, and consider a point $w_0(\neq \infty)$ in the w-plane which does not lie on the curve β. By the argument principle, we have

$$\begin{aligned}\frac{1}{2\pi}\Delta_\gamma \arg\left[f(z) - w_0\right] &= \frac{1}{2\pi i}\int_\gamma \frac{f'(z)}{f(z) - w_0}dz \\ &= \frac{1}{2\pi i}\int_a^b \frac{f'[z(t)]z'(t)}{f[z(t)] - w_0}dt = \frac{\pm 1}{2\pi i}\int_\beta \frac{dw}{w - w_0}\end{aligned}$$

where the integration is taken in the positive orientation, so that if z traverses γ in the positive orientation, and so does $w = f(z)$ along β, we take the $+$ sign in the preceding equation; otherwise we take the $-$ sign. Moreover, the existence of $f'(z)$ on γ (with the possible exception of a finite number of points) follows from Theorem 13.4.2. When w_0 is outside of β, it is clear that

$$\frac{1}{2\pi}\Delta_\gamma \arg\left[f(z) - w_0\right] = \frac{\pm 1}{2\pi i}\int_L \frac{dw}{w - w_0} = 0$$

Hence, $f(z) - w_0$ has no zeros inside U. When w_0 is in the domain V in the interior of β, we have

$$\frac{1}{2\pi}\Delta_\gamma \arg\left[f(z) - w_0\right] = \frac{\pm 1}{2\pi i}\int_L \frac{dw}{w - w_0} = \pm 1$$

where ± 1 represents the number of zeros of $f(z) - w_0$ inside U. Since -1 is meaningless in this context, $f(z) = w_0$ has only one root inside U. At the same time, we also know that when γ is traversed by z in the positive orientation, $w = f(z)$ also moves in a positive orientation along β. Now let us suppose that w_1 is a point on β; we shall prove that $f(z) = w_1$ has no roots in U. If there exists a point z_1 such that $f(z_1) = w_1$, we can then find a disc γ centered at w_1 such that for any point w_2 in γ, $f(z) = w_2$ has roots in U. In particular, we can choose a point w_2 inside γ which is outside V. This, however, contradicts the case we discussed earlier. Therefore, $f(z)$ is univalent in U and conformally maps U onto V. □

13.5 Homotopic Path

Let X be a topological space and let $a, b \in X$. Recall that a path in X from a to b is a map $\gamma : [0, 1] \to X$ such that $\gamma(0) = a$ and $\gamma(1) = b$. We call $[0, 1]$ the parameter interval of γ and denote the range of γ by $\gamma^\star$. Thus γ is a mapping, and $\gamma^\star$ is the set of all points $\gamma(t)$, for $0 \leq t \leq 1$. Two paths γ_0 and γ_1 from a to b are homotopic with endpoints fixed, written $\gamma_0 \simeq \gamma_1$ rel $\{0, 1\}$ if there is a map

$$F : [0, 1] \times [0, 1] \to X$$

such that

$$\begin{aligned} F(s,0) &= \gamma_0(s), & 0 \le s \le 1, \\ F(s,1) &= \gamma_1(s), & 0 \le s \le 1, \\ F(0,t) &= a, & 0 \le t \le 1, \\ F(1,t) &= b, & 0 \le t \le 1. \end{aligned}$$

The map F is referred to as a homotopy of γ_0 and γ_1. For each $t \in [0,1]$, the map $\gamma_t : [0,1] \to X$ defined by

$$\gamma_t(s) = F(s,t), \quad 0 \le s \le 1, \tag{13.3}$$

is then a path from a to b. The variable t is to be regarded as a parameter, and the paths γ_t move "continuously" with t. As the parameter t moves from 0 to 1, the path γ_0 is continuously deformed to the path γ_1 through the γ_t's. Often we shall refer to $\{\gamma_t\}_{0 \le t \le 1}$ as a homotopy of γ_0 and γ_1, understanding that the homotopy F can be recovered from the γ_t's by (13.3).

The situation may be represented by a square, representing the parameter space, with left edge labeled "a" to indicate that the restriction of the homotopy F to the left edge is the constant path at a, with bottom edge labeled "γ_0" to indicate that the restriction of F to the bottom edge coincides with γ_0, etc. (see the diagram Fig. 13.4).

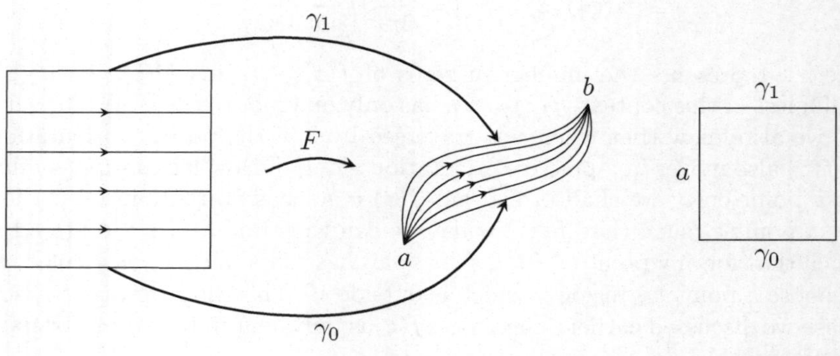

FIGURE 13.4

So two paths are homotopic if one can be deformed into the other continuously. We say that a subset A of a space X separates X if $X - A$ has more than one component. A subspace $J \subseteq \mathbb{C}$ homeomorphic to the circle is normally called a *Jordan curve*, or a simple closed curve. A subspace $A \subseteq \mathbb{C}$ homeomorphic to [0,1] is called an arc. If $J \subseteq \mathbb{C}$ is a Jordan curve, then $\mathbb{C} - J$ has (as one would expect) exactly two components, one bounded, the other unbounded, and J is the frontier of each. This is the famous Jordan curve theorem, a detailed discussion of which can be found in Munkres [158] and Wall [207]. A function f from one metric space to another is a *homeomorphism* if f is continuous, one-to-one, and onto and if, moreover, the inverse function f^{-1} is continuous. A topological space X is called *simply connected* if and only if it is path-connected, and whenever $p : [0,1] \to X$ and $q : [0,1] \to X$ are two paths (that is, continuous maps) with the same start and endpoint ($p(0) = q(0)$ and $p(1) = q(1)$), then p can be continuously deformed into q while keeping both endpoints fixed. Explicitly, there exists a homotopy $F : [0,1] \times [0,1] \to X$ such that $F(x,0) = p(x)$ and $F(x,1) = q(x)$.

13.6 Functions Satisfying a Hölder Condition

Let β be a smooth contour and $\varphi(t)$ a function of position on it. The function $\varphi(t)$ is said to satisfy on the curve the Hölder condition (H condition), if for two arbitrary points of this curve

$$|\varphi(t_2) - \varphi(t_1)| < A\,|t_2 - t_1|^{\lambda} \tag{13.4}$$

where A and λ are positive numbers. A is called the Hölder constant and λ the Hölder index. If λ were greater than unity, then the condition (13.4) would imply that the derivative $\varphi'(t)$ vanishes everywhere, the function $\varphi(t)$, therefore, would identically be equal to a constant. Consequently, we assume that

$$0 < \lambda \leqq 1$$

If $\lambda = 1$ the Hölder condition is identical to the familiar Lipschitz condition. (Observe that in many books the Hölder condition is called the Lipschitz condition of order λ.)

If t_1, t_2 are sufficiently close and the Hölder condition is satisfied for some value of the index λ_1, then it is evident that it is satisfied for every index $\lambda < \lambda_1$. The converse, in general, is not true. Thus, smaller λ is associated with a wider class of functions. The narrowest class is the class of functions obeying the Lipschitz condition of order 1. On the basis of the latter property it is easy to find that if the functions $\varphi_1(t), \varphi_2(t)$ satisfy the Hölder condition with the indices λ_1, λ_2, respectively, then their sum, product and quotient, under the condition that the denominator does not vanish, satisfy the Hölder condition with the index $\lambda = \min(\lambda_1, \lambda_2)$.

If $\varphi(t)$ is differentiable and has a finite derivative, it satisfies the Lipschitz condition. This statement follows from the theorem of finite increment. The converse, in general, is not true; this is shown by the following example of a function given on the real axis:

$$\varphi(x) = |x|$$

The function $\varphi(x)$ satisfies the Lipschitz condition, but its derivative does not exist at the origin of the coordinate system, since the left and right derivatives there are equal to $-1, +1$, respectively.

For a complex function, with superposition of functional relationships of different classes, the resultant function as a rule belongs to the widest (worst) class of all circuit elements. One often comes across analytic functions which satisfy a Hölder condition. Such functions satisfy Hölder's condition with the same indices. A proof is obtained by direct verification with regard to the fact that the analytic function satisfies the Lipschitz condition.

13.6.1 Hölder Transformations

Let $f(x)$ and $K(x)$ be two functions integrable over the interval $(-\infty, +\infty)$. It is very well known that their composition

$$\int_{-\infty}^{+\infty} f(t)K(x-t)dt$$

exists, as an absolutely convergent integral, for almost every x. The integral can, however, exist almost everywhere even if K is not absolutely integrable. The most interesting special case is that of $K(x) = 1/x$. Let us set

$$\tilde{f}(x) = \frac{1}{\pi}\int_{-\infty}^{+\infty} \frac{f(t)}{x-t}dt$$

The function $\hat{f}$ is called the Hilbert transform of f.

It exists for almost every value of x in the Principal Value sense:

$$\tilde{f}(x) = \lim_{\varepsilon\to 0} \frac{1}{\pi}\left(\int_{-\infty}^{x-\varepsilon} + \int_{x+\varepsilon}^{\infty}\right) \frac{f(t)}{x-t}dt$$

Moreover, it is known (See [182] or [212], p. 317) to satisfy the M. Riesz inequality

$$\left[\int_{-\infty}^{+\infty} |\tilde{f}|^p dx\right]^{1/p} \leq A_p \left[\int_{-\infty}^{+\infty} |f|^p dx\right]^{1/p}, \quad 1 < p < \infty$$

where A_p depends on p only. There are substitute results for $p = 1$ and $p = \infty$. The limit $\tilde{f}$ exists almost everywhere also in the case when $f(t)dt$ is replaced there by $dF(t)$, where $F(t)$ is any function of bounded variation over the whole interval $(-\infty, +\infty)$. (For all this, see e.g. [212], Chapters VII and XI, where also bibliographical references can be found).

13.6.2 Riesz Transformation

In the harmonic analysis, the Riesz transforms are a family of generalizations of the Hilbert transform to Euclidean spaces of dimension $d > 1$. They are a type of singular integral operator, meaning that they are given by a convolution of one function with another function having a singularity at the origin.

Specifically, the Riesz transforms of a complex-valued function f on $\mathbb{R}^d$ are defined by

$$R_j f(x) = c_d \lim_{\epsilon \to 0} \int_{\mathbf{R}^d \setminus B_\epsilon(x)} \frac{(x_j - t_j) f(t)}{|x - t|^{d+1}} dt$$

for $j = 1, 2, \ldots, d$. The constant c_d is a dimensional normalization given by

$$c_d = \frac{1}{\pi \omega_{d-1}} = \frac{\Gamma[(d+1)/2]}{\pi^{(d+1)/2}}$$

where ω_{d-1} is the volume of the unit $d-1$-ball. The limit is written as a principal value.

The Riesz transform arises in the study of differentiability properties of harmonic potentials in potential theory and harmonic analysis. In particular, they arise in the proof of the Calderón-Zygmund inequality (Gilbarg & Trudinger 1983, §9.4).

In the Cartesian coordinate system $\mathbb{R}^d$ with coordinates $(x_1, \ldots, x_d)$ and standard basis $\{\mathbf{e}_1, \ldots, \mathbf{e}_d\}$, del is a vector operator whose $x_1, \ldots, x_d$ components are the partial derivative operators $\frac{\partial}{\partial x_1}, \ldots, \frac{\partial}{\partial x_d}$; that is,

$$\nabla = \sum_{i=1}^{d} \mathbf{e}_i \frac{\partial}{\partial x_i} = \left(\frac{\partial}{\partial x_1}, \ldots, \frac{\partial}{\partial x_d} \right)$$

Where the expression in parentheses is a row vector.

The Laplace operator is a second-order differential operator in the d-dimensional Euclidean space, defined as the divergence $(\nabla \cdot)$ of the gradient (∇f). Thus if f is a twice-differentiable real-valued function, then the Laplacian of f is the real-valued function defined by:

$$\Delta f = \nabla^2 f = \nabla \cdot \nabla f$$

Explicitly, the Laplacian of f is thus the sum of all the unmixed second partial derivatives in the Cartesian coordinates x_i:

$$\Delta f = \sum_{i=1}^{d} \frac{\partial^2 f}{\partial x_i^2}$$

As a second-order differential operator, the Laplace operator maps C^k functions to C^{k-2} functions for $k \geq 2$. It is a linear operator $\Delta : C^k\left(\mathbf{R}^d\right) \to C^{k-2}\left(\mathbf{R}^d\right)$, or more generally, an operator $\Delta : C^k(\Omega) \to C^{k-2}(\Omega)$ for any open set $\Omega \subseteq \mathbf{R}^d$.

The Riesz transforms of f give the first partial derivatives of a solution of the equation

$$(-\Delta)^{\frac{1}{2}} u = f$$

where Δ is the Laplacian. Thus the Riesz transform of f can be written as:

$$Rf = \nabla (-\Delta)^{-\frac{1}{2}} f$$

In particular, one should also have

$$R_i R_j \Delta u = -\frac{\partial^2 u}{\partial x_i \partial x_j}$$

so that the Riesz transforms give a way of recovering information about the entire Hessian of a function from knowledge of only its Laplacian.

Bibliography

[1] Lars Ahlfors and Arne Beurling. Conformal invariants and function-theoretic null-sets. *Acta Math.*, 83:101–129, 1950.

[2] Lars V. Ahlfors. Bounded analytic functions. *Duke Math. J.*, 14:1–11, 1947.

[3] Lars V. Ahlfors. *Complex analysis.* International Series in Pure and Applied Mathematics. McGraw-Hill Book Co., New York, third edition, 1978. An introduction to the theory of analytic functions of one complex variable.

[4] Aldo Andreotti and C. Denson Hill. E. E. Levi convexity and the Hans Lewy problem. II. Vanishing theorems. *Ann. Scuola Norm. Sup. Pisa Cl. Sci. (3)*, 26:747–806, 1972.

[5] Lynn Suzanne Apfel. *Localization properties and boundary behavior of the Bergman kernel.* ProQuest LLC, Ann Arbor, MI, 2003. Thesis (Ph.D.)–Washington University in St. Louis.

[6] George B. Arfken, Hans J. Weber, and Frank E. Harris. *Mathematical methods for physicists. A comprehensive guide.* Amsterdam: Elsevier/Academic Press, 7th ed. edition, 2013.

[7] N. Aronszajn. Theory of reproducing kernels. *Trans. Amer. Math. Soc.*, 68:337–404, 1950.

[8] Arzelà. A theorem about series of functions. *Rom. Acc. L. Rend. (4)*, 1:262–267, 1885.

[9] C. Arzelà. On functions of lines. Bologna Mem. (5) V. 225-244 (1895)., 1895.

[10] G. Ascoli. The limit curve of a given family of curves. *Rom. Acc. L. Mem. (3)*, 18:521–586, 1884.

[11] R. Baire. On functions of real variables. *Annali di Mat. (3)*, 3:1–123, 1899.

[12] David E. Barrett. Irregularity of the Bergman projection on a smooth bounded domain in C^2. *Ann. of Math. (2)*, 119(2):431–436, 1984.

[13] David E. Barrett. Behavior of the Bergman projection on the Diederich-Fornæss worm. *Acta Math.*, 168(1-2):1–10, 1992.

[14] Theodore J. Barth. Families of holomorphic maps into Riemann surfaces. *Trans. Amer. Math. Soc.*, 207:175–187, 1975.

[15] Heinrich Behnke and Friedrich Sommer. *Theorie der analytischen Funktionen einer komplexen Veränderlichen*, volume 77 of *Grundlehren Math. Wiss.* Springer, Cham, 1962.

[16] S. Bell. Mapping problems in complex analysis and the $\overline{\partial}$-problem. *Bull. Amer. Math. Soc. (N.S.)*, 22(2):233–259, 1990.

[17] Steve Bell and Ewa Ligocka. A simplification and extension of Fefferman's theorem on biholomorphic mappings. *Invent. Math.*, 57(3):283–289, 1980.

[18] Steve Bell and Ewa Ligocka. A simplification and extension of Fefferman's theorem on biholomorphic mappings. *Invent. Math.*, 57(3):283–289, 1980.

[19] Steven R. Bell. Biholomorphic mappings and the $\bar{\partial}$-problem. *Ann. of Math. (2)*, 114(1):103–113, 1981.

[20] Steven R. Bell. *The Cauchy transform, potential theory and conformal mapping.* Chapman & Hall/CRC, Boca Raton, FL, second edition, 2016.

[21] Steven R. Bell and Harold P. Boas. Regularity of the Bergman projection in weakly pseudoconvex domains. *Math. Ann.*, 257(1):23–30, 1981.

[22] Steven R. Bell and Steven G. Krantz. Smoothness to the boundary of conformal maps. *Rocky Mountain J. Math.*, 17(1):23–40, 1987.

[23] Oleg V. Besov, Valentin P. Il'in, and Sergey M. Nikol'skii. *Integral representations of functions and imbedding theorems. Vol. I.* Scripta Series in Mathematics. V. H. Winston & Sons, Washington, DC; Halsted Press [John Wiley & Sons], New York-Toronto-London, 1978. Translated from the Russian.

[24] Oleg V. Besov, Valentin P. Ilin, and Sergey M. Nikolskii. *Integral representations of functions and imbedding theorems. Vol. II.* Scripta Series in Mathematics. V. H. Winston & Sons, Washington, DC; Halsted Press [John Wiley & Sons], New York-Toronto-London, 1979.

[25] Brian E. Blank and Steven G. Krantz. *Calculus. Single variable.* Emeryville, CA: Key College Publishing, 2006.

[26] Thomas Bloom and Ian Graham. A geometric characterization of points of type m on real submanifolds of C^n. *J. Differential Geometry*, 12(2):171–182, 1977.

[27] Harold P. Boas. Small sets of infinite type are benign for the $\overline{\partial}$-Neumann problem. *Proc. Amer. Math. Soc.*, 103(2):569–578, 1988.

[28] Harold P. Boas and Emil J. Straube. Equivalence of regularity for the Bergman projection and the $\overline{\partial}$-Neumann operator. *Manuscripta Math.*, 67(1):25–33, 1990.

[29] Harold P. Boas and Emil J. Straube. Sobolev estimates for the $\overline{\partial}$-Neumann operator on domains in C^n admitting a defining function that is plurisubharmonic on the boundary. *Math. Z.*, 206(1):81–88, 1991.

[30] Harold P. Boas and Emil J. Straube. Global regularity of the $\overline{\partial}$-Neumann problem: a survey of the L^2-Sobolev theory. In *Several complex variables (Berkeley, CA, 1995–1996)*, volume 37 of *Math. Sci. Res. Inst. Publ.*, pages 79–111. Cambridge Univ. Press, Cambridge, 1999.

[31] S. Bochner. Über orthogonale Systeme analytischer Funktionen. *Math. Z.*, 14(1):180–207, 1922.

[32] Albert Boggess. *CR manifolds and the tangential Cauchy-Riemann complex.* Studies in Advanced Mathematics. CRC Press, Boca Raton, FL, 1991.

[33] Raoul Bott and Loring W. Tu. *Differential forms in algebraic topology*, volume 82 of *Graduate Texts in Mathematics.* Springer-Verlag, New York-Berlin, 1982.

[34] N. Bourbaki. Éléments de mathématique. X. 1re Pt. Les structures fondamentales de l'analyse. Livre III: Topologie générale. Chapitre X: Espaces fonctionnels. Dictionnaire. Paris: Hermann & Cie. 101 p. (1949)., 1949.

[35] L. Boutet de Monvel and J. Sjöstrand. Sur la singularité des noyaux de Bergman et de Szegő. In *Journées: Équations aux Dérivées Partielles de Rennes (1975)*, volume No. 34–35 of *Astérisque*, pages 123–164. Soc. Math. France, Paris, 1976.

[36] Renato Caccioppoli. Sulle famiglie normali di funzioni analitiche di due variabili. *Rend. Semin. Mat. Univ. Padova*, 4:111–121, 1933.

[37] C. Carathéodory. Untersuchungen über die konformen Abbildungen von festen und veränderlichen Gebieten. *Math. Ann.*, 72(1):107–144, 1912.

[38] C. Carathéodory. Über das Schwarzsche Lemma bei analytischen Funktionen von zwei komplexen Veränderlichen. *Math. Ann.*, 97(1):76–98, 1927.

[39] Henri Cartan and Peter Thullen. Zur Theorie der Singularitäten der Funktionen mehrerer komplexen Veränderlichen. *Math. Ann.*, 106(1):617–647, 1932.

[40] David Catlin. Necessary conditions for subellipticity of the $\bar{\partial}$-Neumann problem. *Ann. of Math. (2)*, 117(1):147–171, 1983.

[41] David Catlin. Boundary invariants of pseudoconvex domains. *Ann. of Math. (2)*, 120(3):529–586, 1984.

[42] David Catlin. Subelliptic estimates for the $\overline{\partial}$-Neumann problem on pseudoconvex domains. *Ann. of Math. (2)*, 126(1):131–191, 1987.

[43] David W. Catlin. Global regularity of the $\bar{\partial}$-Neumann problem. In *Complex analysis of several variables (Madison, Wis., 1982)*, volume 41 of *Proc. Sympos. Pure Math.*, pages 39–49. Amer. Math. Soc., Providence, RI, 1984.

[44] David W. Catlin. Estimates of invariant metrics on pseudoconvex domains of dimension two. *Math. Z.*, 200(3):429–466, 1989.

[45] So-Chin Chen. Global regularity of the $\overline{\partial}$-Neumann problem: a sufficient condition. *Bull. Inst. Math. Acad. Sinica*, 19(3):191–205, 1991.

[46] So-Chin Chen. Global regularity of the $\overline{\partial}$-Neumann problem in dimension two. In *Several complex variables and complex geometry, Part 3 (Santa Cruz, CA, 1989)*, volume 52, Part 3 of *Proc. Sympos. Pure Math.*, pages 55–61. Amer. Math. Soc., Providence, RI, 1991.

[47] So-Chin Chen. On the Darboux-Picard theorem in $\mathbb{C}^n$. *Michigan Math. J.*, 40(3):605–608, 1993.

[48] So-Chin Chen and Mei-Chi Shaw. *Partial differential equations in several complex variables*, volume 19 of *AMS/IP Studies in Advanced Mathematics.* American Mathematical Society, Providence, RI; International Press, Boston, MA, 2001.

[49] S. S. Chern and J. K. Moser. Real hypersurfaces in complex manifolds. *Acta Math.*, 133:219–271, 1974.

[50] Michael Christ. *Lectures on singular integral operators*, volume 77 of *CBMS Regional Conference Series in Mathematics.* Conference Board of the Mathematical Sciences, Washington, DC; by the American Mathematical Society, Providence, RI, 1990.

[51] Michael Christ. Global C^∞ irregularity of the $\overline{\partial}$-Neumann problem for worm domains. *J. Amer. Math. Soc.*, 9(4):1171–1185, 1996.

[52] Joseph A. Cima and Steven G. Krantz. The Lindelöf principle and normal functions of several complex variables. *Duke Math. J.*, 50(1):303–328, 1983.

[53] Ronald R. Coifman and Guido Weiss. *Analyse harmonique non-commutative sur certains espaces homogènes*, volume Vol. 242 of *Lecture Notes in Mathematics*. Springer-Verlag, Berlin-New York, 1971. Étude de certaines intégrales singulières.

[54] E. F. Collingwood and A. J. Lohwater. *The theory of cluster sets*, volume No. 56 of *Cambridge Tracts in Mathematics and Mathematical Physics*. Cambridge University Press, Cambridge, 1966.

[55] John B. Conway. *Functions of one complex variable*, volume 11 of *Graduate Texts in Mathematics*. Springer-Verlag, New York-Heidelberg, 1973.

[56] Jiri Dadok and Reese Harvey. The fundamental solution for the Kohn-Laplacian $\Box_b$ on the sphere in $\mathbb{C}^n$. *Math. Ann.*, 244(2):89–104, 1979.

[57] John P. D'Angelo. Real hypersurfaces, orders of contact, and applications. *Ann. of Math. (2)*, 115(3):615–637, 1982.

[58] John P. D'Angelo. *Several complex variables and the geometry of real hypersurfaces*. Studies in Advanced Mathematics. CRC Press, Boca Raton, FL, 1993.

[59] Georges de Rham. *Variétés différentiables. Formes, courants, formes harmoniques*, volume III of *Publications de l'Institut de Mathématique de l'Université de Nancago*. Hermann, Paris, 1973. Troisième édition revue et augmentée, Actualités Scientifiques et Industrielles, No. 1222b. [Current Scientific and Industrial Topics].

[60] Klas Diederich and John E. Fornaess. Pseudoconvex domains with real-analytic boundary. *Ann. of Math. (2)*, 107(2):371–384, 1978.

[61] Klas Diederich and John Erik Fornaess. Pseudoconvex domains: an example with nontrivial Nebenhülle. *Math. Ann.*, 225(3):275–292, 1977.

[62] Alexander Dinghas. *Vorlesungen über Funktionentheorie*, volume Band 110 of *Die Grundlehren der mathematischen Wissenschaften*. Springer-Verlag, Berlin-Göttingen-Heidelberg, 1961.

[63] Peter V. Dovbush and Steven G. Krantz. *Normal families and normal functions*. Boca Raton, FL: CRC Press, 2024.

[64] James Dugundji. *Topology*. Allyn and Bacon, Inc., Boston, MA, 1966.

[65] Donald A. Eisenman. *Intrinsic measures on complex manifolds and holomorphic mappings*, volume No. 96 of *Memoirs of the American Mathematical Society*. American Mathematical Society, Providence, RI, 1970.

[66] P. Erdös, C. D. Godsil, S. G. Krantz, and T. D. Parsons. Intersection graphs for families of balls in $\mathbb{R}^n$. *European J. Combin.*, 9(5):501–505, 1988.

[67] Herbert Federer. *Geometric measure theory*, volume Band 153 of *Die Grundlehren der mathematischen Wissenschaften*. Springer-Verlag New York, Inc., New York, 1969.

[68] Herbert Federer. *Geometric measure theory*, volume Band 153 of *Die Grundlehren der mathematischen Wissenschaften*. Springer-Verlag New York, Inc., New York, 1969.

[69] Charles Fefferman. The Bergman kernel and biholomorphic mappings of pseudoconvex domains. *Invent. Math.*, 26:1–65, 1974.

[70] Charles Fefferman. The Bergman kernel and biholomorphic mappings of pseudoconvex domains. *Invent. Math.*, 26:1–65, 1974.

[71] Stephen D. Fisher. *Function theory on planar domains*. Pure and Applied Mathematics (New York). John Wiley & Sons, Inc., New York, 1983. A second course in complex analysis, A Wiley-Interscience Publication.

[72] G. B. Folland and J. J. Kohn. *The Neumann problem for the Cauchy-Riemann complex*, volume No. 75 of *Annals of Mathematics Studies*. Princeton University Press, Princeton, NJ; University of Tokyo Press, Tokyo, 1972.

[73] G. B. Folland and E. M. Stein. Estimates for the $\bar{\partial}_b$ complex and analysis on the Heisenberg group. *Comm. Pure Appl. Math.*, 27:429–522, 1974.

[74] Frank Forelli. Pluriharmonicity in terms of harmonic slices. *Math. Scand.*, 41(2):358–364, 1977.

[75] P. R. Garabedian. *Partial differential equations*. John Wiley & Sons, Inc., New York-London-Sydney, 1964.

[76] John B. Garnett. *Bounded analytic functions*, volume 96 of *Pure and Applied Mathematics*. Academic Press, Inc. [Harcourt Brace Jovanovich, Publishers], New York-London, 1981.

[77] John B. Garnett. *Bounded analytic functions*, volume 96 of *Pure and Applied Mathematics*. Academic Press, Inc. [Harcourt Brace Jovanovich, Publishers], New York-London, 1981.

[78] Daryl Geller. The Laplacian and the Kohn Laplacian for the sphere. *J. Differential Geometry*, 15(3):417–435, 1980.

[79] John E. Gilbert. Nikišin-Stein theory and factorization with applications. In *Harmonic analysis in Euclidean spaces (Proc. Sympos. Pure Math., Williams Coll., Williamstown, Mass., 1978), Part 2*, volume XXXV, Part 2 of *Proc. Sympos. Pure Math.*, pages 233–267. Amer. Math. Soc., Providence, RI, 1979.

[80] G. M. Goluzin. *Geometric theory of functions of a complex variable*, volume Vol. 26 of *Translations of Mathematical Monographs*. American Mathematical Society, Providence, RI, 1969.

[81] Ian Graham. Boundary behavior of the Carathéodory and Kobayashi metrics on strongly pseudoconvex domains in C^n with smooth boundary. *Trans. Amer. Math. Soc.*, 207:219–240, 1975.

[82] Hans Grauert and Ingo Lieb. Das Ramirezsche Integral und die Lösung der Gleichung $\bar{\partial} f = \alpha$ im Bereich der beschränkten Formen. *Rice Univ. Stud.*, 56(2):29–50, 1970.

[83] Hans Grauert and Ingo Lieb. Das Ramirezsche Integral und die Lösung der Gleichung $\bar{\partial} f = \alpha$ im Bereich der beschränkten Formen. *Rice Univ. Stud.*, 56(2):29–50, 1970.

[84] Robert E. Greene and Steven G. Krantz. Deformation of complex structures, estimates for the $\bar{\partial}$ equation, and stability of the Bergman kernel. *Adv. in Math.*, 43(1):1–86, 1982.

[85] Robert E. Greene and Steven G. Krantz. *Function theory of one complex variable*. Pure and Applied Mathematics (New York). John Wiley & Sons, Inc., New York, 1997. A Wiley-Interscience Publication.

[86] Robert E. Greene and Steven G. Krantz. *Function theory of one complex variable*, volume 40 of *Graduate Studies in Mathematics*. American Mathematical Society, Providence, RI, second edition, 2002.

[87] Robert E. Greene and Steven G. Krantz. *Function theory of one complex variable*, volume 40 of *Graduate Studies in Mathematics*. American Mathematical Society, Providence, RI, third edition, 2006.

[88] P. C. Greiner, J. J. Kohn, and E. M. Stein. Necessary and sufficient conditions for solvability of the Lewy equation. *Proc. Nat. Acad. Sci. U.S.A.*, 72(9):3287–3289, 1975.

[89] P. C. Greiner, J. J. Kohn, and E. M. Stein. Necessary and sufficient conditions for solvability of the Lewy equation. *Proc. Nat. Acad. Sci. U.S.A.*, 72(9):3287–3289, 1975.

[90] Peter Greiner. Subelliptic estimates for the $\bar{\delta}$-Neumann problem in C^2. *J. Differential Geometry*, 9:239–250, 1974.

[91] Helmut Grunsky. *Lectures on theory of functions in multiply connected domains*, volume 4 of *Studia Mathematica: Skript*. Vandenhoeck Ruprecht, Göttingen, 1978.

[92] Nicholas Hanges. Explicit formulas for the Szegő kernel for some domains in $\mathbb{C}^2$. *J. Funct. Anal.*, 88(1):153–165, 1990.

[93] G. H. Hardy. *Divergent Series.* Oxford, at the Clarendon Press,, 1949.

[94] Frank E. Harris. *Mathematics for physical science and engineering. Symbolic computing applications in Maple and Mathematica.* Amsterdam: Elsevier/Academic Press, 2014.

[95] Fritz Hartogs. Zur Theorie der analytischen Funktionen mehrerer unabhängiger Veränderlichen, insbesondere über die Darstellung derselben durch Reihen, welche nach Potenzen einer Veränderlichen fortschreiten. *Math. Ann.*, 62(1):1–88, 1906.

[96] F. Reese Harvey. Integral formulae connected by Dolbeault's isomorphism. *Rice Univ. Stud.*, 56(2):77–97, 1970.

[97] F. Reese Harvey and H. Blaine Lawson, Jr. On boundaries of complex analytic varieties. I. *Ann. of Math. (2)*, 102(2):223–290, 1975.

[98] F. Reese Harvey and H. Blaine Lawson, Jr. On boundaries of complex analytic varieties. II. *Ann. of Math. (2)*, 106(2):213–238, 1977.

[99] Reese Harvey and John Polking. Fundamental solutions in complex analysis. I. The Cauchy-Riemann operator. *Duke Math. J.*, 46(2):253–300, 1979.

[100] Reese Harvey and John Polking. Fundamental solutions in complex analysis. II. The induced Cauchy-Riemann operator. *Duke Math. J.*, 46(2):301–340, 1979.

[101] Joel R. Hass, Maurice D. Weir, and George B. jun. Thomas. *University calculus, early transcendentals.* Boston, MA: Pearson, 2nd revised ed. of University calculus: elements with early transcendentals. International edition edition, 2012.

[102] L. L. Helms. *Introduction to potential theory*, volume Vol. XXII of *Pure and Applied Mathematics.* Wiley-Interscience [A division of John Wiley & Sons, Inc.], New York-London-Sydney, 1969.

[103] G. M. Henkin. Integral representation of functions which are holomorphic in strictly pseudoconvex regions, and some applications. *Mat. Sb. (N.S.)*, 78(120):611–632, 1969.

[104] G. M. Henkin. Integral representation of functions in strongly pseudoconvex regions, and applications to the $\overline{\partial}$-problem. *Mat. Sb. (N.S.)*, 82(124):300–308, 1970.

[105] G. M. Henkin. H. Lewy's equation and analysis on pseudoconvex manifolds. *Uspehi Mat. Nauk*, 32(3(195)):57–118, 247, 1977.

[106] Michel Hervé. *Analytic and plurisubharmonic functions in finite and infinite dimensional spaces*, volume Vol. 198 of *Lecture Notes in Mathematics*. Springer-Verlag, Berlin-New York, 1971. Course given at the University of Maryland, College Park, Md., Spring 1970.

[107] Einar Hille. *Analytic function theory. Vol. 1*. Introductions to Higher Mathematics. Ginn and Company, Boston, MA, 1959.

[108] Kenneth Hoffman. *Banach spaces of analytic functions*. Prentice-Hall Series in Modern Analysis. Prentice-Hall, Inc., Englewood Cliffs, NJ, 1962.

[109] Lars Hörmander. L^2 estimates and existence theorems for the $\bar{\partial}$ operator. *Acta Math.*, 113:89–152, 1965.

[110] Lars Hörmander. Fourier integral operators. I. *Acta Math.*, 127(1-2):79–183, 1971.

[111] Lars Hörmander. *An introduction to complex analysis in several variables*, volume Vol. 7 of *North-Holland Mathematical Library*. North-Holland Publishing Co., Amsterdam-London; American Elsevier Publishing Co., Inc., New York, revised edition, 1973.

[112] Michael Hortmann. Über die Lösbarkeit der $\bar{\partial}$-Gleichung auf Ringgebieten mit Hilfe von L^p-, $\mathcal{C}^k$- und $\mathcal{D}$-stetigen Integraloperatoren. *Math. Ann.*, 223(2):139–156, 1976.

[113] Stefan Bron Albert Jakobsson. *Weighted Bergman kernels and biharmonic Green functions*. ProQuest LLC, Ann Arbor, MI, 2000. Thesis (Ph.D.)–Lunds Universitet (Sweden).

[114] Jae-Cheon Joo, Kang-Tae Kim, and Gerd Schmalz. On the generalization of Forelli's theorem. *Math. Ann.*, 365(3-4):1187–1200, 2016.

[115] Jean-Pierre Kahane. Baire's category theorem and trigonometric series. *J. Anal. Math.*, 80:143–182, 2000.

[116] Yitzhak Katznelson. *An introduction to harmonic analysis*. John Wiley & Sons, Inc., New York-London-Sydney, 1968.

[117] Yitzhak Katznelson. *An introduction to harmonic analysis*. John Wiley & Sons, Inc., New York-London-Sydney, 1968.

[118] John L. Kelley. *Topología general*. Editorial Universitaria de Buenos Aires, Buenos Aires, 1962. Translated from the English and revised by Oscar A. Varsavsky.

[119] N. Kerzman and E. M. Stein. The Szegő kernel in terms of Cauchy-Fantappiè kernels. *Duke Math. J.*, 45(2):197–224, 1978.

[120] Christer O. Kiselman. A study of the Bergman projection in certain Hartogs domains. In *Several complex variables and complex geometry, Part 3 (Santa Cruz, CA, 1989)*, volume 52, Part 3 of *Proc. Sympos. Pure Math.*, pages 219–231. Amer. Math. Soc., Providence, RI, 1991.

[121] Shoshichi Kobayashi. Invariant distances on complex manifolds and holomorphic mappings. *J. Math. Soc. Japan*, 19:460–480, 1967.

[122] J. J. Kohn. Harmonic integrals on strongly pseudo-convex manifolds. I. *Ann. of Math. (2)*, 78:112–148, 1963.

[123] J. J. Kohn. Harmonic integrals on strongly pseudo-convex manifolds. I. *Ann. of Math. (2)*, 78:112–148, 1963.

[124] J. J. Kohn. Boundary behavior of δ on weakly pseudo-convex manifolds of dimension two. *J. Differential Geometry*, 6:523–542, 1972.

[125] J. J. Kohn. Global regularity for $\bar{\partial}$ on weakly pseudo-convex manifolds. *Trans. Amer. Math. Soc.*, 181:273–292, 1973.

[126] J. J. Kohn. Global regularity for $\bar{\partial}$ on weakly pseudo-convex manifolds. *Trans. Amer. Math. Soc.*, 181:273–292, 1973.

[127] J. J. Kohn. Methods of partial differential equations in complex analysis. In *Several complex variables (Proc. Sympos. Pure Math., Vol. XXX, Part 1, Williams Coll., Williamstown, Mass., 1975)*, volume Vol. XXX, Part 1 of *Proc. Sympos. Pure Math.*, pages 215–237. Amer. Math. Soc., Providence, RI, 1977.

[128] J. J. Kohn. Subellipticity of the $\bar{\partial}$-Neumann problem on pseudo-convex domains: sufficient conditions. *Acta Math.*, 142(1-2):79–122, 1979.

[129] J. J. Kohn. A survey of the $\bar{\partial}$-Neumann problem. In *Complex analysis of several variables (Madison, Wis., 1982)*, volume 41 of *Proc. Sympos. Pure Math.*, pages 137–145. Amer. Math. Soc., Providence, RI, 1984.

[130] J. J. Kohn and L. Nirenberg. Non-coercive boundary value problems. *Comm. Pure Appl. Math.*, 18:443–492, 1965.

[131] J. J. Kohn and Hugo Rossi. On the extension of holomorphic functions from the boundary of a complex manifold. *Ann. of Math. (2)*, 81:451–472, 1965.

[132] Walter Koppelman. The Cauchy integral for differential forms. *Bull. Amer. Math. Soc.*, 73:554–556, 1967.

[133] Walter Koppelman. The Cauchy integral for functions of several complex variables. *Bull. Amer. Math. Soc.*, 73:373–377, 1967.

[134] A. Korányi and S. Vági. Singular integrals on homogeneous spaces and some problems of classical analysis. *Ann. Scuola Norm. Sup. Pisa Cl. Sci. (3)*, 25:575–648 (1972), 1971.

[135] Adam Korányi. Harmonic functions on Hermitian hyperbolic space. *Trans. Amer. Math. Soc.*, 135:507–516, 1969.

[136] Steven G. Krantz. Structure and interpolation theorems for certain Lipschitz spaces and estimates for the $\overline{\partial}$ equation. *Duke Math. J.*, 43(2):417–439, 1976.

[137] Steven G. Krantz. On a theorem of Stein. *Trans. Amer. Math. Soc.*, 320(2):625–642, 1990.

[138] Steven G. Krantz. The boundary behavior of the Kobayashi metric. *Rocky Mountain J. Math.*, 22(1):227–233, 1992.

[139] Steven G. Krantz. *Function theory of several complex variables.* The Wadsworth & Brooks/Cole Mathematics Series. Wadsworth & Brooks/Cole Advanced Books & Software, Pacific Grove, CA, second edition, 1992.

[140] Steven G. Krantz. *Partial differential equations and complex analysis.* Studies in Advanced Mathematics. CRC Press, Boca Raton, FL, 1992. Lecture notes prepared by Estela A. Gavosto and Marco M. Peloso.

[141] Steven G. Krantz. *Function theory of several complex variables.* AMS Chelsea Publishing, Providence, RI, 2001. Reprint of the 1992 edition.

[142] Steven G. Krantz. *Geometric function theory.* Cornerstones. Birkhäuser Boston, Inc., Boston, MA, 2006. Explorations in complex analysis.

[143] Steven G. Krantz. On a theorem of F. Forelli and a result of Hartogs. *Complex Var. Elliptic Equ.*, 63(4):591–597, 2018.

[144] Steven G. Krantz. Canonical kernels versus constructible kernels. *Rocky Mountain J. Math.*, 49(6):1931–1959, 2019.

[145] Steven G. Krantz and Harold R. Parks. *A primer of real analytic functions*, volume 4 of *Basler Lehrbücher [Basel Textbooks]*. Birkhäuser Verlag, Basel, 1992.

[146] Steven G. Krantz and T. D. Parsons. Antisocial subcovers of self-centered coverings. *Amer. Math. Monthly*, 93(1):45–48, 1986.

[147] R. Kühnau. Bibliography of geometric function theory. In *Handbook of complex analysis: geometric function theory. Vol. 2*, pages 809–828. Elsevier Sci. B. V., Amsterdam, 2005.

[148] Olli Lehto and K. I. Virtanen. Boundary behaviour and normal meromorphic functions. *Acta Math.*, 97:47–65, 1957.

[149] Jean Leray. Le calcul différentiel et intégral sur une variété analytique complexe. (Problème de Cauchy. III). *Bull. Soc. Math. France*, 87:81–180, 1959.

[150] Hans Lewy. On the local character of the solutions of an atypical linear differential equation in three variables and a related theorem for regular functions of two complex variables. *Ann. of Math. (2)*, 64:514–522, 1956.

[151] Hans Lewy. An example of a smooth linear partial differential equation without solution. *Ann. of Math. (2)*, 66:155–158, 1957.

[152] Ingo Lieb. Die Cauchy-Riemannschen Differentialgleichungen auf streng pseudokonvexen Gebieten. Beschränkte Lösungen. *Math. Ann.*, 190:6–44, 1970.

[153] Ewa Ligocka. The Hölder continuity of the Bergman projection and proper holomorphic mappings. *Studia Math.*, 80(2):89–107, 1984.

[154] E. Lindelöf. Sur un principe général de l'analyse et ses applications à la théorie de la représentation conforme. Acta Soc. Fennicae 46, Nr. 4, 35 S. (1915)., 1915.

[155] S. N. Mergelyan. Uniform approximations of functions of a complex variable. *Uspehi Matem. Nauk (N.S.)*, 7(2(48)):31–122, 1952.

[156] David Minda and Glenn Schober. Another elementary approach to the theorems of Landau, Montel, Picard and Schottky. *Complex Variables Theory Appl.*, 2(2):157–164, 1983.

[157] Dietrich Morgenstern. Unendlich oft differenzierbare nicht-analytische Funktionen. *Math. Nachr.*, 12:74, 1954.

[158] James R. Munkres. *Topology: a first course.* Prentice-Hall, Inc., Englewood Cliffs, NJ, 1975.

[159] Alexander Nagel and Walter Rudin. Local boundary behavior of bounded holomorphic functions. *Canadian J. Math.*, 30(3):583–592, 1978.

[160] Alexander Nagel, Elias M. Stein, and Stephen Wainger. Boundary behavior of functions holomorphic in domains of finite type. *Proc. Nat. Acad. Sci. U.S.A.*, 78(11):6596–6599, 1981.

[161] Raghavan Narasimhan. *Several complex variables.* Chicago Lectures in Mathematics. University of Chicago Press, Chicago, Ill.-London, 1971.

[162] Raghavan Narasimhan. *Several complex variables.* Chicago Lectures in Mathematics. University of Chicago Press, Chicago, Ill.-London, 1971.

[163] Zeev Nehari. *Conformal mapping.* McGraw-Hill Book Co., Inc., New York-Toronto-London, 1952.

[164] Toshio Nishino. Sur une propriété des familles de fonctions analytiques de deux variables complexes. *J. Math. Kyoto Univ.*, 4:255–282, 1965.

[165] Nils Ø vrelid. Integral representation formulas and L^p-estimates for the $\bar{\partial}$-equation. *Math. Scand.*, 29:137–160, 1971.

[166] W. F. Osgood. Some points in the elements of the theory of functions. *Bull. Amer. Math. Soc.*, 2(9):296–302, 1896.

[167] William F. Osgood. Non-Uniform Convergence and the Integration of Series Term by Term. *Amer. J. Math.*, 19(2):155–190, 1897.

[168] N. Ovrelid. Integral representation formulae for differential forms, and solutions of the $\bar{\partial}$-equation. In *Fonctions analytiques de plusieurs variables et analyse complexe (Colloq. Internat. CNRS, No. 208, Paris, 1972)*, volume No. 1 of *Agora Math.*, pages 180–198. Gauthier-Villars Éditeur, Paris-Brussels-Montreal, Que., 1974.

[169] John C. Oxtoby. *Measure and category*, volume 2 of *Graduate Texts in Mathematics*. Springer-Verlag, New York-Berlin, second edition, 1980. A survey of the analogies between topological and measure spaces.

[170] Paul Painlevé. Sur les lignes singulières des fonctions analytiques. *Ann. Fac. Sci. Toulouse Sci. Math. Sci. Phys.*, 2:B1–B130, 1888.

[171] Josip Plemelj. *Problems in the sense of Riemann and Klein*, volume No. 16 of *Interscience Tracts in Pure and Applied Mathematics*. Interscience Publishers John Wiley & Sons, Inc., New York-London-Sydney, 1964. Edited and translated by J. R. M. Radok.

[172] H. Poincaré. Les fonctions analytiques de deux variables et la représentation conforme. *Rend. Circ. Mat. Palermo*, 23:185–220, 1907.

[173] John C. Polking and R. O. Wells, Jr. Boundary values of Dolbeault cohomology classes and a generalized Bochner-Hartogs theorem. *Abh. Math. Sem. Univ. Hamburg*, 47:3–24, 1978.

[174] Wade C. Ramey. Local boundary behavior of pluriharmonic functions along curves. *Amer. J. Math.*, 108(1):175–191, 1986.

[175] Enrique Ramirez de Arellano. Ein Divisionsproblem und Randintegraldarstellungen in der komplexen Analysis. *Math. Ann.*, 184:172–187, 1969.

[176] R. Michael Range. On Hölder estimates for $\bar{\partial}u = f$ on weakly pseudoconvex domains. In *Several complex variables (Cortona, 1976/1977)*, pages 247–267. Scuola Norm. Sup., Pisa, 1978.

[177] R. Michael Range. *Holomorphic functions and integral representations in several complex variables*, volume 108 of *Graduate Texts in Mathematics*. Springer-Verlag, New York, 1986.

[178] R. Michael Range. *Holomorphic functions and integral representations in several complex variables*, volume 108 of *Graduate Texts in Mathematics*. Springer-Verlag, New York, 1986.

[179] Murali Rao, Henrik Stetkæ r, Sø ren Fournais, and Jacob Schach Mø ller. *Complex analysis*. World Scientific Publishing Co. Pte. Ltd., Hackensack, NJ, second edition, 2015. An invitation.

[180] Reinhold Remmert and Karl Stein. Eigentliche holomorphe Abbildungen. *Math. Z.*, 73:159–189, 1960.

[181] Frederic Riesz and Marcel Riesz. On the boundary values of an analytic function. Quatrième congrès des math. scand. 1916, 27-44 (1916)., 1916.

[182] Marcel Riesz. Sur les fonctions conjuguées. *Math. Z.*, 27(1):218–244, 1928.

[183] A. V. Romanov. A formula and estimates for the solution of the tangential Cauchy-Riemann equation. *Dokl. Akad. Nauk SSSR*, 220:532–535, 1975.

[184] A. V. Romanov. A formula and estimates for the solutions of the tangential Cauchy-Riemann equation. *Mat. Sb. (N.S.)*, 99(141)(1):58–83, 135, 1976.

[185] Guy Roos. L'intégration partielle de la formule de Cauchy-Fantappié. *C. R. Acad. Sci. Paris Sér. A-B*, 287(8):A615–A618, 1978.

[186] H. L. Royden. *Real analysis*. The Macmillan Company, New York; Collier Macmillan Ltd., London, 1963.

[187] Walter Rudin. *Principles of mathematical analysis*. International Series in Pure and Applied Mathematics. McGraw-Hill Book Co., New York-Auckland-Düsseldorf, third edition, 1976.

[188] Walter Rudin. *Principles of mathematical analysis*. International Series in Pure and Applied Mathematics. McGraw-Hill Book Co., New York-Auckland-Düsseldorf, third edition, 1976.

[189] Walter Rudin. *Functional analysis*. International Series in Pure and Applied Mathematics. McGraw-Hill, Inc., New York, second edition, 1991.

[190] Helmut Salzmann and Karl Zeller. Singularitäten unendlich oft differenzierbarer Funktionen. *Math. Z.*, 62:354–367, 1955.

[191] Mikio Sato. Regularity of hyperfunctions solutions of partial differential equations. In *Actes du Congrès International des Mathématiciens (Nice, 1970), Tome 2*, pages 785–794. Gauthier-Villars Éditeur, Paris, 1971.

[192] S. Sawyer. Maximal inequalities of weak type. *Ann. of Math. (2)*, 84:157–174, 1966.

[193] Francesco Severi. Contributi alla teoria delle funzioni biarmoniche. *Mem. Accad. Ital. [Spec.]*, 2:1–59, 1931.

[194] Nessim Sibony. Une classe de domaines pseudoconvexes. *Duke Math. J.*, 55(2):299–319, 1987.

[195] Yum Tong Siu. The $\bar{\partial}$ problem with uniform bounds on derivatives. *Math. Ann.*, 207:163–176, 1974.

[196] Henri Skoda. Valeurs au bord pour les solutions de l'opérateur d'', et caractérisation des zéros des fonctions de la classe de Nevanlinna. *Bull. Soc. Math. France*, 104(3):225–299, 1976.

[197] E. M. Stein. On limits of seqences of operators. *Ann. of Math. (2)*, 74:140–170, 1961.

[198] E. M. Stein. *Boundary behavior of holomorphic functions of several complex variables*, volume No. 11 of *Mathematical Notes*. Princeton University Press, Princeton, NJ; University of Tokyo Press, Tokyo, 1972.

[199] E. M. Stein. *Boundary behavior of holomorphic functions of several complex variables*, volume No. 11 of *Mathematical Notes*. Princeton University Press, Princeton, NJ; University of Tokyo Press, Tokyo, 1972.

[200] Elias M. Stein. *Singular integrals and differentiability properties of functions*, volume No. 30 of *Princeton Mathematical Series*. Princeton University Press, Princeton, NJ, 1970.

[201] Elias M. Stein and Guido Weiss. *Introduction to Fourier analysis on Euclidean spaces*, volume No. 32 of *Princeton Mathematical Series*. Princeton University Press, Princeton, NJ, 1971.

[202] Toshiaki Terada. Analyticités relatives à chaque variable. Analogies du théorème de Hartogs. *J. Math. Kyoto Univ.*, 12:263–296, 1972.

[203] E. C. Titchmarsh. *The theory of functions*. Oxford University Press, Oxford, second edition, 1939.

[204] M. Tsuji. *Potential theory in modern function theory*. Maruzen Co. Ltd., Tokyo, 1959.

[205] M. Tsuji. *Potential theory in modern function theory*. Chelsea Publishing Co., New York, 1975. Reprinting of the 1959 original.

[206] E. M. Čirka. The Lindelöf and Fatou theorems in $\mathbb{C}^n$. *Mat. Sb. (N.S.)*, 92(134):622–644, 648, 1973.

[207] C. T. C. Wall. *A geometric introduction to topology.* Addison-Wesley Publishing Co., Reading, Mass.-London-Don Mills, Ont., 1972.

[208] John Wermer. *Seminar über Funktionen-Algebren*, volume 1 of *Lecture Notes in Mathematics.* Springer-Verlag, Berlin-Göttingen-Heidelberg, 1964. Winter-Semester 1963/64.

[209] Gordon Thomas Whyburn. *Topological analysis*, volume No. 23 of *Princeton Mathematical Series.* Princeton University Press, Princeton, NJ, 1958.

[210] Jan Wiegerinck. Separately subharmonic functions need not be subharmonic. *Proc. Amer. Math. Soc.*, 104(3):770–771, 1988.

[211] Giuseppe Zampieri. *Complex analysis and CR geometry*, volume 43 of *University Lecture Series.* American Mathematical Society, Providence, RI, 2008.

[212] A. Zygmund. *Trigonometric series: Vols. I, II.* Cambridge University Press, London-New York, second edition, 1968.

Index